CAMBRIDGE TRACTS IN MATHEMATICS

General Editors

B. BOLLOBÁS, W. FULTON, A. KATOK,
F. KIRWAN, P. SARNAK, B. SIMON, B. TOTARO

179   Dynamics of Linear Operators

T0299092

DYNAMICS OF LINEAR OPERATORS

FRÉDÉRIC BAYART
Université de Clermont-Ferrand, France

ÉTIENNE MATHERON
Université d'Artois, France

CAMBRIDGE
UNIVERSITY PRESS

CAMBRIDGE
UNIVERSITY PRESS

University Printing House, Cambridge CB2 8BS, United Kingdom

Cambridge University Press is part of the University of Cambridge.

It furthers the University's mission by disseminating knowledge in the pursuit of education, learning and research at the highest international levels of excellence.

www.cambridge.org
Information on this title: www.cambridge.org/9780521514965

© F. Bayart and É. Matheron 2009

First published 2009

A catalogue record for this publication is available from the British Library

ISBN 978-0-521-51496-5 Hardback

Contents

Introduction

Linear dynamics is a young and rapidly evolving branch of functional analysis, which was probably born in 1982 with the Toronto Ph.D. thesis of C. Kitai [158]. It has become rather popular, thanks to the efforts of many mathematicians. In particular, the seminal paper [123] by G. Godefroy and J. H. Shapiro, the authoritative survey [133] by K.-G. Grosse-Erdmann and the beautiful notes [222] by J. H. Shapiro have had a considerable influence on both its internal development and its diffusion within the mathematical community. After more than two decades of active research, this would seem to be the proper time to write a book about it.

As the name indicates, linear dynamics is mainly concerned with the behaviour of iterates of linear transformations. On finite-dimensional spaces, things are rather well understood since linear transformations are completely described by their Jordan canonical form. However, a new phenomenon appears in an infinite-dimensional setting: linear operators may have dense orbits. In fact, quite a lot of natural operators have this property.

To settle some terminology, let us recall that if T is a continuous linear operator acting on some topological vector space X, the T-**orbit** of a vector $x \in X$ is the set $O(x,T) := \{x, T(x), T^2(x), \dots\}$. The operator T is said to be **hypercyclic** if there exists some vector $x \in X$ whose T-orbit is dense in X. Such a vector x is said to be hypercyclic for T. Hypercyclicity is the main topic of the present book.

From the definition of hypercyclicity, it is immediately apparent that linear dynamics lies at the intersection of at least three different domains of mathematics.

1. **Topological dynamics** The definition of hypercyclicity does not require any linear structure. It makes sense for an arbitrary continuous map $T : X \to X$ acting on a topological space X and, in fact, continuous maps with dense orbits are in some sense the main objects of study in topological dynamics.

However, the usual setting of topological dynamics is that of *compact* topological spaces, and compactness is essential at many points in the discussion. In linear dynamics, the underlying space is never compact or even locally compact because hypercyclicity turns out to be a purely infinite-dimensional property. Thus, it would seem to be hard to use sophisticated tools from topological dynamics. Nevertheless, when the linear structure is added interesting phenomena appear.

2. **Operator theory**   The word "hypercyclic" comes from the much older notion of a **cyclic** operator. An operator $T \in \mathfrak{L}(X)$ is said to be cyclic if there exists a vector $x \in X$ such that the *linear span* of $O(x, T)$ is dense in X. This notion is of course related to the famous **invariant subspace problem**: given an operator $T \in \mathfrak{L}(X)$, is it possible to find a non-trivial closed subspace $F \subset X$ which is T-invariant (i.e. for which $T(F) \subset F$)? Here, non-trivial means that $F \neq \{0\}$ and $F \neq X$. Clearly, the closed linear span of any T-orbit is an invariant subspace for T; hence, T lacks non-trivial invariant closed subspaces if and only if (iff) every non-zero vector $x \in X$ is cyclic for T.

Similarly, the **invariant subset problem** asks whether any operator $T \in \mathfrak{L}(X)$ has a non-trivial closed invariant subset. Since the closure of any T-orbit is a T-invariant closed set, an operator T lacks non-trivial invariant closed sets iff all non-zero vectors $x \in X$ are hypercyclic for T. In the language of topological dynamics, this means that $(X \setminus \{0\}, T)$ is a *minimal* dynamical system.

Despite considerable efforts, the invariant subspace problem remains largely open, most notably for Hilbert space operators. Since P. Enflo's negative solution on a rather peculiar Banach space [104], the most impressive achievement has been C. J. Read's construction of an operator T on $\ell^1(\mathbb{N})$ for which every non-zero vector $x \in \ell^1(\mathbb{N})$ is hypercyclic ([202]). This means that the invariant *subset* problem has a negative solution on the space ℓ^1.

3. **Universality**   Let $(T_i)_{i \in I}$ be a family of continuous maps $T_i : X \to Y$ between two fixed topological spaces X and Y. The family (T_i) is said to be **universal** if there exists $x \in X$ such that the set $\{T_i(x); \ i \in I\}$ is dense in Y. The first example of universality seems to go back to M. Fekete in 1914 (quoted in [191]) who discovered the existence of a *universal Taylor series* $\sum_{n \geq 1} a_n t^n$: for any continuous function g on $[-1, 1]$ with $g(0) = 0$, there exists an increasing sequence of integers (n_k) such that $\sum_{n=1}^{n_k} a_n t^n \to g(t)$ uniformly as $k \to \infty$. (Here $X = \mathbb{C}^{\mathbb{N}}$, Y is the space of all continuous functions on $[-1, 1]$ vanishing at 0, and $T_i((a_n)) = \sum_{n=1}^{i} a_n t^n, i \geq 1$). Since then, universal families have been exhibited in a huge number of situations; see [133].

Hypercyclicity is of course a particular instance of universality, in which $X = Y$ is a topological vector space and $(T_i)_{i \in \mathbb{N}}$ is the sequence of iterates of a single linear operator $T \in \mathfrak{L}(X)$. Nevertheless, it is worth keeping in mind that a number of results pertaining to hypercyclic operators can be formulated (and proved!) in the more general setting of universal families. When working with the iterates of an operator, however, more specific tools can be used. In particular, spectral theory is often helpful.

One particularly seductive feature of linear dynamics is the diversity of ideas and techniques that are involved in its study, owing to its strong connections with a number of distinct branches of mathematics. For some of them, e.g. topology, operator

theory, and approximation theory, this is rather obvious. More unexpectedly, Banach
space geometry and probability theory also play quite an important role. Even number theory may be useful at times! For that reason, we believe that linear dynamics is
an extremely attractive area, where many beautiful results are still to be discovered.
We hope that the present book will give some substance to this affirmation.

It is now time to describe the contents of the book in more detail.

Chapter 1 contains the basics of linear dynamics. We introduce hypercyclicity and
the weaker (typically linear) notion of supercyclicity. An operator $T \in \mathcal{L}(X)$ is said
to be **supercyclic** if there is some vector $x \in X$ such that the *cone* generated by
$O(x, T)$ is dense in X. Our approach is based on the Baire category theorem. We
start with the well-known equivalence between hypercyclicity and **topological transitivity**: an operator T acting on some separable completely metrizable space X is
hypercyclic iff for each pair of non-empty open sets (U, V) in X, one can find $n \in \mathbb{N}$
such that $T^n(U) \cap V \neq \varnothing$; in this case, there is in fact a residual set of hypercyclic
vectors. From this, one gets immediately the so-called **Hypercyclicity Criterion**, a
sufficient set of conditions for hypercyclicity with a remarkably wide range of application. The analogous Supercyclicity Criterion is proved along the same lines. Next,
we show that hypercyclicity and supercyclicity induce noteworthy spectral properties. Then we discuss the algebraic and topological properties of $HC(T)$, the set of
all hypercyclic vectors for a given hypercyclic operator $T \in \mathcal{L}(X)$. We show that
$HC(T)$ always contains a dense linear subspace of X (except 0) and that $HC(T)$ is
homeomorphic to X when X is a Fréchet space and hence to the separable Hilbert
space. Finally, several fundamental examples are treated in detail: weighted shifts
on ℓ^p spaces, operators commuting with translations on the space of entire functions
$H(\mathbb{C})$, and composition operators acting on the Hardy space $H^2(\mathbb{D})$. We will come
back to these examples several times in the book.

Chapter 2 contains some rather impressive results showing that hypercyclicity is
not a mere curiosity. We first prove that hypercyclic operators can be found in any
infinite-dimensional separable Fréchet space. The key point here is that operators of
the form "identity plus a backward shift" are always hypercyclic, and even **topologically mixing**; that is, for each pair of non-empty open sets (U, V), all but finitely
many $n \in \mathbb{N}$ satisfy $T^n(U) \cap V \neq \varnothing$. Then we show that any countable, dense,
linearly independent set in a separable infinite-dimensional Banach space is an orbit
of some hypercyclic operator. Next, we discuss the size of the set of all hypercyclic
operators on some given infinite-dimensional separable Banach space X. This set
is always dense in $\mathcal{L}(X)$ with respect to the strong operator topology, but nowhere
dense with respect to the norm topology (at least when X is a Hilbert space). Then,
we show that linear dynamics provides a universal model for topological (non-linear)
dynamics: there exists a single hypercyclic operator T acting on the separable Hilbert
space H such that any continuous self-map of a compact metric space is topologically conjugate to the restriction of T to some invariant compact set $K \subset H$. We
conclude the chapter by showing that any Hilbert space operator is the sum of two
hypercyclic operators.

In Chapter 3, we present several elegant and useful results that point out a kind of "rigidity" in linear dynamics: the powers and rotations of hypercyclic operators remain hypercyclic; every single operator in a hypercyclic C_0-semigroup is already hypercyclic; and any orbit of an arbitrary operator is either nowhere dense or everywhere dense in the underlying topological vector space. Besides obvious formal similarities, these results have another interesting common feature: the proof of each ultimately relies on some suitable *connectedness* argument.

Chapter 4 is devoted to the Hypercyclicity Criterion. It turns out that a linear operator $T \in \mathfrak{L}(X)$ satisfies the Hypercyclicity Criterion iff it is topologically **weakly mixing**, which means that the product operator $T \times T$ is hypercyclic on $X \times X$. The **Hypercyclicity Criterion Problem** asks whether every hypercyclic operator has to be weakly mixing. In a non-linear context, very simple examples show that the answer is negative. However, the linear problem proves to be much more difficult. It was solved only recently by C. J. Read and M. De La Rosa, who showed that a counterexample exists in some suitably manufactured Banach space. We present in Chapter 4 a variant of their construction which allows us to exhibit counterexamples in a large class of separable Banach spaces, including separable Hilbert spaces. The chapter also contains various characterizations of the weak mixing property involving the sets of natural numbers $\mathbf{N}(U, V) := \{n \in \mathbb{N}; \ T^n(U) \cap V \neq \varnothing\}$, where U, V are non-empty open sets in X. These characterizations are undoubtly quite well known to people working in topological dynamics, but perhaps less so to the operator theory community.

In Chapter 5, we give a rather detailed account of the connections between linear dynamics and **measurable dynamics**, i.e. ergodic theory. The basic idea is the following: if an operator T turns out to be ergodic with respect to some measure with full support then T is hypercyclic by Birkhoff's ergodic theorem. Accordingly, it is desirable to find conditions ensuring the existence of such an ergodic measure. We concentrate on **Gaussian** measures only, since they are by far the best understood infinite-dimensional measures. We start with a general and essentially self-contained discussion of Gaussian measures and covariance operators on Banach spaces. Then we show how one can construct an ergodic Gaussian measure for an operator T provided that T has "sufficiently many" eigenvectors associated with unimodular eigenvalues. The geometry of the underlying Banach space turns out to be quite important here, which should not be too surprising to anyone who has heard about probability in Banach spaces.

In Chapter 6, we discuss some variants or strengthenings of hypercyclicity. We first show that an operator is hypercyclic whenever it has an orbit passing "not too far" from any point of the underlying space. Then, we consider **chaotic** and **frequently hypercyclic** operators (the latter being implicitly present in Chapter 5). Chaoticity and frequent hypercyclicity are qualitative strengthenings of hypercyclicity, both strictly stronger because operators with one or the other property are shown to be weakly mixing. There are interesting similarities and differences between hypercyclicity and these two variants. For example, on the one hand any rotation and any power of a chaotic or frequently hypercyclic operator has the same property; on the other hand, some separable Banach spaces do not support any chaotic or frequently hypercyclic operator. Moreover, we show the existence of frequently hypercyclic operators which are not chaotic and Hilbert space operators which are both chaotic and frequently hypercyclic but not topologically mixing.

In Chapter 7, we discuss in some detail the problem of the existence of **common hypercyclic vectors** for uncountable families of operators. By the Baire category theorem, any *countable* family of hypercyclic operators has a residual set of common hypercyclic vectors, but there is no obvious result of that kind for uncountable families. We present several positive criteria which prove to be efficient in various situations. These criteria may be viewed as kinds of "uncountable Baire category theorems", applying, of course, to very special families of open sets. Then we consider the particular case of weighted shifts and show that continuous paths of weighted shifts may or may not admit common hypercyclic vectors.

Chapter 8 is centred around the following question: when does a given hypercyclic operator admit a **hypercyclic subspace**, i.e. when is it possible to find an infinite-dimensional *closed* subspace of the underlying space consisting entirely of hypercyclic vectors (except 0)? If the operator T acts on a complex Banach space and satisfies the Hypercyclicity Criterion then there is a complete and very simple characterization: such a subspace can be found iff the essential spectrum of T intersects the closed unit disk. We prove this result in two different ways and then give several natural examples. We also prove some results related to the existence of non-trivial *algebras* of hypercyclic vectors.

Chapter 9 is entirely devoted to supercyclicity. We prove the so-called **Angle Criterion**, a geometrical result which is often useful for showing that a given operator is *not* supercyclic. Then, we illustrate this criterion with two nice examples: the composition operators on $H^2(\mathbb{D})$ associated with non-automorphic parabolic maps of the disk and the classical Volterra operator acting on $L^2([0,1])$.

In Chapter 10, we consider hypercyclicity or supercyclicity with respect to the weak topology of a given Banach space X. We start with a detailed discussion of weakly dense sequences which are not dense with respect to the norm topology. Then we concentrate on weak hypercyclicity or supercyclicity for bilateral weighted shifts acting on $\ell^p(\mathbb{Z})$. In particular we show that there exist bilateral weighted shifts which are weakly hypercyclic but not hypercyclic, that weak hypercyclicity or supercyclicity of a weighted shift really depends on the exponent p (unlike norm hypercyclicity and supercyclicity), and that the unweighted shift is weakly supercyclic on $\ell^p(\mathbb{Z})$ iff $p > 2$. Then we consider *unitary* operators. We show that, surprisingly enough, there exist Borel probability measures μ on $\mathbb{T}$ for which the "multiplication by the variable" operator M_z is weakly supercyclic on $L^2(\mu)$. This holds if the support of μ is "very small" but it is also possible to require that the Fourier coefficients of μ vanish at infinity, in which case the support of μ is rather "large". We conclude the chapter by discussing the notions of weak *sequential* hypercyclicity and supercyclicity, which are still not well understood.

Chapter 11 is devoted to the universality properties of the Riemann zeta function. We show that any holomorphic function in the strip $\{1/2 < \mathrm{Re}(s) < 1\}$ and without zeros can be uniformly approximated on compact sets by imaginary translates of the zeta function. This remarkable result is due to S. M. Voronin. The proof is very much in the spirit of the whole book, being a mixture of analytic number theory, function theory, Hilbert space geometry, and ergodic theory.

In Chapter 12, we try to give a reader-friendly description of one of the many operators constructed by C. J. Read in connection with the invariant subspace problem. We concentrate on the simplest example: an operator without non-trivial invariant subspaces on the space $X = \ell^1(\mathbb{N})$. Even so, the construction is quite involved but is presented at a relatively slow pace in a reasonable number of pages. When working on that chapter, our hope was, of course, to be able to solve the problem on a separable Hilbert space! The final result is much less impressive; nevertheless, we hope that the chapter will be useful to some people. Typically, a non-expert interested in the invariant subspace problem (just like us) may find our exposition convenient.

From this outline, it should be clear that we have not written an encyclopaedic treatise on linear dynamics. This book is rather a selection of results and ideas made mostly according to our personal tastes but also because they fit together to give a reasonably accurate global picture of the subject. As a result, the chapters have few overlaps and can be read more or less independently.

At the end of the book, we have added four appendices on complex analysis, function spaces, Banach space theory, and spectral theory. The reader will find there only definitions and results that are explicitly needed in the main body of the book. Several proofs are given. One reason is that the reader may find it more convenient to have grouped together the proofs of important results that are used several times, rather than to look for them in various sources. Another reason is that some results definitely need to be proved, but it seemed better to postpone the proofs to the appendices in order to keep the reading of the book reasonably fluent. Concerning the appendix on spectral theory, we must confess that the first reason for including these proofs was that this was useful for *us*, since we are very far from being experts in that area.

As a rule, we have tried to give the simplest and most natural proofs that we were able to produce. However, this does not mean that we have refrained from stating a result in great generality whenever this seemed to be both possible and desirable. We hope that various kinds of readers will find our book useful, e.g. Ph.D. students, specialists in the area and non-specialists wanting to get a flavour of the subject. We felt that it should be accessible to a rather large audience, including graduate students with an interest in functional analysis. Perhaps ambitiously, we hope that some of these people actually *enjoy* reading the book!

Each chapter ends with some comments and a set of exercises. Some exercises are quite easy, but these are not necessarily the less interesting ones. Some others outline proofs of useful published results that could have been included in the main body of the book but were relegated to exercises owing to the lack of space. A few exercises have the aim of proving new results which will probably not have been published elsewhere. Finally, some exercises are devoted to results which are used in the text but for which we did not give full proofs in order to make the presentation more digestible. We have worked out each exercise rather carefully and included a number of explicit hints. In that way, we believe that any motivated reader will succeed in finding solutions without excessive effort.

Just like any other, this book cannot pretend to be perfect and is bound to suffer from flaws. The authors take responsibility for these, including any mathematical errors. In particular, if some result has not been included this by no means indicates that it did not deserve to be mentioned. Most likely the omission was due to a lack of space; and if the result does not even appear in the comments this simply means that the authors were not aware of it. We would be very grateful to anyone pointing out unfortunate omissions, mathematical inaccuracies or troublesome typos to us.

Acknowledgements We are very honoured that Cambridge University Press accepted our book for publication. For this many thanks are due to Roger Astley, whose quite positive reception of a first, very preliminary, draft of the manuscript and always optimistic emails were extremely encouraging for us. We also thank Susan Parkinson and Anna-Marie Lovett for their help during the production process.

Many colleagues and friends, one student and one relative have read parts of the manuscript and/or made a number of useful suggestions for improving the text. In alphabetical order, our warmest thanks to Richard Aron, Gilles Bailly-Maitre, Juan Bès, Josée Besenger, Monsieur Bony, Isabelle Chalendar, Kit Chan, Éric Charpentier, Bernard Chevreau, George Costakis, Sylvain Delpech, Robert Deville, Eva Gallardo, Frédéric Gaunard, Gilles Godefroy, Sophie Grivaux, Andreas Hartmann, General Mickael F. Jourdan, Alexandre Matheron, Jonathan Partington, Hervé Queffélec, Elizabeth Strouse and Frédérique Watbled.

Finally, a more personal reading was done by, in decreasing size order, Véronique, Armelle, Mathilde, Juliette, Solène and Émile. For that, and so much more, we dedicate the book to them.

1

Hypercyclic and supercyclic operators

Introduction

The aim of this first chapter is twofold: to give a reasonably short, yet significant and hopefully appetizing, sample of the type of questions with which we will be concerned and also to introduce some definitions and prove some basic facts that will be used throughout the whole book.

Let X be a topological vector space over $\mathbb{K} = \mathbb{R}$ or $\mathbb{C}$. We denote by $\mathfrak{L}(X)$ the set of all continuous linear operators on X. If $T \in \mathfrak{L}(X)$, the T-**orbit** of a vector $x \in X$ is the set

$$O(x, T) := \{T^n(x); \ n \in \mathbb{N}\}.$$

The operator T is said to be **hypercyclic** if there is some vector $x \in X$ such that $O(x, T)$ is dense in X. Such a vector x is said to be hypercyclic for T (or T-*hypercyclic*), and the set of all hypercyclic vectors for T is denoted by $HC(T)$. Similarly, T is said to be **supercyclic** if there exists a vector $x \in X$ whose **projective orbit**

$$\mathbb{K} \cdot O(x, T) := \{\lambda T^n(x); \ n \in \mathbb{N}, \ \lambda \in \mathbb{K}\}$$

is dense in X; the set of all supercyclic vectors for T is denoted by $SC(T)$. Finally, we recall that T is said to be **cyclic** if there exists $x \in X$ such that

$$\mathbb{K}[T]x := \operatorname{span} O(x, T) = \{P(T)x; \ P \text{ polynomial}\}$$

is dense in X.

Of course, these notions make sense only if the space X is *separable*. Moreover, hypercyclicity turns out to be a purely infinite-dimensional phenomenon ([206]):

PROPOSITION 1.1 *There are no hypercyclic operators on a finite-dimensional space $X \neq \{0\}$.*

PROOF Suppose on the contrary that T is a hypercyclic operator on $\mathbb{K}^N$, $N \geq 1$. Pick $x \in HC(T)$ and observe that $(x, T(x), \dots, T^{N-1}(x))$ is a linearly independent family and hence is a basis of $\mathbb{K}^N$. Indeed, otherwise the linear span of $O(x, T)$ would have dimension less than N and hence could not be dense in $\mathbb{K}^N$. For any $\alpha \in \mathbb{R}_+$, one can find a sequence of integers (n_k) such that $T^{n_k}(x) \to \alpha x$. Then $T^{n_k}(T^i x) = T^i(T^{n_k} x) \to \alpha T^i x$ for each $i < N$, and hence $T^{n_k}(z) \to \alpha z$ for any $z \in \mathbb{K}^N$. It follows that $\det(T^{n_k}) \to \alpha^N$, i.e. $\det(T)^{n_k} \to \alpha^N$. Thus, putting $a := |\det(T)|$, we see that the set $\{a^n; \ n \in \mathbb{N}\}$ is dense in $\mathbb{R}_+$. This is clearly impossible. $\square$

The most general setting for linear dynamics is that of an arbitrary (separable) topological vector space X. However, we will usually assume that X is an F-**space**, i.e. a complete and metrizable topological vector space. Then X has a translation-invariant compatible metric (see [210]) and (X, d) is complete for any such metric d. In fact, in most cases X will be a **Fréchet space**, i.e. a locally convex F-space. Equivalently, a Fréchet space is a complete topological vector space whose topology is generated by a countable family of seminorms.

An attractive feature of F-spaces is that one can make use of the Baire category theorem. This will be *very* important for us. Incidentally, we note that the Banach–Steinhaus theorem and Banach's isomorphism theorem are valid in F-spaces, and if local convexity is added then one can also use the Hahn–Banach theorem and its consequences. If the reader feels uncomfortable with F-spaces and Fréchet spaces, he or she may safely assume that the underlying space X is a Banach space, keeping in mind that several natural examples live outside this context.

The chapter is organized as follows. We start by explaining how one can show that a given operator is hypercyclic or supercyclic. In particular, we prove the so-called *Hypercyclicity Criterion*, and the analogous Supercyclicity Criterion. Then we show that hypercyclicity and supercyclicity both entail certain spectral restrictions on the operator and its adjoint. Next, we discuss the "largeness" and the topological properties of the set of all hypercyclic vectors for a given operator T. Finally, we treat in some detail several specific examples: weighted shifts on ℓ^p spaces, composition operators on the Hardy space $H^2(\mathbb{D})$, and operators commuting with translations on the space of entire functions $H(\mathbb{C})$.

1.1 How to prove that an operator is hypercyclic

Our first characterization of hypercyclicity is a direct application of the Baire category theorem. This result was proved by G. D. Birkhoff in [53], and it is often referred to as **Birkhoff's transitivity theorem**.

THEOREM 1.2 (BIRKHOFF'S TRANSITIVITY THEOREM) *Let X be a separable F-space and let $T \in \mathfrak{L}(X)$. The following are equivalent:*

(i) *T is hypercyclic;*

(ii) *T is **topologically transitive**; that is, for each pair of non-empty open sets $(U, V) \subset X$ there exists $n \in \mathbb{N}$ such that $T^n(U) \cap V \neq \varnothing$.*

In that case, $HC(T)$ is a dense G_δ subset of X.

PROOF First, observe that if x is a hypercyclic vector for T then $O(x, T) \subset HC(T)$. Indeed, since X has no isolated points, any dense set $A \subset X$ remains dense after the removal of a finite number of points. Applying this to $A := O(x, T)$, and since $O(T^p(x), T) = O(x, T) \backslash \{x, T(x), \dots, T^{p-1}(x)\}$, we see that $T^p(x) \in HC(T)$ for every positive integer p. Thus $HC(T)$ is either empty or dense in X. From this, it is clear that (i) $\implies$ (ii). Indeed, if (i) holds and the open sets U, V are given, we can pick $x \in U \cap HC(T)$ and then $n \in \mathbb{N}$ such that $T^n(x) \in V$.

To prove the converse, we note that since the space X is metrizable and separable, it is second-countable, i.e. it admits a countable basis of open sets. Let $(V_j)_{j \in \mathbb{N}}$ be such a basis. A vector $x \in X$ is hypercyclic for T iff its T-orbit visits each open set V_j, that is, iff for any $j \in \mathbb{N}$ there exists an integer $n \geq 0$ such that $T^n(x) \in V_j$. Thus one can describe $HC(T)$ as follows:

$$HC(T) = \bigcap_{j \in \mathbb{N}} \bigcup_{n \geq 0} T^{-n}(V_j).$$

This shows in particular that $HC(T)$ is a G_δ set. Moreover, it follows from the Baire category theorem that $HC(T)$ is dense in X iff each open set $W_j :=$ $\bigcup_{n \geq 0} T^{-n}(V_j)$ is dense; in other words, iff for each non-empty open set $U \subset X$ and any $j \in \mathbb{N}$ one can find n such that

$$U \cap T^{-n}(V_j) \neq \varnothing \quad \text{or, equivalently,} \quad T^n(U) \cap V_j \neq \varnothing.$$

Since (V_j) is a basis for the topology of X, this is equivalent to the topological transitivity of T. $\qquad \square$

REMARK The implication (hypercyclic) $\implies$ (topologically transitive) does not require the space X to be metrizable or Baire: it holds for an arbitrary topological vector space X. Indeed, the only thing we use is that $HC(T)$ is dense in X whenever it is non-empty, since X has no isolated points. Moreover, what is really needed for the converse implication (topologically transitive) $\implies$ (hypercyclic) is that X is a Baire space with a countable basis of open sets. We also point out that Theorem 1.2 has nothing to do with linearity, since the definitions of hypercyclicity and topological transitivity do not require any linear structure. Accordingly, Theorem 1.2 holds as stated for an arbitrary continuous map $T : X \to X$ acting on some second-countable Baire space X without isolated points.

When the operator T is invertible, it is readily seen that T is topologically transitive iff T^{-1} is. Thus, we can state

COROLLARY 1.3 *Let X be a separable F-space, and let $T \in \mathfrak{L}(X)$. Assume that T is invertible. Then T is hypercyclic if and only if T^{-1} is hypercyclic.*

It is worth noting that T and T^{-1} do not necessarily share the same hypercyclic vectors; see Exercise 1.11.

We illustrate Theorem 1.2 with the following historic example, also due to Birkhoff [54].

EXAMPLE 1.4 (G. D. BIRKHOFF, 1929) Let $H(\mathbb{C})$ be the space of all entire functions on $\mathbb{C}$ endowed with the topology of uniform convergence on compact sets. For any non-zero complex number a, let $T_a : H(\mathbb{C}) \to H(\mathbb{C})$ be the **translation operator** defined by $T_a(f)(z) = f(z + a)$. Then T_a is hypercyclic on $H(\mathbb{C})$.

PROOF   The space $H(\mathbb{C})$ is a separable Fréchet space, so it is enough to show that T_a is topologically transitive. If $u \in H(\mathbb{C})$ and $E \subset \mathbb{C}$ is compact, we set

$$\|u\|_E := \sup\{|u(z)|; \ z \in E\}.$$

Let U, V be two non-empty open subsets of $H(\mathbb{C})$. There exist $\varepsilon > 0$, two closed disks $K, L \subset \mathbb{C}$ and two functions $f, g \in H(\mathbb{C})$ such that

$$U \supset \{h \in H(\mathbb{C}); \ \|h - f\|_K < \varepsilon\},$$

$$V \supset \{h \in H(\mathbb{C}); \ \|h - g\|_L < \varepsilon\}.$$

Let n be any positive integer such that $K \cap (L + an) = \varnothing$. Since $\mathbb{C} \backslash (K \cup (L + an))$ is connected, one can find $h \in H(\mathbb{C})$ such that

$$\|h - f\|_K < \varepsilon \quad \text{and} \quad \|h - g(\cdot - na)\|_{L + an} < \varepsilon;$$

this follows from Runge's approximation theorem (see e.g. [209] or Appendix A). Thus $h \in U$ and $T_a^n(h) \in V$, which shows that T_a is topologically transitive. $\square$

Topologically transitive maps are far from being exotic objects. For example, the map $x \mapsto 4x(1 - x)$ is transitive on the interval $[0, 1]$ and the map $\lambda \mapsto \lambda^2$ is transitive on the circle $\mathbb{T}$ (see e.g. R. L. Devaney's classical book [94]). However, in a topological setting one often needs a specific argument to show that a given map is transitive.

Nevertheless, in a linear setting an extremely useful general criterion for hypercyclicity does exist. This criterion was isolated by C. Kitai in a restricted form [158] and then by R. Gethner and J. H. Shapiro in a form close to that given below, [119]. The version we use appears in the Ph.D. thesis of J. Bès [45].

DEFINITION 1.5   *Let X be a topological vector space, and let $T \in \mathfrak{L}(X)$. We say that T satisfies the **Hypercyclicity Criterion** if there exist an increasing sequence of integers (n_k), two dense sets $\mathcal{D}_1, \mathcal{D}_2 \subset X$ and a sequence of maps $S_{n_k} : \mathcal{D}_2 \to X$ such that:*

(1) $T^{n_k}(x) \to 0$ *for any $x \in \mathcal{D}_1$;*
(2) $S_{n_k}(y) \to 0$ *for any $y \in \mathcal{D}_2$;*
(3) $T^{n_k} S_{n_k}(y) \to y$ *for each $y \in \mathcal{D}_2$.*

We will sometimes say that T satisfies the Hypercyclicity Criterion *with respect to the sequence* (n_k). When it is possible to take $n_k = k$ and $\mathcal{D}_1 = \mathcal{D}_2$, it is usually said that T satisfies **Kitai's Criterion**. We point out that in the above definition, the maps S_{n_k} are not assumed to be linear or continuous.

THEOREM 1.6   *Let $T \in \mathfrak{L}(X)$, where X is a separable F-space. Assume that T satisfies the Hypercyclicity Criterion. Then T is hypercyclic.*

FIRST PROOF We show that T is topologically transitive. Let U, V be two non-empty open subsets of X and pick $x \in \mathcal{D}_1 \cap U$, $y \in \mathcal{D}_2 \cap V$. Then $x + S_{n_k}(y) \to x \in U$ as $k \to \infty$ whereas $T^{n_k}(x + S_{n_k}(y)) = T^{n_k}(x) + T^{n_k} S_{n_k}(y) \to y \in V$. Thus, $T^{n_k}(U) \cap V \neq \emptyset$ if k is large enough. $\qquad\square$

SECOND PROOF This second proof consists in replacing the Baire category theorem by a suitable series; this was the original idea of Kitai. We may assume that the set $\mathcal{D}_2$ is countable, and we enumerate it as a sequence $(y_l)_{l \in \mathbb{N}}$. Let us also fix some translation-invariant (necessarily complete) metric d for X. For the sake of visual clarity, we write $\|x\|$ instead of $d(x, 0)$.

We construct by induction a subsequence (m_k) of (n_k), a sequence $(x_k) \subset \mathcal{D}_1$, and a decreasing sequence of positive numbers (ε_k) with $\varepsilon_k \leq 2^{-k}$ such that the following properties hold for each $k \in \mathbb{N}$:

(i) $\|x_k\| < \varepsilon_k$;
(ii) $\|T^{m_k}(x_k) - y_k\| < \varepsilon_k$;
(iii) $\|T^{m_k}(x_i)\| < \varepsilon_k$ for all $i < k$;
(iv) if $u \in X$ satisfies $\|u\| < \varepsilon_k$ then $\|T^{m_i}(u)\| < 2^{-k}$ for all $i < k$.

Starting with $\varepsilon_0 := 1$, we use (2) and (3) of Definition 1.5 to find m_0 such that $\|S_{m_0}(y_0)\| < \varepsilon_0$ and $\|T^{m_0} S_{m_0}(y_0) - y_0\| < \varepsilon_0$ and then pick some $x_0 \in \mathcal{D}_1$ close to $S_{m_0}(y_0)$, in order to ensure (i) and (ii). The inductive step is likewise easy: having defined everything up to step k, one can first choose ε_{k+1} such that (iv) holds for $k + 1$ and then m_{k+1} such that (iii) holds, $\|S_{m_{k+1}}(y_{k+1})\| < \varepsilon_{k+1}$ and $\|T^{m_{k+1}} S_{m_{k+1}}(y_{k+1}) - y_{k+1}\| < \varepsilon_{k+1}$, and after that $x_{k+1} \in \mathcal{D}_1$ close enough to $S_{m_{k+1}}(y_{k+1})$ to satisfy (i) and (ii).

By (i) and the completeness of (X, d), the series $\sum x_j$ is convergent in X. We claim that

$$x := \sum_{j=0}^{\infty} x_j$$

is a hypercyclic vector for T. Indeed, for any $l \in \mathbb{N}$ one may write

$$\|T^{m_l}(x) - y_l\| \leq \sum_{j < l} \|T^{m_l}(x_j)\| + \|T^{m_l}(x_l) - y_l\| + \sum_{j > l} \|T^{m_l}(x_j)\|$$
$$\leq l\varepsilon_l + \varepsilon_l + \sum_{j > l} 2^{-j}$$

where we have used (iii), (ii), and (iv). Thus $T^{m_l}(x) - y_l \to 0$ as $l \to \infty$, which concludes the proof. $\qquad\square$

REMARK 1.7 We have in fact proved the following more precise result: if $T \in \mathfrak{L}(X)$ satisfies the Hypercyclicity Criterion with respect to some sequence $(n_k)_{k \geq 0}$ then the family $(T^{n_k})_{k \geq 0}$ is *universal*, i.e. there exists some vector $x \in X$ such that the set $\{T^{n_k}(x); \ k \geq 0\}$ is dense in X. In fact, for *any* subsequence (n'_k) of (n_k), the family $(T^{n'_k})_{k \geq 0}$ is universal: this is apparent from the above proofs.

Theorem 1.6 will ensure the hypercyclicity of almost (!) all the hypercyclic operators in this book. We give two historical examples, due to G. R. MacLane [176] and to S. Rolewicz [206]. The latter was the first example of a hypercyclic operator that acts on a Banach space.

EXAMPLE 1.8 (G. R. MACLANE, 1951)   The **derivative operator** $D : f \mapsto f'$ is hypercyclic on $H(\mathbb{C})$.

PROOF   We apply the Hypercyclicity Criterion to the whole sequence of integers $(n_k) := (k)$, the *same* dense set $\mathcal{D}_1 = \mathcal{D}_2$ made up of all complex polynomials, and the maps $S_k := S^k$, where $Sf(z) = \int_0^z f(\xi)\,d\xi$. It is easy to check that conditions (1), (2) and (3) of Definition 1.5 are satisfied. Indeed, (1) holds because $D^k(P)$ tends to zero for any polynomial P, and (3) holds because $DS = I$ on $\mathcal{D}_2$. To prove (2), it is enough to check that $S_k(z^p) \to 0$ uniformly on compact subsets of $\mathbb{C}$, for any fixed $p \in \mathbb{N}$ (then we conclude using linearity). This, in turn, follows at once from the identity

$$S_k(z^p) = \frac{p!}{(p+k)!} z^{p+k}. \qquad \square$$

EXAMPLE 1.9 (S. ROLEWICZ, 1969)   Let $B : \ell^2(\mathbb{N}) \to \ell^2(\mathbb{N})$ be the **backward shift** operator, defined by $B(x_0, x_1, \dots) = (x_1, x_2, \dots)$. Then λB is hypercyclic for any scalar λ such that $|\lambda| > 1$.

Observe that B itself cannot be hypercyclic since $\|B\| = 1$. Indeed, if T is a hypercyclic Banach space operator then $\|T\| > 1$ (otherwise any T-orbit would be bounded).

PROOF   We apply the Hypercyclicity Criterion to the whole sequence of integers $(n_k) := (k)$, the dense set $\mathcal{D}_1 = \mathcal{D}_2 := c_{00}(\mathbb{N})$ made up of all finitely supported sequences and the maps $S_k := \lambda^{-k} S^k$, where S is the forward shift operator, $S(x_0, x_1, \dots) = (0, x_0, x_1, \dots)$. It is easy to check that conditions (1), (2) and (3) of Definition 1.5 are satisfied. For (1) and (3) the arguments are the same as in Example 1.8, and (2) follows immediately from the estimate $\|S_k\| \le |\lambda|^{-k}$. $\qquad \square$

As a consequence of Theorem 1.6 we get the following result, according to which an operator having a large supply of eigenvectors is hypercyclic. This is the so-called **Godefroy–Shapiro Criterion**, which was exhibited by G. Godefroy and J. H. Shapiro in [123].

COROLLARY 1.10 (GODEFROY–SHAPIRO CRITERION)   *Let $T \in \mathfrak{L}(X)$ where X is a separable F-space. Suppose that $\bigcup_{|\lambda|<1} \mathrm{Ker}(T - \lambda)$ and $\bigcup_{|\lambda|>1} \mathrm{Ker}(T - \lambda)$ both span a dense subspace of X. Then T is hypercyclic.*

PROOF We show that T satisfies the Hypercyclicity Criterion with $(n_k) := (k)$ and

$$\mathcal{D}_1 := \mathrm{span}\left(\bigcup_{|\lambda|<1} \mathrm{Ker}(T-\lambda)\right), \qquad \mathcal{D}_2 := \mathrm{span}\left(\bigcup_{|\lambda|>1} \mathrm{Ker}(T-\lambda)\right).$$

The maps $S_k : \mathcal{D}_2 \to X$ are defined as follows: we set $S_k(y) := \lambda^{-k}y$ if $T(y) = \lambda y$ with $|\lambda| > 1$, and we extend S_k to $\mathcal{D}_2$ by linearity. This definition makes sense because the subspaces $\mathrm{Ker}(T-\lambda)$, $|\lambda| > 1$, are linearly independent. Thus, any non-zero $y \in \mathcal{D}_2$ may be uniquely written as $y = y_1 + \cdots + y_p$, with $y_i \in \mathrm{Ker}(T-\lambda_i)\backslash\{0\}$ and $|\lambda_i| > 1$. Having said that, it is clear that the assumptions of the Hypercyclicity Criterion are satisfied. $\square$

REMARK In Corollary 1.10 we see for the first time that hypercyclicity can be inferred from the existence of a large supply of eigenvectors. This will be a recurrent theme in the book.

To illustrate the Godefroy–Shapiro Criterion, we are now going to establish the hypercyclicity of a certain classical operator defined on a Hilbert space of holomorphic functions. Let us first introduce some terminology. In what follows, $\mathbb{T}$ is the unit circle $\{z \in \mathbb{C}; |z| = 1\}$ and $\mathbb{D}$ is the open unit disk $\{z \in \mathbb{C}; |z| < 1\}$.

We denote by $H^2(\mathbb{D})$ the classical **Hardy space** on $\mathbb{D}$. By definition, $H^2(\mathbb{D})$ is the space of all holomorphic functions $f : \mathbb{D} \to \mathbb{C}$ such that

$$\|f\|_{H^2}^2 := \sup_{r<1} \int_{-\pi}^{\pi} |f(re^{i\theta})|^2 \frac{d\theta}{2\pi} < \infty. \qquad (1.1)$$

The Hardy space will appear several times in the book, and we assume that the reader is more or less familiar with it. Moreover, very few properties of $H^2(\mathbb{D})$ will be needed in our discussion (see Appendix B). We refer to e.g. [101] for an in-depth study of Hardy spaces.

We recall here that $H^2(\mathbb{D})$ can also be defined in terms of Taylor expansions. Any holomorphic function $f : \mathbb{D} \to \mathbb{C}$ can be (uniquely) written as $f(z) = \sum_0^\infty c_n(f)z^n$. Then f is in $H^2(\mathbb{D})$ if and only if $\sum_0^\infty |c_n(f)|^2 < \infty$, and in that case $\|f\|_{H^2}^2 = \sum_0^\infty |c_n(f)|^2$. This shows that $H^2(\mathbb{D})$ is canonically isometric to the sequence space $\ell^2(\mathbb{N})$ and also to the closed subspace of $L^2(\mathbb{T})$ defined by

$$H^2(\mathbb{T}) := \{\varphi \in L^2(\mathbb{T}); \widehat{\varphi}(n) = 0 \text{ for all } n < 0\},$$

where the $\widehat{\varphi}(n)$ are the Fourier coefficients of φ. The function of $H^2(\mathbb{T})$ associated with a given $f \in H^2(\mathbb{D})$ is called the **boundary value** of f and will be denoted by f^*.

We summarize these elementary facts as follows: $H^2(\mathbb{D})$ is a Hilbert space whose norm can be defined in two equally useful ways:

$$\|f\|_{H^2}^2 = \sum_{n=0}^{\infty} |c_n(f)|^2 = \|f^*\|_{L^2(\mathbb{T})}^2.$$

Finally, we recall that convergence in $H^2(\mathbb{D})$ entails uniform convergence on compact sets. In particular the point evaluations $f \mapsto f(z)$ are continuous linear functionals on $H^2(\mathbb{D})$, so that for each $z \in \mathbb{D}$ there is a well-defined **reproducing kernel** k_z at z. By definition, k_z is the unique function in $H^2(\mathbb{D})$ satisfying

$$\forall f \in H^2(\mathbb{D}) \ : \ f(z) = \langle f, k_z \rangle_{H^2}. \tag{1.2}$$

In the present case, k_z is given explicitly by the formula

$$k_z(s) = \frac{1}{1 - \bar{z}s},$$

and (1.2) is just a rephrasing of Cauchy's formula.

The set of all bounded holomorphic functions of $\mathbb{D}$ will be denoted by $H^\infty(\mathbb{D})$. It is a non-separable Banach space when endowed with the norm

$$\|u\|_\infty = \sup\{|u(z)|; \ |z| < 1\}.$$

If ϕ is a function in $H^\infty(\mathbb{D})$, the **multiplication operator** M_ϕ associated with ϕ is defined on $H^2(\mathbb{D})$ by $M_\phi(f) = \phi f$. From formula (1.1), it is readily seen that

$$\inf_{z \in \mathbb{D}} |\phi(z)| \times \|f\|_2 \le \|M_\phi(f)\|_2 \le \sup_{z \in \mathbb{D}} |\phi(z)| \times \|f\|_2$$

for any $f \in H^2(\mathbb{D})$. This shows in particular that M_ϕ is a bounded operator on $H^2(\mathbb{D})$, with $\|M_\phi\| \le \|\phi\|_\infty$.

It is not hard to see that a multiplication operator M_ϕ cannot be hypercyclic (for example, for $f \in H^2(\mathbb{D})$, try to approximate $2f$ by functions of the form $\phi^n f$). As the next example shows, things are quite different for the adjoint operator M_ϕ^*.

EXAMPLE 1.11   Let $\phi \in H^\infty(\mathbb{D})$ and let $M_\phi : H^2(\mathbb{D}) \to H^2(\mathbb{D})$ be the associated multiplication operator. The adjoint multiplier M_ϕ^* is hypercyclic if and only if ϕ is non-constant and $\phi(\mathbb{D}) \cap \mathbb{T} \neq \varnothing$.

PROOF   For any $z \in \mathbb{D}$, let $k_z \in H^2(\mathbb{D})$ be the reproducing kernel at z. Then k_z is an eigenvector of M_ϕ^*, with associated eigenvalue $\lambda(z) := \overline{\phi(z)}$. Indeed, we have

$$\langle f, M_\phi^*(k_z) \rangle_{H^2} = \langle \phi f, k_z \rangle_{H^2} = \phi(z)f(z) = \langle f, \overline{\phi(z)}k_z \rangle_{H^2}$$

for all $f \in H^2(\mathbb{D})$, so that $M_\phi^*(k_z) = \overline{\phi(z)}\,k_z$. Let $U := \{z \in \mathbb{D}; \ |\phi(z)| < 1\}$ and $V := \{z \in \mathbb{D}; \ |\phi(z)| > 1\}$. If ϕ is non-constant and $\phi(\mathbb{D}) \cap \mathbb{T} \neq \varnothing$, the open sets U and V are both non-empty by the open mapping theorem for analytic functions. In view of Corollary 1.10, it is enough to show that $\mathrm{span}\{k_z; \ z \in U\}$ and $\mathrm{span}\{k_z; \ z \in V\}$ are dense in $H^2(\mathbb{D})$. But this is clear, since if $f \in H^2(\mathbb{D})$ is orthogonal to k_z either for all $z \in U$ or for all $z \in V$ then f vanishes on a non-empty open set and hence is identically zero.

Conversely, assume that M_ϕ^* is hypercyclic (so that ϕ is certainly non-constant). Then $\|M_\phi\| = \|M_\phi^*\| > 1$, hence $\sup_{z \in \mathbb{D}} |\phi(z)| > 1$. Moreover, we also have $\inf_{z \in \mathbb{D}} |\phi(z)| < 1$. Indeed, if we assume that $\inf_{z \in \mathbb{D}} |\phi(z)| \ge 1$ then $1/\phi \in H^\infty$ and

$M^*_{1/\phi}$ is not hypercyclic since $\|M^*_{1/\phi}\| = \|M_{1/\phi}\| \leq 1$; and since $M^*_{1/\phi} = (M^*_\phi)^{-1}$, Corollary 1.3 shows that M^*_ϕ is not hypercyclic either. Thus, we get

$$\inf_{z \in \mathbb{D}} |\phi(z)| < 1 < \sup_{z \in \mathbb{D}} |\phi(z)|,$$

which yields $\phi(\mathbb{D}) \cap \mathbb{T} \neq \varnothing$ by a simple connectedness argument.   □

We now turn to the "supercyclic" analogues of Theorems 1.2 and 1.6. This is essentially a matter of including a multiplicative factor, so the proofs will be rather sketchy.

THEOREM 1.12   *Let X be a separable F-space, and let $T \in \mathfrak{L}(X)$. The following are equivalent:*

(i) *T is supercyclic;*
(ii) *For each pair of non-empty open sets $(U, V) \subset X$, there exist $n \in \mathbb{N}$ and $\lambda \in \mathbb{K}$ such that $\lambda T^n(U) \cap V \neq \varnothing$.*

In that case, $SC(T)$ is a dense G_δ subset of X.

PROOF   As before, let $(V_j)_{j \in \mathbb{N}}$ be a countable basis of open sets for X. Then one can write $SC(T) = \bigcap_j \bigcup_{\lambda, n} (\lambda T^n)^{-1}(V_j)$, and the proof is completed exactly as that of Theorem 1.2.   □

The following definition and the theorem below are due to H. N. Salas [216].

DEFINITION 1.13   *Let X be a Banach space, and let $T \in \mathfrak{L}(X)$. We say that T satisfies the **Supercyclicity Criterion** if there exist an increasing sequence of integers (n_k), two dense sets $\mathcal{D}_1, \mathcal{D}_2 \subset X$ and a sequence of maps $S_{n_k} : \mathcal{D}_2 \to X$ such that:*

(1) *$\|T^{n_k}(x)\| \, \|S_{n_k}(y)\| \to 0$ for any $x \in \mathcal{D}_1$ and any $y \in \mathcal{D}_2$;*
(2) *$T^{n_k} S_{n_k}(y) \to y$ for each $y \in \mathcal{D}_2$.*

THEOREM 1.14   *Let $T \in \mathfrak{L}(X)$, where X is a separable Banach space. Assume that T satisfies the Supercyclicity Criterion. Then T is supercyclic.*

PROOF   Let U and V be two non-empty open subsets of X. Pick $x \in \mathcal{D}_1 \cap U$ and $y \in \mathcal{D}_2 \cap V$. It follows from part (1) of Definition 1.13 that we can find a sequence of non-zero scalars (λ_k) such that $\lambda_k T^{n_k}(x) \to 0$ and $\lambda_k^{-1} S_{n_k}(y) \to 0$. (Assume that $\alpha_k := \|T^{n_k}(x)\|$ and $\beta_k := \|S_{n_k}(y)\|$ are not both 0. If $\alpha_k \beta_k \neq 0$, put $\lambda_k := \beta_k^{1/2} \alpha_k^{-1/2}$. Otherwise, take $\lambda_k := 2^k \beta_k$ if $\alpha_k = 0$ and $\lambda_k := 2^{-k} \alpha_k^{-1}$ if $\beta_k = 0$). Then, for large enough k, the vector $z := x + \lambda_k^{-1} S_{n_k}(y)$ belongs to U and $\lambda_k T^{n_k}(z)$ belongs to V. By Theorem 1.12, this shows that T is supercyclic.   □

We illustrate the Supercyclicity Criterion with the following important example.

EXAMPLE 1.15   Let $B_\mathbf{w}$ be a **weighted backward shift** on $\ell^2(\mathbb{N})$; $B_\mathbf{w}$ is the operator defined by $B_\mathbf{w}(e_0) = 0$ and $B_\mathbf{w}(e_n) = w_n e_{n-1}$ for $n \geq 1$, where $(e_n)_{n \in \mathbb{N}}$ is the canonical basis of $\ell^2(\mathbb{N})$ and $\mathbf{w} = (w_n)_{n \geq 1}$ is a bounded sequence of positive numbers. Then $B_{\mathbf{w}}$ is supercyclic. In particular, if e.g. $w_n \to 0$ as $n \to \infty$ then $B_{\mathbf{w}}$ is a supercyclic operator which has no hypercyclic multiple.

PROOF   Let $\mathcal{D}_1 = \mathcal{D}_2 := c_{00}(\mathbb{N})$ be the set of all finitely supported sequences. Let $S_{\mathbf{w}}$ be the linear map defined on $\mathcal{D}_2$ by $S_{\mathbf{w}}(e_n) = w_{n+1}^{-1} e_{n+1}$ and, for each $k \in \mathbb{N}$, set $S_k := S_{\mathbf{w}}^k$. Then, the Supercyclicity Criterion is satisfied with respect to $(n_k) := (k)$ because $\|B_{\mathbf{w}}^k(x)\| = 0$ for large enough k and $B_{\mathbf{w}}^k S_k = I$ on $\mathcal{D}_2$.

If $w_n \to 0$ then $\|(\lambda B_{\mathbf{w}})^n\| = |\lambda|^n \sup_{i \in \mathbb{N}}(w_{i+1} \cdots w_{i+n}) \to 0$ as $n \to \infty$, for each fixed $\lambda \in \mathbb{C}$. Hence, no multiple of $B_{\mathbf{w}}$ can be hypercyclic. $\square$

1.1.1 The hypercyclic comparison principle

We conclude this section by introducing the following well-known concepts; they will be used several times in the book.

DEFINITION 1.16   *Let $T_0 : X_0 \to X_0$ and $T : X \to X$ be two continuous maps acting on topological spaces X_0 and X. The map T is said to be a **quasi-factor** of T_0 if there exists a continuous map with dense range $J : X_0 \to X$ such that the diagram*

$$\begin{array}{ccc} X_0 & \xrightarrow{\ T_0\ } & X_0 \\ {\scriptstyle J}\downarrow & & \downarrow{\scriptstyle J} \\ X & \xrightarrow{\ T\ } & X \end{array}$$

*commutes, i.e. $TJ = JT_0$. When this can be achieved with a homeomorphism $J : X_0 \to X$ (so that $T = JT_0 J^{-1}$), we say that T_0 and T are **topologically conjugate**. Finally, when T_0 and T are linear operators and the factoring map (resp. the homeomorphism) J can be taken as linear, we say that T is a **linear** quasi-factor of T_0 (resp. that T_0 and T are **linearly conjugate**).*

The usefulness of these definitions comes from the following simple but important observation: *hypercyclicity is preserved by quasi-factors and supercyclicity as well as the Hypercyclicity Criterion are preserved by linear quasi-factors. Moreover, any factoring map J sends hypercyclic points to hypercyclic points.* This is indeed obvious since, with the above notation, we have $O(J(x_0), T) = J(O(x_0, T_0))$ for any $x_0 \in X_0$. In J. H. Shapiro's book [220], this observation is called the **hypercyclic comparison principle**. See Exercise 1.14 for a simple illustration.

A particular instance of the hypercyclic comparison principle is the following useful remark: *if $T \in \mathfrak{L}(X)$ is hypercyclic and if $J \in \mathfrak{L}(X)$ has dense range and commutes with T then $HC(T)$ is invariant under J.*

1.2 Some spectral properties

In this section, we show that hypercyclic and supercyclic operators have some noteworthy spectral properties. We start with the following simple observation. Here and elsewhere in the book, we denote by $\sigma_p(R)$ the **point spectrum** of an operator R, i.e. the set of all eigenvalues of R.

PROPOSITION 1.17 *Let $T \in \mathfrak{L}(X)$ be hypercyclic. Then $\sigma_p(T^*) = \varnothing$.*

PROOF Suppose that $T^*(x^*) = \mu x^*$ for some $\mu \in \mathbb{K}$ and some $x^* \in X^*$, $x^* \neq 0$. If $x \in X$ then $\langle x^*, T^n(x) \rangle = \langle T^{*n}(x^*), x \rangle = \mu^n \langle x^*, x \rangle$ for all $n \in \mathbb{N}$. Since the set $\{\mu^n \langle x^*, x \rangle; \ n \in \mathbb{N}\}$ is obviously not dense in $\mathbb{K}$, it follows that no vector $x \in X$ can be hypercyclic for T. □

REMARK When X is locally convex, the statement $\sigma_p(T^*) = \varnothing$ means that for each $\mu \in \mathbb{K}$ the operator $T - \mu$ has dense range. The latter property always holds for a hypercyclic operator, even if X is not locally convex. A more general result will be proved in the next section (Lemma 1.31).

We have already observed that a contractive operator ($\|T\| \leq 1$) cannot be hypercyclic. Likewise, an expansive operator ($\|Tx\| \geq \|x\|$ for all x) cannot be hypercyclic either, since the orbit of any non-zero vector stays away from 0. The following theorem, due to C. Kitai [158], says that a hypercyclic operator cannot be "partly" contractive or expansive.

THEOREM 1.18 *Let X be a complex Banach space, and let $T \in \mathfrak{L}(X)$ be hypercyclic. Then every connected component of the spectrum of T intersects the unit circle.*

For the proof, we need two well-known results from spectral theory, which are stated in the next two lemmas. In what follows, X is a complex Banach space.

LEMMA 1.19 (RIESZ DECOMPOSITION THEOREM) *Let $T \in \mathfrak{L}(X)$, and assume that the spectrum of T can be decomposed as $\sigma(T) = \sigma_1 \cup \cdots \cup \sigma_N$, where the sets σ_i are closed and pairwise disjoint. Then one can write $X = X_1 \oplus \cdots \oplus X_N$, where each X_i is a closed T-invariant subspace and $\sigma(T_{|X_i}) = \sigma_i$ for each $i \in \{1, \ldots, N\}$.*

For a proof, see your favourite functional analysis book or Appendix D.

LEMMA 1.20 *Let $T \in \mathfrak{L}(X)$.*

(1) *Suppose that $\sigma(T) \subset \mathbb{D}$. Then there exist $a < 1$ and $N \in \mathbb{N}$ such that $\|T^n(x)\| \leq a^n \|x\|$ for any $x \in X$ and all $n \geq N$.*
(2) *Suppose that $\sigma(T) \subset \mathbb{C} \backslash \overline{\mathbb{D}}$. Then there exist $A > 1$ and $N \in \mathbb{N}$ such that $\|T^n(x)\| \geq A^n \|x\|$ for any $x \in X$ and all $n \geq N$.*

PROOF This follows from the spectral radius formula, applied to T in case (1) and to T^{-1} in case (2). □

Finally, we need an elementary topological lemma.

LEMMA 1.21 *Let K be a compact subset of $\mathbb{C}$, and let C be a connected component of K. Assume that C is contained in some open set $\Omega \subset \mathbb{C}$. Then one can find a clopen (i.e. closed and open) subset σ of K such that $C \subset \sigma \subset \Omega$.*

PROOF   As in any compact Hausdorff space, the component C is the intersection of all clopen subsets of K containing it, say $C = \bigcap_{O \in \mathcal{F}} O$ (see [145, Theorem 2-14]). By compactness, one can find $O_1, \ldots, O_n \in \mathcal{F}$ such that $O_1 \cap \cdots \cap O_n \cap (\mathbb{C} \backslash \Omega) = \varnothing$. Then $\sigma := O_1 \cap \cdots \cap O_n$ is the clopen set we are looking for.   □

PROOF OF THEOREM 1.18   Assume that some component C_1 of $\sigma(T)$ does not intersect the unit circle, so that $C_1 \subset \mathbb{D}$ or $C_1 \subset \mathbb{C} \backslash \overline{\mathbb{D}}$. By Lemma 1.21, one can find a clopen set $\sigma_1 \subset \sigma(T)$ such that $C_1 \subset \sigma_1 \subset \mathbb{D}$ or $C_1 \subset \sigma_1 \subset \mathbb{C} \backslash \overline{\mathbb{D}}$. By Lemma 1.19 applied for σ_1 and $\sigma_2 := \sigma(T) \backslash \sigma_1$, one can write $T = T_1 \oplus T_2$, where $\sigma(T_1) = \sigma_1$. It is easy to check that if T is hypercyclic then both T_1 and T_2 are hypercyclic (in fact, they are quasi-factors of T via the associated projections onto X_1 and X_2). Now, it follows from Lemma 1.20 that $\|T_1^n(x)\|$ tends to 0 or ∞ for any $x \in X_1 \backslash \{0\}$, so that T_1 cannot be hypercyclic. Hence, T itself is not hypercyclic.   □

COROLLARY 1.22   *No compact operator on a complex Banach space $X \neq \{0\}$ can be hypercyclic.*

PROOF   By Proposition 1.1, we may assume that X is infinite-dimensional. Let $T \in \mathfrak{L}(X)$ be a compact operator. Then the spectrum of T is countable and contains 0; since any countable subset of $\mathbb{C}$ is totally disconnected, it follows that $\{0\}$ is a connected component of $\sigma(T)$. By Theorem 1.18, T cannot be hypercyclic.   □

REMARK 1.23   The same result holds for *real* Banach spaces; see Exercise 1.7. However, it is worth noting that compact operators *can* be supercyclic. For example, consider the backward shift $B_{\mathbf{w}}$ of Example 1.15, where $\mathbf{w} = (w_n)_{n \geq 1}$ is a weight sequence tending to 0.

In the supercyclic case, a result similar to Theorem 1.18 was obtained by N. S. Feldman, V. G. Miller, and T. L. Miller in [109].

THEOREM 1.24   *Let X be a complex Banach space, and let $T \in \mathfrak{L}(X)$ be supercyclic. Then one can find $R \geq 0$ such that the (possibly degenerate) circle $\{|z| = R\}$ intersects each component of the spectrum of T.*

The circle $\{|z| = R\}$ will be called a **supercyclicity circle** for T. For the proof, we need the following simple lemma.

LEMMA 1.25   *Let $\mathcal{C}$ be a collection of connected compact subsets of $\mathbb{C}$. Then either there exists $R \geq 0$ such that the circle $\{|z| = R\}$ intersects every member of $\mathcal{C}$, or there exist $R > 0$ and $C_1, C_2 \in \mathcal{C}$ such that $C_1 \subset R\mathbb{D}$ and $C_2 \subset \mathbb{C} \backslash \overline{R\mathbb{D}}$.*

PROOF   Let us denote by $m : \mathbb{C} \rightarrow \mathbb{R}_+$ the function $z \mapsto |z|$. Then $m(\mathcal{C})$ is a collection of compact intervals in $\mathbb{R}_+$. Thus, either these intervals have a non-empty intersection or there exist two pairwise disjoint intervals in $m(\mathcal{C})$. This proves the lemma.   □

PROOF OF THEOREM 1.24   Assume that T is supercyclic and that no supercyclicity circle exists. By Lemma 1.25, we get $R > 0$ and two components C_1, C_2 of $\sigma(T)$

such that $C_1 \subset R\mathbb{D}$ and $C_2 \subset \mathbb{C}\backslash\overline{R\mathbb{D}}$. Since T is supercyclic if and only if $R^{-1}T$ is, we may, and will, assume that $R = 1$.

One can decompose $\sigma(T)$ as $\sigma(T) = \sigma_1 \cup \sigma_2 \cup \sigma_3$, where the sets σ_i are closed and pairwise disjoint, $C_1 \subset \sigma_1 \subset \mathbb{D}$ and $C_2 \subset \sigma_2 \subset \mathbb{C} \setminus \overline{\mathbb{D}}$. Indeed, by Lemma 1.21 one can find two clopen sets $\sigma_1, \sigma_2 \subset \sigma(T)$ such that $C_1 \subset \sigma_1 \subset \mathbb{D}$ and $C_2 \subset \sigma_2 \subset \mathbb{C} \setminus \overline{\mathbb{D}}$. Set $\sigma_3 := \sigma(T) \setminus (\sigma_1 \cup \sigma_2)$.

By the Riesz decomposition theorem, one can write $T = T_1 \oplus T_2 \oplus T_3$, where $\sigma(T_i) = \sigma_i$. Now, let $x = x_1 \oplus x_2 \oplus x_3$ be a supercyclic vector for T. Then T_1 and T_2 are supercyclic with supercyclic vectors x_1 and x_2, respectively (they are quasi-factors of T via the associated projections onto X_1 and X_2); in particular, x_1 and x_2 are non-zero. We then choose a sequence of integers (n_k) and a sequence of scalars (λ_k) such that

$$\lambda_k T_1^{n_k}(x_1) \oplus \lambda_k T_2^{n_k}(x_2) \oplus \lambda_k T_3^{n_k}(x_3) \to x_1 \oplus 0 \oplus 0.$$

Since $\sigma(T_2) \subset \mathbb{C}\backslash\overline{\mathbb{D}}$ and $x_2 \neq 0$, it follows from Lemma 1.20 that $\lambda_k \to 0$. By Lemma 1.20 again, this implies that $\lambda_k T_1^{n_k}(x_1) \to 0$, which contradicts $x_1 \neq 0$. $\square$

To conclude this section, we now show that, unlike in the case of hypercyclic operators, the adjoint of a supercyclic operator T can have an eigenvalue. However, T^* cannot have more than *one* eigenvalue and if it does have one then the operator T is "almost" hypercyclic. This is the content of the next result.

PROPOSITION 1.26 *Let X be a locally convex space, and let $T \in \mathfrak{L}(X)$ be super-cyclic. Then either $\sigma_p(T^*) = \varnothing$ or $\sigma_p(T^*) = \{\lambda\}$ for some $\lambda \neq 0$. In the latter case, $\mathrm{Ker}(T^* - \lambda)$ has dimension 1 and $\mathrm{Ker}(T^* - \lambda)^n = \mathrm{Ker}(T^* - \lambda)$ for all $n \geq 1$. Moreover, there exists a (closed) T-invariant hyperplane $X_0 \subset X$ such that $T_0 := \lambda^{-1}T_{|X_0}$ is hypercyclic on X_0.*

The proof relies on the following elementary lemma.

LEMMA 1.27 *Let $a, b, c, \lambda, \mu \in \mathbb{K}$. Then the sets $\mathbb{K} \cdot \{(a\lambda^n, b\mu^n);\ n \in \mathbb{N}\}$ and $\mathbb{K} \cdot \{(a, cn + b);\ n \in \mathbb{N}\}$ are not dense in $\mathbb{K}^2$.*

PROOF To prove the first assertion, we may assume that a, b, λ and μ are non-zero and that $|\lambda| \geq |\mu|$. If a sequence $\theta_k(a\lambda^{n_k}, b\mu^{n_k})$ converges to $(0, 1)$ then

$$\frac{|a|}{|b|} \leq \left|\frac{\theta_k a\lambda^{n_k}}{\theta_k b\mu^{n_k}}\right| \to 0,$$

a contradiction. The proof of the second assertion is left to the reader. $\square$

PROOF OF PROPOSITION 1.26 Let us fix some supercyclic vector x for T. We note that if $x^*, y^* \in X^*$ are linearly independent then the set

$$A_{x^*,y^*} := \mathbb{K} \cdot \{((\langle T^{*n}(x^*), x\rangle, \langle T^{*n}(y^*), x\rangle);\ n \in \mathbb{N}\}$$
$$= \mathbb{K} \cdot \{((\langle x^*, T^n(x)\rangle, \langle y^*, T^n(x)\rangle);\ n \in \mathbb{N}\}$$

is dense in $\mathbb{K}^2$. Indeed, the linear map $\Phi_{x^*, y^*} : X \to \mathbb{K}^2$ defined by $\Phi_{x^*, y^*}(z) = (\langle x^*, z \rangle, \langle y^*, z \rangle)$ is continuous and onto, so it maps the dense set $\mathbb{K} \cdot O(x, T)$ onto a dense subset of $\mathbb{K}^2$.

If x^* and y^* are any two eigenvectors of T^*, say $T^*(x^*) = \lambda x^*$ and $T^*(y^*) = \mu x^*$, then $A_{x^*, y^*} = \mathbb{K} \cdot \{(\langle x^*, x \rangle \lambda^n, \langle y^*, x \rangle \mu^n); \ n \in \mathbb{N}\}$, so that A_{x^*, y^*} is not dense in $\mathbb{K}^2$ by Lemma 1.27. Hence, any two eigenvectors of T^* (if there are any) are linearly dependent. It follows that T^* has at most one eigenvalue, and that if it does have one then the corresponding eigenspace has dimension 1. Moreover, $0 \notin \sigma_p(T^*)$ since a supercyclic operator has dense range.

From now on, we assume that $\sigma_p(T^*) = \{\lambda\}$, where $\lambda \neq 0$. Since T is supercyclic iff aT is supercyclic for any $a \neq 0$, we may in fact assume that $\lambda = 1$. Choose $x_0^* \neq 0$ with $T^*(x_0^*) = x_0^*$.

To prove the statement concerning $\mathrm{Ker}(T^* - I)^n$, $n \geq 1$, it is enough to show that $\mathrm{Ker}(T^* - I)^2 = \mathrm{Ker}(T^* - I)$. Then we get $\mathrm{Ker}(T^* - I)^n = \mathrm{Ker}(T^* - I)$ for all $n \geq 1$ by straightforward induction.

Let $y^* \in \mathrm{Ker}(T^* - I)^2$ be arbitrary. Then $(T^* - I)y^* \in \mathrm{Ker}(T^* - I) = \mathbb{K}x_0^*$, so one can write $T^*(y^*) = y^* + \alpha x_0^*$ where $\alpha \in \mathbb{K}$. By induction, we get $T^{*n}(y^*) = y^* + n\alpha x_0^*$ for all $n \in \mathbb{N}$, hence $A_{x_0^*, y^*} = \mathbb{K} \cdot \{(\langle x_0^*, x \rangle, \alpha \langle x_0^*, x \rangle n + \langle y^*, x \rangle); \ n \in \mathbb{N}\}$. By Lemma 1.27, it follows that $A_{x_0^*, y^*}$ is not dense in $\mathbb{K}^2$, so that $y^* \in \mathbb{K}x_0^*$. Thus we get $\mathrm{Ker}(T^* - I)^2 = \mathrm{Ker}(T^* - I)$, as required.

Finally, let $X_0 := \mathrm{Ker}(x_0^*)$. Then X_0 is a T-invariant hyperplane since x_0^* is an eigenvector of T^*. Setting $T_0 := T_{|X_0}$, it remains to be shown that T_0 is hypercyclic on X_0. With a slight abuse of notation, we write $X = \mathbb{K} \oplus X_0$ and assume that the supercyclic vector $x \in SC(T)$ has the form $x = 1 \oplus x_0$. Since $T^*(x_0^*) = x_0^*$, we may then write $T(1 \oplus 0) = 1 \oplus f$ where $f \in X_0$, and it follows that

$$T(1 \oplus u) = 1 \oplus (f + T_0(u)) \tag{1.3}$$

for all $u \in X_0$.

We first note that $\mathrm{Ker}(T_0^* - I) = \{0\}$. Indeed, let $z_0^* \in X_0^*$ satisfy $T_0^*(z_0^*) = z_0^*$, and let $z^* \in X^*$ be any extension of z_0^* to X. Since X_0 is T-invariant, we have $T^*(z^*)_{|X_0} = T_0^*(z_0^*) = z_0^* = z_{|X_0}^*$, so that $X_0 \subset \mathrm{Ker}((T^* - I)(z^*))$. Since $X_0 = \mathrm{Ker}(x_0^*)$, this means that $(T^* - I)(z^*) \in \mathbb{K}x_0^* = \mathrm{Ker}(T^* - I)$. Thus, we get $z^* \in \mathrm{Ker}(T^* - I)^2 = \mathrm{Ker}(T^* - I) = \mathbb{K}x_0^*$, and hence $z_0^* = z_{|\mathrm{Ker}(x_0^*)}^* = 0$. Since X_0 is locally convex, it follows that $T_0 - I$ has dense range.

Now, we show that T_0 is hypercyclic. By (1.3) and a straightforward induction, we have

$$T^n(1 \oplus x_0) = 1 \oplus (f + T_0(f) + \cdots + T_0^{n-1}(f) + T_0^n(x_0)) \tag{1.4}$$

for all $n \geq 1$. Since $1 \oplus x_0$ is a supercyclic vector for T we may find, for any $z \in X_0$, a net $(n_i, \lambda_i) \subset \mathbb{N} \times \mathbb{K}$ such that $\lambda_i T^{n_i}(1 \oplus x_0) \to 1 \oplus z$. By (1.4), the first coordinate gives $\lambda_i \to 1$, and inserting that into the second coordinate, this yields

$$f + T_0(f) + \cdots + T_0^{n_i - 1}(f) + T_0^{n_i}(x_0) \to z.$$

Applying $(T_0 - I)$, we get $T_0^{n_i}(f + (T_0 - I)x_0) \to f + (T_0 - I)z$. Since $z \in X_0$ is arbitrary, we have proved that the T_0-orbit of the vector $u_0 := f + (T_0 - I)x_0$ contains $f + \text{Ran}(T_0 - I)$ in its closure. Since $T_0 - I$ has dense range, this shows that T_0 is hypercyclic with hypercyclic vector u_0. $\qquad\square$

REMARK 1.28 Conversely, assume that $X = \mathbb{K} \oplus X_0$ and let $T_0 \in \mathfrak{L}(X_0)$ be hypercyclic. Then $T := I \oplus T_0$ is readily seen to be supercyclic, with $\sigma_p(T^*) = \{1\}$. However, there exist supercyclic operators T with $\sigma_p(T^*) = \{1\}$ which cannot be written in that form; see Exercise 1.9.

1.3 What does the set of hypercyclic vectors look like?

We have already observed that if T is a hypercyclic operator on some F-space X then the set of all hypercyclic vectors for T is large in a topological sense: it is a dense G_δ subset of X. As is well known, this implies largeness in an algebraic sense.

PROPOSITION 1.29 *Let $T \in \mathfrak{L}(X)$ be hypercyclic on the F-space X. Then every $x \in X$ is the sum of two hypercyclic vectors.*

PROOF Both the sets $HC(T)$ and $x - HC(T)$ are residual in X, so they have a non-empty intersection by the Baire category theorem. $\qquad\square$

The next theorem highlights another kind of "algebraic largeness" for the set of hypercyclic vectors. Here the underlying space X does not need to be complete. We say that a linear subspace $E \subset X$ is a **hypercyclic manifold** for T if $E \setminus \{0\}$ consists entirely of hypercyclic vectors.

THEOREM 1.30 *Let X be a topological vector space, and let $T \in \mathfrak{L}(X)$ be hypercyclic. If $x \in HC(T)$ then $\mathbb{K}[T]x := \{P(T)x;\ P \text{ polynomial}\}$ is a hypercyclic manifold for T. In particular, T admits a dense hypercyclic manifold.*

The proof relies on the following lemma.

LEMMA 1.31 *Let $T \in \mathfrak{L}(X)$ be hypercyclic. For any non-zero polynomial P, the operator $P(T)$ has dense range.*

PROOF We start with the following

FACT If $L \subset X$ is a closed T-invariant subspace then either $L = X$ or L has infinite codimension in X.

PROOF OF THE FACT Assume that L has finite codimension in X, i.e. the quotient space X/L is finite-dimensional. Let $q : X \to X/L$ be the canonical quotient map. The T-invariance of L means that $\text{Ker}(q) \subset \text{Ker}(qT)$. Therefore $qT : X \to X/L$ factors through the quotient map q, i.e. one can find an operator $A \in \mathfrak{L}(X/L)$ such that $Aq = qT$. Since q is continuous and onto, the operator A is a quasi-factor of T.

Hence A is hypercyclic on the finite-dimensional space X/L. By Proposition 1.1, it follows that $X/L = \{0\}$, i.e. $L = X$. ☐

Now let P be a non-zero polynomial and set $L := \overline{\mathrm{Ran}(P(T))}$. Since T commutes with $P(T)$, it is readily seen that L is T-invariant. Hence, by the above fact it is enough to show that L has finite codimension in X.

Let $x \in HC(T)$ and let $q : X \to X/L$ be the canonical quotient map. By the division algorithm and the commutativity of the algebra $\mathbb{K}[T]$, we see that

$$\mathbb{K}[T]x \subset \mathrm{Ran}(P(T)) + \mathrm{span}\,\{T^i(x);\ i < \deg(P)\}.$$

Thus $q(\mathbb{K}[T]x)$ is finite-dimensional, and so is $X/L = q(X)$ since x is a cyclic vector for T. ☐

REMARK   When the underlying space X is complex and locally convex, Lemma 1.31 has a much simpler proof. Indeed, if the polynomial P is non-constant then one can factor $P(T)$ as $P(T) = a(T - \mu_1) \ldots (T - \mu_d)$, where $a \neq 0$ and $\mu_1, \ldots, \mu_d \in \mathbb{C}$. Since $\sigma_p(T^*) = \varnothing$, each operator $T - \mu_i$ has dense range; hence $P(T)$ has dense range as well.

PROOF OF THEOREM 1.30   Let us fix $x \in HC(T)$. For any non-zero polynomial P, the operator $P(T)$ commutes with T and has dense range by Lemma 1.31. By the hypercyclic comparison principle (see p. 10), it follows that $P(T)x \in HC(T)$. Thus $\mathbb{K}[T]x$ is a hypercyclic manifold for T. Finally, $\mathbb{K}[T]x$ is dense in X since it contains $O(x, T)$. ☐

Theorem 1.30 is really a *linear* statement. Several interesting properties of hypercyclic operators can be deduced from this result. Here we just point out one consequence, on which we shall elaborate in Chapter 3.

COROLLARY 1.32   *If $T \in \mathfrak{L}(X)$ is hypercyclic then $HC(T)$ is connected.*

PROOF   The set $HC(T)$ lies between the two connected sets $\mathbb{K}[T]x$ and X, the smaller of which is dense in the larger. ☐

When the topological vector space X is a Fréchet space, much more can be said. The following theorem is inspired by a result of A. Fathi [105].

THEOREM 1.33   *Let X be a separable Fréchet space, and let $T \in \mathfrak{L}(X)$ be hypercyclic. Then $HC(T)$ is homeomorphic to X.*

REMARK 1.34   By a famous result of R. D. Anderson [5], every separable infinite-dimensional Fréchet space is homeomorphic to the separable Hilbert space. Hence the topological structure of $HC(T)$ is in some sense trivial.

REMARK 1.35   Let H be the separable infinite-dimensional Hilbert space. It follows from Theorem 1.33 that there exists a continuous map $f : H \to H$ such that the dynamical system (H, f) is **minimal**, i.e. the orbit of every point $x \in H$ under f is dense in H: to see this just put $f := \phi \circ T \circ \phi^{-1}$, where $T \in \mathfrak{L}(H)$ is hypercyclic and ϕ is a homeomorphism from $HC(T)$ onto H. This is Fathi's result mentioned above.

The proof of Theorem 1.33 requires a definition and a non-trivial result from infinite-dimensional topology.

DEFINITION 1.36 *Let X be a separable Fréchet space. A closed set $A \subset X$ is said to be a **Z-set** if, for any compact metric space K, the set of continuous functions $\mathcal{C}(K, X \backslash A)$ is dense in $\mathcal{C}(K, X)$ (with respect to the topology of uniform convergence).*

Thus, Z-sets are small enough to have in some sense no influence on the function spaces $\mathcal{C}(K, X)$. The result we need is as follows. For a proof, the reader should refer to [52, Theorem V.6.4].

LEMMA 1.37 *Let X be a separable Fréchet space. If $A \subset X$ is a countable union of Z-sets then $X \backslash A$ is homeomorphic to X.*

PROOF OF THEOREM 1.33 Recall that if $(V_j)_{j \in \mathbb{N}}$ is a countable basis of open sets for X then one can write $X \backslash HC(T) = \bigcup_{j \in \mathbb{N}} \bigcap_{n \geq 0} X \backslash T^{-n}(V_j)$. Thus, it suffices to prove that, for each non-empty open set $V \subset X$, the closed set $\bigcap_{n \geq 0} X \backslash T^{-n}(V)$ is a Z-set. In other words, given any $f \in \mathcal{C}(K, X)$ (where K is a compact metric space) and any open neighbourhood O of 0 in X, we need to find a $g \in \mathcal{C}(K, X)$ such that $g(K) \subset \bigcup_{n \geq 0} T^{-n}(V)$ and $(g - f)(K) \subset O$. Moreover, since X is locally convex we may assume that O is convex.

Let us choose $x \in HC(T)$. With each point $t \in K$ one may associate a natural number m_t such that $T^{m_t}(x) - f(t) \in O$ and also an open set $W_t \subset K$ such that $t \in W_t$ and $T^{m_t}(x) - f(s) \in O$ for all $s \in W_t$. By compactness, we may find a finite partition of unity $(\phi_i)_{1 \leq i \leq p}$ subordinate to the open covering (W_t); that is, the functions ϕ_i are continuous with $0 \leq \phi_i \leq 1$, $\sum_i \phi_i = 1$, and $\text{supp}(\phi_i) \subset W_{t_i}$ for some $t_i \in K$. Setting $m_i := m_{t_i}$ and $g := \sum_{i=1}^{p} \phi_i T^{m_i}(x)$, we then have $(g - f)(K) \subset O$, by the convexity of O: to see this just write $g(s) - f(s) = \sum_i \phi_i(s)[T^{m_i}(x) - f(s)]$. Now, for every $a \in K$ one can write $g(a) = P_a(T)x$, where P_a is a non-zero polynomial. Since $P_a(T)x$ is itself a hypercyclic vector for T, by Theorem 1.30, one can find an integer n_a such that $T^{n_a}(g(a)) \in V$. Hence $g(K) \subset \bigcup_{n \geq 0} T^{-n}(V)$, as required. $\qquad \square$

REMARK As observed by G. Godefroy, the proof yields in fact the following more general result: *if X is a separable Fréchet space and if $G \subset X$ is a G_δ set containing a dense linear subspace of X then G is homeomorphic to X.*

1.4 Three examples

We conclude this chapter by studying hypercyclicity in three classical families of operators. Our aim is to illustrate some of the beautiful features of concrete operator theory that appear in linear dynamics.

1.4.1 Weighted shifts

In this subsection we characterize the hypercyclicity and the supercyclicity of a weighted backward shift in terms of its weight sequence. All results are due to H. N. Salas [215], [216].

We first consider **bilateral** weighted backward shifts $B_{\mathbf{w}}$ acting on $\ell^2(\mathbb{Z})$. Thus $B_{\mathbf{w}}$ is defined by $B_{\mathbf{w}}(e_n) = w_n e_{n-1}$, where $(e_n)_{n \in \mathbb{Z}}$ is the canonical basis of $\ell^2(\mathbb{Z})$ and $\mathbf{w} = (w_n)_{n \in \mathbb{Z}}$ is a bounded sequence of positive real numbers. Any such sequence $\mathbf{w}$ will be called a **weight sequence**. It is clear that $B_{\mathbf{w}}$ is bounded (with $\|B_{\mathbf{w}}\| \le \|\mathbf{w}\|_\infty$) and that $B_{\mathbf{w}}$ is invertible iff $\mathbf{w}$ is bounded below, i.e. $\inf_n w_n > 0$.

Instead of weighted shifts acting on unweighted sequence spaces, it is often convenient to consider unweighted shifts acting on weighted spaces. Accordingly, with any weight sequence $\mathbf{w} = (w_n)_{n \in \mathbb{Z}}$ we associate the sequence $\omega = (\omega_n)_{n \in \mathbb{Z}}$ defined by $\omega_0 = 1$ and $\omega_n / \omega_{n+1} = w_{n+1}$ (the reader should take care to distinguish between ω and w). We introduce the weighted space

$$\ell^2(\mathbb{Z}, \omega) := \left\{ x \in \mathbb{C}^{\mathbb{Z}}; \ \|x\|^2 := \sum_{n \in \mathbb{Z}} \omega_n^2 |x_n|^2 < \infty \right\}.$$

The operator $B_{\mathbf{w}} : \ell^2(\mathbb{Z}) \to \ell^2(\mathbb{Z})$ is unitarily equivalent to the bilateral *unweighted* backward shift $B : \ell^2(\mathbb{Z}, \omega) \to \ell^2(\mathbb{Z}, \omega)$. Indeed, the operator $U : \ell^2(\mathbb{Z}) \to \ell^2(\mathbb{Z}, \omega)$ defined by $U(x_n) = (x_n / \omega_n)$ is unitary, and clearly $U B_{\mathbf{w}} = B U$. In particular, $B_{\mathbf{w}}$ is linearly conjugate with B, so that the hypercyclicity or supercyclicity of $B_{\mathbf{w}}$ is equivalent to that of B. Since the results for B are easier to state, we first prove the following theorem.

THEOREM 1.38   *Let $\omega = (\omega_n)_{n \in \mathbb{Z}}$ be a sequence of positive numbers such that $\sup_n \omega_n / \omega_{n+1} < \infty$, and let B be the unweighted backward shift acting on $\ell^2(\mathbb{Z}, \omega)$.*

(a) *The operator B is hypercyclic if and only if*

$$\forall q \in \mathbb{N} : \ \liminf_{n \to +\infty} \omega_{\pm n + q} = 0. \tag{1.5}$$

(b) *The operator B is supercyclic if and only if*

$$\forall q \in \mathbb{N}: \ \liminf_{n \to +\infty} \omega_{n+q} \omega_{-n+q} = 0. \tag{1.6}$$

PROOF   We start with the hypercyclic case. Suppose first that (1.5) holds, and let us show that B satisfies the Hypercyclicity Criterion. Choose a positive number $C > \max(1, \sup_n(\omega_n / \omega_{n+1}))$. By (1.5), one can find an increasing sequence $(n_k) \subset \mathbb{N}$ such that $\omega_{\pm n_k - k} \le C^{-3k}$ for all k in $\mathbb{N}$. Then $\omega_{\pm n_k + i} \xrightarrow{k \to \infty} 0$ for every $i \in \mathbb{Z}$. Indeed, if i is fixed and $k \ge |i|$ then $\omega_{\pm n_k + i} \le C^{i+k} \omega_{\pm n_k + k} \le C^{-k}$. Now, let $\mathcal{D}_1 = \mathcal{D}_2 := c_{00}(\mathbb{Z})$ (the linear span of the basis vectors e_i) and let S be the forward shift, defined on $\mathcal{D}_2$ by $S(e_i) = e_{i+1}$. Since $BS = I$ on $\mathcal{D}_2$, we just need to prove that $B^{n_k}(e_i)$ and $S^{n_k}(e_i)$ both tend to 0 for any $i \in \mathbb{Z}$ (then we conclude by using linearity). But this is clear since

$$\|B^{n_k}(e_i)\| = \omega_{-n_k + i} \quad \text{and} \quad \|S^{n_k}(e_i)\| = \omega_{n_k + i}.$$

Conversely, suppose that B is hypercyclic and fix $q \in \mathbb{N}$. Since the hypercyclic vectors are dense in $\ell^2(\mathbb{Z}, \omega)$, for any $\delta \in (0,1)$ one may find $x \in \ell^2(\mathbb{Z}, \omega)$ and an integer $n > 2q$ such that

$$\left\| x - e_q \right\| < \delta \quad \text{and} \quad \left\| B^n(x) - e_q \right\| < \delta.$$

Looking at the qth and the $(n + q)$th coordinate in the first inequality we get $|\omega_q(x_q - 1)| < \delta$ and $|\omega_{n+q} x_{n+q}| < \delta$. Likewise, looking at the qth and the $(-n + q)$th coordinate in the second inequality, we get $|\omega_q(x_{n+q} - 1)| < \delta$ and $|\omega_{-n+q} x_q| < \delta$. Putting all this together, using the triangle inequality and assuming $\delta < \omega_q$, we conclude that

$$\omega_{\pm n + q} < \frac{\delta \omega_q}{\omega_q - \delta}.$$

Since δ is arbitrary, this gives the converse implication.

Now we turn to the supercyclic case. If condition (1.6) holds then we can find as above an increasing sequence $(n_k) \subset \mathbb{N}$ such that, for any $i, j \in \mathbb{Z}$,

$$\omega_{n_k + i}\, \omega_{-n_k + j} \xrightarrow{k \to \infty} 0.$$

This means exactly that the Supercyclicity Criterion is satisfied for $\mathcal{D}_1 = \mathcal{D}_2 := c_{00}(\mathbb{Z})$ and the forward shift S, since $\| B^{n_k}(e_j) \| \| S_{n_k}(e_i) \| = \omega_{-n_k + j} \omega_{n_k + i}$.

Conversely, suppose that B is supercyclic and fix $q \in \mathbb{N}$. Since the supercyclic vectors are dense in $\ell^2(\mathbb{Z})$, for each $\delta \in (0, 1)$ one may find $x \in \ell^2(\mathbb{Z}, \omega)$, $\lambda \in \mathbb{K} \setminus \{0\}$ and $n > 2q$ such that

$$\left\| x - e_q \right\| < \delta \quad \text{and} \quad \left\| \lambda B^n(x) - e_q \right\| < \delta.$$

Then, as above, we get $|\omega_q(x_q - 1)| < \delta$, $|\omega_{n+q} x_{n+q}| < \delta$ from the first inequality and $|\omega_q(\lambda x_{n+q} - 1)| < \delta$, $|\lambda \omega_{-n+q} x_q| < \delta$ from the second. This leads to

$$\omega_{-n+q} \omega_{n+q} < \frac{\delta^2}{|\lambda|\, |x_q|\, |x_{n+q}|} < \frac{(\omega_q \delta)^2}{(\omega_q - \delta)^2},$$

and the result follows. $\qquad \square$

Returning to the discussion preceding Theorem 1.38, the result can be reformulated as a characterization of hypercyclicity and supercyclicity for weighted bilateral shifts acting on the unweighted space $\ell^2(\mathbb{Z})$.

COROLLARY 1.39 *Let $B_{\mathbf{w}}$ be a bilateral weighted backward shift on $\ell^2(\mathbb{Z})$, with weight sequence $\mathbf{w} = (w_n)_{n \in \mathbb{Z}}$. Then $B_{\mathbf{w}}$ is hypercyclic if and only if, for any $q \in \mathbb{N}$,*

$$\liminf_{n \to +\infty} \max\left\{ (w_1 \cdots w_{n+q})^{-1}, (w_0 \cdots w_{-n+q+1}) \right\} = 0.$$

Likewise, $B_{\mathbf{w}}$ is supercyclic if and only if, for any $q \in \mathbb{N}$,

$$\liminf_{n \to +\infty} (w_1 \cdots w_{n+q})^{-1} \times (w_0 \cdots w_{-n+q+1}) = 0.$$

PROOF   This follows at once from Theorem 1.38, since the associated sequence (ω_n) satisfies $\omega_{n+q} = (w_1 \cdots w_{n+q})^{-1}$ and $\omega_{-n+q} = w_0 \cdots w_{-n+q+1}$. $\square$

REMARK   When the shift $B_{\mathbf{w}}$ is invertible, these characterizations can be stated in a simpler way. For example, $B_{\mathbf{w}}$ is hypercyclic if and only if

$$\liminf_{n \to +\infty} \max \left\{ (w_1 \cdots w_n)^{-1}, (w_{-1} \cdots w_{-n}) \right\} = 0.$$

Indeed, since the weights w_p are bounded above and below, the products $(w_1 \cdots w_{n+q})^{-1}$ and $w_0 \cdots w_{-n+q+1}$ are equivalent to $(w_1 \cdots w_n)^{-1}$ and $w_{-1} \cdots w_{-n}$, up to constants depending only on q.

We now turn to **unilateral** weighted backward shifts $B_{\mathbf{w}}$ acting on $\ell^2(\mathbb{N})$. Thus $B_{\mathbf{w}}$ is defined by $B_{\mathbf{w}}(e_0) = 0$ and $B_{\mathbf{w}}(e_n) = w_n e_{n-1}$ for all $n \geq 1$, where $(e_n)_{n \in \mathbb{N}}$ is the canonical basis of $\ell^2(\mathbb{N})$ and $\mathbf{w} = (w_n)_{n \geq 1}$ is a bounded sequence of positive numbers. We saw in Example 1.15 that $B_{\mathbf{w}}$ is always supercyclic, regardless of the weight sequence $\mathbf{w}$. Concerning hypercyclicity, we have the following characterization [215], which is both simple and intuitive.

THEOREM 1.40   *Let $B_{\mathbf{w}}$ be a unilateral weighted backward shift acting on $\ell^2(\mathbb{N})$. Then $B_{\mathbf{w}}$ is hypercyclic if and only if $\limsup_{n \to \infty} (w_1 \cdots w_n) = \infty$.*

The proof of the "if" part is quite similar to that given for Example 1.9. For the "only if" part, one argues as in the proof of Theorem 1.38. We leave the details as a useful exercise for the reader.

REMARK 1.41   We have stated the above results for shifts acting on $\ell^2(\mathbb{Z})$ or $\ell^2(\mathbb{N})$. The same characterizations are valid for shifts acting on c_0 or ℓ^p, $1 \leq p < \infty$, with exactly the same proofs. For similar statements in a large class of F-spaces, we refer to [134].

1.4.2 Operators commuting with translations

At first sight, one might think that Birkhoff's and MacLane's examples 1.4 and 1.8 are completely unrelated results. On the contrary, however, they are particular instances of a beautiful general theorem due to G. Godefroy and J. H. Shapiro [123]. Recall that the translation operators $T_a : H(\mathbb{C}) \to H(\mathbb{C})$ are defined by $T_a f(z) = f(z + a)$.

THEOREM 1.42   *Let T be a continuous linear operator on $H(\mathbb{C})$. Assume that T commutes with every translation operator T_a and is not a scalar multiple of the identity. Then T is hypercyclic.*

The proof of Theorem 1.42 requires two steps. We shall first show that T has a representation of the form $T = \phi(D)$, where ϕ is an entire function and D is the derivative operator. This representation will help us to find many eigenvectors for T, and an application of the Godefroy–Shapiro Criterion (Corollary 1.10) will give the conclusion.

LEMMA 1.43   *Let T be a continuous operator on $H(\mathbb{C})$, and assume that T commutes with every translation operator T_a. Then there exists an entire function of exponential type $\phi : \mathbb{C} \to \mathbb{C}$ such that $T = \phi(D)$.*

PROOF   We first explain the meaning of $\phi(D)$. We recall that an entire function ϕ is said to be of **exponential type** if there exist finite constants A and B such that

$$|\phi(z)| \le A e^{B|z|}$$

for all $z \in \mathbb{C}$. Writing $\phi(z) = \sum_{n \ge 0} c_n z^n$, it follows from Cauchy's inequalities that ϕ is of exponential type iff $|c_n| \le C R^n / n!$ for some finite constants C, R and all $n \in \mathbb{N}$ (see Appendix A). Using Cauchy's inequalities again, we see that for every $f \in H(\mathbb{C})$, the series $\sum c_n D^n(f)$ converges uniformly on compact subsets of $\mathbb{C}$. The sum of this series we denote by $\phi(D)f$.

We now turn to the proof of Lemma 1.43. Let $L : H(\mathbb{C}) \to \mathbb{C}$ be the linear functional defined by

$$L(f) = (Tf)(0).$$

By the assumption on T, we may write

$$Tf(z) = (T_z(Tf))(0) = (T(T_z f))(0) = L(T_z f)$$

for any $z \in \mathbb{C}$. Moreover, the linear functional L is continuous, so one can find $C < \infty$ and a closed disk $K = \overline{D}(0, R)$ such that $|L(f)| \le C \|f\|_K$ for any $f \in H(\mathbb{C})$. Appealing to the Hahn–Banach theorem and the Riesz representation theorem, we get a complex measure μ supported on K such that

$$L(f) = \int_K f \, d\mu$$

for every $f \in H(\mathbb{C})$. Thus we obtain

$$Tf(z) = \int_K T_z f \, d\mu = \int_K f(z + w) \, d\mu(w).$$

Since K is compact, the power series expansion of the function f at z is uniformly convergent on K. Hence, we may write

$$Tf(z) = \int_K \sum_{n=0}^{\infty} \frac{f^{(n)}(z)}{n!} w^n \, d\mu(w)$$

$$= \sum_{n=0}^{\infty} c_n D^n f(z)$$

where $c_n = (1/n!) \int_K w^n d\mu(w)$. Since $|c_n| \le (R^n/n!) \|\mu\|$ for all n, the formula $\phi(z) := \sum_0^{\infty} c_n z^n$ defines an entire function of exponential type, which concludes the proof.   □

In order to apply the Godefroy–Shapiro Criterion, we also need the following density lemma. For any $\lambda \in \mathbb{C}$, let us denote by $e_\lambda \in H(\mathbb{C})$ the function $z \mapsto e^{\lambda z}$.

LEMMA 1.44   *If $\Lambda \subset \mathbb{C}$ has an accumulation point in $\mathbb{C}$ then* span $\{e_\lambda; \ \lambda \in \Lambda\}$ *is dense in $H(\mathbb{C})$.*

PROOF   Let us fix Λ. Since $H(\mathbb{C})$ is locally convex, one can use the Hahn–Banach theorem. Let L be a continuous linear functional on $H(\mathbb{C})$ which is orthogonal to span$\{e_\lambda; \ \lambda \in \Lambda\}$. As above, we find a complex measure μ with compact support $K \subset \mathbb{C}$ such that

$$L(f) = \int_K f \, d\mu$$

for all $f \in H(\mathbb{C})$. Let $F : \mathbb{C} \to \mathbb{C}$ be defined by $F(\lambda) = L(e_\lambda) = \int_K e^{\lambda z} d\mu(z)$. Differentiation under the integral sign shows that F is an entire function; moreover, $F \equiv 0$ on Λ by assumption. Since Λ has an accumulation point, F is identically zero. Thus we get

$$0 = F^{(n)}(0) = \int_K z^n d\mu = L(z^n)$$

for all $n \in \mathbb{N}$. Since the polynomials are dense in $H(\mathbb{C})$, it follows that $L = 0$. This concludes the proof.   □

PROOF OF THEOREM 1.42   Since T commutes with translations and is not a multiple of the identity, it can be written as $T = \phi(D)$ where ϕ is a *non-constant* entire function of exponential type. Then

$$T(e_\lambda) = \sum_{n \geq 0} a_n D^n(e_\lambda) = \sum_{n \geq 0} a_n \lambda^n e_\lambda = \phi(\lambda) e_\lambda$$

for every $\lambda \in \mathbb{C}$. In other words, each e_λ is an eigenvector of T with associated eigenvalue $\phi(\lambda)$. Moreover, it follows from Liouville's theorem applied to ϕ (and to $1/\phi$ if necessary) that the open sets $V := \{\lambda \in \mathbb{C}; \ |\phi(\lambda)| > 1\}$ and $U := \{\lambda \in \mathbb{C}; \ |\phi(\lambda)| < 1\}$ are both non-empty. By Lemma 1.44, we conclude that the assumptions of Corollary 1.10 are satisfied. Hence T is hypercyclic.   □

REMARK   It is not hard to show that an operator on $H(\mathbb{C})$ commutes with all translation operators if and only if it commutes with the derivative operator D. The "only if" part is easy since $Df = \lim_{h \to 0} h^{-1} (T_h f - f)$, and the "if" part follows from the identity $T_a = e^{aD}$.

1.4.3 Composition operators

Let $\phi : \mathbb{D} \to \mathbb{D}$ be a holomorphic self-map of $\mathbb{D}$. The **composition operator** induced by ϕ is defined on $H(\mathbb{D})$ by $C_\phi(f) = f \circ \phi$. Obviously, C_ϕ sends $H(\mathbb{D})$ continuously into itself. It is also true, but non-trivial, that C_ϕ sends the Hardy space $H^2(\mathbb{D})$ continuously into itself. The shortest way to prove this is by observing that a function $f \in H(\mathbb{D})$ is in H^2 iff the function $|f|^2$ has a *harmonic majorant*; see [90] or Exercise 1.12. For an elementary proof, see [220].

The study of composition operators consists in the comparison between the properties of an operator C_ϕ and that of the function ϕ itself. For useful accounts of this very interesting subject, we refer again to [220] and to [90]. In this section, we will focus of course on the hypercyclicity of a composition operator C_ϕ acting on $H^2(\mathbb{D})$.

We start with the following result concerning the fixed points of ϕ.

PROPOSITION 1.45   *Let ϕ be a holomorphic self-map of $\mathbb{D}$. If C_ϕ is hypercyclic on $H^2(\mathbb{D})$ then ϕ has no fixed points in $\mathbb{D}$.*

PROOF   First we note that if α is any point of $\mathbb{D}$ then

$$C_\phi^*(k_\alpha) = k_{\phi(\alpha)}, \qquad (1.7)$$

where $k_z \in H^2(\mathbb{D})$ is the reproducing kernel at $z \in \mathbb{D}$. This follows from the identities

$$\langle f, C_\phi^*(k_\alpha)\rangle_{H^2} = \langle f \circ \phi, k_\alpha\rangle_{H^2} = f(\phi(\alpha)) = \langle f, k_{\phi(\alpha)}\rangle_{H^2}.$$

Thus if ϕ has a fixed point $\alpha \in \mathbb{D}$ then $C_\phi^*(k_\alpha) = k_\alpha$, so that k_α is an eigenvector for C_ϕ^*. By Proposition 1.17, C_ϕ cannot be hypercyclic.   □

Thus, we are reduced to maps ϕ without fixed points in $\mathbb{D}$. We will in fact restrict ourselves to a particular class of maps, the so-called linear fractional maps.

DEFINITION 1.46   *A non-constant map $\phi : \mathbb{D} \to \mathbb{D}$ is called a **linear fractional map** if it can be written as*

$$\phi(z) = \frac{az + b}{cz + d},$$

where $a, b, c, d \in \mathbb{C}$.

Equivalently, a linear fractional map $\phi : \mathbb{D} \to \mathbb{D}$ is the restriction to $\mathbb{D}$ of an automorphism of the Riemann sphere $\widehat{\mathbb{C}} = \mathbb{C} \cup \{\infty\}$ (still denoted by ϕ) mapping the unit disk $\mathbb{D}$ into itself. We shall denote by $LFM(\mathbb{D})$ the set of all linear fractional maps of $\mathbb{D}$ and by $Aut(\mathbb{D})$ the set of all automorphisms of $\mathbb{D}$. Then $Aut(\mathbb{D}) \subset LFM(\mathbb{D})$, and if we conjugate a function $\phi \in LFM(\mathbb{D})$ by an automorphism of $\mathbb{D}$ then we still get a linear fractional map.

If $\phi \in LFM(\mathbb{D})$ then ϕ has at least one and at most two fixed points in $\widehat{\mathbb{C}}$. Moreover, it is not too hard to show that if ϕ has no fixed points in $\mathbb{D}$ then ϕ has an attractive fixed point on $\mathbb{T} = \partial\mathbb{D}$ (see [220] or [90]). Thus, the linear fractional maps of $\mathbb{D}$ without fixed points in $\mathbb{D}$ fall into one of the following two classes:

- **parabolic** maps: those having a unique attractive fixed point $\alpha \in \mathbb{T}$;
- **hyperbolic** maps: those having an attractive fixed point $\alpha \in \mathbb{T}$ and a second fixed point $\beta \in \widehat{\mathbb{C}}\backslash\mathbb{D}$.

In particular, if $\alpha \in \mathbb{T}$ is the attractive fixed point of $\phi \in LFM(\mathbb{D})$, the iterates

$$\phi_n(z) := \phi \circ \cdots \circ \phi(z)$$

converge to α for any $z \in \overline{\mathbb{D}}$, except possibly the repulsive fixed point $\beta \in \mathbb{T}$. Moreover, the convergence is uniform on compact subsets of $\overline{\mathbb{D}} \backslash \{\beta\}$.

At this point, it is useful to observe that if u is any automorphism of $\mathbb{D}$ then the associated composition operator C_u is invertible on $H^2(\mathbb{D})$ and $C_{u \circ \phi \circ u^{-1}} = (C_u)^{-1} \circ C_\phi \circ C_u$. Thus, we do not change the dynamical properties of C_ϕ if we conjugate ϕ by a disk automorphism. Since the group $Aut(\mathbb{D})$ acts doubly transitively on $\mathbb{T}$, we may therefore assume, by conjugating with a suitable automorphism if necessary, that the fixed points on $\mathbb{T}$ of the linear fractional map at which we are looking are located exactly where we want. For example, we may assume that the attractive fixed point is $\alpha = +1$, and (if necessary) that the repulsive fixed point is $\beta = -1$.

Finally, we point out that linear fractional maps are easier to visualize when considered as transformations of the upper half-plane $\mathbb{P}_+ := \{s \in \mathbb{C}; \ \mathrm{Im}(s) > 0\}$. This is done via the so-called **Cayley map**

$$\omega(z) := i \frac{1+z}{1-z},$$

which maps $\mathbb{D}$ conformally onto $\mathbb{P}_+$ with $\omega(0) = i$ and $\omega(1) = \infty$. Thus, if $\phi : \mathbb{D} \to \mathbb{D}$ is a holomorphic self-map of $\mathbb{D}$ then the map $\psi := \omega \circ \phi \circ \omega^{-1}$ is a holomorphic self-map of $\mathbb{P}_+$.

Let $\phi \in LFM(\mathbb{D})$ have no fixed points within $\mathbb{D}$, and let it have its attractive fixed point $\alpha = +1$ on the boundary. When ϕ is parabolic, the associated map ψ turns out to be a translation $\psi(s) = s + ia$, where $\mathrm{Re}(a) \geq 0$. It is an automorphism of $\mathbb{P}_+$ if and only if $\mathrm{Re}(a) = 0$. When ϕ is hyperbolic the map ψ is the dilation $\psi(s) = \lambda(s - s_0) + s_0$, where $\lambda > 1$ and $\mathrm{Im}(s_0) \leq 0$. It is an automorphism if and only if $\mathrm{Im}(s_0) = 0$, which means that the second fixed point of ϕ lies on $\mathbb{T}$.

We now have the following characterization of hypercyclicity for composition operators induced by linear fractional maps.

THEOREM 1.47 *Let $\phi \in LFM(\mathbb{D})$ have no fixed points in $\mathbb{D}$. Then C_ϕ is hypercyclic on $H^2(\mathbb{D})$ if and only if ϕ is either hyperbolic or a parabolic automorphism of $\mathbb{D}$.*

For the proof, we need the following elementary density lemma.

LEMMA 1.48 *Let $\alpha \in \mathbb{C} \backslash \mathbb{D}$, and set $\mathcal{P}_\alpha := \{P \text{ polynomial}; \ P(\alpha) = 0\}$. Then $\mathcal{P}_\alpha$ is dense in $H^2(\mathbb{D})$.*

PROOF Let $f = \sum_0^\infty c_n z^n \in H^2(\mathbb{D})$ be orthogonal to $\mathcal{P}_\alpha$. Then

$$0 = \langle f, z^{p+1} - \alpha z^p \rangle_{H^2} = c_{p+1} - \alpha c_p$$

for all $p \in \mathbb{N}$. This gives $c_p = \alpha^p c_0$ for all p, and since $|\alpha| \geq 1$ it follows that f can belong to $H^2(\mathbb{D})$ only if it is zero. $\qquad \square$

PROOF OF THEOREM 1.47 In what follows, we denote by $\psi = \omega \circ \phi \circ \omega^{-1}$ the sister of ϕ living on $\mathbb{P}_+$. Moreover, we assume that the attractive fixed point of ϕ is $\alpha = +1$ and we denote by β the other fixed point of ϕ, setting $\beta := \alpha$ if ϕ is parabolic.

We first suppose that ϕ is an automorphism of $\mathbb{D}$. Then β is the attractive fixed point of ϕ^{-1} (this is particularly clear when considering ψ). We set $\mathcal{D}_1 := \mathcal{P}_\alpha, \mathcal{D}_2 := \mathcal{P}_\beta$ and $S_k = C_{\phi^{-1}}^k = C_\phi^{-k}$, $k \in \mathbb{N}$. For any $P \in \mathcal{D}_1$ and any $z \in \mathbb{T} \setminus \{\beta\}$, one has

$$\lim_{k \to \infty} P \circ \phi_k(z) = P(\alpha) = 0.$$

Since

$$\|C_\phi^k(P)\|_{H^2}^2 = \int_{-\pi}^{\pi} |P \circ \phi_k(e^{i\theta})|^2 \frac{d\theta}{2\pi},$$

it follows from Lebesgue's convergence theorem that $C_\phi^k(P) \to 0$ in $H^2(\mathbb{D})$. Similarly, $S_k(Q) = C_{\phi^{-1}}^k(Q) \to 0$ in $H^2(\mathbb{D})$ for any $Q \in \mathcal{D}_2$. Thus we see that C_ϕ satisfies the Hypercyclicity Criterion.

Suppose now that ϕ is hyperbolic and not an automorphism. Then ψ is given by $\psi(s) = \lambda(s - s_0) + s_0$, where $\lambda > 1$ and $\text{Im}(s_0) < 0$. If we conjugate ψ by an automorphism of $\mathbb{P}_+$ of the form $u(z) = \varepsilon z$, $\varepsilon \in (0,1)$, we get $u \circ \psi \circ u^{-1}(s) = \lambda(s - \varepsilon s_0) + \varepsilon s_0$. Thus we may assume that $\text{Im}(s_0) > -1$. Then the domain $\Delta := \omega^{-1}(\{\text{Im}(s - s_0) > 0\})$ is *bounded* (and contains $\mathbb{D}$) because $\omega^{-1}(s) = (s - i)/(s + i)$ is bounded outside any neighbourhood of $-i$; and ϕ induces an automorphism of Δ because ψ induces an automorphism of the half-plane $\{\text{Im}(s - s_0) > 0\}$.

Let us denote by ϕ^{-1} the inverse of $\phi_{|\Delta}$, whose attractive fixed point is β, the repulsive fixed point of ϕ. We set $\mathcal{D}_1 := \mathcal{P}_\alpha$, $\mathcal{D}_2 := \mathcal{P}_\beta$, and $S_k Q(z) = Q \circ \phi^{-1} \circ \cdots \circ \phi^{-1}(z)$ for any $Q \in \mathcal{D}_2$. As above, the assumptions of the Hypercyclicity Criterion are satisfied; the boundedness of Δ ensures that one may apply Lebesgue's theorem even to the sequence $(S_k Q)$.

Finally, suppose that ϕ is parabolic and not an automorphism. To show that C_ϕ is not hypercyclic in this case, it is clearly enough to check that, for any $f \in H^2(\mathbb{D})$, one can find some non-zero function $g \in H^2(\mathbb{D})$ such that $\langle C_\phi^n(f), g \rangle_{H^2} \to 0$ as $n \to \infty$. We shall in fact prove that

$$\langle C_\phi^n(f), z \rangle_{H^2} \to 0$$

for every $f \in H^2(\mathbb{D})$. So let us fix the function f. As usual, we set $\phi_n := \phi \circ \cdots \circ \phi$. We first note that

$$\langle C_\phi^n(f), z \rangle_{H^2} = f'(\phi_n(0)) \, \phi_n'(0),$$

since $\langle h, z \rangle_{H^2} = h'(0)$ for any $h \in H^2(\mathbb{D})$. Moreover, writing $f(w) = \sum_0^\infty a_n w^n$ and using Schwarz's inequality, one can estimate f' as follows:

$$|f'(w)|^2 \le \left(\sum_{n \ge 1} |a_n|^2 \right) \left(\sum_{n \ge 1} n^2 |w|^{2(n-1)} \right) \le \frac{2 \|f\|_{H^2}^2}{(1 - |w|^2)^3}$$

for any $w \in \mathbb{D}$. Thus, we get

$$|\langle C_\phi^n(f), z \rangle_{H^2}| \le C \frac{|\phi_n'(0)|}{(1 - |\phi_n(0)|^2)^{3/2}}. \tag{1.8}$$

Here and below, $C = C(f, \phi)$ is a constant which does not depend on n and may change from line to line.

We now have to make some calculations. Since ϕ is parabolic and not an automorphism, the associated map ψ is a translation, $\psi(s) = s + ia$, with $\mathrm{Re}(a) > 0$. An easy computation yields

$$\phi(z) = \frac{(2-a)z + a}{-az + 2 + a}.$$

Set $\psi_n := \psi \circ \cdots \circ \psi$ $(n \geq 1)$. Then $\psi_n(s) = s + ina$, so that

$$\phi_n(z) = \frac{(2-na)z + na}{-naz + 2 + na}.$$

Differentiating and evaluating at zero, we find

$$\phi_n'(0) = \frac{4}{(2+na)^2}. \tag{1.9}$$

Moreover, it is readily seen that

$$1 - \phi_n(0) \sim \frac{2}{na} \quad \text{as} \quad n \to \infty. \tag{1.10}$$

At this point, we use the fact that ϕ is not an automorphism, i.e. $\mathrm{Re}(a) > 0$. Since $\psi_n(i) = i + ina$ for all $n \in \mathbb{N}$, we see that $\psi_n(i)$ goes to ∞ along a line in the half-plane $\mathbb{P}_+$ which is not parallel to the real axis. Going back to the unit disk, this forces $\phi_n(0)$ to approach the attractive fixed point $\alpha = 1$ *non-tangentially*; in other words $|1 - \phi_n(0)|$ is comparable with $1 - |\phi_n(0)|$ and hence with $1 - |\phi_n(0)|^2$, since $1 - |\phi_n(0)| \leq 1 - |\phi_n(0)|^2 \leq 2(1 - |\phi_n(0)|)$. In view of (1.8), (1.9) and (1.10), it follows that

$$|\langle C_\phi^n(f), z \rangle_{H^2}| \leq \frac{Cn^{3/2}}{n^2} \leq \frac{C}{\sqrt{n}}.$$

This concludes the proof. $\qquad\qquad\qquad\qquad\qquad\qquad\qquad\qquad\qquad\qquad\qquad\square$

REMARK   With more sophisticated arguments, it can be shown that C_ϕ is in fact not even *supercyclic* when ϕ is a parabolic non-automorphism. This will be done in Chapter 9.

1.5 Comments and exercises

The word "hypercyclic" seems to have been introduced by B. Beauzamy; see [34]. Hypercyclic vectors have also been called **universal** (e.g. in [119]) or **orbital** [158]. The word "supercyclic" is older. It was first used by H. M. Hilden and L. J. Wallen in [144].

Birkhoff's transitivity theorem appears in [53]. One may ask whether, besides being a set of the first Baire category, the set of all non-hypercyclic vectors for a given hypercyclic operator is small in some other natural sense. In fact this is not always the case, at least for the two natural notions of σ-*porosity* and *Haar negligibility* [21], [32]. The set of non-hypercyclic vectors may be extremely small, however, even for a Hilbert space operator; see the recent paper [131] by S. Grivaux and M. Roginskaya.

The first version of the Hypercyclicity Criterion goes back to the unpublished Ph.D. thesis of C. Kitai [158]. It was formulated there for the full sequence of integers ($n_k = k$) and for a single dense set $\mathcal{D} = \mathcal{D}_1 = \mathcal{D}_2$. Kitai's proof was constructive, whereas R. M. Gethner and J. H. Shapiro used the Baire category theorem in [119]. However, Kitai's method is still interesting for several extensions where one cannot hope to use Baire category arguments, e.g. *frequent* hypercyclicity (see Chapter 6) or unbounded operators (see [49] and [93]). The version of the Hypercyclicity Criterion given here is due to J. Bès [45].

Looking at the proof of Theorem 1.6, it is apparent that the assumptions in the Hypercyclicity Criterion can be weakened. Namely, instead of two dense sets $\mathcal{D}_1$ and $\mathcal{D}_2$ we may assume that we have one dense set $\mathcal{D}_1$ and, for each $x \in \mathcal{D}_1$, a sequence (n_k), a dense set $\mathcal{D}_{2,x}$ and maps $S_{n_k,x}$ that may depend on x. This sharpened Hypercyclicity Criterion is in fact equivalent to the original version (see e.g. [80]). This can be shown by making use of the so-called *three open sets conditions*; see Chapter 4. However, there may be cases where the improved criterion can easily be applied, while this is not so obvious for the "standard" version (see Chapter 7).

It is important to note that when an operator T satisfies the Hypercyclicity Criterion then so does $T \oplus T$; hence, $T \oplus T$ is automatically hypercyclic. We will come back to this in Chapter 4.

The role played by the eigenvalues in the dynamics of an operator was pointed out by K. F. Clancey and D. D. Rogers in [78] (see also Exercise 1.4 below). This is a very interesting topic, which will be investigated further in Chapters 2, 5 and 6. In particular, we will see that the existence of many *unimodular* eigenvalues often ensures a strong form of hypercyclicity.

The Supercyclicity Criterion is due to A. Montes-Rodríguez and H. N. Salas [184]. Some other supercyclicity criteria were given by N. S. Feldman, V. G. Miller and T. L. Miller in [109]. All these criteria are in fact equivalent (see [39]). However, one variant may be more tractable than the others for a specific application.

Many of the ideas of Section 1.2 are contained in D. A. Herrero's paper [143]. This paper contains a remarkable result: a complete spectral characterization of Hilbert space operators which are limits of hypercyclic operators with respect to the operator norm topology (and likewise for the supercyclic case).

Theorem 1.30 is due to several authors: [143] and [63] for complex locally convex spaces, [46] for the real case and finally [236] for non-locally convex spaces. Regarding Theorem 1.33, A. Fathi's paper [105] contains the result for an operator satisfying the Hypercyclicity Criterion. This theorem was obtained independently by A. Fathi and G. Godefroy.

EXERCISE 1.1 **Universal families** ([133])
Let $\mathbf{T} = (T_i)_{i \in I}$ be a family of continuous maps between two topological spaces X and Y. The family $\mathbf{T}$ is said to be **universal** if there exists some point $x \in X$ such that the set $\{T_i(x); \; i \in I\}$ is dense in Y. Such a point x is said to be universal for $\mathbf{T}$, and the set of universal points for $\mathbf{T}$ is denoted by $Univ(\mathbf{T})$.

1. Assume that X is a Baire space and that Y has a countable basis of open sets. Show that the following are equivalent:

    (i) $Univ(\mathbf{T})$ is a dense G_δ subset of X;
    (ii) $Univ(\mathbf{T})$ is dense in X;
    (iii) for each pair of non-empty open sets $(U, V) \subset X \times Y$, one can find $i \in I$ such that $T_i(U) \cap V \neq \varnothing$.

2. Assume that $X = Y$ is a second-countable Baire space. Moreover, assume that each map T_i has dense range and that the family $\mathbf{T}$ commutes, i.e. $T_i \circ T_j = T_j \circ T_i$ for any $i, j \in I$. Show that $Univ(\mathbf{T})$ is either empty, or dense in X.

3. *The Universality Criterion* Suppose now that X is a Baire topological vector space, that Y is a separable and metrizable topological vector space, that $I = \mathbb{N}$ and that each $T_n : X \to Y$ is a continuous linear operator. Assume that there exist dense sets $\mathcal{D}_1 \subset X$ and $\mathcal{D}_2 \subset Y$, a sequence of integers (n_k) and a sequence of maps $S_{n_k} : \mathcal{D}_2 \to X$ such that (see Definition 1.5):

(1) $T_{n_k}(x) \to 0$ for any $x \in \mathcal{D}_1$;
(2) $S_{n_k}(y) \to 0$ for any $y \in \mathcal{D}_2$;
(3) $T_{n_k} S_{n_k}(y) \to y$ for each $y \in \mathcal{D}_2$.

Show that $Univ(\mathbf{T})$ is a dense G_δ subset of X.

4. Assume that X is a separable F-space. Let $T \in \mathfrak{L}(X)$ and assume that $\bigcup_{|\lambda|>1} \mathrm{Ker}(T-\lambda)$ and $\bigcup_{\lambda \in e^{2i\pi\mathbb{Q}}\setminus\{1\}} \mathrm{Ker}(T - \lambda)$ both span a dense linear subspace of X. For each $n \geq 1$, set $T_n := I + T + \cdots + T^{n-1}$. Show that the sequence $(T_n)_{n\geq 1}$ is universal, with a dense G_δ set of universal vectors. (*Hint:* Apply the Universality Criterion with $n_k = k!$.)

EXERCISE 1.2   **The Cesàro operator on L^p**

Let $p \in [1,\infty)$, and let $T : L^p(0,1) \to L^p(0,1)$ be the operator defined by

$$Tf(x) = \frac{1}{x} \int_0^x f(t)dt.$$

The operator T is bounded on L^p by the classical **Hardy inequality**.

1. For any $\alpha \in (-\infty, 1/p)$, let $\phi_\alpha \in L^p(0,1)$ be defined by $\phi_\alpha(x) = x^{-\alpha}$. Show that if $I \subset (-\infty, 1/p)$ is a non-empty open interval then $\mathrm{span}\{\phi_\alpha; \alpha \in I\}$ is dense in $L^p(0,1)$. (*Hint:* Use the Hahn–Banach theorem, a suitable change of variable and some properties of the Laplace transform.)
2. Find the eigenvectors of T and the corresponding eigenvalues.
3. Show that T is hypercyclic on $L^p(0,1)$. (*Hint:* Use the Godefroy–Shapiro Criterion).

EXERCISE 1.3 ([39])   Let X be a separable Banach space. The **generalized kernel** of an operator $T \in \mathfrak{L}(X)$ is the subspace $\mathrm{Ker}^*(T) := \bigcup_{n\geq 1} \mathrm{Ker}(T^n)$. Prove that an operator T with a dense generalized kernel is supercyclic if and only if it has dense range.

EXERCISE 1.4   **Eigenvalues and cyclicity** ([78], [6], [129])

1. *A cyclicity criterion*   Let X be a separable F-space and let $T \in \mathfrak{L}(X)$. Assume that there exist two dense sets $\mathcal{D}_1$ and $\mathcal{D}_2$, a sequence of polynomials (P_k) and a sequence of maps $S_k : \mathcal{D}_2 \to X$ such that
   (i) $\|P_k(T)x\|\|S_k(y)\| \to 0$ for any $x \in \mathcal{D}_1$ and any $y \in \mathcal{D}_2$;
   (ii) $P_k(T)S_k(y) \to y$ for each $y \in \mathcal{D}_2$.
   Prove that T is cyclic, with a dense G_δ set of cyclic vectors.
2. Let $T \in \mathfrak{L}(X)$ be an F-space operator. Assume that $\bigcup_{P\in\mathcal{P}} \mathrm{Ker}\,(P(T))$ is dense in X, where $\mathcal{P}$ is the set of all polynomials $P \in \mathbb{K}[X]$ such that $P(T)$ has dense range. Show that T is cyclic. (In particular, if X is a Fréchet space then T is cyclic provided that $\sigma_p(T^*) = \varnothing$ and the eigenvectors of T span a dense subspace of X.)

EXERCISE 1.5   **Classes of operators which are not supercyclic**

1. *Unitary operators* ([144])   Show that a unitary operator on a complex Hilbert space cannot be supercyclic. (*Hint:* By the spectral theorem, it is enough to consider a multiplication operator $T = M_\phi$ acting on $L^2(\mu)$, for some probability measure μ. Assume that $f \in L^2(\mu)$ is a supercyclic vector for M_ϕ and try to approximate a function of the form fg, where $|g|$ is not constant.)
2. *Normal operators* ([144])   Show that a normal operator on a complex Hilbert space cannot be supercyclic. (*Hint:* Proceed as above and approximate f to show that $|\phi|$ is constant.)
3. *Isometries* ([8])   We recall that any Banach space isometry has a non-trivial invariant subspace; this result is due to R. Godement [124, Théorème J]. Let X be a complex Banach space and let $T \in \mathfrak{L}(X)$ be an isometry.
   (a) Let $x_0, x, z \in X$, $\alpha, \beta \in \mathbb{C} \setminus \{0\}$ and $m < n \in \mathbb{N}$. Show that

$$\left\| \frac{\beta}{\alpha} T^{n-m}(x) - z \right\| \leq \left| \frac{\beta}{\alpha} \right| \|x - \alpha T^m(x_0)\| + \|\beta T^n(x_0) - z\|.$$

(b) Show that if T is supercyclic then any non-zero vector $x \in X$ is supercyclic for T.
(c) Conclude that T is not supercyclic.

4. *Hyponormal operators* ([65]) Let T be a **hyponormal** operator on the Hilbert space H, namely $\|T(h)\| \geq \|T^*(h)\|$ for every $h \in H$. Equivalently, $T^*T - TT^*$ is a positive operator.

   (a) Prove that for every $f \in H$, the sequence $(\|T^n(f)\|)$ is log-convex, i.e. for all $n \in \mathbb{N}$,
   $$\|T^n(f)\|^2 \leq \|T^{n+1}(f)\| \|T^{n-1}(f)\|.$$

   (b) Deduce that if $f \in H$ is such that $\|T(f)\| \geq \|f\|$ then the sequence $(\|T^n(f)\|)$ is non-decreasing. Why does this imply that T is not hypercyclic (see [158])?

   (c) Suppose that T is supercyclic. For any $\varepsilon > 0$, let $A_\varepsilon := (1 + \varepsilon) T/\|T\|$. Show that there exists $f \in SC(T)$ such that $\|A_\varepsilon(f)\| \geq \|f\|$. Deduce that $\|A_\varepsilon(h)\| \geq \|h\|$ for any $h \in H$. Conclude that $A = T/\|T\|$ is a unitary operator and obtain a contradiction.

EXERCISE 1.6 Let X be a non-zero finite-dimensional vector space. Show that X supports a supercyclic operator iff $\dim(X) = 1$ in the complex case, or $\dim(X) \in \{1, 2\}$ in the real case.

EXERCISE 1.7 ([59]) **Compact operators on real Banach spaces**
Let X be a real Banach space, $X \neq \{0\}$. The aim of the exercise is to show that no compact operator on X can be hypercyclic. So we fix $T \in \mathfrak{L}(X)$ and, towards a contradiction, we assume that T is hypercyclic and compact. Let $Y = X \oplus iX$ and $S = T \oplus iT$ be the complexifications of X and T.

1. Show that S is compact and that S^* has no eigenvalues.
2. Deduce that $\sigma(S) = \{0\}$ and obtain a contradiction.

EXERCISE 1.8 Let X be an infinite-dimensional Banach space and let $T \in \mathfrak{L}(X)$ be compact and supercyclic. Show that $\sigma(T) = \{0\}$. (*Hint*: First show that $\sigma_p(T^*) = \varnothing$.)

EXERCISE 1.9 **A strange supercyclic operator**

1. Let $\phi \in H^\infty(\mathbb{D})$ be defined by $\phi(z) = 1 + i + z$. Let M_ϕ be the associated multiplication operator on $H^2(\mathbb{D})$ and set $R = M_\phi^*$. For each $n \geq 1$, put $R_n = I + \cdots + R^{n-1}$. Show that the sequence (R_n) is universal, with a dense G_δ set of universal vectors. (*Hint*: Look at Exercise 1.1.)
2. Show that $R - I$ is not onto. (*Hint*: 1 is a non-isolated point of the boundary of $\sigma(R)$.) Deduce that one can find $u \in H^2(\mathbb{D})$ such that u is a universal vector for (R_n) and $u \notin \text{Ran}(R - I)$. (*Hint*: A proper linear subspace of $H^2(\mathbb{D})$ cannot be comeager; see the proof of Proposition 1.29.)
3. Let $X := \mathbb{C} \oplus H^2(\mathbb{D})$ and let $T \in \mathfrak{L}(X)$ have the matrix representation
   $$T = \begin{pmatrix} 1 & 0 \\ u & R \end{pmatrix}.$$

   (a) Show that T is supercyclic with supercyclic vector $1 \oplus 0$.
   (b) Show that $\sigma_p(T^*) = \{1\}$ and that T is not similar to an operator of the form $I \oplus T_0$. (*Hint*: 1 is not an eigenvalue of T.)

EXERCISE 1.10 *J*-**sets** ([84])
Let X be a topological space and let $T : X \to X$ be a continuous map. For any $x \in X$, we denote by $J(x, T)$ the set of all $z \in X$ such that $(x, z) \in \overline{\bigcup_{n \in \mathbb{N}} Gr(T^n)}$, where $Gr(T^n) \subset X \times X$ is the graph of T^n. When X is metrizable, a point $z \in X$ is in $J(x, T)$ iff there exist a sequence of integers (n_k) and a sequence $(x_k) \subset X$ such that $x_k \to x$ and $T^{n_k}(x_k) \to z$.

1. Show that $J(x, T)$ is a closed T-invariant subset of X containing $O(x, T)$.
2. Observe that if $x \in X$ then a point $z \in X$ is in $J(x, T)$ iff for any neighbourhood U of x and any neighbourhood V of z, one can find $n \in \mathbb{N}$ such that $T^n(U) \cap V \neq \varnothing$. Deduce that the following are equivalent.

   (i) T is topologically transitive;
   (ii) $J(x, T) = X$ for all $x \in X$;
   (iii) the set $\{x \in X; \ J(x, T) = X\}$ is dense in X.

We now assume that X is a complex separable Banach space and that $T \in \mathfrak{L}(X)$.

3. Let $x \in X$. Show that if z is an interior point of $J(x, T)$ then one can find a sequence $(x_k) \subset X$ and a sequence (n_k) *tending to infinity* such that $x_k \to x$ and $T^{n_k}(x_k) \to z$.
4. Let $x^* \in X^*$ and $\lambda \in \mathbb{C}$ satisfy $T^*(x^*) = \lambda x^*$. Also let $x \in X$, and assume that $J(x, T)$ has non-empty interior. Prove that the following holds (where we set $r := |\langle x^*, x \rangle|$):

if $|\lambda| < 1$ then $\langle x^*, z \rangle = 0$ for every z in the interior of $J(x, T)$;
if $|\lambda| = 1$ then $|\langle x^*, z \rangle| = r$ for every z in the interior of $J(x, T)$;
if $|\lambda| > 1$ then $\langle x^*, Q(T)x \rangle = 0$ for any polynomial Q.

5. Show that if T is cyclic and $J(x, T)$ has non-empty interior for some T-cyclic vector x then $P(T)$ has dense range for any non-zero polynomial P. (*Hint:* Use part 4 of the exercise.)
6. Show that the following are equivalent:

   (i) T is hypercyclic;
   (ii) T is cyclic and $J(x, T) = X$ for some T-cyclic vector x.

(*Hint:* Show first that if $A \in \mathfrak{L}(X)$ commutes with T then $A(J(x, T)) \subset J(Ax, T)$.)

EXERCISE 1.11   Let $\omega = (\omega_n)_{n \in \mathbb{Z}}$ be a weight sequence such that the backward shift B acting on $\ell^2(\mathbb{Z}, \omega)$ is invertible and hypercyclic. Show that there exists an $x \in HC(B)$ which is not hypercyclic for B^{-1}.

EXERCISE 1.12 ([215])   Let $B_{\mathbf{w}_1}, \dots, B_{\mathbf{w}_N}$ be unilateral backward weighted shifts acting on $\ell^2(\mathbb{N})$. Find a necessary and sufficient condition for the operator $T := B_{\mathbf{w}_1} \oplus \cdots \oplus B_{\mathbf{w}_N}$ to be hypercyclic on $\ell^2 \oplus \cdots \oplus \ell^2$.

EXERCISE 1.13   Let $p \in [1, \infty)$ and let $H^p(\mathbb{D})$ be the usual Hardy space on the disk.

1. Show that a function $f \in H(\mathbb{D})$ is H^p if and only if $|f|^p$ admits a harmonic majorant, that is, there exists a (non-negative) harmonic function u on $\mathbb{D}$ such that $|f|^p \leq u$. (*Hint:* For the "only if" part, consider the Poisson integral of $|f^*|^p$; for the "if" part, use the mean-value formula for harmonic functions).
2. Deduce that if $\phi : \mathbb{D} \to \mathbb{D}$ is a holomorphic self-map of $\mathbb{D}$ then the composition operator C_ϕ maps $H^p(\mathbb{D})$ continuously into itself.

EXERCISE 1.14 **Composition operators on $H(\mathbb{D})$** ([219])
Let $\phi \in H(\mathbb{D})$ be an automorphism of $\mathbb{D}$ without fixed points in $\mathbb{D}$. Show that the composition operator C_ϕ acting on $H(\mathbb{D})$ is hypercyclic. (*Hint:* Use the hypercyclic comparison principle.)

2

Hypercyclicity everywhere

Introduction

In this chapter, we prove several striking results showing that hypercyclicity is far from being an exotic phenomenon. In particular, we show that hypercyclic operators exist on any infinite-dimensional separable Fréchet space and that one can construct hypercyclic operators with "arbitrary" orbits. In the same spirit, we show that linear dynamics is in some sense as complicated as topological dynamics. We also discuss the size of the set of all hypercyclic operators on some given space X. Finally, we show that any Hilbert space operator may be written as the sum of two hypercyclic operators.

2.1 Mixing operators

When considering a strategy for constructing hypercyclic operators on an arbitrary "abstract" separable Banach space, one certainly has to keep in mind that some spaces have very few operators. Indeed, it is now well known that there exist (infinite-dimensional) Banach spaces on which any operator has the form $\lambda I + S$, where S is strictly singular (for example, the so-called *hereditarily indecomposable* spaces; see Chapter 6 for some results concerning hypercyclic operators on such spaces). By a recent result of S. A. Argyros and R. Haydon [9], one can even encounter Banach spaces on which every operator has the form $\lambda I + K$, where K is a *compact* operator.

The examples of hypercyclic operators encountered in Chapter 1 are very far from being small perturbations of λI. Yet, if hypercyclic operators are supposed to exist on any separable Banach space, it *must* be true that some operators of the form $\lambda I + K$ are hypercyclic. And indeed, it was shown by H. N. Salas that any operator of the form "identity plus a backward shift" is hypercyclic [215].

As should be clear from the above, Salas' theorem is in some sense central in the theory. The aim of this section is to prove a generalization of this result, which is due to S. Grivaux and S. Shkarin [132].

The theorem we want to prove is in fact a statement about *topologically mixing* operators; so first we have to introduce the mixing property. This property has nothing to do with linearity. It is quite well known and has been extensively studied in topological dynamics.

DEFINITION 2.1 *Let X be a topological vector space. An operator $T \in \mathfrak{L}(X)$ is said to be (topologically) **mixing** if the following property holds true: for any pair (U, V) of non-empty open subsets of X, one can find an $N \in \mathbb{N}$ such that $T^n(U) \cap V \neq \varnothing$ for all $n \geq N$.*

By its very definition, the mixing property is a strong form of topological transitivity; in particular, mixing operators are hypercyclic if the underlying space X is a separable F-space. In fact, in this case it is easy to see that an operator $T \in \mathfrak{L}(X)$ is mixing if and only if it is **hereditarily hypercyclic**, which means that for any infinite set $\mathbf{N} \subset \mathbb{N}$ the family $\{T^n;\ n \in \mathbf{N}\}$ is universal. This happens in particular if T satisfies the Hypercyclicity Criterion with respect to the full sequence $(n_k) = (k)$.

The following theorem, which is a slight improvement of a result from [132], gives a general and easily verifiable sufficient condition for a linear operator to be mixing. It applies in particular to operators of the form $T = I + B$, where B is any weighted backward shift on $\ell^p(\mathbb{N})$, and this yields a rather transparent proof of Salas' result mentioned above. It can also be seen as an extension of the Godefroy–Shapiro Criterion (see Chapter 1).

THEOREM 2.2   *Let X be a topological vector space, and let $T \in \mathfrak{L}(X)$. Set*

$$\Lambda_1(T) := \mathrm{span}\left(\bigcup_{|\lambda|=1, N \in \mathbb{N}} \mathrm{Ker}(T-\lambda)^N \cap \mathrm{Ran}(T-\lambda)^N \right),$$

$$\Lambda^+(T) := \mathrm{span}\left(\Lambda_1(T) \cup \bigcup_{|\lambda|>1, N \in \mathbb{N}} \mathrm{Ker}(T-\lambda)^N \right),$$

$$\Lambda^-(T) := \mathrm{span}\left(\Lambda_1(T) \cup \bigcup_{|\lambda|<1, N \in \mathbb{N}} \mathrm{Ker}(T-\lambda)^N \right).$$

If $\Lambda^+(T)$ and $\Lambda^-(T)$ are both dense in X then T is mixing.

Applying this result to operators of the form $T = I + B$, we see that if $B \in \mathfrak{L}(X)$ and span $\left(\bigcup_{N \in \mathbb{N}} \mathrm{Ker}(B^N) \cap \mathrm{Ran}(B^N) \right)$ is dense in X then $I + B$ is mixing. This leads to the following widely applicable corollary. Part (1) is Salas' theorem [215].

COROLLARY 2.3   *Let $B \in \mathfrak{L}(X)$, where X is a topological vector space. Assume that the **generalized kernel** $\mathrm{Ker}^*(B) := \bigcup_{N \in \mathbb{N}} \mathrm{Ker}(B^N)$ is dense in X, and that $\mathrm{Ker}^*(B) \subset \mathrm{Ran}(B)$. Then $I + B$ is mixing. This happens in the following two cases.*

(a) *$X = c_0(\mathbb{N})$ or $\ell^p(\mathbb{N})$, and B is any weighted backward shift with non-zero weights.*

(b) *$\dim X = \infty$ and B is a **generalized backward shift** in the sense of Godefroy–Shapiro [123]; that is, $\mathrm{Ker}(B)$ is one-dimensional and $\mathrm{Ker}^*(B)$ is dense in X.*

The sequence $(\mathrm{Ker}(B^k))_{k \geq 0}$ is called the **sequence of iterated kernels** of B. We shall use without comment the following well-known and easily proved facts:

- the sequence of iterated kernels is non-decreasing, and if $\mathrm{Ker}(B^k) = \mathrm{Ker}(B^{k+1})$ for some $k \geq 0$ then $\mathrm{Ker}(B^j) = \mathrm{Ker}(B^k)$ for all $j \geq k$;

- if $\mathrm{Ker}(B)$ is finite-dimensional then $\mathrm{Ker}(B^k)$ is finite-dimensional for each $k \geq 1$ and $\dim(\mathrm{Ker}(B^k)) - \dim(\mathrm{Ker}(B^{k-1}))$ is non-increasing with respect to k.

PROOF OF COROLLARY 2.3   Without any assumption on the operator B, we have $\mathrm{Ker}(B^N) \cap \mathrm{Ran}(B) = B(\mathrm{Ker}B^{N+1})$ for all $N \geq 1$. Therefore, if $\mathrm{Ker}(B^N) \subset \mathrm{Ran}(B)$ for all N then in fact

$$\mathrm{Ker}(B^N) = B(\mathrm{Ker}(B^{N+1})) = B(\mathrm{Ker}(B^{N+1}) \cap \mathrm{Ran}(B))$$
$$= B^2(\mathrm{Ker}(B^{N+2})) = \cdots$$

so that $\mathrm{Ker}(B^N) \subset \mathrm{Ran}(B^k)$ for any choice of $N, k \geq 1$. Thus, $I + B$ is mixing if $\mathrm{Ker}^*(B)$ is dense and contained in $\mathrm{Ran}(B)$.

Case (a) is clear since $\mathrm{Ker}^*(B) = c_{00}(\mathbb{N})$ (the set of all finitely supported sequences) and c_{00} is obviously contained in $\mathrm{Ran}(B)$. Of course, one may also observe that (a) is a particular case of (b).

In case (b), we have $\dim(\mathrm{Ker}(B^k)) - \dim(\mathrm{Ker}(B^{k-1})) \leq 1$ for each positive integer k, so that $\dim(\mathrm{Ker}(B^N)) \leq N$ for all $N \in \mathbb{N}$. Moreover, if the inequality is strict for some N then the sequence $(\mathrm{Ker}(B^k))$ is stationary. This is impossible since $\mathrm{Ker}^*(B)$ is dense in X and hence infinite-dimensional. Thus $\dim(\mathrm{Ker}(B^N)) = N$ for all $N \in \mathbb{N}$.

We now apply the rank-nullity theorem to the operator $B_{|\mathrm{Ker}(B^{N+1})}$. This gives

$$\dim\left(B(\mathrm{Ker}(B^{N+1}))\right) + \dim\left(\mathrm{Ker}(B_{|\mathrm{Ker}(B^{N+1})})\right) = \dim\left(\mathrm{Ker}(B^{N+1})\right),$$

so that $\dim\left(B(\mathrm{Ker}(B^{N+1}))\right) \geq N + 1 - \dim\left(\mathrm{Ker}(B)\right) \geq N$. Since $B(\mathrm{Ker}(B^{N+1}))$ is contained in $\mathrm{Ker}(B^N)$, it follows that $B(\mathrm{Ker}(B^{N+1})) = \mathrm{Ker}(B^N)$ for all N, hence $\mathrm{Ker}^*(B) \subset \mathrm{Ran}(B)$. $\qquad\square$

The following definition will be useful for the proof of Theorem 2.2. The notation J^{mix} comes from the paper [84] by G. Costakis and A. Manoussos.

DEFINITION 2.4   *If R is a continuous linear operator acting on some topological vector space E, we put*

$$J^{\mathrm{mix}}(R) := \{(u,v) \in E \times E; \, \exists (x_n)_{n \in \mathbb{N}} \subset X \, : \, x_n \to u \text{ and } R^n(x_n) \to v\}.$$

Notice that the space E may be finite-dimensional. The usefulness of this definition comes from the following simple lemma.

LEMMA 2.5   *Let $T \in \mathfrak{L}(X)$. If $J^{\mathrm{mix}}(T)$ is dense in $X \times X$ then T is mixing.*

PROOF   Assume that $J^{\mathrm{mix}}(T)$ is dense in X. If (U,V) is any pair of non-empty open subsets of X, one can find a point $(u,v) \in (U \times V) \cap J^{\mathrm{mix}}(T)$ to which corresponds a sequence $(x_n)_{n \in \mathbb{N}}$ given by the definition of $J^{\mathrm{mix}}(T)$. Then, for large enough $n \in \mathbb{N}$, we have $x_n \in U$ and $T^n(x_n) \in V$, whence $T^n(U) \cap V \neq \varnothing$. This shows that T is mixing. $\qquad\square$

REMARK   When the space X is metrizable, it can be shown that T is mixing if and only if $J^{\mathrm{mix}}(T) = X \times X$.

In view of this lemma, Theorem 2.2 follows at once from the next proposition.

PROPOSITION 2.6   *If $T \in \mathfrak{L}(X)$ then $\Lambda^-(T) \times \Lambda^+(T) \subset J^{\mathrm{mix}}(T)$.*

We now turn to the proof of Proposition 2.6. Perhaps surprisingly, this proof is finite-dimensional in essence, the main step being the following lemma.

LEMMA 2.7   *Let N be a positive integer and let E be a $2N$-dimensional topological vector space with basis $(e_1, \ldots, e_{2N})$. Set $E_0 := \mathrm{span}\,(e_1, \ldots, e_N)$. If $B \in \mathfrak{L}(E)$ is the backward shift built on the basis $(e_1, \ldots, e_{2N})$ then $E_0 \times E_0 \subset J^{\mathrm{mix}}(I + B)$.*

PROOF   We may assume that $E = \mathbb{K}^{2N}$ and that $(e_1, \ldots, e_{2N})$ is the canonical basis of $\mathbb{K}^{2N}$. Then each vector $x \in E$ can be written as

$$x = \begin{pmatrix} u \\ \widetilde{u} \end{pmatrix},$$

where $u, \widetilde{u} \in \mathbb{K}^N$. Finally, we identify any operator $R \in \mathfrak{L}(\mathbb{K}^{2N})$ with its matrix in the canonical basis of $\mathbb{K}^{2N}$. Thus, B can be identified with the standard $2N \times 2N$ Jordan matrix:

$$B = \begin{pmatrix} 0 & 1 & 0 & \cdots & 0 \\ 0 & \ddots & \ddots & \ddots & \vdots \\ \vdots & \ddots & \ddots & \ddots & 0 \\ \vdots & & \ddots & \ddots & 1 \\ 0 & \cdots & \cdots & 0 & 0 \end{pmatrix}$$

We have to show that, given $(u, v) \in \mathbb{K}^N \times \mathbb{K}^N$, one can find a sequence $(x_n)_{n \in \mathbb{N}}$ in $\mathbb{K}^{2N}$ such that

$$x_n \to \begin{pmatrix} u \\ 0 \end{pmatrix} \quad \text{and} \quad (I + B)^n x_n \to \begin{pmatrix} v \\ 0 \end{pmatrix}. \tag{2.1}$$

A straightforward computation shows that

$$(I + B)^n = \begin{pmatrix} K_n & L_n \\ 0 & K_n \end{pmatrix}$$

for any $n \geq 2N$, where

$$K_n = \begin{pmatrix} 1 & \binom{n}{1} & \cdots & \binom{n}{N-1} \\ 0 & 1 & \ddots & \vdots \\ \vdots & \ddots & \ddots & \binom{n}{1} \\ 0 & \cdots & 0 & 1 \end{pmatrix}$$

and

$$L_n = \begin{pmatrix} \binom{n}{N} & \binom{n}{N+1} & \cdots & \binom{n}{2N-1} \\ \binom{n}{N-1} & \binom{n}{N} & \ddots & \vdots \\ \vdots & \ddots & \ddots & \binom{n}{N+1} \\ \binom{n}{1} & \cdots & \binom{n}{N-1} & \binom{n}{N} \end{pmatrix}$$

Thus, we have to find two sequences (u_n), $(\widetilde{u}_n)$ in $\mathbb{K}^N$ such that

$$\begin{cases} u_n \to u \quad \text{and} \quad \widetilde{u}_n \to 0; \\ K_n u_n + L_n \widetilde{u}_n \to v \quad \text{and} \quad K_n \widetilde{u}_n \to 0. \end{cases}$$

This will be achieved as follows. First we show that the matrix L_n is invertible for large enough n. Putting $u_n := u$ and $\widetilde{u}_n := L_n^{-1}(v - K_n u)$, we will then get the required result if we are able to prove that

$$\|L_n^{-1}\| + \|L_n^{-1} K_n\| \to 0 \quad \text{and} \quad \|K_n L_n^{-1}\| + \|K_n L_n^{-1} K_n\| \to 0. \qquad (2.2)$$

Heuristically speaking, the matrix L_n behaves like

$$L_n' := \begin{pmatrix} \frac{n^N}{N!} & \frac{n^{N+1}}{(N+1)!} & \cdots & \frac{n^{2N-1}}{(2N-1)!} \\ \frac{n^{N-1}}{(N-1)!} & \frac{n^N}{N!} & \ddots & \vdots \\ \vdots & \ddots & \ddots & \frac{n^{N+1}}{(N+1)!} \\ n & \cdots & \frac{n^{N-1}}{(N-1)!} & \frac{n^N}{N!} \end{pmatrix}$$

This latter matrix is more tractable because it admits a useful factorization: if we denote by D_n the diagonal matrix $\mathrm{diag}\,(1,\ldots,n^{N-1})$ and by $\check{D}_n$ the matrix $\mathrm{diag}\,(n^{N-1},\ldots,1)$ then

$$L_n' = n\check{D}_n L D_n,$$

where

$$L := \begin{pmatrix} \frac{1}{N!} & \frac{1}{(N+1)!} & \cdots & \frac{1}{(2N-1)!} \\ \frac{1}{(N-1)!} & \frac{1}{N!} & \ddots & \vdots \\ \vdots & \ddots & \ddots & \frac{1}{(N+1)!} \\ 1 & \cdots & \frac{1}{(N-1)!} & \frac{1}{N!} \end{pmatrix}$$

Having exhibited the matrix L, we now see that

$$(n^{-1}\check{D}_n^{-1} L_n D_n^{-1})_{i,j} = \frac{\binom{n}{N+j-i}}{n^{N+j-i}} = \frac{1}{(N+j-i)!} + o(1).$$

This means that $n^{-1}\check{D}_n^{-1} L_n D_n^{-1} = L + M_n$, where (M_n) is a sequence of matrices tending to 0. Thus the invertibility of L_n reduces to that of L for large enough n.

Let us denote by $\mathcal{E}$ the N-dimensional vector space made up of all complex polynomials of the form $P(t) = \sum_{i=N}^{2N-1} c_i t^i$, and let $\mathbf{L} : \mathcal{E} \to \mathbb{C}^N$ be the linear map defined by $\mathbf{L}(P) = (P(1), \ldots, P^{(N-1)}(1))$. The linear map $\mathbf{L}$ is clearly invertible. Moreover, its matrix with respect to the basis

$$\left(\frac{t^N}{N!}, \ldots, \frac{t^{2N-1}}{(2N-1)!} \right)$$

of $\mathcal{E}$ and the canonical basis of $\mathbb{C}^N$ is precisely the matrix L. Hence L is invertible† and so is L_n for large enough n, say $n \geq n_0$.

It remains to check that (2.2) holds true. Let us denote by $a_{i,j}(n)$ the entries of the matrix $(L + M_n)^{-1}$. Then each sequence $(a_{i,j}(n))_{n \geq n_0}$ is bounded. Since $L_n^{-1} = n^{-1} D_n^{-1} (L + M_n)^{-1} \check{D}_n^{-1}$ we see that $(L_n^{-1})_{i,j} = a_{i,j}(n)/n^{N+i-j}$, hence $\|L_n^{-1}\| \to 0$. Next, writing $K_n = (b_{i,j}(n))_{i,j}$ we get $(\check{D}_n^{-1} K_n)_{i,j} = b_{i,j}(n)/n^{N-i}$. Since $b_{i,j}(n) = O(n^{j-i})$, it follows that the sequence $(\check{D}_n^{-1} K_n)$ is bounded, and it may be checked in the same way that $(K_n D_n^{-1})$ is bounded as well. Since D_n^{-1}, $(L + M_n)^{-1}$ and $\check{D}_n^{-1}$ are also bounded, we now get the second part of (2.2) by considering the expression for L_n^{-1}. This concludes the proof of the lemma. □

REMARK   It can be shown that the matrix L_n introduced in the above proof is in fact invertible for every $n \in \mathbb{N}$; see [132].

PROOF OF PROPOSITION 2.6   Let $T \in \mathfrak{L}(X)$. Since $J^{\mathrm{mix}}(T)$ is obviously a vector space, it is enough to check that the following properties hold for each $N \in \mathbb{N}$ and any $\lambda \in \mathbb{C}$:

  (i) $\{0\} \times (\mathrm{Ker}(T-\lambda)^N \cap \mathrm{Ran}(T-\lambda)^N)$ and $(\mathrm{Ker}(T-\lambda)^N \cap \mathrm{Ran}(T-\lambda)^N) \times \{0\}$
      are contained in $J^{\mathrm{mix}}(T)$ when $|\lambda| = 1$;
 (ii) $\mathrm{Ker}(T-\lambda)^N \times \{0\} \subset J^{\mathrm{mix}}(T)$ when $|\lambda| < 1$;
(iii) $\{0\} \times \mathrm{Ker}(T-\lambda)^N \subset J^{\mathrm{mix}}(T)$ when $|\lambda| > 1$.

Let us fix N and $\lambda \in \mathbb{T}$, and let us also fix $z \in \mathrm{Ker}(T-\lambda)^N \cap \mathrm{Ran}(T-\lambda)^N$. Without loss of generality, we may assume that N is the smallest integer such that $z \in \mathrm{Ker}(T-\lambda)^N \cap \mathrm{Ran}(T-\lambda)^N$.

Set $R := \lambda^{-1} T - I$. Then $z \in \mathrm{Ker}(R^N) \cap \mathrm{Ran}(R^N)$, and N is the smallest integer with that property. Write $z = R^N(e)$, so that e satisfies $R^{2N}(e) = 0$ and $R^k(e) \neq 0$ for all $k < 2N$. Setting $E := \mathrm{span}\{R^i(e); \ 0 \leq i < 2N\}$ and $e_j = R^{2N-j}(e)$ for $j = 1, \ldots, 2N$, we see that E is a $2N$-dimensional R-invariant subspace of X with basis $(e_1, \ldots, e_{2N})$ and that $B := R_{|E}$ is the backward shift built on the basis $(e_1, \ldots, e_{2N})$. Moreover, we have $z = e_N$. By Lemma 2.7 and since $I + R = \lambda^{-1} T$, it follows that $(0, z)$ and $(z, 0)$ both belong to $J^{\mathrm{mix}}(\lambda^{-1} T)$. Since $|\lambda| = 1$, this amounts to saying that $(0, z)$ and $(z, 0)$ belong to $J^{\mathrm{mix}}(T)$.

Suppose now that $\lambda \in \mathbb{C}$ and let $z \in \mathrm{Ker}(T-\lambda)^N$, where N is the smallest integer with that property. Let $E := \mathrm{span}(z, T(z), \ldots, T^{N-1}(z))$. Then E is T-invariant and $(T - \lambda)_{|E}$ is a nilpotent operator. If $|\lambda| < 1$, it is thereby clear that $T^n(z) \to 0$ as $n \to \infty$, so that $(z, 0) \in J^{\mathrm{mix}}(T)$. If $|\lambda| > 1$, then $x_n := S^n(z) \to 0$, where S is the inverse of $T_{|E}$. Since $T^n(x_n) = z$ for all n, it follows that $(0, z) \in J^{\mathrm{mix}}(T)$. □

---

† This proof of the invertibility of L was suggested to us by S. Grivaux.

2.2 Existence of hypercyclic operators

In this section, we show that every infinite-dimensional separable Fréchet space supports a hypercyclic operator, and even a mixing operator. Before going into any detail, let us explain the main ideas. Suppose first that X is a Banach space having a normalized Schauder basis $(e_n)_{n \in \mathbb{N}}$. Let $\mathbf{w} = (w_n)_{n \geq 1}$ be a weight sequence with $\sum_1^\infty |w_n| < \infty$, and let $B_{\mathbf{w}}$ be the backward shift defined on span$\{e_n; \ n \in \mathbb{N}\}$ by $B_{\mathbf{w}}(e_0) = 0$ and $B_{\mathbf{w}}(e_n) = w_n e_{n-1}$, $n \geq 1$. Then $B_{\mathbf{w}}$ extends to a bounded operator on X. Indeed, one can write

$$\left\| B_{\mathbf{w}} \left(\sum_k \alpha_k e_k \right) \right\| \leq \sum_{k \geq 1} |\alpha_k| |w_k| \leq 2M \|\mathbf{w}\|_{\ell^1} \left\| \sum_k \alpha_k e_k \right\|, \quad (2.3)$$

where M is the basis constant of (e_n). By Theorem 2.2, $I + B_{\mathbf{w}}$ is the mixing operator on X for which we are looking. Incidentally, there may be no hypercyclic weighted shift on X. Indeed, the weights of a hypercyclic weighted shift cannot be too small; but if the weights are not small, then the shift has no reason for being continuous!

Unfortunately there exist Banach spaces with no Schauder basis. Yet, it is always possible to find a sequence which behaves almost like a basis and hence to construct an operator resembling $I + B_{\mathbf{w}}$. This is exactly what we are going to do. As may be guessed from (2.3), the Banach space $\ell^1(\mathbb{N})$ will play an extremal role in the construction.

The result that we intend to prove is given in the theorem below. In the Fréchet space setting, this theorem is due to J. Bonet and A. Peris [59]. The Banach space case was obtained earlier by S. I. Ansari [7] and L. Bernal-González [40]. Recall that an operator $T \in \mathfrak{L}(X)$ is said to be a (linear) **quasi-factor** of some operator $T_0 \in \mathfrak{L}(X_0)$ if there exists a continuous (linear) map with dense range $J : X_0 \to X$ such that $TJ = JT_0$.

THEOREM 2.8 *If X is an infinite-dimensional separable Fréchet space then any weighted backward shift on $\ell^1(\mathbb{N})$ with sufficiently small weights has a linear quasi-factor $B \in \mathfrak{L}(X)$.*

COROLLARY 2.9 *Any infinite-dimensional separable Fréchet space supports a mixing operator.*

PROOF If $B \in \mathfrak{L}(X)$ is a quasi-factor of some weighted backward shift B_0 on $\ell^1(\mathbb{N})$ then $I + B$ is mixing, since it is a quasi-factor of the mixing operator $I + B_0$. □

The proof of Theorem 2.8 relies on the next two lemmas. Only the second is explicitly needed, but its proof uses the following result of G. Metafune and V. B. Moscatelli [182], which is of independent interest. This is a purely "non-Banach" statement, so a reader interested in Banach spaces only should skip this part of the proof.

LEMMA 2.10 *Let X be an infinite-dimensional separable Fréchet space, and assume that X is not isomorphic to $\mathbb{K}^{\mathbb{N}}$. Then there exists a dense subspace $M \subset X$ on which one can define a continuous norm.*

PROOF   We start with the following

CLAIM   There exists a continuous seminorm p on X such that $\mathrm{Ker}(p)$ has infinite codimension.

PROOF OF THE CLAIM   Let $(p_i)_{i \geq 1}$ be non-decreasing sequence of continuous seminorms generating the topology of X. Towards a contradiction, assume that $\mathrm{Ker}(p_i)$ has finite codimension for each $i \geq 1$. Since X is Hausdorff, the non-increasing sequence $(\mathrm{Ker}(p_i))$ is not stationary; by passing to a subsequence we may assume that this sequence is decreasing. Let $(E_i)_{i \geq 1}$ be an increasing sequence of finite-dimensional subspaces of X such that $X = E_i \oplus \mathrm{Ker}(p_i)$ for all $i \in \mathbb{N}$, and denote by π_i the corresponding projection on E_i. The projections π_i satisfy the compatibility conditions $\pi_j \pi_i = \pi_j$, $j \leq i$. Moreover, since $p_i(x - \pi_i(x)) = 0$ for all i, it follows from the choice of the sequence (p_i) that $\pi_i(x) \to x$ for all $x \in X$ as $i \to \infty$.

One can find a sequence $(e_n)_{n \in \mathbb{N}} \subset X$, and an increasing sequence of integers $(n_i)_{i \in \mathbb{N}}$ with $n_0 = 0$, such that $(e_0, \ldots, e_{n_i - 1})$ is a basis of E_i for each $i \geq 1$ and $e_n \in \mathrm{Ker}(p_i)$ whenever $n \geq n_i$. Since $\pi_j \pi_i = \pi_j$ when $i \geq j$, one can associate with each $x \in X$ a uniquely defined sequence of scalars $(\lambda_n(x))_{n \in \mathbb{N}}$ such that $\pi_i(x) = \sum_{n < n_i} \lambda_n(x) e_n$ for all $i \geq 1$. Moreover, each map $\lambda_n : X \to \mathbb{K}$ is a continuous linear functional on X, since one can write $\lambda_n(x) = \langle e_n^*, \pi_i(x) \rangle$ for any i such that $n_i > n$, where $(e_0^*, \ldots, e_{n_i - 1}^*)$ is the dual basis of $(e_0, \ldots, e_{n_i - 1})$. Thus, one defines a continuous linear operator $J : X \to \mathbb{K}^{\mathbb{N}}$ by setting $J(x) := (\lambda_n(x))_{n \in \mathbb{N}}$.

Since $\pi_i(x) \to x$ for all $x \in X$, the operator J is one-to-one. Moreover, J is also onto. Indeed, if $(a_n) \in \mathbb{K}^{\mathbb{N}}$ then the partial sums of the series $\sum a_n e_n$ form a Cauchy sequence in X, since for any $i \geq 1$ we have $p_i(\sum_N^{N'} a_n e_n) = 0$ if $n_i \leq N < N'$. Thus $x := \sum_0^{\infty} a_n e_n$ is well defined and $J(x) = (a_n)$. Therefore, we have shown that J is a continuous linear isomorphism from X onto $\mathbb{K}^{\mathbb{N}}$ and hence a topological isomorphism, since the two spaces are Fréchet. This is a contradiction.   □

Now we return to the proof of Lemma 2.10. Let us fix a continuous seminorm p on X such that $K := \mathrm{Ker}(p)$ has infinite codimension. To conclude the proof it is enough to find a dense subspace $M \subset X$ such that $M \cap K = \{0\}$. The required norm will be the restriction of p to the subspace M.

For any finite-dimensional subspace $Z \subset X$, the linear subspace $Z + K$ is not the whole space X, hence $X \setminus (Z + K)$ is dense in X. Therefore, having fixed a countable basis (V_n) for the topology of X, one can construct by induction a sequence $(z_n) \subset X$ such that $z_n \in V_n$ and $z_n \notin K + \mathrm{span}\{z_i; \ i < n\}$ for all $n \in \mathbb{N}$. Then $M := \mathrm{span}\{z_n; \ n \in \mathbb{N}\}$ is the subspace we are looking for.   □

The next lemma is crucial. As mentioned at the beginning of this section, it says that in any infinite-dimensional separable Fréchet space it is always possible to find a sequence which behaves more or less like a Schauder basis. Stronger results are available (see Remark 2.13 below), but this lemma is easy to prove and sufficient for our purpose. Recall that a sequence $(e_n) \subset X$ is said to be **bounded** if $\sup_n q(e_n) < \infty$ for every continuous seminorm q on X.

LEMMA 2.11 *Let X be an infinite-dimensional separable Fréchet space and assume that X is not isomorphic to $\mathbb{K}^{\mathbb{N}}$. Then one can find two sequences $(e_n)_{n \in \mathbb{N}} \subset X$ and $(e_n^*)_{n \in \mathbb{N}} \subset X^*$ such that*

(1) *(e_n) is bounded and $\overline{\text{span}}\{e_n; \ n \in \mathbb{N}\} = X$;*
(2) *(e_n^*) is equicontinuous;*
(3) *$\langle e_n^*, e_m \rangle = 0$ if $n \neq m$, and $\langle e_n^*, e_n \rangle \in (0, 1]$ for all $n \in \mathbb{N}$.*

PROOF   Let M be a dense subspace of X having a continuous norm $\| \cdot \|$. Then M is infinite-dimensional. Since X has a countable basis of open sets, one can find a linearly independent set $\{z_n; \ n \in \mathbb{N}\}$ contained in M and dense in X (see the end of the proof of Lemma 2.10).

Using the Hahn–Banach theorem, one can construct by induction two sequences $(x_n) \subset M$ and $(x_n^*) \subset (M, \| \cdot \|)^*$ such that $\text{span}\{x_0, \dots, x_n\} = \text{span}\{z_0, \dots, z_n\}$ for all $n \in \mathbb{N}$, and $\langle x_n^*, x_m \rangle = \delta_{n,m}$ for all $n, m \in \mathbb{N}$. Indeed, if x_i and x_i^* have been constructed for all $i < n$, one can choose first a non-zero vector $x_n \in \text{span}\{z_0, \dots, z_n\} \cap \bigcap_{i<n} \text{Ker}(x_i^*)$ (because $\bigcap_{i<n} \text{Ker}(x_i^*)$ has codimension n) and then the linear functional x_n^*. For each $n \in \mathbb{N}$, we denote by C_n the norm of the linear functional $x_n^* \in (M, \| \cdot \|)^*$.

Since M is dense in X, there is a unique continuous seminorm q on X extending $\| \cdot \|$, and each linear functional x_n^* can be uniquely extended to a continuous linear functional on X, still denoted by x_n^*. Then we have $|\langle x_n^*, x \rangle| \leq C_n \, q(x)$ for each $n \in \mathbb{N}$ and all $x \in X$, and hence the sequence $(C_n^{-1} x_n^*)$ is equicontinuous. Finally, since X is metrizable, one can find a sequence $(\alpha_n) \subset (0, 1)$ such that the sequence $(\alpha_n x_n)$ is bounded in X. Setting $e_n := \alpha_n x_n$ and $e_n^* = (1 + C_n)^{-1} x_n^*$, the sequences (e_n) and (e_n^*) have the required properties. $\square$

PROOF OF THEOREM 2.8   For any pair $(u, u^*) \in X \times X^*$ we will denote by $u \otimes u^* \in \mathfrak{L}(X)$ the operator defined by $u \otimes u^*(x) = \langle u^*, x \rangle u$.

If $X = \mathbb{K}^{\mathbb{N}}$, we denote by (e_n) the "canonical basis" of X and by (e_n^*) the associated sequence of coordinate functionals. If X is not isomorphic to $\mathbb{K}^{\mathbb{N}}$, let $(e_n) \subset X$ and $(e_n^*) \subset X^*$ be given by Lemma 2.11. In both cases the sequence (e_n) is bounded in X and the sequence of linear operators $(e_{n-1} \otimes e_n^*)_{n \geq 1}$ is equicontinuous (when $X = \mathbb{K}^{\mathbb{N}}$, this is temporarily left as an exercise; see Remark 2.14 below). We show that if $\mathbf{w} = (w_n)_{n \geq 1}$ is a sequence of positive numbers such that

$$\sum_{n=1}^{\infty} \frac{w_n}{\langle e_n^*, e_n \rangle} < \infty$$

then the associated backward shift $B_{\mathbf{w}}$ on $\ell^1(\mathbb{N})$ has a linear quasi-factor $B \in \mathfrak{L}(X)$.

For any sequence $(\alpha_n) \in \ell^1(\mathbb{N})$, the partial sums of the series $\sum \alpha_n e_n$ form a Cauchy sequence in X by the boundedness of the sequence (e_n). Thus, one can define a linear operator $J : \ell^1(\mathbb{N}) \to X$ by

$$J((\alpha_n)) = \sum_{n=0}^{\infty} \alpha_n e_n.$$

The operator J is continuous, as can be seen by a direct computation or by applying the Banach–Steinhaus theorem. Moreover, $\mathrm{Ran}(J)$ contains $\mathrm{span}\{e_n;\ n \in \mathbb{N}\}$, so that J has dense range by (1) of Lemma 2.11.

Similarly, for any $x \in X$ the partial sums of the series

$$\sum \frac{w_n}{\langle e_n^*, e_n \rangle}\, e_{n-1} \otimes e_n^*$$

form a Cauchy sequence in $\mathfrak{L}(X)$, by the equicontinuity of the sequence $(e_{n-1} \otimes e_n^*)$. Therefore, the formula

$$B := \sum_{n=1}^{\infty} \frac{w_n}{\langle e_n^*, e_n \rangle}\, e_{n-1} \otimes e_n^*$$

makes sense and defines a continuous linear operator $B : X \to X$. An immediate computation shows that $BJ = JB_{\mathbf{w}}$. Thus, the operator B is a linear quasi-factor of $B_{\mathbf{w}}$. $\qquad\square$

REMARK 2.12   The statement of Corollary 2.9 can be sharpened as follows: there exists a mixing operator $T \in \mathfrak{L}(X)$ of the form $T = I + K$, where K is a nuclear operator. Moreover, one may require that $\|I - T\|$ is arbitrarily small in the Banach space case. This is apparent from the above proof.

REMARK 2.13   Lemma 2.11 is all that was needed to prove Theorem 2.8. When X is a *Banach* space there is, however, a stronger result: the sequences (e_n) and (e_n^*) can be chosen in such a way that $\langle e_n^*, e_n \rangle = 1$ for all $n \in \mathbb{N}$; in other words, (e_n, e_n^*) is a true biorthogonal system, with both sequences (e_n) and (e_n^*) bounded and the sequence (e_n) spanning a dense subspace of X. This is a classical and non-trivial result due to R. I. Ovsepian and A. Pełczyński ([190]; see [174, Vol. 1, Theorem 1.f.4]). From this and the above proof, it follows that any weighted backward shift on $\ell^1(\mathbb{N})$ with summable weight sequence has a linear quasi-factor $B \in \mathfrak{L}(X)$.

REMARK 2.14   Let us come back to one particular point of the above proof, the only point at which the space $\mathbb{K}^{\mathbb{N}}$ has to be treated separately. If (e_n) is the canonical basis of $\mathbb{K}^{\mathbb{N}}$ then (e_n^*) is not equicontinuous. Nevertheless, the sequence $(e_{n-1} \otimes e_n^*)$ *is* equicontinuous! Indeed, let (p_k) be the sequence of seminorms defined by $p_k((a_j)) = \max_{0 \le j \le k} |a_j|$. If $x = (x_j) \in \mathbb{K}^{\mathbb{N}}$ and $n \ge 1$ then $e_{n-1} \otimes e_n^*(x) = x_n e_{n-1}$, so that $p_k(e_{n-1} \otimes e_n^*(x))$ is equal to $|x_n|$ if $k \ge n-1$ and to 0 if $k < n-1$. In any case, we get

$$p_k(e_{n-1} \otimes e_n^*(x)) \le p_{k+1}(x)$$

for each k and all $n \ge 1$. Since the seminorms p_k generate the topology of $\mathbb{K}^{\mathbb{N}}$, this shows that the sequence $(e_{n-1} \otimes e_n^*)$ is indeed equicontinuous.

2.3 Operators with prescribed orbits

If $T \in \mathfrak{L}(X)$ is hypercyclic with hypercyclic vector x then the orbit $O(x, T)$ is dense in X, by definition, and it must also be a linearly independent family otherwise $\mathrm{span}\{T^n(x); \; n \in \mathbb{N}\}$ would be finite-dimensional. The following theorem shows that these obvious necessary conditions are in fact the only restrictions needed for a subset of X to be a dense orbit of some hypercyclic operator. This result is due to S. Grivaux [127].

THEOREM 2.15 *Let X be a separable Banach space, and let $A = \{a_n; \; n \in \mathbb{N}\}$ be a countable dense subset of X consisting of linearly independent vectors. Then one can find an operator $T \in \mathfrak{L}(X)$ such that the T-orbit of a_0 is exactly the set A. Moreover, given any hypercyclic operator $S \in \mathfrak{L}(X)$, one may require that T has the form $T = J^{-1}SJ$, where $J : X \to X$ is an isomorphism such that $\|J - I\|$ is arbitrarily small.*

The heart of the proof is the following lemma, which is reminiscent of a classical result of G. Cantor on countable dense linear orderings ([68]; see [140] or Exercise 2.4). The **back and forth method** used in the proof is a ubiquitous tool in set theory and logic.

LEMMA 2.16 *Let X be a separable infinite-dimensional Banach space, and let $A = \{a_n; \; n \in \mathbb{N}\}$, $B = \{b_n; \; n \in \mathbb{N}\}$ be two countable dense and linearly independent subsets of X. Then, for any given $\varepsilon \in (0, 1)$, there exists an invertible operator $J \in \mathfrak{L}(X)$ such that $\|J - I\| < \varepsilon$ and $J(A) = B$. Moreover, if $\|b_0 - a_0\| < \varepsilon \|a_0\|$ then one may require that $J(a_0) = b_0$.*

PROOF Let us fix $\varepsilon \in (0, 1)$ and assume that $\|b_0 - a_0\| < \varepsilon \|a_0\|$. Let us also fix a sequence of positive numbers $(\varepsilon_n)_{\in \mathbb{N}}$ such that $\varepsilon_0 \|a_0\| > \|b_0 - a_0\|$ and $\sum_0^\infty \varepsilon_n < \varepsilon$.

Set $J_{-1} := I$. We construct by induction a sequence of invertible operators $(J_n)_{n \in \mathbb{N}}$ and two increasing sequences of finite sets $(\mathfrak{a}_n)$, $(\mathfrak{b}_n)$ such that the following properties hold true for each $n \in \mathbb{N}$:

(i) $\|J_n - J_{n-1}\| < \varepsilon_n$;
(ii) $\{a_0, \dots, a_n\} \subset \mathfrak{a}_n \subset A$ and $\{b_0, \dots, b_n\} \subset \mathfrak{b}_n \subset B$;
(iii) $J_n(\mathfrak{a}_n) = \mathfrak{b}_n$;
(iv) $J_n \equiv J_{n-1}$ on $\mathfrak{a}_{n-1}$ if $n \geq 1$.

Assume that this has been done. By (i), the sequence (J_n) is norm-convergent to some operator $J \in \mathfrak{L}(X)$ satisfying $\|J - I\| < \varepsilon$. Since $\varepsilon < 1$, this operator is invertible. By (iii) and (iv), we have $J(\mathfrak{a}_k) = \mathfrak{b}_k$ for all $k \in \mathbb{N}$, whence $J(A) = B$ by (ii).

The operator J_0 is defined as follows. We choose a linear functional $a_0^* \in X^*$ such that $\langle a_0^*, a_0 \rangle = 1 = \|a_0^*\| \, \|a_0\|$, and we set $J_0 := I + (b_0 - a_0) \otimes a_0^*$. Since $\|b_0 - a_0\| < \varepsilon_0 \|a_0\|$, we have $\|J_0 - I\| < \varepsilon_0$ and of course $J_0(a_0) = b_0$. Then (i) is satisfied, as well as (ii).

For the inductive step, it is clearly enough to prove the following

FACT   Let $\mathfrak{a}$, $\mathfrak{b}$ be two finite subsets of A, B respectively and let $R_0 \in \mathcal{L}(X)$ be an invertible operator such that $R_0(\mathfrak{a}) = \mathfrak{b}$. Finally, let $(a, b) \in A \times B$ and let $\eta > 0$. Then one can find an invertible operator $R_1 \in \mathcal{L}(X)$ and $(a', b') \in A \times B$ such that

- $\|R_1 - R\| < \eta$;
- $R_1 \equiv R_0$ on $\mathfrak{a}$;
- $R_1(\mathfrak{a} \cup \{a, a'\}) = \mathfrak{b} \cup \{b, b'\}$.

PROOF OF THE FACT   Let α be a small positive number. We will first define an invertible operator $R_{1/2} \in \mathcal{L}(X)$ such that $b' := R_{1/2}(a) \in B$, $R_{1/2} \equiv R_0$ on $\mathfrak{a}$ and $\|R_{1/2} - R_0\| < \alpha$. Then we will define the operator R_1 and $a' := R_1^{-1}(b)$.

If $a \in \mathfrak{a}$, we set $R_{1/2} = R_0$. Otherwise the vector a does not belong to span($\mathfrak{a}$), so one can find a linear functional $a^* \in X^*$ such that $a^* \equiv 0$ on $\mathfrak{a}$ and $\langle a^*, a \rangle = 1$. Since B is dense in X, one can find $b' \in B$ such that $\|a^*\| \, \|b' - R_0(a)\| < \alpha$. Then the operator $R_{1/2} := R_0 + (b' - R_0(a)) \otimes a^*$ satisfies $R_{1/2}(a) = b'$, $R_{1/2} \equiv R_0$ on $\mathfrak{a}$ and $\|R_{1/2} - R_0\| < \alpha$.

Since R_0 is invertible, the operator $R_{1/2}$ is invertible if α is small enough. Repeating the above procedure with R_0 replaced by $(R_{1/2})^{-1}$, one gets an invertible operator R_1 such that $a' := (R_1)^{-1}(b) \in A$, $(R_1)^{-1} \equiv R_{1/2}^{-1}$ on $\mathfrak{b} \cup \{b'\}$ and $\|(R_1)^{-1} - (R_{1/2})^{-1}\| < \alpha$. Then $\|R_1 - R_0\| < \eta$ if α is small enough, and so R_1 has the required properties.

We point out that it is essential to proceed in two steps (back and forth). If we stopped the construction at $R_{1/2}$, we would just get an invertible operator J such that $J(A) \subset B$. □

At this point, we have proved Lemma 2.16 under the assumption $\|b_0 - a_0\| < \varepsilon$. Since one can always relabel the sets A and B to ensure that $\|b_0 - a_0\| < \varepsilon$, this concludes the whole proof. □

PROOF OF THEOREM 2.15   Let S be any hypercyclic operator on X. Then the hypercyclic vectors for S are dense in X. Given $\varepsilon > 0$, choose $z \in HC(S)$ such that $\|z - a_0\| < \varepsilon \|a_0\|$. By Lemma 2.16, one can find an isomorphism $J : X \to X$ such that $\|J - I\| < \varepsilon$, $J(a_0) = z$ and $J(A) = O(z, S)$. Then $T := J^{-1}SJ$ has the required properties. □

We now point out several interesting consequences of Theorem 2.15, also taken from [127].

COROLLARY 2.17   *Let X be a separable Banach space, and let $M \subset X$ be a dense linear subspace of X with countably infinite algebraic dimension. For any non-zero vector $x \in M$, one can find a hypercyclic operator $T \in \mathcal{L}(X)$ such that $x \in HC(T)$ and $\mathbb{K}[T]x = M$. Moreover, one may require that T has the form $J^{-1}SJ$, where $S \in \mathcal{L}(X)$ is any given hypercyclic operator and J is arbitrarily close to I.*

PROOF   It is not hard to construct an algebraic basis $(e_n)_{n \in \mathbb{N}}$ of M such that the set $A := \{e_n; \ n \in \mathbb{N}\}$ is dense in X and $e_0 = x$, as follows. Fix a countable basis of open sets (V_i) for X with $x \in V_0$, construct inductively a linearly independent sequence $(f_i) \subset M$ such that $f_0 = x$ and $f_i \in V_i$ for all i (see the end of the proof of Lemma 2.10), and add more vectors, if necessary, to obtain a basis for M. Then Theorem 2.15 provides an operator $T \in \mathfrak{L}(X)$ of the required form such that $O(x,T) = A$. $\square$

COROLLARY 2.18   *Let M be a normed space with countably infinite algebraic dimension. Then there exists a bounded operator R on M which has no non-trivial invariant closed set.*

PROOF   Let X be the completion of M, fix $x \in M$ and let $T \in \mathfrak{L}(X)$ be given by Corollary 2.17. Then M is T-invariant and $M \setminus \{0\} = (\mathbb{K}[T]x) \setminus \{0\} \subset HC(T)$, so that every non-zero vector $z \in M$ is hypercyclic for $R := T_{|M}$. $\square$

Of course, a normed space with countably infinite algebraic dimension is never complete, so Corollary 2.18 does not help much in solving the invariant subset problem on a Banach space!

COROLLARY 2.19   *Let X be a separable Banach space, and let $(T_\lambda)_{\lambda \in \Lambda}$ be a countable family of hypercyclic operators on X. Then one can find a dense linear subspace $Z \subset X$ which is a* **common hypercyclic manifold** *for the operators T_λ, that is, every non-zero vector $z \in Z$ is hypercyclic for all operators T_λ.*

PROOF   We first note that, for any compact set $K \subset X$, the set

$$\mathcal{O}(K,V) := \{T \in \mathfrak{L}(X); \forall x \in K \ \exists n \in \mathbb{N} : T^n(x) \in V\}$$

is open in $\mathfrak{L}(X)$. Indeed, its complement is the projection of a closed subset of $X \times \mathfrak{L}(X)$ along the compact set K (see [100, Chapter XI, Theorem 2.5]).

Now fix a countable basis (V_j) for the topology of X and a dense linear subspace $M \subset X$ with countable algebraic dimension. Then $M \setminus \{0\}$ is a K_σ subset of X, since M can be written as a countable union of finite-dimensional subspaces of X. Writing $M \setminus \{0\} = \bigcup_{m \geq 1} K_m$ (where each K_m is compact) it follows that

$$\mathcal{G} := \{T \in \mathfrak{L}(X); \ M \setminus \{0\} \subset HC(T)\}$$
$$= \bigcap_m \bigcap_j \mathcal{O}(K_m, V_j)$$

is a G_δ subset of the Banach space $\mathfrak{L}(X)$. Therefore, all sets

$$\mathcal{G}_\lambda := \{V \in \mathcal{B}_{1/2}; \ M \setminus \{0\} \subset HC(V^{-1}T_\lambda V)\}$$

are G_δ in the open ball $\mathcal{B}_{1/2} := B(I, 1/2) \subset \mathfrak{L}(X)$. Moreover, each set $\mathcal{G}_\lambda$ is dense in $\mathcal{B}_{1/2}$. Indeed, let us fix $\lambda \in \Lambda$ and let $V_0 \in \mathcal{B}_{1/2}$ be arbitrary. Then $S := V_0^{-1}T_\lambda V_0$ is hypercyclic. By Corollary 2.17, one can find an invertible operator $J \in \mathfrak{L}(X)$

arbitrarily close to I such that M is a hypercyclic manifold for $T := J^{-1}SJ$. Then $V_1 := V_0 J$ is close to V_0 and belongs to $\mathcal{G}_\lambda$.

By the Baire category theorem, one can find an operator $V \in \bigcap_{\lambda \in \Lambda} \mathcal{G}_\lambda$. This means that M is a hypercyclic manifold for each operator $V^{-1}T_\lambda V$; equivalently, $Z := V(M)$ is a common hypercyclic manifold for the operators T_λ. $\qquad \square$

REMARK   This last result has a very simple proof if the operators T_λ commute with each other: just put $Z := \mathbb{K}(T_{\lambda_0})x$, where $\lambda_0 \in \Lambda_0$ is arbitrary and $x \in \bigcap_\lambda HC(T_\lambda)$. This works equally well if X is an F-space.

2.4 There are many hypercyclic operators

We now discuss the size of the set of hypercyclic operators. In what follows, we denote by $\mathfrak{L}_{HC}(X)$ the set of all hypercyclic operators on a topological vector space X. From the outset it is clear that $\mathfrak{L}_{HC}(X)$ cannot be dense in $\mathfrak{L}(X)$ with respect to the norm topology when X is a Banach space. Indeed, every hypercyclic operator has norm greater than 1.

Now, the norm topology is not always the most natural topology on $\mathfrak{L}(X)$. It is often more useful to consider the **strong operator topology** (in short, **SOT**). Recall that the strong operator topology is the weakest topology on $\mathfrak{L}(X)$ for which the evaluation maps $T \mapsto T(x)$, $x \in X$, are continuous. With respect to this topology, any $A \in \mathfrak{L}(X)$ has a neighbourhood basis consisting of sets of the form

$$\mathcal{U} = \{S \in \mathfrak{L}(X); \ S(e_1) - A(e_1) \in U, \dots, S(e_N) - A(e_N) \in U\},$$

where $e_1, \dots, e_N \in X$ are *linearly independent* and U is a neighbourhood of 0 in X.

The following result shows that, whenever one hypercyclic operator exists, there are in fact many such operators. This was observed in [47]. Recall that the **similarity orbit** of an operator $T \in \mathfrak{L}(X)$ is defined by

$$\mathrm{Sim}(T) := \{J^{-1}TJ; \ J \in GL(X)\},$$

where $GL(X)$ is the set of all invertible operators.

PROPOSITION 2.20   *Let X be a locally convex topological vector space. If there is at least one hypercyclic operator on X then $\mathfrak{L}_{HC}(X)$ is dense in $\mathfrak{L}(X)$ with respect to the strong operator topology. In fact, the similarity orbit of any hypercyclic operator $T \in \mathfrak{L}(X)$ is SOT-dense in $\mathfrak{L}(X)$.*

The local convexity of X arises through the following simple lemma.

LEMMA 2.21   *Let X be a locally convex topological vector space, and let E, F be two finite-dimensional subspaces of X with the same dimension. Then there exists a closed subspace $Z \subset X$ such that $E \oplus Z = X = F \oplus Z$. Moreover, the associated projections are continuous.*

PROOF   Let us write $E = \mathrm{span}(x_1, \ldots, x_p)$ and $F = \mathrm{span}(y_1, \ldots, y_p)$, where $p = \dim(E) = \dim(F)$. For any $i \in \{1, \ldots, p\}$, one can find a linear functional x_i^* on $\mathrm{span}(x_i, y_i)$ such that $\langle x_i^*, x_i \rangle \neq 0$ and $\langle x_i^*, y_i \rangle \neq 0$. By the Hahn–Banach theorem, we may extend each x_i^* to a continuous linear functional on X, still denoted by x_i^*. Then $Z = \mathrm{Ker}(x_1^*) \cap \cdots \cap \mathrm{Ker}(x_p^*)$ is the desired subspace. The continuity of the associated projections follows from the fact that $\pi_E : X \to E$ and $\pi_F : X \to F$ factor through the canonical quotient map $q : X \to X/Z$. $\qquad\square$

Since hypercyclicity is preserved under similarity, it is sufficient to prove the last statement in Proposition 2.20, namely, that the similarity orbit of any hypercyclic operator $T \in \mathfrak{L}(X)$ is already SOT-dense in X. This in turn is an easy consequence of the following observation, due to D. W. Hadwin, E. A. Nordgren, H. Radjavi and P. Rosenthal [137].

LEMMA 2.22   *Let X be a locally convex topological vector space, and let $T \in \mathfrak{L}(X)$. Assume that for each positive integer N one can find $x_1, \ldots, x_N \in X$ such that the vectors $x_1, \ldots, x_N, T(x_1), \ldots, T(x_N)$ are linearly independent. Then $\mathrm{Sim}(T)$ is SOT-dense in $\mathfrak{L}(X)$.*

PROOF   Let $A \in \mathfrak{L}(X)$ be arbitrary and let $\mathcal{U}$ be a SOT-neighbourhood of A. We may assume that $\mathcal{U}$ has the form

$$\mathcal{U} = \{S \in \mathfrak{L}(X); \; S(e_1) - A(e_1) \in U, \ldots, S(e_N) - A(e_N) \in U\},$$

where $e_1, \ldots, e_N \in X$ are linearly independent and U is a neighbourhood of 0 in X. Since the space X is clearly infinite-dimensional one can find $f_1, \ldots, f_N \in X$ such that $f_i - A(e_i) \in U$ for $i = 1, \ldots, N$ and the vectors $e_1, \ldots, e_N, f_1, \ldots, f_N$ are linearly independent. Moreover, one can pick $x_1, \ldots, x_N \in X$ such that the vectors $x_1, \ldots, x_N, T(x_1), \ldots, T(x_N)$ are linearly independent.

Then there exists an invertible operator $J \in \mathfrak{L}(X)$ such that $J(e_i) = x_i$ and $J(f_i) = T(x_i)$ for $i = 1, \ldots, N$: just extend the obvious isomorphism between $E := \mathrm{span}(e_1, \ldots, e_N, f_1, \ldots, f_N)$ and $F := \mathrm{span}(x_1, \ldots, x_N, T(x_1), \ldots, T(x_N))$ by the identity on some closed subspace $Z \subset X$ such that $E \oplus Z = X = F \oplus Z$. By the choice of J, the operator $S := J^{-1}TJ$ satisfies $S(e_i) = f_i$ for all i, hence $S \in \mathcal{U}$. $\qquad\square$

PROOF OF PROPOSITION 2.20   Let $T \in \mathfrak{L}(X)$ be hypercyclic with hypercyclic vector x. Then the vectors $T^n(x)$, $n \in \mathbb{N}$, are linearly independent, so one can apply Lemma 2.22 with $x_i := T^{2i}(x)$. $\qquad\square$

In a Hilbert space context, Proposition 2.20 can be sharpened [128].

PROPOSITION 2.23   *Let H be an infinite-dimensional complex separable Hilbert space. Then, for any $C > 1$, the set of all hypercyclic operators with norm at most C is SOT-dense in the closed ball $\overline{B}(0, C) \subset \mathfrak{L}(H)$.*

PROOF   Let us fix $C > 1$ and an arbitrary operator $A \in \mathfrak{L}(H)$ with $\|A\| < C$ (the open ball is norm-dense, hence SOT-dense in the closed ball). It is enough to show that, for any given finite-dimensional subspace $E \subset H$ and every $\varepsilon > 0$, one can find a hypercyclic operator T such that $\|T\| \leq C$ and $\|T\pi_E - A\pi_E\| < \varepsilon$, where $\pi_E : H \to E$ is the orthogonal projection onto E. Now, since $A\pi_E$ is a finite-rank operator, one can find a finite-dimensional subspace $F \supset E$ such that $\pi_F A\pi_E = A\pi_E$. Therefore, it is enough to prove the following fact: given a finite-dimensional subspace $F \subset H$ and an operator $R \in \mathfrak{L}(F)$ such that $\|R\| < C$, one can find $T \in \mathfrak{L}_{HC}(H)$ such that $\|T\| \leq C$, $T(F) \subset F$ and $T_{|F} = R$. Let us fix F and R.

Set $F_0 := F$ and let $(F_n)_{n \geq 1}$ be a sequence of pairwise orthogonal subspaces of H such that $\dim F_n = \dim F := d$ for all n and $H = \oplus_{n \geq 0} F_n$. By choosing an orthonormal basis in each F_n, we identify H with $\ell^2(\mathbb{N}, \mathbb{C}^d)$ and R with a $d \times d$ matrix.

Let us choose $\varepsilon > 0$ with $(\|R\|^2 + \varepsilon^2)^{1/2} \leq C$ and a positive real number ω such that $\max(1, \|R\|) < \omega \leq C$. We define the weight sequence $\mathbf{w} = (\varepsilon, \omega, \omega, \dots)$ and the associated backward shift $B_{\mathbf{w}}$ on $H = \ell^2(\mathbb{N}, \mathbb{C}^d)$; that is, the operator defined by $B_{\mathbf{w}}(x_0 \oplus x_1 \oplus x_2 \oplus \dots) = (\varepsilon x_1 \oplus \omega x_2 \oplus \omega x_3 \oplus \dots)$. Finally, set $T := R\pi_F + B_{\mathbf{w}}$. Clearly, T has norm at most C, F is T-invariant and $T_{|F} = R$. To conclude the proof, we show that T satisfies the Godefroy–Shapiro Criterion.

An easy calculation reveals that a complex number λ is an eigenvalue of T iff $|\lambda| < \omega$ and that the corresponding eigenvectors can be parametrized analytically. Explicitly, the eigenvectors have the form

$$x_a(\lambda) = a \oplus \bigoplus_{i \geq 1} \frac{\lambda^{i-1}}{\varepsilon \omega^{i-1}} (\lambda - R)(a),$$

where $a \in \mathbb{C}^d$ is an arbitrary non-zero vector.

Suppose that a vector $y \in H$ is orthogonal to $\bigcup_{\lambda \in U} \mathrm{Ker}(T - \lambda)$ for some non-empty open set $U \subset D(0, \omega)$, i.e. $\langle x_a(\lambda), y \rangle = 0$ for any $a \in \mathbb{C}^d \backslash \{0\}$ and any $\lambda \in U$. Since each map $\lambda \mapsto x_a(\lambda)$ is analytic in the disk $D(0, \omega)$, the vector y is then orthogonal to any $x_a(\lambda)$, $|\lambda| < \omega$. Applying this with $U^- = \{|\lambda| < 1\}$ and $U^+ = \{1 < |\lambda| < \omega\}$, we see that the closed linear spans of $\bigcup_{|\lambda| < 1} \mathrm{Ker}(T - \lambda)$ and $\bigcup_{|\lambda| > 1} \mathrm{Ker}(T - \lambda)$ both contain $G := \mathrm{span} \bigcup_{|\lambda| < \omega} \mathrm{Ker}(T - \lambda)$. Therefore it is enough to check that G is dense in H.

Let $x = x_0 \oplus x_1 \oplus \dots \in H$ be any vector orthogonal to G. Then $\langle x_a(\lambda), x \rangle = 0$ for every $\lambda \in D(0, |\omega|)$ and all $a \in \mathbb{C}^d$; in other words,

$$\langle a, x_0 \rangle + \sum_{i=1}^{\infty} \frac{\lambda^{i-1}}{\varepsilon \omega^{i-1}} \langle (\lambda - R)a, x_i \rangle \equiv 0.$$

Looking at the Taylor coefficients of the analytic function appearing on the left-hand side, this leads to the following equations:

$$R^*(x_1) = \varepsilon \, x_0;$$
$$R^*(x_{i+1}) = \omega \, x_i, \quad i \geq 1.$$

Since $\omega \geq \|R^*\| = \|R\|$, it follows that $\|x_{i+1}\| \geq \|x_i\|$ for all $i \geq 1$. This is possible only if $x = 0$. □

2.5 There are few hypercyclic operators

We now return to the operator norm topology of $\mathfrak{L}(X)$. We have observed already that the set of all hypercyclic operators on a Banach space X is not dense in $\mathfrak{L}(X)$ with respect to the norm topology. In the Hilbert space at least, a much stronger result holds true.

THEOREM 2.24 *Let H be an infinite-dimensional Hilbert space. Then the set of all cyclic operators is nowhere dense in $\mathfrak{L}(H)$ with respect to the norm topology.*

For the proof, we need the following nice result of M. Rosenblum ([207]; see also [141, p. 53], or the first chapter of [199]).

LEMMA 2.25 *Let X, Y be complex Banach spaces, and let $S \in \mathfrak{L}(X \oplus Y)$. Assume that S has a block-upper-triangular representation,*

$$S = \begin{pmatrix} A & C \\ 0 & B \end{pmatrix},$$

with $\sigma(A) \cap \sigma(B) = \emptyset$. Then S is similar to $A \oplus B$.

PROOF   We have to find an invertible operator $P \in \mathfrak{L}(X \oplus Y)$ such that $P^{-1}SP = A \oplus B$. Seeking an operator of the form

$$P = \begin{pmatrix} 1 & V \\ 0 & 1 \end{pmatrix},$$

we arrive at the equation $AV - VB = -C$. Thus, it is natural to introduce the operator $\tau_{A,B} : \mathfrak{L}(Y,X) \to \mathfrak{L}(Y,X)$ defined by

$$\tau_{A,B}(V) = AV - VB.$$

The proof will be complete if we are able to show that this operator $\tau_{A,B}$ is onto. We shall in fact prove that $\tau_{A,B}$ is invertible.

Let $\Gamma \subset \mathbb{C}$ be a contour enclosing some bounded domain Ω such that $\sigma(B) \subset \Omega$ and $\sigma(A) \cap \overline{\Omega} = \emptyset$. Then one can define an operator $\tau'_{A,B} : \mathfrak{L}(Y,X) \to \mathfrak{L}(Y,X)$ by the formula

$$\tau'_{A,B}(V') = \frac{1}{2i\pi} \int_\Gamma (A - \xi)^{-1} V'(\xi - B)^{-1} d\xi.$$

We claim that $\tau'_{A,B}$ is the inverse for which we are looking.

A simple calculation reveals that if $V \in \mathfrak{L}(X)$ then

$$(A - \xi)^{-1}\tau_{A,B}(V)(\xi - B)^{-1} = (A - \xi)^{-1}((A - \xi)V - V(B - \xi))(\xi - B)^{-1}$$
$$= (A - \xi)^{-1}V + V(\xi - B)^{-1}$$

for any complex number $\xi \notin \sigma(A) \cup \sigma(B)$. From this, we get

$$\tau'_{A,B}\tau_{A,B}(V) = \left(\frac{1}{2i\pi} \int_\Gamma (A - \xi)^{-1} d\xi \right) V + V \left(\frac{1}{2i\pi} \int_\Gamma (\xi - B)^{-1} d\xi \right).$$

By the choice of Γ, the first integral on the right-hand side is equal to 0 and the second is simply $I \in \mathcal{L}(Y)$. Thus, we have $\tau'_{A,B}\tau_{A,B}(V) = V$ for any $V \in \mathcal{L}(Y, X)$.

Similarly, one can compute $\tau_{A,B}\tau'_{A,B}(V')$ as follows:

$$\begin{aligned}
\tau_{A,B}\tau'_{A,B}(V') &= \frac{1}{2i\pi} \int_\Gamma \tau_{A,B} \left((A - \xi)^{-1} V'(\xi - B)^{-1} \right) d\xi \\
&= \frac{1}{2i\pi} \int_\Gamma \left(V'(\xi - B)^{-1} + (A - \xi)^{-1} V' \right) d\xi \\
&= V'.
\end{aligned}$$

Thus $\tau_{A,B}$ is indeed invertible, with inverse $\tau'_{A,B}$. $\square$

PROOF OF THEOREM 2.24   Let us denote by $\mathcal{L}_C(H)$ the set of all cyclic operators on H. We have to show that, given $T \in \mathcal{L}(H)$ and $\varepsilon > 0$, one can find $S \in \mathcal{L}(H)$ such that $\|S - T\| < \varepsilon$ and S is not in the norm-closure of $\mathcal{L}_C(H)$.

We start with the following simple

FACT   If $R \in \mathcal{L}(H)$ is a Fredholm operator with index -2 then R is not cyclic.

PROOF OF THE FACT   If R is cyclic with cyclic vector e then $H = \mathbb{C}e + \overline{\mathrm{Ran}(R)}$. Indeed, $\mathbb{C}e + \overline{\mathrm{Ran}(R)}$ is a closed subspace of H that contains $\mathrm{span}\{R^n(e); \; n \in \mathbb{N}\}$. It follows that $\mathrm{Ker}(R^*) = \mathrm{Ran}(R)^\perp$ has dimension at most 1, so that R cannot be Fredholm with index ≤ -2. $\square$

Since an operator $S \in \mathcal{L}(H)$ is cyclic if and only if $S - \mu$ is cyclic for any $\mu \in \mathbb{C}$, it follows from the above fact that

$$\mathcal{U} := \{S \in \mathcal{L}(H); \; \exists \mu \in \mathbb{C} \; : \; S - \mu \text{ is Fredholm with index } - 2\}$$

contains no cyclic operator. Moreover, $\mathcal{U}$ is an open subset of $\mathcal{L}(H)$ by the continuity of the Fredholm index. Hence, the proof will be complete if we are able to show that $\mathcal{U}$ is dense in $\mathcal{L}(H)$.

Let us fix an arbitrary operator $T \in \mathcal{L}(H)$ and $\varepsilon > 0$. Choose any complex number $\lambda \in \partial \sigma(T)$ such that the operator $T - \lambda$ is not *left-Fredholm* (see Proposition D.3.6 in Appendix D), and fix $\mu \in \mathbb{C}$ such that $T - \mu$ is invertible and $|\lambda - \mu| < \varepsilon/2$. Since $T - \lambda$ is not left-Fredholm, one can find an operator $T_0 \in \mathcal{L}(H)$ and an infinite-dimensional closed subspace $E \subset H$ such that $\|T_0 - T\| < \alpha$, where $\alpha < \varepsilon/2$ is very small, E is invariant under T_0 and $(T_0)_{|E} = \lambda I_E$; see Corollary D.3.5 in Appendix D. Then T_0 has a block-upper-triangular representation of the form

$$T_0 = \begin{pmatrix} \lambda & C_0 \\ 0 & B_0 \end{pmatrix}$$

with respect to the orthogonal decomposition $H = E \oplus E^\perp$. Moreover $T_0 - \mu$ is invertible if α is small enough, which we now assume to be the case. Hence $\mu \notin \sigma(B_0)$.

Now, let us define $S_0 \in \mathfrak{L}(H)$ by

$$S_0 := \begin{pmatrix} \mu & C_0 \\ 0 & B_0 \end{pmatrix},$$

where the block representation is of course taken with respect to the decomposition $H = E \oplus E^\perp$. Clearly $\|S_0 - T\| < \varepsilon$, and we recall that $\mu \notin \sigma(B_0)$.

Let $A_0 \in \mathfrak{L}(E)$ be any Fredholm operator with index -2. For example, A_0 could be a forward shift with multiplicity 2, that is, the operator defined by $A_0(e_n) = e_{n+2}$ where $(e_n)_{n \in \mathbb{N}}$ is some orthonormal basis of E. If $\delta > 0$ is small enough then $\sigma(\mu + \delta A_0) \cap \sigma(B_0) = \varnothing$. By Lemma 2.25 the operator

$$S_\delta := \begin{pmatrix} \mu + \delta A_0 & C_0 \\ 0 & B_0 \end{pmatrix}$$

is similar to $(\mu + \delta A_0) \oplus B_0$. Since $B := B_0 - \mu$ is invertible, it follows that $S_\delta - \mu \sim \delta A_0 \oplus B$ is Fredhom with index -2. Since S_δ is close to T, this concludes the proof. $\qquad\square$

REMARK 2.26 When the Hilbert space H is finite-dimensional, we have the opposite situation: the cyclic operators are dense in $\mathfrak{L}(H)$. See Exercise 2.10.

2.6 Linear dynamics is complicated

In this short section, we prove a result of N. S. Feldman [106] that states that, in some sense, linear dynamics is as complicated as topological dynamics. Recall that two continuous maps $f : X \to X$ and $g : Y \to Y$ acting on topological spaces X and Y are said to be **topologically conjugate** if there exists a homeomorphism $\phi : X \to Y$ such that $g = \phi \circ f \circ \phi^{-1}$; that is, f and g are the same map up to a change of variable.

THEOREM 2.27 *There exists a hypercyclic operator T acting on a separable Hilbert space $\mathcal{H}$ which has the following property. For any compact metrizable space K and any continuous map $f : K \to K$, there exists a T-invariant compact set $L \subset \mathcal{H}$ such that f and $T_{|L}$ are topologically conjugate.*

PROOF We first observe that any compact metrizable space K is homeomorphic to some compact subset of $\ell^2(\mathbb{N})$. Indeed, let us define $h : K \to \ell^2(\mathbb{N})$ by $h(x) = (2^{-n} d(x, x_n))$, where d is a compatible metric on K and $(x_n)_{n \in \mathbb{N}}$ is a countable dense sequence in K. The map h is indeed well-defined, because the metric d is bounded, and one-to-one because (x_n) is dense in K. It is also continuous, being the uniform limit of the continuous maps $h_n(x) :=$

$(d(x, x_0), 2^{-1}d(x, x_1), \ldots, 2^{-n}d(x, x_n), 0, 0, \ldots)$. Since K is compact, h is a homeomorphism onto its range. Thus, we may and will assume that K is a compact subset of $\ell^2(\mathbb{N})$.

The Hilbert space $\mathcal{H}$ is defined as the countable ℓ^2-direct sum of infinite-dimensional ℓ^2 spaces, i.e. $\mathcal{H} = \ell^2(\mathbb{N}, \ell^2)$. Explicitly,

$$\mathcal{H} = \left\{ x = (x_0, x_1, \ldots); \; x_i \in \ell^2(\mathbb{N}) \text{ and } \|x\|^2 := \sum_{i \geq 0} \|x_i\|^2_{\ell^2(\mathbb{N})} < \infty \right\}.$$

Let T be twice the backward shift on $\mathcal{H}$, i.e. $T(x_0, x_1, \ldots) = (2x_1, 2x_2, \ldots)$. As in the scalar case $\ell^2(\mathbb{N}, \mathbb{K})$, one may show that T is hypercyclic by applying the Hypercyclicity Criterion. We now define $\phi : K \to \mathcal{H}$ by $\phi(x) = (x, f(x)/2, f^2(x)/4, \ldots)$. Since K is bounded, the map ϕ is indeed well defined and clearly one-to-one. Moreover, by the boundedness of K again, ϕ is the uniform limit of the continuous maps ϕ_n defined by $\phi_n(x) = (x, f(x)/2, \ldots, f^n(x)/2^n, 0, 0, \ldots)$. Hence ϕ is continuous.

Since K is compact, ϕ is a homeomorphism from K onto $L := \phi(K)$. Moreover, $\phi \circ f = T \circ \phi$ by the definitions of ϕ and T. This implies that L is T-invariant and that f is topologically conjugate to $T_{|L}$. $\qquad\square$

2.7 Sums of hypercyclic operators

We conclude this chapter by proving the following result, again due to S. Grivaux [128].

THEOREM 2.28 *Let H be a separable infinite-dimensional complex Hilbert space. Then every operator $T \in \mathfrak{L}(H)$ is the sum of two mixing operators.*

We now briefly explain the strategy for proving this theorem. If T is a finite-rank operator then we may proceed as in Proposition 2.23. More precisely, let $F \subset H$ be a finite-dimensional subspace such that $T(H) \subset F$. We identify H with $\ell^2(\mathbb{N}, \mathbb{C}^d)$, $d = \dim(F)$ and we consider the backward shift B on H, i.e. the operator defined by $B(x_0 \oplus x_1 \oplus \cdots) = x_1 \oplus x_2 \oplus \cdots$. Then, for $|\omega|$ large enough, $T - \omega B$ and ωB are both mixing. Therefore it suffices to write $T = (T - \omega B) + \omega B$.

To derive the general case, we will first prove that if T is upper-triangular with respect to some orthonormal basis of H (in a sense to be made precise later), then $T + \omega B$ is still mixing when $|\omega|$ is large enough, where B is a shift constructed on this basis. Theorem 2.28 follows if we are able to decompose T as $T = A_1 + A_2$, where A_1 and A_2 are both upper-triangular in the same basis: to see this just write $T = (A_1 + \omega B) + (A_2 - \omega B)$.

We now proceed to the details, which are not so simple! A number of lemmas will be needed. The first consists of well-known results from the perturbation theory of Fredholm operators.

LEMMA 2.29 *Let X be a complex Banach space, and let $B \in \mathfrak{L}(X)$ be onto and Fredholm with index 1, that is, B is onto and $\mathrm{Ker}(B)$ has dimension 1. Set*

$$\gamma(B) := \inf\{\|B(x)\|; \ \mathrm{dist}(x, \mathrm{Ker}(B)) = 1\}$$

and let $c \in (0, \gamma(B)/2)$. Finally, let A be any bounded operator on X and let $\omega \in \mathbb{C}$ with $|\omega| > (\gamma(B)/2 - c)^{-1}\|A\|$. Then the following properties hold.

(a) *For any complex number $\lambda \in D(0, c|\omega|)$, the operator $A + \omega B - \lambda$ is onto and Fredholm with index 1.*

(b) *There is an analytic function $\lambda \mapsto x(\lambda)$ defined on $D(0, c|\omega|)$ such that $x(\lambda)$ is an eigenvector for $A + \omega B$ associated with the eigenvalue λ.*

(c) *For any non-empty open set $U \subset D(0, c|\omega|)$, the closed linear span of the set $\bigcup_{\lambda \in U} \mathrm{Ker}(A + \omega B - \lambda)$ contains $\bigcup_{\lambda \in D(0,c|\omega|)} \bigcup_{k \geq 1} \mathrm{Ker}(A + \omega B - \lambda)^k$.*

PROOF   Set $K := \mathrm{Ker}(B)$ and let $\widetilde{B} : X/K \to X$ be the operator canonically induced by B. Then $\widetilde{B}$ is invertible and $\|\widetilde{B}^{-1}\| = \gamma(B)^{-1}$. Write $K = \mathbb{C}e$, where $\|e\| = 1$, choose a linear functional $e^* \in X^*$ such that $\|e^*\| = 1 = \langle e^*, e \rangle$ and set $M := \mathrm{Ker}(e^*)$. Then $X = \mathbb{C}e \oplus M$ and the corresponding projection $\pi : X \to M$ has norm at most 2. Thus, the operator $B_{|M} : M \to X$ is invertible and $\|(B_{|M})^{-1}\| \leq 2\gamma(B)^{-1}$.

If $\lambda \in D(0, c|\omega|)$ then $\|(A - \lambda)_{|M}\| \leq \|A\| + |\lambda| < \gamma(B)|\omega|/2$. Since $\|(\omega B_{|M})^{-1}\| \leq (\gamma(B)|\omega|/2)^{-1}$, it follows that the operator $(A + \omega B - \lambda)_{|M}$ is invertible from M onto X. Since M has codimension 1 in X, this shows that $A + \omega B - \lambda$ is Fredholm with index ≤ 1. Equality will follow from (b).

To prove (b), we set $R := A + \omega B$. Write $X = K \oplus M = \mathbb{C}e \oplus \mathrm{Ker}(e^*)$ as above, and define $\mathcal{R}_\lambda : X \to X \times \mathbb{C}$ by $\mathcal{R}_\lambda(x) = ((R - \lambda)x, \langle e^*, x \rangle)$. Then $\mathcal{R}_\lambda$ is invertible if $\lambda \in D(0, c|\omega|)$, with $\mathcal{R}_\lambda^{-1}(z, \alpha) = (R - \lambda)_{|M}^{-1}(z) + \alpha e$. Moreover, $\mathcal{R}_\lambda^{-1}$ depends analytically on λ because $\mathcal{R}_\lambda$ does. Therefore, $x(\lambda) := \mathcal{R}_\lambda^{-1}(0, 1)$ is a nowhere zero analytic function of λ, which satisfies $(R - \lambda)x(\lambda) \equiv 0$ by definition.

To prove (c), set $R := A + \omega B$ and $E := \overline{\mathrm{span}} \bigcup_{\lambda \in U} \mathrm{Ker}(R - \lambda)$. If $x^* \in X^*$ is any linear functional orthogonal to E then $\langle z, x(\lambda) \rangle = 0$ in the open set U; since $x(\lambda)$ is analytic, this is in fact true for each $\lambda \in D(0, c|\omega|)$. Thus, we have $E = \overline{\mathrm{span}} \bigcup_{\lambda \in D(0,c|\omega|)} \mathrm{Ker}(R - \lambda)$.

By (a), $\mathrm{Ker}(R - \lambda)$ has dimension 1 for any $\lambda \in D(0, c|\omega|)$. By the properties of the sequence of iterated kernels, it follows that for each $k \geq 1$ the kernel $\mathrm{Ker}(R-\lambda)^k$ has dimension at most k (in fact, it has dimension exactly k because $(R - \lambda)^k$ is onto and Fredholm with index k). Moreover, differentiating the equation $(R-\lambda)x(\lambda) = 0$ we get $(R-\lambda)x^{(k)}(\lambda) = kx^{(k-1)}(\lambda)$ for all $k \geq 1$, from which we infer by induction that $x^{(k)}(\lambda) \in \mathrm{Ker}(R-\lambda)^{k+1} \setminus \mathrm{Ker}(R-\lambda)^k$. Thus, $\bigcup_{k \geq 1} \mathrm{Ker}(R-\lambda)^k$ is spanned by the vectors $x^{(k)}(\lambda)$, $k \geq 0$. Now, $x'(\lambda) = \lim_{h \to 0} (x(\lambda + h) - x(\lambda))/h$ belongs to E and so does $x^{(k)}(\lambda)$ for all $k \geq 1$ by straightforward induction. This concludes the proof. $\square$

To state the next two lemmas, the following terminology will be convenient. Let $(e_i)_{i \in \mathbb{N}}$ be an orthonormal basis of H and let N be a positive integer. We say that an operator $A \in \mathfrak{L}(H)$ is N-**triangular** with respect to (e_i) if A has a block-upper-triangular representation with respect to (e_i), with blocks of size at most N. When $N = 1$, this means that the matrix of T with respect to (e_i) is upper-triangular; in that case we say that T is **triangular**.

LEMMA 2.30   *Let T be a triangular operator on H, and let $(e_i)_{i \geq 1}$ be an orthonormal basis with respect to which the matrix $(t_{i,j})$ of T is upper-triangular. Then the space spanned by the kernels $\mathrm{Ker}(T - t_{i,i}I)^k$, $i \geq 1$, $k \geq 1$, is dense in H.*

PROOF   This is easy. For each $n \geq 1$, set $E_n := \mathrm{span}(e_1, \dots, e_n)$ and $T_n := T_{|E_n}$. Then T_n is triangular with respect to $(e_1, \dots, e_n)$, with diagonal coefficients $(t_{i,i})_{1 \leq i \leq n}$. Thus, $\mathrm{span}\left[\bigcup_{i=1}^{n} \bigcup_{k \geq 1} \mathrm{Ker}(T_n - t_{i,i}I)^k \right] = E_n$. Since T and T_n coincide on E_n and $\bigcup_n E_n$ is dense in H, the result follows.   □

LEMMA 2.31   *Let $A \in \mathfrak{L}(H)$. Assume that there exist some positive integer N and a finite or infinite orthogonal decomposition $H = \oplus_{\alpha \geq 1} H_\alpha$ such that the following properties hold.*

(1) *Each H_α is infinite-dimensional.*
(2) *The operator A has a block-upper-triangular representation $(A_{\alpha\beta})$ with respect to the decomposition $H = \oplus_\alpha H_\alpha$.*
(3) *Each diagonal block $A_{\alpha\alpha}$ is N-triangular with respect to some orthonormal basis $(e_{\alpha i})_{i \in \mathbb{N}}$ of H_α.*

For each $\alpha \geq 1$, let us denote by B_α the backward shift constructed on the basis $(e_{\alpha i})_{i \in \mathbb{N}}$ and set $B := \oplus_\alpha B_\alpha$. Then, if $\omega \in \mathbb{C}$ and $|\omega|$ is larger than some constant $C(N, \|A\|)$, the operator $A + \omega B$ is mixing. More precisely, $A + \omega B$ satisfies the Godefroy–Shapiro Criterion.

PROOF   We have to show that if $|\omega|$ is larger than some suitable constant $C = C(N, \|A\|)$ then $\bigcup_{|\lambda| < 1} \mathrm{Ker}(A + \omega B - \lambda)$ and $\bigcup_{|\lambda| > 1} \mathrm{Ker}(A + \omega B - \lambda)$ both span a dense subspace of H. We divide the proof into three steps using Facts 1–3 given below. The conclusion of the first two steps will be that if $|\omega|$ is large enough then

$$\bigcup_{|\lambda| < 1} \mathrm{Ker}(A_{\alpha\alpha} + \omega B_\alpha - \lambda) \quad \text{and} \quad \bigcup_{|\lambda| > 1} \mathrm{Ker}(A_{\alpha\alpha} + \omega B_\alpha - \lambda)$$

both span a dense subspace of H_α, for each $\alpha \geq 1$. The proof has some similarity to that of Proposition 2.23. The last step allows to collect the information obtained in each block.

FACT 1   We fix $\alpha \geq 1$ and let $c > 0$. Then we can find some constant $K_{N,c}$ depending only on N and c such that $\mathrm{span}[\bigcup_{\lambda \in D(0, c|\omega|)} \bigcup_{k \geq 1} \mathrm{Ker}(A_{\alpha\alpha} + \omega B_\alpha - \lambda)^k]$ is dense in H_α whenever $|\omega| > K_{N,c} \|A_{\alpha\alpha}\|$.

PROOF OF FACT 1   We denote by $A_1, A_2, \ldots$ the diagonal blocks of the matrix representation of $A_{\alpha\alpha}$, which are of size N. Let $B_1, B_2, \ldots$ be the relevant finite-dimensional shift. Then $A_{\alpha\alpha} + \omega B_\alpha$ has a block-upper-triangular representation with diagonal blocks $A_1 + \omega B_1, A_2 + \omega B_2, \ldots$

Since the finite-dimensional operators $A_p + \omega B_p$ are triangularizable, the operator $A_{\alpha\alpha} + \omega B_\alpha$ is also triangularizable, in some suitable orthonormal basis of H_α, and the coefficients appearing on the diagonal of the corresponding matrix are the eigenvalues of the operators $A_p + \omega B_p$. Setting $\Lambda := \bigcup_p \sigma(A_p + \omega B_p)$, it follows from Lemma 2.30 that $\bigcup_{\lambda \in \Lambda} \bigcup_{k \geq 1} \mathrm{Ker}(A_{\alpha\alpha} + \omega B_\alpha - \lambda)^k$ spans a dense subspace of H_α for any complex number ω. The proof will be complete if we can find some constant $K_{N,c}$ such that $\sigma(A_p + \omega B_p) \subset D(0, c|\omega|)$ for all $p \geq 1$ if $|\omega| > K_{N,c}\|A_{\alpha\alpha}\|$.

Let us denote by $r(S)$ the spectral radius of an operator S. Then $r(A_p + \omega B_p) = |\omega| r(A_p/\omega + B_p)$ and $r(B_p) = 0$ for all $p \geq 1$. Since the shifts B_p act on spaces with dimension at most N (so that $B_p^N = 0$) and $\|A_p\| \leq \|A_{\alpha\alpha}\|$ for all p, this yields easily the desired conclusion. Indeed $B_p - \lambda$ is invertible for each $\lambda \neq 0$, with inverse $(B_p - \lambda)^{-1} = -\lambda^{-1} \sum_{i=0}^{N-1} (B_p/\lambda)^i$. Since $\|B_p\| = 1$, it follows that $\|(B_p - \lambda)^{-1}\|$ does not exceed some constant $K_{N,c}$ if $|\lambda| \geq c$. Therefore, $A_p/\omega + B_p - \lambda$ is invertible if $|\lambda| \geq c$ and $|\omega| > K_{N,c}\|A_{\alpha\alpha}\|$; in other words, $r(A_p/\omega + B_p) < c$ if $|\omega| > K_{N,c}\|A_{\alpha\alpha}\|$, so that $r(A_p + \omega B_p) < c|\omega|$.   □

FACT 2   If $|\omega|$ is greater than some constant $C = C(N, \|A\|)$ then each operator $A_{\alpha\alpha} + \omega B_\alpha$ has the following properties: span $\left[\bigcup_{|\lambda| < 1} \mathrm{Ker}(A_{\alpha\alpha} + \omega B_\alpha - \lambda) \right]$ and span $\left[\bigcup_{1 < |\lambda| < 2} \mathrm{Ker}(A_{\alpha\alpha} + \omega B_\alpha - \lambda) \right]$ are dense in H_α and $A_{\alpha\alpha} + \omega B_\alpha - \lambda$ is onto whenever $|\lambda| < 2$.

PROOF OF FACT 2   For any $\alpha \geq 1$, the operator B_α is onto and Fredholm with index 1. With the notation of Lemma 2.29, we have $\gamma(B_\alpha) = 1$, so the value $c := 1/4$ is allowed. If $|\omega| > 2c^{-1}$ then $U^+ := \{1 < |\lambda| < 2\}$ and $U^- := \{|\lambda| < 1\}$ are non-empty open subsets of $D(0, c|\omega|)$. By Lemma 2.29 and Fact 1, we know that $\bigcup_{\lambda \in U^-} \mathrm{Ker}(A_{\alpha\alpha} + \omega B_\alpha - \lambda)$ and $\bigcup_{\lambda \in U^+} \mathrm{Ker}(A_{\alpha\alpha} + \omega B_\alpha - \lambda)$ both span a dense subspace of H_α if $|\omega| > C_\alpha := (K_{N,c} + (1/2 - c)^{-1})\|A_{\alpha\alpha}\| + 2c^{-1}$. Moreover, $A_{\alpha\alpha} + \omega B_\alpha - \lambda$ is onto for every $\lambda \in D(0, c|\omega|)$, by Lemma 2.29 again. Since we clearly have $\|A_{\alpha\alpha}\| \leq \|A\|$ for every $\alpha \geq 1$, the result follows if we take $C := (K_{N,c} + (1/2 - c)^{-1})\|A\| + 2c^{-1}$.   □

The third step is the following simple observation.

FACT 3   Let $(R(\lambda))_{\lambda \in \Lambda} \subset \mathfrak{L}(H)$ be a family of operators and, for each $\lambda \in \Lambda$, let us denote by $(R_{\alpha\beta}(\lambda))$ the block decomposition of $R(\lambda)$ associated with the decomposition $H = \oplus_\alpha H_\alpha$. Assume that the following properties hold.

(i) For each $\lambda \in \Lambda$, the decomposition $(R_{\alpha\beta}(\lambda))$ is upper-triangular.
(ii) For each $\alpha \geq 1$, the operators $R_{\alpha\alpha}(\lambda)$ are onto and the span of $\bigcup_{\lambda \in \Lambda} \mathrm{Ker}(R_{\alpha\alpha}(\lambda))$ is dense in H_α.

Then the linear span of $\bigcup_{\lambda \in \Lambda} \mathrm{Ker}(R(\lambda))$ is dense in H.

PROOF OF FACT 3   Since all operators $R_{\alpha\alpha}(\lambda)$ are onto and the decompositions $(R_{\alpha\beta}(\lambda))$ are upper-triangular, it is easily checked that

$$R(\lambda)(H_1 \oplus \cdots \oplus H_{\alpha-1} \oplus \{x_\alpha\}) = H_1 \oplus \cdots \oplus H_{\alpha-1} \oplus \{R_{\alpha\alpha}(\lambda)x_\alpha\}$$

for every $\alpha > 1$, every $\lambda \in \Lambda$ and any vector $x_\alpha \in H_\alpha$. In particular, for any vector $x_\alpha \in \mathrm{Ker}(R_{\alpha\alpha}(\lambda))$ one can find $(u_1, \ldots, u_{\alpha-1}) \in H_1 \times \cdots \times H_{\alpha-1}$ such that the vector $x = u_1 \oplus \cdots \oplus u_{\alpha-1} \oplus x_\alpha \oplus 0 \oplus 0 \oplus \cdots$ belongs to $\mathrm{Ker}(R(\lambda))$.

Now, let $z = z_1 \oplus z_2 \oplus \cdots$ be any vector orthogonal to $\bigcup_\lambda \mathrm{Ker}(R(\lambda))$. Then z_1 is orthogonal to $\bigcup_\lambda \mathrm{Ker}(R_{11}(\lambda))$ since $\mathrm{Ker}(R_{11}(\lambda)) \oplus 0 \oplus 0 \oplus \cdots \subset \mathrm{Ker}(R(\lambda))$ for every λ; hence $z_1 = 0$. Using the above observation with $\alpha = 2$, it follows that z_2 is orthogonal to any vector $x_2 \in \bigcup_\lambda \mathrm{Ker}(R_{22}(\lambda))$, whence $z_2 = 0$. Continuing in this way, we can show by induction that $z_\alpha = 0$ for every α.   □

To conclude the proof of Lemma 2.31 we now apply Facts 2 and 3, with $R(\lambda) := A + \omega B - \lambda$ and $\Lambda := \{|\lambda| < 1\}$ or $\Lambda := \{1 < |\lambda| < 2\}$.   □

The next lemma shows that any operator $T \in \mathfrak{L}(H)$ can be put into a useful block-upper-triangular form. Only part (2) is explicitly needed for our purpose, but the proof of (2) uses (1) in an essential way. The result stated in (1) is due to R. Douglas and C. Pearcy [99]. Part (2) comes from [128].

LEMMA 2.32   *Let $T \in \mathfrak{L}(H)$ be an arbitrary operator.*

(1) *One can find a finite or infinite orthogonal decomposition $H = \oplus_{k\geq1}\mathcal{H}_k$ such that the associated block decomposition (T_{kl}) of T is upper-triangular and each operator T_{kk} is cyclic.*

(2) *One can find a positive integer r and a finite or infinite orthogonal decomposition $H = \oplus_{\alpha\geq1}\mathcal{H}_\alpha$ such that the following properties hold, where $(T_{\alpha\beta})$ is the associated block decomposition of T.*

    (a) *Each space $\mathcal{H}_\alpha$ is infinite-dimensional.*

    (b) *The decomposition $(T_{\alpha\beta})$ is upper-triangular.*

    (c) *For each $\alpha \geq 1$, one can find an orthonormal basis $(e_{\alpha i})_{i \in \mathbb{N}}$ of $\mathcal{H}_\alpha$ such that the matrix of the operator $T_{\alpha\alpha}$ in this basis is "upper-triangular plus r", which means that all coefficents below the rth subdiagonal are 0.*

PROOF   (1) Let $(f_j)_{j\geq1}$ be an orthonormal basis of H. We shall say that a (possibly empty) finite or infinite sequence $(\mathcal{H}_k)_{1\leq k<K}$ consisting of pairwise orthogonal non-zero closed subspaces of H is *admissible* if it has the following properties:

(i) $\mathcal{H} := \oplus_{1\leq j<K}\mathcal{H}_k$ is T-invariant and the associated block decomposition of $T_{|\mathcal{H}}$ is upper-triangular, with cyclic diagonal blocks;

(ii) $f_k \in \oplus_{1\leq j\leq k}\mathcal{H}_j$ for each $k \in [1, K)$.   □

Obviously, part (1) will be proved if we can find an admissible sequence with $\oplus_k\mathcal{H}_k = H$; since this is true for any infinite admissible sequence (thanks to the above property (ii)), we simply need to show that any finite admissible sequence $(\mathcal{H}_k)_{1 \le k < K}$ with $\oplus_k \mathcal{H}_k \neq H$ can be extended to an admissible sequence $(\mathcal{H}_k)_{1 \le k < K+1}$ by the addition of one more subspace $\mathcal{H}_K$. Let us fix such a sequence $(\mathcal{H}_k)_{1 \le k < K}$.

Set $\mathcal{H} := \oplus_{1 \le k < K} \mathcal{H}_k$ and $j_K := \min\{j \ge 1; \ f_j \notin \mathcal{H}\}$. Then $j_K \ge K$ by property (ii). Write $f_{j_K} = h + z$, where $h \in \mathcal{H}$ and $z \in \mathcal{H}^\perp$ is non-zero. Denote by $\pi : H \to \mathcal{H}^\perp$ the orthogonal projection onto $\mathcal{H}^\perp$ and let $\mathcal{H}_K$ be the closed linear span of the vectors $\tilde{T}^n(z)$, $n \in \mathbb{N}$, where $\tilde{T} \in \mathfrak{L}(\mathcal{H}^\perp)$ is the operator $\pi T_{|\mathcal{H}^\perp}$. The space $\mathcal{H} \oplus \mathcal{H}_K$ is T-invariant, since $\pi T(\pi T_{|\mathcal{H}^\perp})^n(z) = (\pi T_{|\mathcal{H}^\perp})^{n+1}(z) \in \mathcal{H}_K$ for every $n \in \mathbb{N}$. Moreover, we have $f_j \in \mathcal{H} \oplus \mathcal{H}_K$ for all $j \le j_K$ by the choice of j_K, whence $f_k \in \mathcal{H} \oplus \mathcal{H}_K$ for all $k \le K$. Finally, the Kth triangular block is cyclic because $\mathcal{H}_K$ is spanned by the vectors $\tilde{T}^n(z)$, $n \in \mathbb{N}$. Therefore the sequence $(\mathcal{H}_k)_{1 \le k < K+1}$ is admissible. This proves (1).

To prove (2), we first note that if R is a cyclic operator acting on some separable Hilbert space E then one can find an orthonormal basis (e_i) of E with respect to which the matrix of R is triangular plus 1: starting with a cyclic vector $e_0 \in E$, just apply the Gram–Schmidt orthonormalization process to the sequence $(R^i(e_0))_{0 \le i < \dim E}$.

Now let $H = \oplus_{1 \le k < K} \mathcal{H}_k$ be the decomposition of H given by (1). By the above remark, each space $\mathcal{H}_k$ has an orthonormal basis with respect to which the matrix of T_{kk} is triangular plus 1. Thus, there would be nothing to do (except to change the notation) if all spaces $\mathcal{H}_k$ were infinite-dimensional. However, the situation is less simple if some $\mathcal{H}_k$ are finite-dimensional.

We say that an interval of integers $I \subset [1, K)$ is *manageable* if either I is infinite and $\mathcal{H}_k$ is finite-dimensional for all $k \in I$ or I is finite, $\mathcal{H}_k$ is finite-dimensional for all $k < \max I$ and $\mathcal{H}_{\max I}$ is infinite-dimensional. Then we are in a favourable situation if the interval $[1, K)$ can be written as $[1, K) = \bigcup_{\alpha \ge 1} I_\alpha$, where (I_α) is a finite or infinite sequence of consecutive manageable intervals. Indeed, all spaces $H_\alpha := \oplus_{k \in I_\alpha} \mathcal{H}_k$ are infinite-dimensional and $H = \oplus_\alpha H_\alpha$. Let $(T_{\alpha\beta})$ be the associated block decomposition of T. For each $\alpha \ge 1$, one can construct suitable orthonormal bases of the spaces $\mathcal{H}_k$, $k \in I_\alpha$, to get an orthonormal basis of H_α with respect to which the matrix of $T_{\alpha\alpha}$ is triangular plus 1.

The only remaining case is when $[1, K)$ is finite and the last space $\mathcal{H}_{K-1}$ is finite-dimensional. In this case one can still write $[1, K) = \bigcup_{1 \le \alpha \le \alpha_0} I_\alpha \cup [k_0, K)$, where the I_α are consecutive manageable intervals and $\mathcal{H}_k$ is finite-dimensional for all $k \ge k_0$. We set $H_\alpha := \oplus_{k \in I_\alpha} \mathcal{H}_k$ if $\alpha < \alpha_0$, and $H_{\alpha_0} := \oplus_{k \in I_{\alpha_0}} \mathcal{H}_k \bigoplus \oplus_{k \ge k_0} \mathcal{H}_k$. Then each H_α is infinite-dimensional, the associated block decomposition $T_{\alpha\beta}$ is upper-triangular and $T_{\alpha\alpha}$ is triangular plus 1 (in some suitable orthonormal basis of H_α) for each $\alpha < \alpha_0$, by what was stated above. To conclude the proof, we need to find $r \in \mathbb{N}^*$ and some orthonormal basis H_{α_0} with respect to which the matrix of $T_{\alpha_0\alpha_0}$ is triangular plus r.

For notational simplicity, we put $R := T_{\alpha_0\alpha_0}$ and $\mathcal{H} := \mathcal{H}_{\max I_{\alpha_0}}$. Then we have an orthogonal decomposition $H_{\alpha_0} = E \oplus \mathcal{H} \oplus F$, where E and F are finite-dimensional, E is R-invariant and $R_{|E}$ can be made triangular plus 1 in some suitable orthonormal basis of E (because it has a block-diagonal decomposition with cyclic diagonal blocks). Therefore, we just have to check that the operator $S := \pi R_{|\mathcal{H} \oplus F} \in \mathfrak{L}(\mathcal{H} \oplus F)$ can be made triangular plus r for some r, where $\pi : H_{\alpha_0} \to \mathcal{H} \oplus F$ is the orthogonal projection onto $\mathcal{H} \oplus F$.

Let us denote by $\pi_{\mathcal{H}} : \mathcal{H} \oplus F \to \mathcal{H}$ the orthogonal projection onto $\mathcal{H}$. Then the operator $\pi_{\mathcal{H}} S_{|\mathcal{H}} \in \mathfrak{L}(\mathcal{H})$ is cyclic, since it is a diagonal block of the original decomposition (T_{kl}); let $z \in \mathcal{H}$ be a cyclic vector for $\pi_{\mathcal{H}} S_{|\mathcal{H}}$. Also let $(f_1, \dots, f_p)$ be an orthonormal basis of F. Now, consider the sequence

$$(v_i)_{i \in \mathbb{N}} := (z, f_1, \dots, f_p, S(z), S(f_1), \dots, S(f_p), S^2(z), S^2(f_1), \dots).$$

By the choice of z, the linear span of the sequence (v_i) is dense in $\mathcal{H} \oplus F$. Moreover, $S(v_i) \in \mathrm{span}(v_0, \dots, v_{i+p+1})$ for all $i \in \mathbb{N}$. Cancelling (from left to right) any vector of the sequence (v_i) which is a linear combination of the preceding ones, we obtain a linearly independent sequence $(u_i)_{i \in \mathbb{N}}$ with the same two properties as those just mentioned and, applying the Gram–Schmidt orthonormalization procedure to that sequence (u_i), we finally arrive at an orthonormal basis of $\mathcal{H} \oplus F$ with respect to which the matrix of S is $(p+1)$-triangular. This concludes the proof of (2).

PROOF OF THEOREM 2.28   It is enough to show that any operator $T \in \mathfrak{L}(H)$ can be written as $T = A_1 + A_2$, where A_1, A_2 satisfy the assumptions of Lemma 2.31 with the same decomposition $H = \oplus_\alpha H_\alpha$ and the same orthonormal bases $(e_{\alpha i})_{i \in \mathbb{N}}$, $\alpha \geq 1$. Indeed, once this is done one may apply Lemma 2.31 to choose a complex number ω such that $T_1 := A_1 + \omega B$ and $T_2 := A_2 - \omega B$ are mixing and then write $T = T_1 + T_2$.

Let $T \in \mathfrak{L}(H)$ and let $r \in \mathbb{N}^*$ and $(H_\alpha)_{\alpha \geq 1}$ be given by Lemma 2.32(2). In what follows, we denote by the same symbol an operator $R \in \mathfrak{L}(H_\alpha, H_\beta)$ and its matrix with respect to the bases $(e_{\alpha i})_{i \in \mathbb{N}}$, $(e_{\beta i})_{i \in \mathbb{N}}$.

Let us define infinite matrices $(A_1)_{\alpha\beta}$ and $(A_2)_{\alpha\beta}$ as follows.

- $(A_1)_{\alpha\beta} = T_{\alpha\beta}$ if $\alpha \neq \beta$.
- $(A_1)_{\alpha\alpha}$ is the block-upper-triangular matrix with blocks of size $2r$ obtained from $T_{\alpha\alpha}$ by making a decomposition of $T_{\alpha\alpha}$ into blocks of size $2r$ and replacing every coefficient below the diagonal by 0.
- $(A_2)_{\alpha\beta} = T_{\alpha\beta} - (A_1)_{\alpha\beta}$.

By definition we have $(A_2)_{\alpha\beta} = 0$ if $\alpha \neq \beta$. Moreover, since $T_{\alpha\alpha}$ is triangular plus r, each matrix $(A_2)_{\alpha\alpha}$ is easily seen to be block-diagonal, with one diagonal block of size r at the top and then diagonal blocks of size $2r$. Finally, each coefficient of $(A_2)_{\alpha\alpha}$ is either 0 or equal to the corresponding coefficient of $T_{\alpha\alpha}$, so that the moduli of these coefficients never exceed $\|T\|$. Since the blocks have size at most $2r$, it follows that each matrix $(A_2)_{\alpha\alpha}$ defines a bounded operator on H_α with $\|(A_2)_{\alpha\alpha}\| \leq 2r\|T\|$. Therefore, the diagonal matrix of operators $((A_2)_{\alpha\beta})_{\alpha,\beta \geq 1}$ defines a bounded operator $A_2 \in \mathfrak{L}(H)$ and consequently the matrix $((A_1)_{\alpha\beta})_{\alpha,\beta \geq 1}$

defines a bounded operator $A_1 \in \mathfrak{L}(H)$, with $A_1 + A_2 = T$. Both A_1 and A_2 are block-upper-triangular with respect to the decomposition $H = \oplus_\alpha H_\alpha$, and each diagonal block $(A_j)_{\alpha\alpha}$ is $2r$-triangular with respect to the orthonormal basis $(e_{\alpha i})_{i \in \mathbb{N}}$. This concludes the proof. □

REMARK 2.33   Theorem 2.28 does not hold if the Hilbert space H is replaced by an arbitrary (complex, separable, infinite-dimensional) Banach space. See Exercise 2.5.

REMARK 2.34   By writing down the definition of topological transitivity, it is not difficult to check that $\mathfrak{L}_{HC}(X)$ is SOT-G_δ in $\mathfrak{L}(X)$ if X is a separable F-space. In view of Proposition 2.23 this may have some importance, since any closed ball of $\mathfrak{L}(H)$ is completely metrizable in the strong operator topology. In fact, if one remembers the classical Pettis' lemma (see Exercise 2.9), it is very tempting to believe that Theorem 2.28 could be deduced from Proposition 2.23 by some category-like argument. However, we have been unable to find such an argument.

2.8 Comments and exercises

Salas' result on the hypercyclicity of "identity plus a shift" plays a central role in [7] and [59], and the proof in [40] uses the same ideas. The finite-dimensional arguments needed to prove Theorem 2.2 are already present in Salas' original proof, where the same kind of matrices appear.

It is interesting to note that the Hypercyclicity Criterion is used neither in the proof of Theorem 2.2 nor in Salas' original proof. That operators of the form "identity plus shift" do satisfy the Hypercyclicity Criterion is proved in [166]. More generally, every mixing operator satisfies the Hypercyclicity Criterion: this follows from the *Bès–Peris theorem*, to be proved in Chapter 4. However, operators of the form "identity plus shift" need not satisfy **Kitai's Criterion**, i.e. the Hypercyclicity Criterion with respect to the full sequence $(n_k) = (k)$. Some of them do, however, as shown by S. Shkarin in [225], who used it to prove that any infinite-dimensional separable Banach space supports an operator satisfying Kitai's Criterion.

The existence of hypercyclic operators on any separable Banach space was proved by S. I. Ansari in [7] and independently by L. Bernal-González in [40]. In [7], the result is stated for general Fréchet spaces, but there is a gap in the proof, which is filled in [59]. The corresponding result for hypercyclic *semigroups* was obtained in [38] (see also [43]).

Some complete, locally convex, separable, infinite-dimensional topological vector spaces fail to support any hypercyclic operator; see Exercise 2.3. There are also examples of (non-locally convex) separable F-spaces with this property. In fact, there exist separable F-spaces on which every continuous linear operator is a scalar multiple of the identity (the so-called **rigid** F-spaces; see [153]).

In [132] it is shown that a separable infinite-dimensional (complex) Fréchet space X is isomorphic to $\mathbb{C}^\mathbb{N}$ if and only if every hypercyclic operator on X is mixing.

Recently, H. N. Salas proved that on any separable Banach space with a separable dual there exists an operator T such that T and T^* are both hypercyclic [218]. Such operators are called **dual hypercyclic operators**. His idea was to consider operators of the form "identity plus *bilateral* shift".

The denseness of the set of hypercyclic operators with respect to the strong operator topology was first proved by K. Chan [71], in a Hilbert space setting. The idea of using Lemma 2.22 appears in [47] and [198].

We have stated Proposition 2.23 in a Hilbert space setting, but the result holds at least in any ℓ^p space (see Exercise 2.7). It would be of interest to know whether the validity of the result depends on some property of the underlying Banach space.

Theorem 2.24 seems to be a "folklore" result. The proof we give is an elaboration of the ideas of [110], where it is proved that the non-cyclic operators are dense in $\mathfrak{L}(H)$; it was extracted from D. A. Herrero's book [141]. See Exercise 2.11 for an abstract version of the result. As mentioned in Chapter 1, the norm-closure of the set of all hypercyclic operators was completely described by D. A. Herrero in [143]. Several other results of the same type were obtained by the same author; see e.g. [142].

Theorem 2.28 answers a question from [47], where it is shown that any operator $T \in \mathfrak{L}(H)$ is the sum of two hypercyclic operators and a compact operator of arbitrarily small norm. It was shown earlier by P. Y. Wu [237] that any Hilbert space operator is the sum of two *cyclic* operators. The result of Douglas and Pearcy stated in Lemma 2.32 is also used in Wu's paper. As already mentioned, Theorem 2.28 does not hold for an arbitrary separable Banach space X. It would be quite interesting to isolate some general property of X ensuring the validity of Theorem 2.28 (let alone a complete characterization).

EXERCISE 2.1   For any bounded weight sequence $\mathbf{w} = (w(n))_{n \geq 1}$, let us denote by $B_{\mathbf{w}}$ the associated weighted backward shift acting on $X = c_0(\mathbb{N})$ or $\ell^p(\mathbb{N})$. Characterize those weight sequences $\mathbf{w}$ for which $B_{\mathbf{w}}$ is mixing.

EXERCISE 2.2   Prove the assertions concerning iterated kernels stated just after Corollary 2.3.

EXERCISE 2.3   **When supercyclic operators do not exist** ([59])
Let $X = c_{00}(\mathbb{N})$ be the vector space of all finitely supported sequences of scalars. For each $N \in \mathbb{N}$, set $E_N := \mathrm{span}(e_0, \dots, e_N)$, where $(e_i)_{i \in \mathbb{N}}$ is the canonical basis of X, and equip E_N with its unique vector space topology. Finally, let τ be the inductive limit topology on $X = \bigcup_N E_N$, i.e. the finest locally convex topology on X such that the natural inclusions $j_N : E_N \to X$ are continuous. In what follows, we assume that X is equipped with the topology τ.

1. Show that the topological vector space X is complete and separable.
2. Show that every linear subspace of X is closed in X.
3. Let $T \in \mathfrak{L}(X)$, and assume that T is supercyclic with supercyclic vector x. Show that $X = \mathrm{span}\{T^n(x); \ n \geq 1\}$.
4. Show that X does not support any supercyclic operators.

EXERCISE 2.4   **Countable dense linear orderings**
A **dense linear ordering** is a linearly ordered set $(D, \leq)$ with the following property: whenever $u, v \in D$ satisfy $u < v$, one can find a w such that $u < w < v$. Show that any two countable dense linear orderings without endpoints are isomorphic: i.e. if $(D_1, \leq_1)$ and $(D_2, \leq_2)$ are two such orderings then there exists an increasing bijection $J : D_1 \to D_2$. (*Hint*: Use the same back and forth method as in the proof of Lemma 2.16.)

EXERCISE 2.5   Let X be a separable hereditarily indecomposable Banach space. Show that $3I$ cannot be written as the sum of two hypercyclic operators. (*Hint*: Read Chapter 6.)

EXERCISE 2.6   Prove the converse of Lemma 2.22.

EXERCISE 2.7   Show that Proposition 2.23 remains true if the Hilbert space H is replaced by $X = c_0(\mathbb{N})$ or $\ell^p(\mathbb{N})$, $1 \leq p < \infty$. (*Hint*: Follow the proof of Proposition 2.23. Show first that it is enough to consider subspaces of the form $F = \mathrm{span}(e_0, \dots, e_N)$, where $(e_i)_{i \in \mathbb{N}}$ is the canonical basis of X.)

EXERCISE 2.8   **Topologically transitive extensions** ([129])
Let X be a separable Banach space, and let $T \in \mathfrak{L}(X)$. Show that T has a hypercyclic extension $\tilde{T}$ acting on some larger Banach space $\tilde{X}$. More accurately, prove the following result: there exists an isometric embedding $J : X \to \tilde{X}$ into some separable Banach space $\tilde{X}$

and a hypercyclic operator $\widetilde{T} \in \mathfrak{L}(\widetilde{X})$ such that $\widetilde{T}J = JT$. (*Hint*: Consider $\widetilde{X} = \ell^p(\mathbb{N}, X)$ for any $p \in [1, \infty)$, and look at the proof of Proposition 2.23.)

EXERCISE 2.9 **Pettis' lemma**
Let $(G, \cdot)$ be a Baire topological group, i.e. a topological group which is also a Baire space.

1. Let U be an open subset of G, and let A be a comeager subset of U. Also let V be a non-empty open subset of U. Show that if $x \in G$ and $x \cdot V \subset U$ then $(x \cdot A) \cap A \neq \varnothing$.
2. Let A be a subset of G with the Baire property, and assume that A is non-meager in G. Show that $A \cdot A^{-1}$ is a neighbourhood of the unit element e.

EXERCISE 2.10 Let H be a finite-dimensional complex Hilbert space.

1. Show that if $T \in \mathfrak{L}(H)$ is diagonal (in some suitable basis, not necessarily orthogonal) and all eigenvalues of T are simple then T is cyclic.
2. Show that the cyclic operators are dense in $\mathfrak{L}(H)$.

EXERCISE 2.11 **Bad properties**
Let (P) be a property of Hilbert space operators. Following [141], (P) is said to be a **bad property** if it satisfies the following "axioms":

- if A has property (P) then so does $\alpha A + \beta$ for any $(\alpha, \beta) \in (\mathbb{C} \setminus \{0\}) \times \mathbb{C}$;
- if A has property (P) and A' is similar to A then A' has (P);
- if A, B have property (P) and $\sigma(A) \cap \sigma(B) = \varnothing$ then $A \oplus B$ has (P).

1. Give examples of bad properties.
2. Let H be an infinite-dimensional Hilbert space. Show that the set of all operators $T \in \mathfrak{L}(H)$ with property (P) is either empty or dense in $\mathfrak{L}(H)$.

3

Connectedness and hypercyclicity

Introduction

The starting point of this chapter is the following observation: if T is a hypercyclic operator acting on some topological vector space X then the set of all hypercyclic vectors for T is connected (Corollary 1.32).

This innocent remark is already sufficient to show that if an operator $T \in \mathfrak{L}(X)$ is hypercyclic then T^2 is hypercyclic as well, with the same hypercyclic vectors. Indeed, let $x \in HC(T)$. Given a non-empty open set $V \subset X$, we would like to find $q \in \mathbb{N}$ such that $T^{2q}(x) \in V$. Let us define

$$F_0 := HC(T) \cap \overline{\{T^{2n}(x); \ n \geq 0\}},$$
$$F_1 := HC(T) \cap \overline{\{T^{2n+1}(x); \ n \geq 0\}},$$

where the overbars indicate the closures of the sets. Then $F_0 \cup F_1 = HC(T)$, since x is a hypercyclic vector for T. Moreover F_0 and F_1 are closed in $HC(T)$, and they are both non-empty since $x \in F_0$ and $T(x) \in F_1$. By the connectedness of $HC(T)$, it follows that $F_0 \cap F_1 \neq \varnothing$. Take any $z \in F_0 \cap F_1$. Since $z \in HC(T)$, one can find $m \in \mathbb{N}$ such that $T^m(z) \in V$. If m is even, we use the fact that $z \in F_0$ to obtain $q \in \mathbb{N}$ such that $T^{2q}(x) \in V$, and likewise if m is odd, since $z \in F_1$.

There is nothing particular here about $p = 2$. Indeed, it was shown by S. A. Ansari [6] that if $T \in \mathfrak{L}(X)$ is hypercyclic then T^p is hypercyclic for any $p \geq 1$, with the same hypercyclic vectors. This interesting result is in fact a special case of a now famous theorem due to P. S. Bourdon and N. S. Feldman [66], according to which the orbits of any operator $T \in \mathfrak{L}(X)$ satisfy the following dichotomy: if $x \in X$ then $O(x, T)$ is either dense or nowhere dense in X.

One may also note that Ansari's theorem has a group-theoretic flavour, since it asserts that any non-trivial sub-semigroup of the discrete hypercyclic semigroup $(T^n)_{n \in \mathbb{N}}$ is itself hypercyclic. Another instance of the same phenomenon is the following result of F. León-Saavedra and V. Müller [168]: if $T \in \mathfrak{L}(X)$ and if the semigroup $\mathcal{T} := \{\lambda T^n; \ n \in \mathbb{N}, \ \lambda \in \mathbb{T}\}$ is hypercyclic then the operator T itself is hypercyclic. A third example is a theorem due to J. A. Conejero, V. Müller and A. Peris stating that if $(T_t)_{t \geq 0}$ is a hypercyclic C_0-semigroup, then each individual operator T_t, $t > 0$, is itself hypercyclic [79].

Connectedness plays an essential role in the proofs of each of the above results, and this is what we would like to emphasize in this chapter. We will first give somewhat unified proofs of the results of Ansari, León and Müller and of Conejero, Müller and Peris, which are based on the ideas of León and Müller [168]. Then we will prove the Bourdon–Feldman theorem, following fairly closely the arguments of [66].

3.1 Connectedness and semigroups

In this section X is a topological vector space, infinite-dimensional but otherwise arbitrary.

We consider the following problem. Let $\mathcal{T}$ be a multiplicative sub-semigroup of $\mathfrak{L}(X)$ and let $\mathcal{T}_0$ be a sub-semigroup of $\mathcal{T}$. We say that a vector $x \in X$ is **hypercyclic** for $\mathcal{T}$, and we write $x \in HC(\mathcal{T})$, if its $\mathcal{T}$-orbit $O(x, \mathcal{T}) := \{S(x);\ S \in \mathcal{T}\}$ is dense in X. We look for conditions ensuring that $HC(\mathcal{T}) = HC(\mathcal{T}_0)$. We find positive results in the following three cases:

(i) $\mathcal{T}$ is the discrete semigroup generated by a single operator $T \in \mathfrak{L}(X)$, i.e. $\mathcal{T} = \{T^n;\ n \in \mathbb{N}\}$, and $\mathcal{T}_0$ is the semigroup generated by some power of T;
(ii) $\mathcal{T}_0$ is a sub-semigroup of $\mathfrak{L}(X)$ (with some additional property) and $\mathcal{T}$ is the semigroup generated by all rotations of operators from $\mathcal{T}_0$, i.e. $\mathcal{T} = \{\lambda S;\ S \in \mathcal{T}_0, \lambda \in \mathbb{T}\}$;
(iii) $\mathcal{T}$ is a C_0-semigroup $(T_t)_{t \geq 0}$, and $\mathcal{T}_0$ is the discrete semigroup generated by a single operator $T_{t_0}, t_0 > 0$.

The first positive result is Ansari's theorem [6].

THEOREM 3.1 *Let $T \in \mathfrak{L}(X)$, and let p be a positive integer. If T is hypercyclic then T^p is also hypercyclic, with the same hypercyclic vectors.*

The second result is the León–Müller theorem [168].

THEOREM 3.2 *Assume that X is a complex topological vector space. Let $\mathcal{T}_0$ be a sub-semigroup of $\mathfrak{L}(X)$ and let $\mathcal{T}$ be the semigroup made up of all rotations of operators from $\mathcal{T}_0$, i.e. $\mathcal{T} = \{\lambda S;\ (S, \lambda) \in \mathcal{T}_0 \times \mathbb{T}\}$. Assume that there exists some operator $T \in \mathfrak{L}(X)$ such that $TS = ST$ for each $S \in \mathcal{T}_0$ and $T - \mu$ has dense range for every complex number μ. If the semigroup $\mathcal{T}$ is hypercyclic then so is $\mathcal{T}_0$, with the same hypercyclic vectors.*

COROLLARY 3.3 *If $T \in \mathfrak{L}(X)$ is hypercyclic then, for any $\lambda \in \mathbb{T}$, the operator λT is also hypercyclic, with the same hypercyclic vectors.*

PROOF Fix $\lambda \in \mathbb{T}$ and apply Theorem 3.2 to the semigroup $\mathcal{T}_0$ generated by λT. Then the associated semigroup $\mathcal{T}$ contains T and the result follows, since T commutes with $\mathcal{T}_0$ and $T - \mu$ has dense range for every $\mu \in \mathbb{C}$ (by Lemma 1.31). $\square$

The next corollary is sometimes called the **positive supercyclicity theorem**.

COROLLARY 3.4 *Let $T \in \mathfrak{L}(X)$, and assume that $T - \mu$ has dense range for every $\mu \in \mathbb{C}$ (if X is locally convex, this means that T^* has no eigenvalues). Then a vector $x \in X$ is a supercyclic vector for T iff it is **positively supercyclic**, i.e. the "positive projective orbit" $(0, \infty) \cdot O(x, T)$ is dense in X.*

PROOF Apply Theorem 3.2 to the semigroup $\mathcal{T}_0 := \{rT^n;\ (n, r) \in \mathbb{N} \times (0, \infty)\}$. Then $O(x, \mathcal{T}_0) = (0, \infty) \cdot O(x, T)$ and $O(x, \mathcal{T}) = \mathbb{C}^* \cdot O(x, T)$. $\square$

REMARK   In Theorem 3.2, one cannot dispense with some assumption on the semigroup $\mathcal{T}_0$. For example, let A be an operator of the form $I \oplus R$, where R is hypercyclic and I acts on a one-dimensional space. Then A is supercyclic but not positively supercyclic. Therefore, the semigroup $\mathcal{T}_0 := \{rA^n;\ (n,r) \in \mathbb{N} \times (0,\infty)\}$ is not hypercyclic even though the associated semigroup $\mathcal{T}$ is. Likewise, the compactness of the group $\mathbb{T}$ is essential even if the semigroup $\mathcal{T}_0$ is generated by a single operator: to see this, consider $\mathcal{T}_0 := \{A^n;\ n \in \mathbb{N}\}$ and $\mathcal{T} := \{\lambda A^n;\ (n,\lambda) \in \mathbb{N} \times \mathbb{C}^*\}$, where A is any supercyclic operator which is not hypercyclic. In the same spirit, if α is an irrational number of $\mathbb{T}$ and A is a supercyclic operator with no hypercyclic multiple then the finitely generated semigroup $\mathcal{T} := \{2^p 3^{-q} \alpha^r A^n;\ (p,q,r,n) \in \mathbb{N}^4\}$ is hypercyclic with no hypercyclic operator in it. These observations are due to N. S. Feldman [108].

The third result is the Conejero–Müller–Peris theorem [79]. We recall that a one-parameter family of operators $(T_t)_{t \geq 0} \subset \mathfrak{L}(X)$ is a C_0-**semigroup** (or **strongly continuous semigroup**) if $T_0 = I$, $T_{t+s} = T_t T_s$ for any $t, s \geq 0$ and $\lim_{t \to s} T_t(x) = T_s(x)$ for each $s \geq 0$ and all $x \in X$. The semigroup (T_t) is said to be **locally equicontinuous** if, for any $s \geq 0$, the family $\{T_t;\ t \in [0, s]\}$ is equicontinuous. This property is often automatic. For example, it is well known that if X is an F-space then every C_0-semigroup on X is locally equicontinuous (see the remark after conditions (H1) and (H2) below).

THEOREM 3.5   *Let $\mathcal{T} = (T_t)_{t \geq 0}$ be a locally equicontinuous C_0-semigroup on X. Assume that $\mathcal{T}$ is hypercyclic. Then, for any $t_0 > 0$, the operator T_{t_0} is hypercyclic, with the same hypercyclic vectors as $\mathcal{T}$.*

We point out that in the above theorems the topological vector space X was not assumed to be metrizable or Baire, nor even locally convex. Moreover, X can be real or complex in Theorems 3.1 and 3.5. If the reader feels uncomfortable with that setting, he or she can safely assume that X is a Banach space and replace everywhere the word "net" (see e.g. the proof of Lemma 3.9) by "sequence".

We will see that the proofs of Theorems 3.1, 3.2 and 3.5 are very similar. We start with some general discussion on the semigroups of operators and then use a specific connectedness argument to conclude in each particular case. The argument is straightforward for Ansari's theorem but more elaborated in the other two cases, since it requires the notion of homotopy. We recall that two continuous maps $f_0, f_1 : E \to F$ between topological spaces E and F are said to be **homotopic in** $\mathcal{C}(E, F)$ if there is a continuous map $H : [0, 1] \times E \to F$ such that $H(0, \cdot) = f_0$ and $H(1, \cdot) = f_1$. When $E = \mathbb{T} = F$, it is well known that two maps are homotopic in $\mathcal{C}(\mathbb{T}, \mathbb{T})$ iff they have the same winding number (see your favourite monograph on complex analysis or algebraic topology).

Let us now describe the general framework. For the sake of clarity it seems better to view our semigroups of operators as abstract semigroups acting on X by continuous linear transformations. Thus we start with a topological semigroup $(\Gamma, \cdot)$ and a homomorphism $\gamma \mapsto T_\gamma$ from Γ into the multiplicative semigroup $\mathfrak{L}(X)$. We will usually write $\gamma \cdot x$ or even γx instead of $T_\gamma(x)$. We assume that the action is continuous in the following sense: for each $x \in X$ the map $\gamma \mapsto \gamma x$ is continuous on Γ. The semigroup Γ is not assumed to be abelian.

A vector $x \in X$ is said to be Γ-**hypercyclic** if $\Gamma \cdot x := \{\gamma x; \ \gamma \in \Gamma\}$ is dense in X; the set of all Γ-hypercyclic vectors is denoted by $HC(\Gamma)$.

Now let Γ_0 be a sub-semigroup of Γ. As already stated above, the general problem we are considering is the following: when do Γ and Γ_0 have the same hypercyclic vectors? Of course, it seems hopeless to give a satisfactory answer at such a level of generality. Our goal will be much more modest. We just want to show that the three positive results stated above can be proved in a very similar way. Thus, we will stay within the general framework as long as it seems natural to do so and then give the specific arguments needed in each case.

We make the following assumptions on the sub-semigroup $\Gamma_0 \subset \Gamma$.

(H1) The sub-semigroup Γ_0 is the kernel of a continuous homomorphism $\rho : \Gamma \to G$ from Γ *onto* some compact abelian group $(G, \cdot)$.

(H2) There exists some compact set $K \subset \Gamma$ such that

- the family $(T_k)_{k \in K}$ is equicontinuous;
- $K \cap \Gamma_0$ is finite, and T_γ has dense range for each $\gamma \in K \cap \Gamma_0$;
- for each $\gamma \in \Gamma$, one can find $k \in K$ such that $k \cdot \gamma \in \Gamma_0$.

REMARK If X is an F-space then the equicontinuity of the family $(T_k)_{k \in K}$ in (H2) follows automatically from the continuity of the action. Indeed, for each $x \in X$ the set $\{T_k(x); \ k \in K\}$ is compact. Therefore, the family $(T_k)_{k \in K}$ is pointwise bounded, hence equicontinuous by the Banach–Steinhaus theorem.

Let us first check that this general framework includes the three examples we have in mind.

- In the case of Ansari's theorem, Γ is the discrete additive semigroup $\mathbb{N}$ acting in an obvious way ($n \cdot x = T^n(x)$) and $\Gamma_0 = p\mathbb{N}$. The sub-semigroup Γ_0 is the kernel of the canonical quotient map $\rho : \mathbb{N} \to \mathbb{Z}/p\mathbb{Z}$, so that (H1) holds. Condition (H2) is satisfied by $K := \{0, \ldots, p-1\}$.
- In the case of the León–Müller theorem, Γ is the semigroup $\mathbb{T} \times \mathcal{T}_0$ with an obvious action, and Γ_0 is the sub-semigroup $\{1\} \times \mathcal{T}_0 \simeq \mathcal{T}_0$. The topology on Γ is the product of the usual topology on $\mathbb{T}$ and the discrete topology on $\mathcal{T}_0$. Then (H1) holds because Γ_0 is the kernel of the canonical projection map $\rho : \mathbb{T} \times \mathcal{T}_0 \to \mathbb{T}$; assuming that $I \in \mathcal{T}_0$, (H2) is satisfied by $K := \mathbb{T} \times \{I\}$.
- In the case of the Conejero–Müller–Peris theorem, Γ is the additive semigroup $\mathbb{R}_+$ and $\Gamma_0 = t_0\mathbb{N}$. Then (H1) is satisfied, thanks to the homomorphism $\rho : \mathbb{R}_+ \to \mathbb{T}$ defined by $\rho(t) = e^{2i\pi t/t_0}$. Condition (H2) holds with $K := [0, t_0]$. Indeed, the family $(T_t)_{t \in [0, t_0]}$ is equicontinuous by the local equicontinuity assumption. Moreover, each operator T_t has dense range. To prove this, let us fix $t \geq 0$ and let

$x \in X$ be any hypercyclic vector for Γ. By the continuity of the map $s \mapsto T_s(x)$, the set $\{T_s(x);\ s \in [0, t]\}$ is compact. Since compact sets have empty interior in X, it follows that the set $\{T_s(x);\ s > t\}$ is dense in X. But the latter set is contained in $\mathrm{Ran}(T_t)$, by the semigroup property.

We now intend to prove the following theorem. Recall that a **character** of the group G is a continuous homomorphism $\chi : G \to \mathbb{T}$. A character is **non-trivial** if it is not identically 1. We denote by $\widehat{G}$ the character group of G. Finally, if F is a closed subgroup of G, we denote by $[g]_{G/F}$ the image of a point $g \in G$ in the quotient group G/F.

THEOREM 3.6   *Assume that Γ and Γ_0 satisfy* (H1), (H2) *above. Let $x \in HC(\Gamma)$, and put $\Gamma_{HC}(x) := \{\gamma \in \Gamma;\ \gamma x \in HC(\Gamma)\}$. Consider the following assertions.*

(i) *The vector x is not Γ_0-hypercyclic.*

(ii) *There exist a proper closed subgroup $F \subset G$ and a continuous map $\phi : HC(\Gamma) \to G/F$ such that $\phi(\gamma x) = [\rho(\gamma)]_{G/F}$ for every $\gamma \in \Gamma_{HC}(x)$.*

(iii) *There exist a non-trivial character $\chi \in \widehat{G}$ and a continuous map $\psi : HC(\Gamma) \to \mathbb{T}$ such that $\psi(\gamma x) = \langle \chi, \rho(\gamma) \rangle$ for every $\gamma \in \Gamma_{HC}(x)$.*

Then (i) $\implies$ (ii) $\implies$ (iii). *Moreover,* (i), (ii) *and* (iii) *are equivalent if* $\Gamma_{HC}(x) = \Gamma$.

REMARK   The condition $\Gamma_{HC}(x) = \Gamma$ is satisfied if the semigroup Γ is abelian and each operator T_γ has dense range.

The statement of Theorem 3.6 is not particularly appealing. To convince the reader that the result is nevertheless useful, we first show how it implies the three theorems we have in mind.

We handle the three proofs separately. In each case, we fix a point $x \in HC(\Gamma)$. We assume that x is not Γ_0-hypercyclic and look for a contradiction. We will keep the notation of Theorem 3.6.

PROOF OF THEOREM 3.1   In this case we have $\Gamma = \mathbb{N}$ and $G = \mathbb{Z}/p\mathbb{Z}$. We also note that $T^n(x) \in HC(T)$ for all $n \in \mathbb{N}$; that is, $\Gamma_{HC}(x) = \Gamma$.

Since G is finite, the continuous map $\phi : HC(\Gamma) \to G/F$ given by Theorem 3.6(ii) has finite range and, since $HC(\Gamma) = HC(T)$ is connected, it follows that ϕ is constant. But ϕ is also onto, since $\rho : \Gamma \to G$ is onto and $\Gamma_{HC}(x) = \Gamma$. Thus the quotient group G/F is trivial, a contradiction.   □

REMARK   The proof works in a more general context. Let Γ have the form $\Gamma = \mathbb{N} \times H$ where H is an abelian semigroup. If we assume that each operator T_γ has dense range and that $HC(\Gamma)$ is connected then $HC(p\mathbb{N} \times H) = HC(\Gamma)$ for any positive integer p. Taking $H := \mathbb{C}^*$, we get that for any $T \in \mathfrak{L}(X)$ (where X is a complex topological vector space), T and T^p share the same *supercyclic* vectors. (See Exercise 3.6 for a more general result.) This works on a *complex* vector space because $SC(T)$ is connected in that case, which is not necessarily true in the real case (see Exercise 3.2).

PROOF OF THEOREM 3.2   In this case we have $\Gamma = \mathbb{T} \times \mathcal{T}_0$ and $G = \mathbb{T}$, and $\rho : \Gamma \to G$ is defined by $\rho(\lambda, S) = \lambda$. Moreover, replacing $\mathcal{T}_0$ by $\mathcal{T}_0 \cup \{I\}$ if necessary, we may assume that $I \in \mathcal{T}_0$. Then $\Gamma_{HC}(x)$ contains $\mathbb{T} \times \{I\}$.

Choose χ and ψ according to Theorem 3.6(iii). The character $\chi \in \widehat{\mathbb{T}}$ is given by $\langle \chi, \lambda \rangle = \lambda^k$ for some non-zero integer k. Since $\mathbb{T} \times \{I\} \subset \Gamma_{HC}(x)$, it follows that the continuous map $\psi : HC(\Gamma) \to \mathbb{T}$ satisfies $\psi(\lambda x) = \lambda^k$ for all $\lambda \in \mathbb{T}$.

By assumption, each operator $T - \mu$ has dense range ($\mu \in \mathbb{C}$). Since operators of this form commute with all T_γ, it follows that $HC(\Gamma)$ contains every vector of the form $P(T)x$, where P is a non-zero polynomial of degree ≤ 1. In particular, putting $z_0 := Tx$ we see that $[z_0, \lambda x] \subset HC(\Gamma)$ for any $\lambda \in \mathbb{T}$. Therefore, one can define a continuous map $H : [0,1] \times \mathbb{T} \to \mathbb{T}$ by setting

$$H(r, \lambda) := \psi((1-r)z_0 + r\lambda x).$$

Then $H(0, \cdot)$ is constant and $H(1, \lambda) \equiv \lambda^k$, so the map $\lambda \mapsto \lambda^k$ is homotopic to a constant map in $\mathcal{C}(\mathbb{T}, \mathbb{T})$. Since $k \neq 0$ this is a contradiction. $\square$

PROOF OF THEOREM 3.5   Without loss of generality, we may assume that $t_0 = 1$. Thus $\Gamma = \mathbb{R}_+$, $\Gamma_0 = \mathbb{N}$, $G = \mathbb{T}$ and $\rho : \Gamma \to \mathbb{T}$ is defined by $\rho(t) = e^{2i\pi t}$. We have already observed that each operator T_t has dense range (see the comments after (H1) and (H2)). Since Γ is abelian, it follows that $\Gamma_{HC}(x) = \Gamma$.

Proceeding exactly as in the León–Müller case, we obtain a non-zero integer k and a continuous map $\psi : HC(\Gamma) \to \mathbb{T}$ such that $\psi(T_t x) = e^{2i\pi kt}$ for all $t \in \mathbb{R}_+$. Replacing ψ by $\overline{\psi}$, we may assume that $k > 0$.

At this point, we need the following fact (as in the León–Müller case), whose proof is quite similar to that of Lemma 1.31.

FACT   We have $(T_s - \mu)x \in HC(\Gamma)$ for any $s > 0$ and every $\mu \in \mathbb{K}$.

PROOF OF THE FACT   Let us fix $s > 0$ and $\mu \in \mathbb{K}$. Since $x \in HC(\Gamma)$ and $T_s - \mu$ commutes with Γ, it is enough to show that $T_s - \mu$ has dense range. Moreover, we know that $\Gamma_{HC}(x) = \Gamma$, i.e. $T_u(x) \in HC(\Gamma)$ for any $u \geq 0$. Thus, it is in fact enough to show that $T_u(x) \in \overline{\mathrm{Ran}(T_s - \mu)}$ for *some* $u \geq 0$: indeed, once this is done we get $\Gamma \cdot T_u(x) \subset \overline{\mathrm{Ran}(T_s - \mu)}$ since (again) $T_s - \mu$ commutes with Γ, and the result follows.

Set $X_0 := \overline{\mathrm{Ran}(T_s - \mu)}$ and let $q : X \to X/X_0$ be the canonical quotient map. Towards a contradiction, assume that $q(T_u x) \neq 0$ for any $u \geq 0$.

Set $M := q(\Gamma \cdot x) = \{\mu^n q(T_u x); \; n \in \mathbb{N}, \; u \in [0,s]\}$. Then M is dense in X/X_0 since $x \in HC(\Gamma)$. This is impossible. Indeed, if $|\mu| \leq 1$ then M is contained in a compact subset of the non-zero vector space X/X_0, and if $|\mu| > 1$ then $\mathcal{U} \cap M = \varnothing$, where $\mathcal{U}$ is any balanced neighbourhood of 0 in X/X_0 which does not intersect the compact set $\{q(T_u x);\ u \in [0, s]\}$.   □

We can now conclude the proof of Theorem 3.5 by a homotopy argument, which is a little trickier than the one used in the León-Müller case.

Put $z_0 := x$. By the above fact, we know that $[z_0, T_t x] \subset HC(\Gamma)$ for any $t \in \mathbb{R}_+$. Thus, it is tempting to define a map $H : [0, 1] \times \mathbb{T} \to \mathbb{T}$ by the formula $H(r, e^{2i\pi t}) = \psi((1 - r)z_0 + rT_t x)$; and it is not immediately obvious why this cannot work. In fact the trouble is that the formula simply does not make sense, because there is no reason why the map $t \mapsto \psi((1 - r)z_0 + rT_t x)$ should be 1-periodic when $r \neq 0$. To get round this problem, we will essentially define $H(1, e^{2i\pi t})$ as $\psi(T_t x)$ on a large subarc of $\mathbb{T}$ and extend this partial map continuously on $[0, 1] \times \mathbb{T}$. This will ensure that $H(1, \cdot)$ has a non-zero winding number and yield the desired contradiction.

Let $\mathcal{E} := \bigcup_{s \in \mathbb{R}_+} [x, T_s x]$. By the above fact, we know that $\mathcal{E} \subset HC(\Gamma)$. Moreover, $\mathcal{E}$ is star-shaped at x, i.e. $[x, z] \subset \mathcal{E}$ for every $z \in \mathcal{E}$. Thus, we may put

$$\mathcal{U} := \{z \in \mathcal{E};\ |\psi(w) - 1| < 1 \ \text{ for all } w \in [x, z]\}.$$

Then $x \in \mathcal{U}$ since $\psi(x) = 1$. Moreover, the set $\mathcal{U}$ is open in $\mathcal{E}$, since its relative complement is the projection of the closed set $\{(z, \lambda);\ |\psi(\lambda x + (1 - \lambda)z) - 1| \geq 1\}$ of $\mathcal{E} \times [0, 1]$ along the compact factor $[0, 1]$ (see e.g. [100, Chapter XI, Theorem 2.5]). Since, as we have already observed, the set $\{T_s x;\ s > 1\}$ is dense in X (and contained in $\mathcal{E}$), it follows that one can find $s > 1$ such that $T_s x \in \mathcal{U}$, i.e.

$$\forall r \in [0, 1]\ :\ |\psi((1 - r)x + rT_s x) - 1| < 1.$$

We then define $h : \mathbb{T} \to X$ by

$$h(e^{2\pi it}) := \begin{cases} T_{2ts} x & \text{if } 0 \leq t < 1/2, \\ (2t - 1)x + (2 - 2t)T_s x & \text{if } 1/2 \leq t < 1. \end{cases}$$

The map h is continuous on $\mathbb{T}$ (one just has to check the continuity at $e^{2i\pi 0} = e^{2i\pi 1}$ and $e^{2i\pi 1/2}$). Moreover, h maps $\mathbb{T}$ into $\mathcal{E}$, hence $rh(\lambda) + (1 - r)x \in HC(\Gamma)$ for each $r \in [0, 1]$ and all $\lambda \in \mathbb{T}$. Thus, one can define a continuous map $H : [0,1] \times \mathbb{T} \to \mathbb{T}$ by setting

$$H(r, \lambda) := \psi((1 - r)x + rh(\lambda)).$$

Then $H(1, e^{2\pi it}) = e^{4\pi istk}$ when $0 \leq t \leq 1/2$ whereas $|H(1, e^{2i\pi t}) - 1| < 1$ when $1/2 \leq t < 1$, by the choice of s. It follows that the map $H(1, \cdot)$ has winding number at least $[s]k$. Since the map $H(0, \cdot)$ is constant, this is the required contradiction. □

We now turn to the proof of Theorem 3.6, which requires some preliminaries. For any $u, v \in X$, we put

$$F_{u,v} := \left\{g \in G;\ (v, g) \in \overline{\{(\gamma u, \rho(\gamma));\ \gamma \in \Gamma\}}\right\}.$$

Thus, a point $g \in G$ is in $F_{u,v}$ iff there exists some net $(\gamma_i) \subset \Gamma$ such that $\rho(\gamma_i) \to g$ and $\gamma_i u \to v$.

The sets $F_{u,v}$ have some useful formal properties, which are collected in the next lemma.

LEMMA 3.7   *The following properties hold for $x, y, z \in X$ and $\gamma \in \Gamma$.*

(1) $F_{x,y}$ *is a closed subset of G;*
(2) *if $x \in HC(\Gamma)$ then $F_{x,y} \neq \varnothing$;*
(3) $F_{x,y} \cdot F_{y,z} \subset F_{x,z}$;
(4) $F_{x,\gamma x} \supset \rho(\gamma) \cdot F_{x,x}$.

PROOF   (1) This is obvious, since the map $g \mapsto (y, g)$ is continuous.

(2) This follows easily from the compactness of G.

(3) Let $(g_1, g_2) \in F_{x,y} \times F_{y,z}$. Then $(z, g_1 g_2) \in \overline{\{(\gamma_2 y, g_1 \rho(\gamma_2)); \ \gamma_2 \in \Gamma\}}$, by the continuity of the map $(u, g) \mapsto (u, g_1 g)$. For each $\gamma_2 \in \Gamma$, we have

$$(\gamma_2 y, g_1 \rho(\gamma_2)) = (\gamma_2 y, \rho(\gamma_2) g_1)$$
$$\in \overline{\{(\gamma_2 \gamma_1 x, \rho(\gamma_2) \rho(\gamma_1)); \ \gamma_1 \in \Gamma\}} \subset \overline{\{(\gamma x, \rho(\gamma)); \ \gamma \in \Gamma\}}.$$

Thus, we get $(z, g_1 g_2) \in \overline{\{(\gamma x, \rho(\gamma)); \ \gamma \in \Gamma\}}$, that is, $g_1 g_2 \in F_{x,z}$.

(4) Let $g_0 \in F_{x,x}$ and $\gamma_0 \in \Gamma$. By the continuity of the map $(u, g) \mapsto (\gamma_0 u, \rho(\gamma_0) g)$, we have

$$(\gamma_0 x, \rho(\gamma_0) g_0) \in \overline{\{(\gamma_0 \gamma x, \rho(\gamma_0) \rho(\gamma)); \ \gamma \in \Gamma\}} \subset \overline{\{(\gamma' x, \rho(\gamma')); \ \gamma' \in \Gamma\}},$$

so that $\rho(\gamma_0) g_0 \in F_{x,\gamma_0 x}$. □

COROLLARY 3.8   *If $x \in HC(\Gamma)$ then $F_{x,x}$ is a closed subgroup of G.*

PROOF   By (1), (2), (3), $F_{x,x}$ is closed, non-empty and stable under the group multiplication. Since G is compact, this implies that $F_{x,x}$ is in fact a subgroup of G, since it is well known that the inverse g^{-1} of any point $g \in G$ is in the closure of the set $\{g^n; \ n \in \mathbb{N}\}$.

To prove the latter fact, let us fix $g \in G$. By compactness, the sequence $(g^p)_{p \in \mathbb{N}}$ has a cluster point $a \in G$. Thus, for any neighbourhood V of the unit element $e \in G$, one can find $p, p' \in \mathbb{N}$ with $p < p'$ such that g^p and $g^{p'}$ both belong to $V \cdot a$, so that $g^{p'-p} \in V \cdot V^{-1}$. It follows that e is in the closure of the set $\{g^m; \ m \geq 1\}$ and hence that $g^{-1} \in \overline{\{g^n; \ n \geq 0\}}$. □

The next lemma justifies the introduction of the sets $F_{u,v}$.

LEMMA 3.9   *If $x \in HC(\Gamma)$ and $F_{x,x} = G$ then $x \in HC(\Gamma_0)$.*

PROOF   We start with the following

FACT   There exists some $\gamma_0 \in \Gamma_0$ such that T_{γ_0} has dense range and the following property holds: if $u, v \in X$ and $e \in F_{u,v}$ then $\gamma_0 v \in \overline{\Gamma_0 u}$.

PROOF OF THE FACT   Let $u, v \in X$ with $e \in F_{u,v}$. Then one can find a net $(\gamma_i)_{i \in I} \subset \Gamma$ such that $\gamma_i u \to v$ and $\rho(\gamma_i) \to e$. For each $i \in I$, property (H2) allows us to pick $k_i \in K$ such that $k_i \gamma_i \in \Gamma_0$. By the compactness of K, we may assume that the net (k_i) converges to some $k \in K$. Since $\rho(k_i \gamma_i) = e$ for all i and $\rho(\gamma_i) \to e$, we get $\rho(k) = e$, that is, $k \in K \cap \Gamma_0$.

We have $(k_i \gamma_i) \cdot u = k_i \cdot (\gamma_i \cdot u)$ for all i. Since $\gamma_i \cdot u \to v$, $k_i \to k$ *and the family* (T_{k_i}) *is equicontinuous*, it follows easily that $(k_i \gamma_i) \cdot u \to k \cdot v$. This shows that $k \cdot v \in \overline{\Gamma_0 u}$.

Thus we have shown that if $u, v \in X$ and $e \in F_{u,v}$ then one can find some $k \in K \cap \Gamma_0$ such that $k \cdot v \in \overline{\Gamma_0 u}$. Now, $K \cap \Gamma_0$ is finite by (H2); let us write $K \cap \Gamma_0 = \{k_1, \ldots, k_p\}$. Setting $\gamma_0 := k_1 \cdots k_p$ and remembering that each operator T_{k_j} has dense range (by (H2) again), we see at once that γ_0 has the required property.   □

Now, let $x \in HC(\Gamma)$ and assume that $F_{x,x} = G$. Since $F_{x,\gamma x} \supset \rho(\gamma) F_{x,x}$, we then have $F_{x,\gamma x} = G$ for any $\gamma \in \Gamma$. By the choice of γ_0 (taking $u = x$ and $v = \gamma x$), it follows that $\gamma_0 \gamma x \in \overline{\Gamma_0 x}$ for every $\gamma \in \Gamma$. Since $x \in HC(\Gamma)$ and T_{γ_0} has dense range, this shows that $x \in HC(\Gamma_0)$.   □

In the next lemma, we fix some vector $x \in HC(\Gamma)$. For any $g \in G$ we denote by $[g]$ the image of g under the canonical quotient map $\pi : G \to G/F_{x,x}$. Likewise, for any set $A \subset G$ we denote by $[A]$ the image of A under π.

LEMMA 3.10   *For any* $y \in HC(\Gamma)$ *the set* $[F_{x,y}]$ *contains only one point. Moreover the map* $\phi : HC(\Gamma) \to G/F_{x,x}$ *defined by* $[F_{x,y}] = \{\phi(y)\}$ *is continuous, and* $\phi(\gamma x) = [\rho(\gamma)]$ *for any* $\gamma \in \Gamma_{HC}(x)$.

PROOF   Let $y \in HC(\Gamma)$. By Lemma 3.7, one can choose some point $a \in F_{y,x}$. Then $g \cdot a \in F_{x,x}$ for any $g \in F_{x,y}$ (by Lemma 3.7 again), so that $[g] = [a^{-1}]$. This shows that $[F_{x,y}]$ contains only one point.

For any closed set $C \subset G/F_{x,x}$, the definition of ϕ gives

$$\phi^{-1}(C) = \{y \in HC(\Gamma); \ F_{x,y} \cap \pi^{-1}(C) \neq \varnothing\}.$$

Thus, to prove the continuity of the map $\phi : HC(\Gamma) \to G/F_{x,x}$, it is enough to check that for any closed set $E \subset G$ the set $\widetilde{E} := \{y \in HC(\Gamma); \ F_{x,y} \cap E \neq \varnothing\}$ is closed in $HC(\Gamma)$. Now, $\widetilde{E}$ is the projection along G of the set $\{(y, g) \in HC(\Gamma) \times G; \ g \in F_{x,y} \cap E\}$, which is readily seen to be closed in $HC(\Gamma) \times G$. Since G is compact, this shows that $\widetilde{E}$ is indeed closed in $HC(\Gamma)$ (see [100, Chapter XI, Theorem 2.5]).

Finally, if $\gamma \in \Gamma_{HC}(x)$ then $\rho(\gamma) \in F_{x,\gamma x}$ by Lemma 3.7 and Corollary 3.8, so that $\phi(\gamma x) = [\rho(\gamma)]$.   □

We can now prove Theorem 3.6.

PROOF OF THEOREM 3.6   The implication (i) $\implies$ (ii) follows from Lemmas 3.9 and 3.10: if x is not Γ_0-hypercyclic then $F_{x,x}$ is a proper closed subgroup of G by Lemma 3.9, and the map $\phi : HC(\Gamma) \to G/F_{x,x}$ given by Lemma 3.10 has the required properties.

The implication (ii) $\implies$ (iii) is easy. Indeed, since F is a proper closed subgroup of G, the character group $\widehat{G/F}$ is non-trivial. If $\theta : G/F \to \mathbb{T}$ is any non-trivial character of G/F then (iii) holds with $\langle \chi, g \rangle := \langle \theta, [g]_{G/F} \rangle$ and $\psi := \theta \circ \phi$.

Finally, the implication (iii) $\implies$ (i) is also easy if $\Gamma_{HC}(x) = \Gamma$. Indeed, if $\chi \in \widehat{G}$ and $\psi : HC(\Gamma) \to \mathbb{T}$ are given by (ii) then $\psi(\gamma x) = 1$ for all $\gamma \in \Gamma_0$; however, ψ is not constant because ρ is onto and the character χ is non-trivial. Since ψ is continuous, it follows that $\Gamma_0 \cdot x$ cannot be dense in X. $\qquad\square$

REMARK It follows from the proof that in (ii) one may take $F = F_{xx}$ and in (iii) one may take as χ any non-trivial character such that $F \subset \mathrm{Ker}(\chi)$, where F is given by (ii).

3.2 Somewhere dense orbits

In this section, we prove the Bourdon–Feldman theorem. We will deduce it from an abstract result, Theorem 3.11 below. For any topological space X, let us denote by $\mathcal{C}(X, X)$ the set of all continuous maps $F : X \to X$. Then $\mathcal{C}(X, X)$ is a semigroup when equipped with the composition product. Let $\mathcal{T} \subset \mathcal{C}(X, X)$ be an abelian sub-semigroup. For $T, U \in \mathcal{T}$, we shall write $U \leq T$ provided that there exists $V \in \mathcal{T}$ such that $T = VU$. For any $z \in X$, we put

$$\Gamma(z, \mathcal{T}) := \bigcap_{U \in \mathcal{T}} \overline{\{Tz; \ T \geq U\}}.$$

In the application to the Bourdon–Feldman theorem, we will of course put $\mathcal{T} :=$ $\{T^n; \ n \geq 0\}$. In that case $T^n \leq T^m$ holds whenever $n \leq m$, and $\Gamma(z, \mathcal{T})$ contains every limit point of $O(z, \mathcal{T})$.

THEOREM 3.11 *Let X be a Hausdorff topological space, and let $\mathcal{T} \subset \mathcal{C}(X, X)$ be an abelian sub-semigroup. Also let $x \in X$. Assume that the following properties hold.*

(1) *$\Gamma(x, \mathcal{T})$ has non-empty interior.*
(2) *There exists a family $\mathcal{S} \subset \mathcal{C}(X, X)$ such that*

- *$\mathcal{T} \subset \mathcal{S}$;*
- *each $S \in \mathcal{S}$ has dense range and commutes with every $T \in \mathcal{T}$;*
- *$\mathcal{S} \cdot x := \{Sx; \ S \in \mathcal{S}\}$ is connected and dense in X.*

Then $\Gamma(x, \mathcal{T}) = X$.

PROOF We will follow the arguments in the paper [66] of P. S. Bourdon and N. S. Feldman. The idea is to show that $\mathcal{S} \cdot x$ does not meet the boundary of $\Gamma(x, \mathcal{T})$. Since $\mathcal{S} \cdot x$ is connected and $(\mathcal{S} \cdot x) \cap \Gamma(x, \mathcal{T}) \neq \varnothing$ (because $\mathcal{S} \cdot x$ is dense in X and $\Gamma(x, \mathcal{T})$ has non-empty interior), it follows that $\mathcal{S} \cdot x \subset \Gamma(x, \mathcal{T})$ and hence $\Gamma(x, \mathcal{T}) = X$ because $\Gamma(x, \mathcal{T})$ is closed and $\mathcal{S} \cdot x$ is dense in X.

The proof that $S \cdot x$ does not meet $\partial\Gamma(x, \mathcal{T})$ is divided into four steps. We will use the following notation: the interior of a set $A \subset X$ is denoted by $\mathring{A}$, and if A has the form $\Gamma(z, \mathcal{T})$ we write $\mathring{\Gamma}(z, \mathcal{T})$. Finally, we say that a point $z \in X$ is $\mathcal{T}$-**recurrent** if $z \in \Gamma(z, \mathcal{T})$.

FACT 1   Let $z \in X$, $A \in \mathcal{T}$ and $S \in \mathcal{S}$. Then the following hold: $\Gamma(z, \mathcal{T}) \subset \Gamma(Az, \mathcal{T})$ and $S(\Gamma(z, \mathcal{T})) \subset \Gamma(Sz, \mathcal{T})$.

PROOF OF FACT 1   Let $U, T \in \mathcal{T}$ with $T \geq AU$. Then we may write $T = VAU = (VU)A$ for some $V \in \mathcal{T}$, so that $\{Tz; \ T \geq AU\} \subset \{TAz; \ T \geq U\}$. This gives the first part. The second part is easy, since any $S \in \mathcal{S}$ is continuous and commutes with $\mathcal{T}$. ☐

FACT 2   The $\mathcal{T}$-recurrent points are dense in X.

PROOF OF FACT 2   Since $\Gamma(x, \mathcal{T})$ has non-empty interior and is contained in the closure of $O(x, \mathcal{T})$, one can find $A \in \mathcal{T}$ such that $Ax \in \Gamma(x, \mathcal{T})$. Then $Ax \in \Gamma(Ax, \mathcal{T})$ by Fact 1, i.e. Ax is $\mathcal{T}$-recurrent. Likewise, if $S \in \mathcal{S}$ then $ASx = SAx \in \Gamma(SAx, \mathcal{T}) = \Gamma(ASx, \mathcal{T})$, by Fact 1 again. Thus, ASx is $\mathcal{T}$-recurrent for any $S \in \mathcal{S}$. Since $\mathcal{S} \cdot x$ is dense and A has dense range, the result follows. ☐

FACT 3   For any closed $\mathcal{T}$-invariant set $\Gamma \subset X$, the set $X \setminus \mathring{\Gamma}$ is $\mathcal{T}$-invariant.

PROOF OF FACT 3   Let $\Gamma \subset X$ be a closed $\mathcal{T}$-invariant set, and let $U \in \mathcal{T}$ be arbitrary. Since $X \setminus \mathring{\Gamma} = \overline{X \setminus \Gamma}$ and U is continuous, it is enough to show that $U(X \setminus \Gamma) \subset X \setminus \mathring{\Gamma}$. Assume that $U(X \setminus \Gamma) \cap \mathring{\Gamma} \neq \varnothing$. Since the $\mathcal{T}$-recurrent points are dense in X and since $X \setminus \Gamma$ and $\mathring{\Gamma}$ are open sets, one can find a $\mathcal{T}$-recurrent point z such that $z \in X \setminus \Gamma$ and $Uz \in \mathring{\Gamma}$. Since Γ is $\mathcal{T}$-invariant, we then have $Tz \in \Gamma$ for all $T \geq U$. By $\mathcal{T}$-recurrence it follows that $z \in \Gamma$, a contradiction. ☐

Applying Fact 3 with $\Gamma := \Gamma(x, \mathcal{T})$, we get the following dichotomy, for any $z \in X$:

- if $z \in \Gamma(x, \mathcal{T})$ then $\Gamma(z, \mathcal{T}) \subset \Gamma(x, \mathcal{T})$;
- if $z \in X \setminus \mathring{\Gamma}(x, \mathcal{T})$ then $\Gamma(z, \mathcal{T}) \subset X \setminus \mathring{\Gamma}(x, \mathcal{T})$.

FACT 4   Let $S \in \mathcal{S}$ and assume that $Sx \in \Gamma(x, \mathcal{T})$. Then $Sx \in \mathring{\Gamma}(x, \mathcal{T})$.

PROOF OF FACT 4   We will assume that $Sx \in X \setminus \mathring{\Gamma}(x, \mathcal{T})$ and then prove that $\text{Ran}(S) \subset X \setminus \mathring{\Gamma}(x, \mathcal{T})$, a contradiction since S has dense range. This will be done by checking separately that $S(\Gamma(x, \mathcal{T}))$ and $S(X \setminus \Gamma(x, \mathcal{T}))$ are both contained in $X \setminus \mathring{\Gamma}(x, \mathcal{T})$.

Since we are assuming that $Sx \in X \setminus \mathring{\Gamma}(x, \mathcal{T})$, on the one hand it follows from Facts 1 and 3 that

$$S(\Gamma(x, \mathcal{T})) \subset \Gamma(Sx, \mathcal{T}) \subset X \setminus \mathring{\Gamma}(x, \mathcal{T}).$$

On the other hand, let $z = S'x$ be an arbitrary point in $(S \cdot x) \cap (X \setminus \Gamma(x, T))$. Since $Sx \in \Gamma(x, T)$, we have

$$S(z) = S'Sx \in S'(\Gamma(x, T)) \subset \Gamma(S'x, T) \subset X \setminus \overset{\circ}{\Gamma}(x, T),$$

again by Fact 3. Thus S maps $(S \cdot x) \cap (X \setminus \Gamma(x, T))$ into $X \setminus \overset{\circ}{\Gamma}(x, T)$. Since $(S \cdot x) \cap (X \setminus \Gamma(x, T))$ is dense in the open set $X \setminus \Gamma(x, T)$, it follows that

$$S(X \setminus \Gamma(x, T)) \subset \overline{X \setminus \overset{\circ}{\Gamma}(x, T)} = X \setminus \overset{\circ}{\Gamma}(x, T),$$

which concludes the proof. □

Fact 4 says exactly that $S \cdot x$ does not meet the boundary of $\Gamma(x, T)$. Thus, we have achieved our goal and the proof of Theorem 3.11 is complete. □

To deduce the Bourdon–Feldman theorem from Theorem 3.11 we need the following lemma, whose proof is inspired by that of Lemma 1.31.

LEMMA 3.12 *Let X be a topological vector space, and let $T \in \mathfrak{L}(X)$. Assume that T has a somewhere dense orbit. Then, for each non-zero polynomial P, the operator $P(T)$ has dense range.*

PROOF By the proof of Lemma 1.31, we just need to show that if X is a finite-dimensional vector space, $X \neq \{0\}$, then $O(x, T)$ is nowhere dense for every $x \in X$. Towards a contradiction, we fix $x \in X$ and assume that $O(x, T)$ is somewhere dense.

Suppose first that X is a complex vector space. Let $x^* \neq 0$ be any eigenvector for T^*, with associated eigenvalue λ (such an eigenvector does exist since X is finite-dimensional). Then the set $\{\langle x^*, T^n(x) \rangle; \ n \geq 0\} = \{\lambda^n \langle x^*, x \rangle; \ n \geq 0\}$ is nowhere dense in $\mathbb{C}$, which contradicts our assumptions on x.

If X is a real vector space, we consider its complexification $\widetilde{X} = X \oplus iX$ and define $\widetilde{T} \in \mathfrak{L}(\widetilde{X})$ by $\widetilde{T}(u \oplus iv) := T(u) \oplus iT(v)$. Then $M := \{T^n(x) \oplus iT^m(x); \ n, m \geq 0\}$ is somewhere dense in $\widetilde{X}$. However, if $\tilde{x}^* \in \widetilde{X}^*$ is an eigenvector for $\widetilde{T}^*$ (with eigenvalue λ) then $\{\langle \tilde{x}^*, T^n(x) \oplus iT^m(x) \rangle; \ n, m \geq 0\} = \{\langle \tilde{x}^*, x \oplus 0 \rangle \lambda^n + \langle \tilde{x}^*, 0 \oplus ix \rangle \lambda^m; \ n, m \geq 0\}$ is nowhere dense in $\mathbb{C}$, so again we have a contradiction. □

We can now prove the Bourdon–Feldman theorem [66].

THEOREM 3.13 *Let X be a topological vector space. If $T \in \mathfrak{L}(X)$ then the T-orbit of any vector $x \in X$ is either nowhere dense, or everywhere dense, in X.*

PROOF Let $x \in X$ and assume that $O(x, T)$ is somewhere dense. We will apply Theorem 3.11 with $\mathcal{T} := \{T^n; \ n \in \mathbb{N}\}$.

The first point is to observe that $\overline{O(x, T)} = \overset{\circ}{\Gamma}(x, T)$: this is easily checked using the fact that X has no isolated points. Next, x is a *cyclic* vector for T, since $\overline{\text{span}}\{T^n(x); \ n \in \mathbb{N}\}$ is a linear subspace of X with non-empty interior. Moreover, if P is any non-zero polynomial then $S := P(T)$ has dense range by Lemma 3.12. Thus, we may apply Theorem 3.11 with $S := \{P(T);\ P \text{ polynomial} \neq 0\}$. □

An immediate corollary is the following result, obtained independently by G. Costakis in [83] and A. Peris in [194]. We note that it contains Ansari's theorem as a special case.

COROLLARY 3.14 *Let X be a topological vector space, and let $T \in \mathfrak{L}(X)$. Assume that there exist $x_1, \ldots, x_p \in X$ such that $O(x_1, T) \cup O(x_2, T) \cup \ldots \cup O(x_p, T)$ is dense in X. Then T is hypercyclic and one of the x_i is a hypercyclic vector for T.*

REMARK 3.15 There is a Bourdon–Feldman theorem for C_0-semigroups, which is due to G. Costakis and A. Peris [85]. The result reads as follows: *if $\mathcal{T} = (T_t)_{t\geq 0}$ is a C_0-semigroup acting on an infinite-dimensional Banach space X then the $\mathcal{T}$-orbit of any vector $x \in X$ is either dense or nowhere dense in X.* We note that some assumption is needed on X, since everything breaks down when $X = \mathbb{R}$ and $T_t(x) := e^t x$!

To prove this result along the same lines as for the Bourdon–Feldman theorem 3.13, the key point is to show that, for each $s > 0$ and every non-zero polynomial P, the operator $P(T_s)$ has dense range. Then one can apply Theorem 3.11 with $S := \{P(T_s)T_t;\ s, t > 0, P \text{ polynomial} \neq 0\}$.

Incidentally, the result can also be deduced from Theorem 3.13 itself by a Baire category argument. Indeed, let $U \subset X$ be a non-empty open set contained in $\overline{O(x, \mathcal{T})}$. Let $(x_j)_{j\geq 1}$ be a dense sequence in U and put

$$E := \bigcap_{j\geq 1} \bigcap_{k\geq 1} \bigcup_{n\geq 0} \{t > 0;\ \|T_{nt}(x) - x_j\| < 1/k\}.$$

For any $j, k \geq 1$, the set $E_{j,k} := \bigcup_{n\geq 0} \{t > 0;\ \|T_{nt}(x) - x_j\| < 1/k\}$ is open and dense in $(0, \infty)$. Indeed, let $(a, b) \subset (0, \infty)$ be a non-trivial open interval. One can find $A > 0$ such that $\bigcup_{n\geq 0}(na, nb) \supset [A, \infty)$. Since X is infinite dimensional, the compact set $\{T_x(x);\ s \in [0, A]\}$ is nowhere dense in X, and hence the set $\{T_s(x);\ s > A\}$ contains U in its closure. Choose $s > A$ such that $\|T_s(x) - x_j\| < 1/k$. Then $s = nt$ with $t \in (a, b)$ and $n \in \mathbb{N}$, and $t \in E_{j,k}$ by the definition of $E_{j,k}$. By the Baire category theorem, the set $E = \bigcap_{j,k} E_{j,k}$ is non-empty. Now, if $t_0 \in E$ then $U \subset \overline{O(x, T_{t_0})}$. By the Bourdon–Feldman theorem 3.13 the operator T_{t_0} is hypercyclic, which concludes the proof.

3.3 Comments and exercises

The first use of a connectedness argument in linear dynamics can be found in S. Ansari's paper [6]. Some similar ideas appear in an earlier work by P. S. Bourdon [64]. The main open question related to the results of this chapter is the validity of the positive supercyclity theorem when the adjoint of the supercyclic operator T has an eigenvalue:

OPEN QUESTION   Let X be a complex Banach space and let $T \in \mathfrak{L}(X)$ be supercyclic with $\sigma_p(T^*) = \{\lambda\}, \lambda \notin e^{2i\pi\mathbb{Q}}$. Is T positively supercyclic?

See Exercise 3.8 for a partial positive result.

EXERCISE 3.1   Find a cyclic operator T such that T^2 is not cyclic.

EXERCISE 3.2   Let B be the usual backward shift on the *real* Hilbert space $\ell^2(\mathbb{N})$, and let $T := I \oplus 2B$ act on $\mathbb{R} \oplus \ell^2$. Show that $SC(T)$ is not connected.

EXERCISE 3.3   It follows from the Conejero–Müller–Peris theorem that there are no hypercyclic C_0-semigroups on any finite-dimensional space $X \neq \{0\}$. Prove this result directly. (*Hint:* If $(T_t)_{t \geq 0}$ were such a semigroup then the set $\{\det(T_t); \ t \in \mathbb{R}^+\}$ would be dense in $\mathbb{K}^*$.)

EXERCISE 3.4   Let $\mathcal{T} = (T_t)_{t \geq 0}$ be a hypercyclic C_0-semigroup (perhaps not locally equicontinuous) acting on a complex topological vector space X. Prove that if P is any non-zero polynomial then $P(T_s)$ has dense range for every $s > 0$. (*Hint:* Look at the proof of the Conejero–Müller–Peris theorem 3.5.)

EXERCISE 3.5   Let X be a topological vector space, and let $\mathcal{T}$ be a multiplicative sub-semigroup of $\mathfrak{L}(X)$. Assume that $\mathcal{T}$ is hypercyclic, with a dense set of hypercyclic vectors. Show that if $x \in X$ has a somewhere dense $\mathcal{T}$-orbit then $x \in HC(\mathcal{T})$.

EXERCISE 3.6   **The supercyclic Bourdon-Feldman theorem** ([66])
Let T be an operator on a complex and locally convex topological vector space X, and let $x \in X$. Assume that $\mathbb{C} \cdot O(x, T)$ is somewhere dense in X.

1. Show that one can find a complex number λ such that $P(T)$ has dense range for any polynomial P with $P(\lambda) \neq 0$.
2. Deduce that $\mathbb{C} \cdot O(x, T)$ is everywhere dense.

EXERCISE 3.7   **Supercyclic C_0-semigroups**
Let X be a complex topological vector space, and let $\mathcal{T} = (T_t)_{t \geq 0}$ be a locally equicontinuous C_0-semigroup on X. Assume that $\mathcal{T}$ is supercyclic, i.e. there exists $x \in X$ such that $\mathbb{C} \cdot O(\mathcal{T}, x) = \{\lambda T_t(x); \ (t, \lambda) \in \mathbb{R}_+ \times \mathbb{C}\}$ is dense in X. Moreover, assume that $T_s - \mu$ has dense range for each $s > 0$ and every $\mu \in \mathbb{C}$. The aim of the exercise is to show that any operator $T_{t_0}, t_0 > 0$ is positively supercyclic and that every supercyclic vector for $\mathcal{T}$ is in fact positively supercyclic for T_{t_0}. For simplicity, we assume that $t_0 = 1$. So we fix some $\mathcal{T}$-supercyclic vector $x \in X$, and we want to show that the set $\{\lambda T_n(x); \ n \in \mathbb{N}, \lambda > 0\}$ is dense in X.

1. Let Γ be the semigroup $(\mathbb{R}_+, +) \times (0, \infty)$ acting on X in the expected way, i.e. $(t, \lambda) \cdot z = \lambda T_t(z)$. Show that $x \in HC(\Gamma)$.
2. Let $\Gamma_0 := \mathbb{N} \times (0, \infty) \subset \Gamma$. We define $\rho : \Gamma \to \mathbb{T}$ by $\rho(t, r) = e^{2\pi i t}$, and we put $K := [0, 1] \times \{1\}$. Show that assumptions (H1) and (H2) are satisfied.
3. Show that $HC(\Gamma)$ contains all vectors of the form $z = (1-r)x + \lambda T_s(x)$, where $r \in [0, 1]$ and $(s, \lambda) \in \Gamma$.
4. Suppose that $x \notin HC(\Gamma_0)$.
   (a) Show that one can find a continuous map $\psi : HC(\Gamma) \to \mathbb{T}$ and a positive integer k such that that $\psi(\lambda T_t(x)) = e^{2\pi i k t}$ for all $(t, \lambda) \in \Gamma$.
   (b) Show that one can find $\lambda \in (0, \infty)$ and $s > 1$ such that $|\psi(z) - 1| < 1$ for all $z \in [x, \lambda T_s(x)]$.
   (c) Let $h : \mathbb{T} \to X$ be defined by

$$h(e^{2\pi i t}) := \begin{cases} T_{3ts}(x) & \text{if } 0 \leq t < 1/3, \\ (3(\lambda - 1)t + 2 - \lambda)T_s(x) & \text{if } 1/3 \leq t < 2/3, \\ (3t - 2)x + (3 - 3t)\lambda T_s(x) & \text{if } 2/3 \leq t < 1. \end{cases}$$

Check that h is continuous, that $h(\mathbb{T}) \subset HC(\Gamma)$ and that $\psi \circ h$ has winding number at least $[s]k$.

5. Prove the desired result.

EXERCISE 3.8 **Positively supercyclic operators** ([92])

1. Let (n_k) be a sequence of natural numbers tending to infinity. Show that there exists $\lambda \in \mathbb{T}$ such that $\{\lambda^{n_k} ; \ k \geq 0\}$ is dense in $\mathbb{T}$. (*Hint*: Put $\phi_n(z) = z^n$. Then the sequence $(\phi_n)_{n \in \mathbb{N}}$ is topologically mixing on $\mathbb{T}$, in a very strong sense. In fact, given a non-empty open arc $I \subset \mathbb{T}$, one can find $N \in \mathbb{N}$ such that $\phi_n(I) = \mathbb{T}$ for every $n \geq N$.)

2. Let $S \in \mathfrak{L}(X)$ be a hypercyclic operator, and let $x \in HC(S)$. Show that there exist a sequence $(n_k) \subset \mathbb{N}$ and a complex number $\lambda \in \mathbb{T}$ such that $S^{n_k}(x) \to x$ and $\{\lambda^{n_k} ; \ k \geq 0\}$ is dense in $\mathbb{T}$.

3. Conclude that $T = \lambda \oplus S$ defined on $\mathbb{C} \oplus X$ is positively supercyclic.

4
Weakly mixing operators

Introduction

In Chapter 3, we saw that hypercyclicity is a rather "rigid" property: if T is hyper-cyclic then so is T^p for any positive integer p and so is λT for any $\lambda \in \mathbb{T}$. In the same spirit, it is natural to ask whether $T \oplus T$ remains hypercyclic. In topological dynamics, this property is quite well known.

DEFINITION   *Let X be a topological space. A continuous map $T : X \to X$ is said to be **(topologically) weakly mixing** if $T \times T$ is topologically transitive on $X \times X$.*

Here, $T \times T : X \times X \to X \times X$ is the map defined by $(T \times T)(x, y) = (T(x), T(y))$. When T is a linear operator, we identify $T \times T$ with the operator $T \oplus T \in \mathfrak{L}(X \oplus X)$. We note that, by Birkhoff's transitivity theorem 1.2 and the remarks following it, one can replace "topologically transitive" by "hypercyclic" in the above definition if the underlying topological space X is a second-countable Baire space with no isolated points. In particular, a linear operator T on a separable F-space is weakly mixing iff $T \oplus T$ is hypercyclic.

By definition, weakly mixing maps are topologically transitive. In the topological setting, it is easy to see that the converse is not true: for example, any irrational rotation of the circle $\mathbb{T}$ is topologically transitive but such a rotation is never weakly mixing. In the linear setting, things become very interesting because weak mixing turns out to be equivalent to the Hypercyclicity Criterion. It is readily seen that if a linear operator T satisfies the Hypercyclicity Criterion then T is weakly mixing because $T \oplus T$ satisfies the Hypercyclicity Criterion as well. By a nice result of J. P. Bès and A. Peris [50], the converse is also true: if X is a separable F-space and if $T \in \mathfrak{L}(X)$ is weakly mixing then T satisfies the Hypercyclicity Criterion.

Thus, the present chapter is centred on the following problem, originally raised by D. A. Herrero in the $T \oplus T$ form [143].

Hypercyclicity Criterion problem   *Does every hypercyclic operator on a sep-arable F-space X satisfy the Hypercyclicity Criterion? Equivalently, is every hypercyclic operator $T \in \mathfrak{L}(X)$ necessarily weakly mixing?*

This problem has been recognized as one of the most exciting questions in linear dynamics (see e.g. [50], [129] or [214]). Since all known "natural" hypercyclic op-erators do satisfy the Hypercyclicity Criterion, it has seemed reasonable to hope that the answer would be positive. Nevertheless, the problem was recently solved in the negative by M. De La Rosa and C. J. Read [91].

This chapter is organized as follows. We start with the Bès–Peris theorem. Then we give several characterizations of weak mixing, most of which are valid for non-linear transformations. Using one of these characterizations, we prove a few positive results showing that hypercyclicity entails weak mixing when combined with some additional "regularity" property. Finally, we present a counter-example which solves the Hypercyclicity Criterion problem in the negative for a large class of Banach spaces, including the Hilbert space.

4.1 Characterizations of weak mixing

4.1.1 Hereditarily hypercyclic operators

In this subsection, we prove the Bès–Peris theorem. The following definition will establish the connection between weak mixing and the Hypercyclicity Criterion.

DEFINITION 4.1   *Let T be a separable F-space. Given an increasing sequence of integers (n_k), an operator $T \in \mathfrak{L}(X)$ is said to be **hereditarily hypercyclic with respect to** (n_k) if, for any subsequence (m_k) of (n_k), the sequence $(T^{m_k})_{k \in \mathbb{N}}$ is universal, i.e. there exists an $x \in X$ such that $\{T^{m_k}(x); \ k \in \mathbb{N}\}$ is dense in X. An operator T is said to be **hereditarily hypercyclic** if it is hereditarily hypercyclic with respect to some sequence (n_k).*

We note that, by the proof of Birkhoff's transitivity theorem 1.2, a sequence (T^{m_k}) is universal iff for any non-empty open sets $U, V \subset X$ one can find a k such that $T^{m_k}(U) \cap V \neq \varnothing$, and if this holds then one can find infinitely many k with that property.

THEOREM 4.2   (BÈS–PERIS THEOREM)   *Let X be a separable F-space, and let $T \in \mathfrak{L}(X)$. The following are equivalent:*

(i) *T satisfies the Hypercyclicity Criterion;*
(ii) *T is hereditarily hypercyclic;*
(iii) *T is weakly mixing.*

PROOF   (i) $\implies$ (ii): If T satisfies the Hypercyclicity Criterion with respect to (n_k) then, for every subsequence (m_k) of (n_k), the sequence (T^{m_k}) is universal by Remark 1.7; that is, T is hereditarily hypercyclic with respect to (n_k).

(ii) $\implies$ (iii): Suppose that T is hereditarily hypercyclic with respect to (n_k), and let U_1, U_2, V_1, V_2 be four non-empty open subsets of X. Since the sequence (T^{n_k}) is universal, one can find an infinite subsequence (m_k) of (n_k) such that $T^{m_k}(U_1) \cap V_1 \neq \varnothing$ for all $k \in \mathbb{N}$. By assumption, the sequence (T^{m_k}) is universal, so there exists at least one $k \in \mathbb{N}$ such that $T^{m_k}(U_2) \cap V_2 \neq \varnothing$. Then $(T \times T)^{m_k}(U_1 \times U_2) \cap (V_1 \times V_2) \neq \varnothing$, which proves that T is weakly mixing.

(iii) $\implies$ (i): Assume that $T \oplus T$ is hypercyclic, with hypercyclic vector $x \oplus y$. We show that the Hypercyclicity Criterion is satisfied by $\mathcal{D}_1 = \mathcal{D}_2 := O(x, T)$.

The key point to observe is that $x \oplus T^n(y) \in HC(T \oplus T)$ for any $n \in \mathbb{N}$, because the operator $I \oplus T^n$ has dense range and commutes with $T \oplus T$. Since $y \in HC(T)$, it follows that, for each non-empty open set $U \subset X$, one can find a $u \in U$ such that $x \oplus u \in HC(T \oplus T)$.

In particular, there exist a u arbitrarily close to 0 and an $n \in \mathbb{N}$ such that $T^n(x) \oplus T^n(u)$ is arbitrarily close to $0 \oplus x$; in other words, one can find a sequence $(u_k) \subset X$ and an increasing sequence of integers (n_k) such that $u_k \to 0$, $T^{n_k}(x) \to 0$ and $T^{n_k}(u_k) \to x$.

Let us define maps $S_{n_k} : \mathcal{D}_2 \to X$ by $S_{n_k}(T^j x) = T^j u_k$, $j \in \mathbb{N}$. This definition is meaningful because the vectors $T^j x$ are pairwise distinct. Then

$$T^{n_k}(T^j x) = T^j T^{n_k}(x) \to 0,$$
$$S_{n_k}(T^j x) = T^j u_k \to 0,$$
$$T^{n_k} S_{n_k}(T^j x) = T^j T^{n_k}(u_k) \to T^j x$$

for all $j \in \mathbb{N}$, so that T satisfies the Hypercyclicity Criterion with respect to (n_k). $\square$

REMARK   The above proof shows that if T satisfies the Hypercyclicity Criterion then in fact it satisfies it with one and the same dense set $\mathcal{D} = \mathcal{D}_1 = \mathcal{D}_2$. This is not always apparent in concrete applications (see e.g. Corollary 1.10). Moreover, since the vectors $T^j x$ are linearly independent, the maps S_{n_k} above can be extended by linearity to the span of $O(x, T)$. Thus, if T satisfies the Hypercyclicity Criterion then it satisfies it for a dense linear subspace $\mathcal{D} = \mathcal{D}_1 = \mathcal{D}_2$ and linear maps S_{n_k}.

4.1.2 Non-linear statements

In this subsection, we give several characterizations of weak mixing. The results have nothing to do with linearity, so we formulate them for an arbitrary continuous map $T : X \to X$ acting on some topological space X. These characterizations have a combinatorial flavour. Let us introduce the following sets of integers, where U, V are non-empty open sets in X and $x \in X$:

$$\mathbf{N}(U, V) := \{n \in \mathbb{N}; \ T^n(U) \cap V \neq \varnothing\},$$
$$\mathbf{N}(x, V) := \{n \in \mathbb{N}; \ T^n(x) \in V\}.$$

By definition, the map T is topologically transitive if and only if $\mathbf{N}(U, V) \neq \varnothing$ for any pair of non-empty open sets (U, V). When X is a second-countable Baire space with no isolated points, this is equivalent to saying that $\mathbf{N}(x, U) \neq \varnothing$ for some $x \in X$ and every non-empty open set U. We will prove below that T is weakly mixing iff all sets $\mathbf{N}(U, V)$ are "large" in the following sense.

DEFINITION 4.3   *A set $A \subset \mathbb{N}$ is said to be **thick** if it contains arbitrarily large intervals.*

Besides the sets $\mathbf{N}(U, V)$ and $\mathbf{N}(x, V)$, we may also define

$$\mathbf{C}(U, V) := \{n \in \mathbb{N}; \ T^n(U) \subset V\}.$$

Several nice formal relations hold between the sets introduced above. We summarize them in the following two lemmas. Lemma 4.4 is needed for the proof of Theorem 4.6 below, which is the main result of this subsection. Lemma 4.5 will be used three times later on: in the proofs of Corollary 4.8 and Lemma 4.17 below, and in Chapter 6 as well.

We recall that if A and B are two subsets of $\mathbb{N}$ then the **difference set** $A - B$ is defined by

$$A - B = \{n - m; \ (n, m) \in A \times B, \ n \geq m\}.$$

The **sum set** $A + B$ is defined in an obvious way.

LEMMA 4.4   *Let U, V, W be non-empty open subsets of X. Then*

$$\mathbf{N}(U, V) + \mathbf{C}(V, W) \subset \mathbf{N}(U, W),$$
$$\mathbf{N}(U, W) - \mathbf{C}(U, V) \subset \mathbf{N}(V, W).$$

PROOF   The first inclusion is trivial. For the second, just note that if $k = n - m \in \mathbf{N}(U, W) - \mathbf{C}(U, V)$ then $T^k(V) \supset T^n(T^{-m}(V)) \supset T^n(U)$, so that $T^k(V) \cap W \neq \varnothing$.   □

LEMMA 4.5   *Assume that X is a second-countable Baire space with no isolated points and that $T : X \to X$ is a continuous hypercyclic map. Also, let x be a T-hypercyclic point. Then, for every pair (U, V) of non-empty open sets, the following relations hold:*

$$\mathbf{N}(U, V) = \mathbf{N}(x, V) - \mathbf{N}(x, U),$$
$$\mathbf{N}(U, V) - \mathbf{N}(U, V) = \mathbf{N}(V, V) - \mathbf{N}(U, U).$$

PROOF   The inclusion $\mathbf{N}(x, V) - \mathbf{N}(x, U) \subset \mathbf{N}(U, V)$ is straightforward. Indeed, if $k = m - n$ with $m \in \mathbf{N}(x, V)$ and $n \in \mathbf{N}(x, U)$ then $T^k(T^n(x)) \in V$, hence $k \in \mathbf{N}(U, V)$. Conversely, let $k \in \mathbf{N}(U, V)$. Then $W := U \cap T^{-k}(V)$ is a non-empty open set, so one can find $n \in \mathbb{N}$ such that $T^n(x) \in W$. Then $n \in \mathbf{N}(x, U)$ and $n + k \in \mathbf{N}(x, V)$, hence $k \in \mathbf{N}(x, V) - \mathbf{N}(x, U)$.

We note that, since X has no isolated points, the above integer n such that $T^n(x) \in W$ can be taken as arbitrarily large. It follows that if $k \in \mathbf{N}(U, V)$ then one can find arbitrarily large $(n, m) \in \mathbf{N}(x, V) \times \mathbf{N}(x, U)$ such that $k = m - n$. This legitimates the following formal derivation of the second identity we are looking at:

$$\mathbf{N}(U, V) - \mathbf{N}(U, V) = [\mathbf{N}(x, V) - \mathbf{N}(x, U)] - [\mathbf{N}(x, V) - \mathbf{N}(x, U)]$$
$$= [\mathbf{N}(x, V) - \mathbf{N}(x, V)] - [\mathbf{N}(x, U) - \mathbf{N}(x, U)]$$
$$= \mathbf{N}(V, V) - \mathbf{N}(U, U).$$ □

Before stating the main result of this subsection, we make two more observations. By definition,

$$T \text{ is topologically transitive} \iff \mathbf{N}(U,V) \neq \varnothing \quad \text{for all } U, V;$$
$$T \text{ is weakly mixing} \iff \mathbf{N}(U_1, V_1) \cap \mathbf{N}(U_2, V_2) \neq \varnothing$$
$$\text{for all } U_1, V_1, U_2, V_2.$$

Thus, topological transitivity is an approximation property involving pairs of open sets whereas weak mixing is an approximation property involving 4-tuples of open sets. Property (6) below states that it is in fact enough to consider 3-tuples of open sets only. Here are now the promised characterizations of weak mixing.

THEOREM 4.6 *Let X be a topological space, and let $T : X \to X$ be a continuous map. The following are equivalent:*

 (i) *T is weakly mixing;*
 (ii) *The sets $\mathbf{N}(U, V)$ form a filter basis, i.e. each $\mathbf{N}(U, V)$ is non-empty and, given U_1, V_1, U_2, V_2, one can find U_3, V_3 such that $\mathbf{N}(U_3, V_3) \subset \mathbf{N}(U_1, V_1) \cap \mathbf{N}(U_2, V_2)$;*
 (iii) *For any $L \geq 1$, the L-fold product map $T \times \cdots \times T$ is topologically transitive;*
 (iv) *All sets $\mathbf{N}(U, V)$ are thick;*
 (v) *$\mathbf{N}(U, V) - \mathbf{N}(U, V) = \mathbb{N}$ for any U, V;*
 (vi) *$\mathbf{N}(U, V) \cap \mathbf{N}(U, V') \neq \varnothing$ for any U, V, V'.*

PROOF OF THEOREM 4.6

(i) $\implies$ (ii): Assume that T is weakly mixing, and fix four non-empty open sets U_1, V_1, U_2, V_2. One can pick an $m \in \mathbf{N}(U_1, U_2) \cap \mathbf{N}(V_1, V_2)$ and hence two non-empty open sets $U_3 \subset U_1$ and $V_3 \subset V_1$ such that $T^m(U_3) \subset U_2$ and $T^m(V_3) \subset V_2$. Then $\mathbf{N}(U_3, V_3) \subset \mathbf{N}(U_1, V_1)$. Moreover, if $n \in \mathbf{N}(U_3, V_3)$ then $n + m \in \mathbf{N}(U_3, V_3) + \mathbf{C}(V_3, V_2) \subset \mathbf{N}(U_3, V_2)$, so that $n = (n + m) - m \in \mathbf{N}(U_3, V_2) - \mathbf{C}(U_3, U_2) \subset \mathbf{N}(U_2, V_2)$ by Lemma 4.4. This proves (ii).

(ii) $\implies$ (iii): If the sets $\mathbf{N}(U, V)$ form a filter basis then any finite intersection of them is non-empty. In other words, $\mathbf{N}(U_1, V_1) \cap \cdots \cap \mathbf{N}(U_L, V_L) \neq \varnothing$ for each $L \geq 1$ and any non-empty open sets $U_1, \ldots, U_L, V_1, \ldots, V_L$. This means that each map $T \times \cdots \times T$ is topologically transitive.

(iii) $\implies$ (iv): Given U, V and a positive integer L, set $V_i := T^{-i}(V)$, $i = 0, \ldots, L$. If (iii) holds then one can find an $n \in \mathbb{N}$ such that $T^n(U) \cap V_i \neq \varnothing$ for all $i \in \{0, \ldots, L\}$. This means that $\mathbf{N}(U, V)$ contains the interval $[n, n + L]$.

(iv) $\implies$ (v): This is trivial.

(v) $\implies$ (vi): Assume that (v) holds, and let U, V, V' be given. Since T is topologically transitive by (v), one can find an $m \in \mathbb{N}$ and a non-empty open set $V_1 \subset V$ such that $T^m(V_1) \subset V'$. By (v), we can choose $k \in \mathbb{N}$ such that $k \in \mathbf{N}(U, V_1)$ and $k + m \in \mathbf{N}(U, V_1)$. Then

$$k + m \in \mathbf{N}(U, V) \cap [\mathbf{N}(U, V_1) + \mathbf{C}(V_1, V')] \subset \mathbf{N}(U, V) \cap \mathbf{N}(U, V').$$

(vi) $\implies$ (i): Assume that (vi) holds, and let U_1, V_1, U_2, V_2 be four non-empty open sets in X. There exists $m \in \mathbb{N}$ and a non-empty open set $U \subset U_1$ such that $T^m(U) \subset$

U_2. Applying (vi) with $U := U$, $V := V_1$ and $V' := T^{-m}(V_2)$, we find a $k \in \mathbb{N}$ such that $T^k(U) \cap V_1 \neq \varnothing$ and $T^{k+m}(U) \cap V_2 \neq \varnothing$. Thus, on the one hand $k \in \mathbf{N}(U, V_1) \subset \mathbf{N}(U_1, V_1)$, and on the other hand $k = k + m - m \in \mathbf{N}(U, V_2) - \mathbf{C}(U, U_2) \subset \mathbf{N}(U_2, V_2)$. This shows that $T \times T$ is topologically transitive. $\qquad\square$

REMARK   When the topological space X is second-countable, there is a countable family of pairs of open sets $(U_i, V_i)_{i \in I}$ such that each set $\mathbf{N}(U, V)$ contains some $\mathbf{N}(U_i, V_i)$. By (iii) and a simple diagonal argument, it follows that T is weakly mixing iff one can find an infinite set $\mathbf{N} \subset \mathbb{N}$ such that each $\mathbf{N}(U, V)$ contains all but finitely many $n \in \mathbf{N}$. In that case, it seems reasonable to say that the map T is **N-mixing**.

In the two corollaries that follow, we quote two interesting consequences of Theorem 4.6.

The first is a result due to A. Peris and L. Saldivia [195] and, independently, to S. Grivaux [129]. Recall that an increasing sequence $(n_k) \subset \mathbb{N}$ is said to be **syndetic** if $\sup_k (n_{k+1} - n_k) < \infty$. An infinite *set* $\mathbf{N} \subset \mathbb{N}$ is syndetic if its increasing enumeration is a syndetic sequence, i.e. $\mathbf{N}$ has bounded gaps.

There is an obvious "duality" between thick sets and syndetic sets: a set is thick iff it intersects any syndetic set, and a set is syndetic iff it intersects any thick set.

COROLLARY 4.7   *Assume that X is a second-countable Baire space with no isolated points. Then a continuous map $T : X \to X$ is weakly mixing iff it is* **syndetically hypercyclic**, *which means that for any syndetic sequence of integers (n_k), the sequence (T^{n_k}) is universal.*

PROOF   By the proof of Birkhoff's transitivity theorem, the statement "T is syndetically hypercyclic" is equivalent to "each set $\mathbf{N}(U, V)$ intersects any syndetic set", i.e. "each set $\mathbf{N}(U, V)$ is thick". $\qquad\square$

The second result we would like to point out is due to S. Grivaux [129].

COROLLARY 4.8   *Assume that X is a second-countable Baire space with no isolated points. Then a continuous map $T : X \to X$ is weakly mixing iff each set $\mathbf{N}(U, V)$ contains two consecutive integers.*

PROOF   Let us denote by $(v)_1$ the statement "Each set $\mathbf{N}(U, V)$ contains two consecutive integers", i.e. "$1 \in \mathbf{N}(U, V) - \mathbf{N}(U, V)$ for any U, V". By Theorem 4.6, it is enough to prove that if $(v)_1$ holds then in fact $\mathbf{N}(U, V) - \mathbf{N}(U, V) = \mathbb{N}$ for any U, V. So we may assume that $(v)_1$ holds.

We first note that T is topologically transitive by $(v)_1$, hence hypercyclic by the assumptions on X. By Lemma 4.5, it follows that we also have

$$1 \in \mathbf{N}(V, V) - \mathbf{N}(U, U) \quad \text{for any } U, V.$$

We now prove by induction on $m \in \mathbb{N}$ that $m \in \mathbf{N}(U, V) - \mathbf{N}(U, V)$ for any U, V. This is trivial for $m = 0$ and we assume that it holds true for m. Using the induction hypothesis, one can find a $k \in \mathbb{N}$ and two non-empty open sets $U_1, U_2 \subset U$ such that

$T^k(U_1) \subset V$ and $T^{k+m}(U_2) \subset V$. By the property above, one can find an l such that $T^l(U_1) \cap U_1 \neq \varnothing$ and $T^{l+1}(U_2) \cap U_2 \neq \varnothing$. Then $l + k \in \mathbf{N}(U_1, U_1) + \mathbf{C}(U_1, V) \subset \mathbf{N}(U, V)$ and $l + k + m + 1 = (l + 1) + (k + m) \in \mathbf{N}(U_2, U_2) + \mathbf{C}(U_2, V) \subset \mathbf{N}(U, V)$, so that $m + 1 \in \mathbf{N}(U, V) - \mathbf{N}(U, V)$. $\qquad\square$

REMARK When X is an F-space and T is linear, a more general result can be proved; see Exercise 4.2.

4.1.3 The three open sets condition

In the linear setting, the neighbourhoods of 0 provide natural examples of thick sets.

LEMMA 4.9 *Let X be a topological vector space, and let $T \in \mathfrak{L}(X)$ be topologically transitive.*

(a) *For any open neighbourhood W of 0 and any non-empty open sets $U, V \subset X$, the sets $\mathbf{N}(U, W)$ and $\mathbf{N}(W, V)$ are thick.*
(b) *Suppose that all the sets $\mathbf{N}(U, W) \cap \mathbf{N}(W, V)$ are non-empty, for U, V, W as above. Then all these sets are thick.*

PROOF To prove (a), let us fix $L \in \mathbb{N}$. Since $T(0) = 0$, one can find an open neighbourhood W' of zero such that $T^k(W') \subset W$ for all $k \in \{0, \ldots, L\}$. Moreover, since T is topologically transitive one can choose $n, n' \in \mathbb{N}$ such that $T^n(U) \cap W' \neq \varnothing$ and $T^{n'}(W') \cap T^{-L}(V) \neq \varnothing$, i.e. $T^{n'+L}(W') \cap V \neq \varnothing$. Then $n + \{0, \ldots, L\} \subset \mathbf{N}(U, W') + \mathbf{C}(W', W) \subset \mathbf{N}(U, W)$ and $n' + \{0, \ldots, L\} = (n' + L) - \{0, \ldots, L\} \subset \mathbf{N}(W', V) - \mathbf{C}(W', W) \subset \mathbf{N}(W, V)$. Since L is arbitrary, this shows that $\mathbf{N}(U, W)$ and $\mathbf{N}(W, V)$ are thick.

The proof of (b) is the same. Indeed, in this case one may take $n = n'$. $\qquad\square$

This lemma provides yet another characterization of weak mixing, the so-called **three open sets condition**.

THEOREM 4.10 (THREE OPEN SETS CONDITION) *Let X be a topological vector space and let $T \in \mathfrak{L}(X)$. The following are equivalent:*

(i) *T is weakly mixing;*
(ii) *$\mathbf{N}(U, W) \cap \mathbf{N}(W, V) \neq \varnothing$ for any non-empty open sets $U, V \subset X$ and any neighbourhood W of 0.*

PROOF It is plain that (i) implies (ii). Conversely, suppose that (ii) holds true. We will show that then all sets $\mathbf{N}(U, V)$ are thick.

Let us fix non-empty open sets $U, V \subset X$. There exist non-empty open sets U_0, V_0 and an open neighbourhood W of 0 such that $U \supset U_0 + W$ and $V \supset V_0 + W$. By (ii) and Lemma 4.9 the set $\mathbf{N}(U_0, W) \cap \mathbf{N}(W, V_0)$ is thick, so it is enough to show that $\mathbf{N}(U_0, W) \cap \mathbf{N}(W, V_0) \subset \mathbf{N}(U, V)$.

Let $n \in \mathbf{N}(U_0, W) \cap \mathbf{N}(W, V_0)$, i.e. $T^n(U_0) \cap W \neq \varnothing$ and $T^n(W) \cap V_0 \neq \varnothing$. By linearity, we get $T^n(U_0 + W) \cap (V_0 + W) \neq \varnothing$, which implies that $n \in \mathbf{N}(U, V)$. This concludes the proof. $\qquad\square$

We now deduce several corollaries on the theme that a hypercyclic operator having many "regular" orbits is in fact weakly mixing. The first may not look very exciting, but the two others (which are immediate consequences of the first) are more appealing. Recall that a set B in a topological vector space X is said to be *bounded* if it is absorbed by any neighbourhood W of 0, i.e. if one can find a $\lambda > 0$ such that $\lambda B \subset W$.

COROLLARY 4.11   *Let $T \in \mathfrak{L}(X)$ be hypercyclic. Suppose that, for each non-empty open set $U \subset X$, there exist an operator $S \in \mathfrak{L}(X)$ which has dense range and commutes with T and a bounded set $B \subset X$ such that $ST^n(U) \cap B \neq \varnothing$ for all $n \geq 0$. Then T is weakly mixing.*

PROOF   In order to apply Theorem 4.10, let us fix non-empty open sets U, V and a neighbourhood W of 0. Let $S_U \in \mathfrak{L}(X)$ and a bounded set B_U be associated with U by the above assumption. Then one can find a $\lambda > 0$ such that $\lambda B_U \subset W$. Putting $S := \lambda S_U$, we get an operator $S \in \mathfrak{L}(X)$ with dense range, such that $TS = ST$ and $ST^n(U) \cap W \neq \varnothing$ for all $n \geq 0$.

Since T is topologically transitive and $S^{-1}(V)$ is a non-empty open set (because S has dense range), one can find an integer $N \geq 1$ such that $ST^N(W) \cap V \neq \varnothing$. We set $A := ST^N$. Then $AT = TA$, $A(W) \cap V \neq \varnothing$ by definition and $A(U) \cap W \neq \varnothing$ by the choice of S.

Let us choose some T-hypercyclic vector $x \in U$ such that $A(x) \in W$ and also $m \in \mathbb{N}$ such that $T^m x \in W$ and $A(T^m x) \in V$, i.e. $T^m(A(x)) \in V$. Now, let (n_i) be a net of integers such that $T^{n_i}(x) \to A(x)$. Then $T^{n_i}(x) \in W$ and $T^{n_i}(T^m x) = T^m(T^{n_i}(x)) \in V$ if i is large enough. Since $x \in U$ and $T^m x \in W$, it follows that $\mathbf{N}(U, W) \cap \mathbf{N}(W, V) \neq \varnothing$. This concludes the proof. $\qquad\square$

For the next corollary, let us say that a vector $x \in X$ is T-**algebraic** if there exists some non-zero polynomial P such that $P(T)x = 0$. In particular, any eigenvector of T is T-algebraic.

COROLLARY 4.12   *Let $T \in \mathfrak{L}(X)$ be hypercyclic, and assume that the T-algebraic vectors are dense in X. Then T is weakly mixing.*

PROOF   Let U be any non-empty open subset of X. By assumption, one can find a non-zero polynomial P and $x \in U$ such that $P(T)x = 0$. Since $P(T)$ has dense range (see Theorem 1.30), the result follows immediately from Corollary 4.11 applied for $S := P(T)$ and $B := \{0\}$. $\qquad\square$

COROLLARY 4.13   *If $T \in \mathfrak{L}(X)$ is hypercyclic and if there exists a dense set of vectors whose T-orbit is bounded then T is weakly mixing.*

PROOF For any non-empty open set $U \subset X$, one can find an $x \in U$ such that $B := O(x, T)$ is bounded. Thus, one can apply Corollary 4.11 with $S = I$. □

4.2 Hypercyclic non-weakly-mixing operators

4.2.1 Results

In 2006, as mentioned at the start of this chapter, M. De La Rosa and C. J. Read solved in the negative the Hypercyclicity Criterion problem [91]. More precisely, they constructed a Banach space X and a hypercyclic operator $T \in \mathfrak{L}(X)$ such that $T \oplus T$ is not hypercyclic.

Their construction may be roughly described as follows. One starts with a hypercyclic operator S on some Banach space $(Z_0, \| \cdot \|_0)$. If e_0 is a hypercyclic vector for S, and if we set $e_i := S^i(e_0)$, then S can be seen as a shift acting on $c_{00} = \text{span}\{e_i; \ i \geq 0\}$. The idea is to define a new norm $\| \cdot \|$ on c_{00} in such a way that S remains continuous and hypercyclic and, moreover, that $\| \cdot \|$ is in some sense maximal with respect to these properties; that is, one wants S to be hypercyclic but nothing more. The desired space X is the completion of c_{00} under this new norm, and T is the extension of S to X.

Although the new norm is not intractably complicated, it is not clear whether the space(s) constructed in [91] can be identified with a "classical" Banach space. Building on the ideas of De La Rosa and Read, the authors of the present book were able to produce hypercyclic, non-weakly-mixing operators on many classical spaces, including the separable Hilbert space. The philosophy of [30] is the following: in De La Rosa and Read's example, the operator is very simple (it is a shift) whereas the norm is rather complicated. Now, if one wants to construct a counter-example on some specified classical space then the norm is given and simple, so one has to transfer the complexity to the operator itself.

Before stating the main result of [30], we settle some terminology. If $(e_i)_{i \in \mathbb{N}}$ is a linearly independent sequence in a Banach space X the **forward shift associated with** (e_i) is the linear map $S : E \to E$ defined by $S(e_i) = e_{i+1}$, where $E = \text{span} \{e_i; \ i \in \mathbb{N}\}$.

A sequence $(e_i)_{i \in \mathbb{N}} \subset X$ is an **unconditional basis** of X if every $x \in X$ can be uniquely written as $x = \sum_0^\infty x_i e_i$ and if the convergence of the latter series is unconditional, i.e. $\sum \pm x_i e_i$ is convergent for any choice of signs $\pm$. Equivalently, (e_i) is a Schauder basis of X and, for any sequence $\lambda \in \ell^\infty(\mathbb{N})$, the linear map $M_\lambda : E \to E$ defined by $M_\lambda(e_i) = \lambda_i e_i$ is continuous. We denote by (e_i^*) the sequence of coordinate functionals associated with (e_i). Then $\sup_i \|e_i\| \|e_i^*\| < \infty$. In particular, if (e_i) is **normalized** (i.e. $\|e_i\| = 1$ for all i) then (e_i^*) is bounded.

We are going to prove the following theorem.

THEOREM 4.14 *Let X be a Banach space. Assume that X has a normalized unconditional basis $(e_i)_{i \in \mathbb{N}}$ whose associated forward shift is continuous. Then there exists a hypercyclic operator $T \in \mathfrak{L}(X)$ which is not weakly mixing.*

From this, we deduce immediately:

COROLLARY 4.15 *There exist hypercyclic operators on $c_0(\mathbb{N})$ or $\ell^p(\mathbb{N})$, $1 \leq p < \infty$, which do not satisfy the Hypercyclicity Criterion. In particular, one can find such operators on the Hilbert space.*

Another direct consequence is the following more general result.

COROLLARY 4.16 *The conclusion of Theorem 4.14 remains true if one assumes only that X has a complemented subspace X_0 admitting a basis with the above properties.*

PROOF Write $X = X_0 \oplus Y$, where Y is a closed subspace. If $\dim(Y) < \infty$ then it is easily checked that one can apply Theorem 4.14 directly to X. Otherwise, Y supports a mixing operator R (see Chapter 2) whereas X_0 supports a hypercyclic non-weakly-mixing operator T_0. Then $T := T_0 \oplus R$ is hypercyclic and non-weakly mixing. $\square$

This result can be applied e.g. to $X = L^1([0,1])$ and $X = \mathcal{C}([0,1])$. Indeed, $L^1([0,1])$ contains a complemented copy of $\ell^1(\mathbb{N})$ and $\mathcal{C}([0,1])$ contains a copy of $c_0(\mathbb{N})$, which is necessarily complemented by a classical result of A. Sobczyk ([229]; see e.g. [96] or [3]).

4.2.2 Strategy

To prove Theorem 4.14, we need some criterion for checking that an operator is *not* weakly mixing. This is the content of the next lemma.

Recall that if T is a linear operator on some topological vector space X and if $e_0 \in X$ then

$$\mathbb{K}[T]e_0 = \{P(T)e_0;\ P \text{ polynomial}\}$$
$$= \text{span}\{T^i(e_0):\ i \in \mathbb{N}\}.$$

By the commutativity of the algebra $\mathbb{K}[T]$, one can unambiguously define a product on $\mathbb{K}[T]e_0$ by setting

$$P(T)e_0 \cdot Q(T)e_0 := PQ(T)e_0.$$

We should perhaps have indicated that this product heavily depends on the operator T, but we prefer to avoid clumsy notation.

LEMMA 4.17 *Let X be a topological vector space, and let $T \in \mathfrak{L}(X)$ be hypercyclic, with hypercyclic vector e_0. Assume there exists a non-zero linear functional $\phi : \mathbb{K}[T]e_0 \to \mathbb{K}$ such that the map $(x,y) \mapsto \phi(x \cdot y)$ is continuous on $\mathbb{K}[T]e_0 \times \mathbb{K}[T]e_0$. Then T is not weakly mixing.*

PROOF   By assumption, $\mathcal{W} := \{(x, y); \, |\phi(x \cdot y)| < 1\}$ and $\mathcal{W}' := \{(x', y'); \, |\phi(x' \cdot y')| > 1\}$ are non-empty open sets in the product $\mathbb{K}[T]e_0 \times \mathbb{K}[T]e_0$, and we have

$$\{x \cdot y; \, (x, y) \in \mathcal{W}\} \cap \{x' \cdot y'; \, (x', y') \in \mathcal{W}'\} = \varnothing. \tag{4.1}$$

Let us choose non-empty open sets $U_1, V_1, U_2, V_2 \subset X$ such that $(U_1 \times V_2) \cap (O(e_0, T) \times O(e_0, T)) \subset \mathcal{W}$ and $(V_1 \times U_2) \cap (O(e_0, T) \times O(e_0, T)) \subset \mathcal{W}'$. Then (4.1) gives

$$(\mathbf{N}(e_0, U_1) + \mathbf{N}(e_0, V_2)) \cap (\mathbf{N}(e_0, V_1) + \mathbf{N}(e_0, U_2)) = \varnothing.$$

This, in turn, yields

$$(\mathbf{N}(e_0, V_1) - \mathbf{N}(e_0, U_1)) \cap (\mathbf{N}(e_0, V_2) - \mathbf{N}(e_0, U_2)) = \varnothing.$$

By Lemma 4.5, it follows that $\mathbf{N}(U_1, V_1) \cap \mathbf{N}(U_2, V_2) = \varnothing$, so that T is not weakly mixing. $\qquad\square$

REMARK 4.18   Linearity plays no role in the above lemma. If $T : X \to X$ is a continuous map acting on some topological space X then T is not weakly mixing provided there exist some T-hypercyclic point $e \in X$ and a non-constant map $\phi : O(e, T) \to \mathbb{K}$ such that the map $(x, y) \mapsto \phi(x \cdot y)$ is continuous on $O(e, T) \times O(e, T)$. One can even replace $\mathbb{K}$ by an arbitrary Hausdorff topological space.

REMARK 4.19   In Lemma 4.17 it is in fact enough to assume that e_0 is a *cyclic* vector for T; this is not apparent from the above proof. See Exercise 4.7.

From now on, we take X to be a Banach space with a normalized unconditional basis $(e_i)_{i \in \mathbb{N}}$ whose associated forward shift is continuous. We put

$$c_{00} := \operatorname{span} \{e_i; \, i \in \mathbb{N}\}.$$

In view of Lemma 4.17, our main result will be proved if we are able to construct a linear operator $T : c_{00} \to c_{00}$ and a non-zero linear functional $\phi : c_{00} \to \mathbb{K}$ such that the following properties hold:

(P1) $\operatorname{span} \{T^i e_0; \, i \in \mathbb{N}\} = \operatorname{span} \{e_i; \, i \in \mathbb{N}\}$, in other words $\mathbb{K}[T]e_0 = c_{00}$;
(P2) the closure of $\{T^i(e_0); \, i \in \mathbb{N}\}$ contains $\mathbb{K}[T]e_0$;
(P3) T is continuous;
(P4) the map $(x, y) \mapsto \phi(x \cdot y)$ is continuous on $c_{00} \times c_{00}$.

Indeed, by (P3) and since c_{00} is dense in X, the linear map T extends to a continuous linear operator on X which is hypercyclic with hypercyclic vector e_0, by (P1) and (P2), and not weakly mixing by (P4) and Lemma 4.17.

The operator T and the linear functional ϕ will be constructed in the next two subsections. From now on, we fix some countable dense set $\mathbf{Q} \subset \mathbb{K}$. We will say that a sequence of polynomials $\mathbf{P} = (P_n)_{n \in \mathbb{N}}$ is **admissible** if $P_0 = 0$ and $\mathbf{P}$ enumerates all polynomials with coefficients in $\mathbf{Q}$, *not* necessarily in a one-to-one way. Finally, if P is a polynomial, we will denote by $\deg(P)$ its degree and by $|P|_1$ its ℓ^1 norm, i.e. the sum of the moduli of its coefficients.

4.2.3 The operator T

In this subsection, we describe a "natural" candidate for solving the Hypercyclicity Criterion problem. The main difficulty is that we want our operator T to be hypercyclic, but we cannot use the Hypercyclicity Criterion to check this property. The solution is to ensure that T is hypercyclic *by definition*.

First, we note that the above property (P1) will be satisfied if T is an upper-triangular perturbation of a forward weighted shift. Thus, we consider a sequence of positive numbers $(w(n))_{n \geq 1}$ and require that each vector $T(e_i)$ has the form

$$T(e_i) = w(i+1)e_{i+1} + \sum_{k \leq i} \alpha_{k,i} e_k.$$

Now, we want to ensure (P2). This will be achieved if we have at hand an admissible sequence of polynomials (P_n), a sequence of positive numbers $(a_n)_{n \geq 1}$ tending to infinity and an increasing sequence of integers $(b_n)_{n \geq 1}$ such that

$$T^{b_n}(e_0) = P_n(T)(e_0) + \frac{1}{a_n} e_{b_n} \quad \text{for all } n \geq 1. \tag{4.2}$$

If we try to give the simplest definition of an operator T subject to the above constraints, it is natural to forget the upper-triangular perturbation almost every time and to put

$$T(e_i) := w(i+1)e_{i+1} \quad \text{for } i \in [b_{n-1}, b_n - 1), \ n \geq 1$$

(we have set $b_0 := 0$). Then only the vectors $T(e_{b_n-1})$ remain to be defined. Now, if (4.2) holds, we write

$$
\begin{aligned}
T^{b_n}(e_0) &= T^{b_n - b_{n-1}} T^{b_{n-1}}(e_0) \\
&= T^{b_n - b_{n-1}} \left(P_{n-1}(T)(e_0) + \frac{1}{a_{n-1}} e_{b_{n-1}} \right) \\
&= T^{b_n - b_{n-1}} P_{n-1}(T)(e_0) + \frac{w(b_{n-1}+1) \cdots w(b_n - 1)}{a_{n-1}} T(e_{b_n-1}).
\end{aligned}
$$

Then, using (4.2) once more to express $T^{b_n}(e_0)$ in a different way, we arrive at the following identity:

$$T(e_{b_n-1}) := \varepsilon_n e_{b_n} + f_n,$$

where

$$\varepsilon_n = \frac{a_{n-1}}{a_n \, w(b_{n-1}+1) \cdots w(b_n - 1)}$$

(we have set $a_0 := 1$) and

$$f_n = \frac{a_{n-1}}{w(b_{n-1}+1) \cdots w(b_n - 1)} \left(P_n(T)e_0 - T^{b_n - b_{n-1}} P_{n-1}(T)e_0 \right). \tag{4.3}$$

Conversely, if the vectors $T(e_{b_n-1})$ are defined in this way then (4.2) is satisfied. This definition is consistent provided that $\deg(P_n) < b_n - 1$ for all n, which we now assume.

Observe that the sequences (a_n), (b_n), (P_n) and $(w(n))$ are parameters in the definition of T. We could adjust them in order to obtain different versions of Theorem 4.14. For the present purpose, we will fix once and for all (a_n), (b_n) and $(w(n))$ by setting $a_0 := 1$, $b_0 := 0$ and, for $n \geq 1$,

$$w(n) := 4\left(1 - \frac{1}{2\sqrt{n}}\right),$$
$$a_n := n + 1,$$
$$b_n := 3^n.$$

Since this will be used several times below, we note that $2 \leq w(n) \leq 4$ for all $n \geq 1$.

By the above discussion, our linear map $T : c_{00} \to c_{00}$ satisfies properties (P1) and (P2) for any choice of the admissible sequence $\mathbf{P} = (P_n)$. What remains to be done is to show that if this last parameter $\mathbf{P}$ is suitably chosen then the operator T is continuous, and a linear functional $\phi : c_{00} \to \mathbb{K}$ with the required continuity property can be defined.

The following terminology will be useful: we will say that an admissible sequence $\mathbf{P}$ is **controlled** by some sequence of positive numbers (c_n) if $\deg(P_n) < c_n$ and $|P_n|_1 \leq c_n$ for all $n \in \mathbb{N}$. Clearly, for any sequence of positive numbers (c_n) such that $\limsup_{n\to\infty} c_n = \infty$, one can find an admissible sequence $\mathbf{P}$ which is controlled by (c_n). Any such sequence (c_n) will be called a **control sequence**.

The ℓ^1 norm on c_{00} will play a crucial role in what follows (essentially because it dominates the given norm on X). We denote this norm by $\| \cdot \|_1$; thus, if $x = \sum_i x_i e_i \in c_{00}$ then $\|x\|_1 = \sum_i |x_i|$.

The next lemma will be our main tool for checking the continuity of T. Here and elsewhere, we put

$$d_n := \deg(P_n).$$

LEMMA 4.20   *The following properties hold:*

(1) $\varepsilon_n \leq 1$ for any $n \geq 1$;
(2) *if $n \geq 1$ and if $\|f_k\|_1 \leq 1$ for all $k < n$ then*

$$\|f_n\|_1 \leq n\, 4^{\max(d_n, d_{n-1})+1}\left(\frac{|P_n|_1}{2^{b_{n-1}}} + |P_{n-1}|_1 \exp\left(-c\sqrt{b_{n-1}}\right)\right),$$

where $c > 0$ is a numerical constant.

PROOF   Part (1) is obvious. To prove (2), let us fix n and assume that $\|f_k\|_1 \leq 1$ for all $k < n$. For each $j \in \mathbb{N}$, set $E_j := \text{span}(e_0, \ldots, e_j)$. Since we are working with the ℓ^1 norm, we have $\|T_{|E_j}\|_1 \leq \max(\|Te_i\|_1; \ i \leq j)$. Moreover, since the sequence $(w(i))$ is increasing and $w(i) \geq 2$ for all i, it follows from our assumption that if $j < b_n - 1$ then $\|T(x)\|_1 \leq w(j+1)\|x\|_1$ for all $x \in E_j$. From this, we deduce that

$$\|T^p(e_0)\|_1 \leq \prod_{i=1}^{p} w(i)$$

for all $p \in [1, b_n)$. This is also true for $p = 0$, if we agree that the value of an empty product is 1. Looking at (4.3), this gives the inequality

$$\|f_n\|_1 \leq \frac{n \left(|P_n|_1 \prod_{i=1}^{d_n} w(i) + |P_{n-1}|_1 \prod_{i=1}^{b_n - b_{n-1} + d_{n-1}} w(i) \right)}{w(b_{n-1} + 1) \cdots w(b_n - 1)};$$

and, since $2 \leq w(i) \leq 4$ for all i, it follows that

$$\|f_n\|_1 \leq n \, 4^{\max(d_n, d_{n-1})+1} \left(\frac{|P_n|_1}{2^{b_n - b_{n-1} - 1}} + |P_{n-1}|_1 \prod_{i=1}^{b_n - b_{n-1} - 1} \frac{w(i)}{w(i + b_{n-1})} \right).$$

Now, it is easy to check that $\log((1-v)/(1-u)) \leq u - v$ whenever $0 \leq u < v < 1$. Thus, we get

$$
\begin{aligned}
\prod_{i=1}^{b_n - b_{n-1} - 1} \frac{w(i)}{w(i + b_{n-1})} &\leq \exp \left[\sum_{i=1}^{b_n - b_{n-1} - 1} \left(\frac{1}{2\sqrt{i + b_{n-1}}} - \frac{1}{2\sqrt{i}} \right) \right] \\
&\leq \exp \left(-\frac{1}{4} \sum_{i=1}^{b_{n-1}} \frac{b_{n-1}}{(i + b_{n-1})^{3/2}} \right) \\
&\leq \exp \left(-\frac{1}{8} \sum_{j=b_{n-1}+1}^{2b_{n-1}} \frac{1}{\sqrt{j}} \right) \\
&\leq \exp \left(-c\sqrt{b_{n-1}} \right),
\end{aligned}
$$

where we have used the inequality $b_n - b_{n-1} - 1 \geq b_{n-1}$ and the mean-value theorem. $\qquad\square$

We can now show that T is continuous if the admissible sequence $\mathbf{P}$ is suitably chosen.

PROPOSITION 4.21 *There exists a control sequence (u_n) tending to infinity such that the following holds: if the sequence $\mathbf{P}$ is controlled by (u_n) then T is continuous on c_{00} with respect to the topology of X.*

PROOF Let (u_n) be an increasing sequence of positive numbers tending to infinity such that

$$n \, 4^{u_n + 1} \left(\frac{u_n}{2^{b_{n-1}}} + u_n \exp\left(- c\sqrt{b_{n-1}} \right) \right) \leq \frac{1}{2^n}$$

for all $n \geq 1$; since $b_n = 3^n$, it is clearly possible to choose such a sequence (u_n). If $\mathbf{P}$ is controlled by (u_n) then Lemma 4.20 and a straightforward induction show that $\|f_n\|_1 \leq 1/2^n$ for all $n \geq 1$.

We decompose T as $T = R + K$, where R is a forward weighted shift associated with a bounded weight sequence and K is defined by $K(e_{b_n - 1}) = f_n$ for all n and $K(e_i) = 0$ if i is not of the form $b_n - 1$.

Since the forward shift associated with (e_i) is continuous and since the basis (e_i) is unconditional, the operator R is continuous. Moreover, the sequence (e_i) is normalized, so $\|K(e_{b_n-1})\|_X \leq \|f_n\|_1$. It follows that $\sum_0^\infty \|K(e_i)\|_X < \infty$. Writing each $x \in c_{00}$ as $\sum_i x_i e_i$ with $|x_i| \leq \|e_i^*\|\|x\|$ and remembering that the sequence (e_i^*) is bounded, we conclude that the operator K is continuous, and even compact. Hence T is continuous. $\qquad\square$

4.2.4 The linear functional

To complete the proof of Theorem 4.14, it remains to construct a non-zero linear functional $\phi : c_{00} \to \mathbb{K}$ such that the map $(x, y) \mapsto \phi(x \cdot y)$ is continuous on $c_{00} \times c_{00}$. The following simple lemma gives us a computational way of checking continuity.

LEMMA 4.22 *Let ϕ be a linear functional on c_{00}. Suppose that $\sum_{p,q} |\phi(e_p \cdot e_q)| < \infty$. Then the map $(x, y) \mapsto \phi(x \cdot y)$ is continuous on $c_{00} \times c_{00}$.*

PROOF Writing $x = \sum_p x_p e_p$ and $y = \sum_q y_q e_q$, we get

$$|\phi(x \cdot y)| \leq \sum_{p,q} |x_p| |y_q| |\phi(e_p \cdot e_q)| \leq C^2 \sum_{p,q} |\phi(e_p \cdot e_q)| \|x\| \|y\|$$

for all $(x, y) \in c_{00} \times c_{00}$, where $C = \sup_i \|e_i^*\|$. $\qquad\square$

So, we are left with an estimate of the coefficients $\phi(e_p \cdot e_q)$. At this point, one difficulty shows up: products like $e_p \cdot e_q$ are not easy to handle because, in order to investigate them, we have to express e_p and e_q in the basis $(T^i(e_0); i \in \mathbb{N})$. Indeed, the basis $(T^i(e_0))$ is the natural one for dealing with the product that we have defined on $c_{00} = \mathbb{K}[T]e_0$. Accordingly, we will say that a vector $x \in c_{00}$ is supported on some set $I \subset \mathbb{N}$ if $x \in \operatorname{span}\{T^i(e_0); i \in I\}$.

Fix $p \leq q$ and write $p = b_k + u$, $q = b_l + v$ with $u \in [0, b_{k+1} - b_k)$ and $v \in [0, b_{l+1} - b_l)$. The definition of T gives

$$e_p = \frac{k+1}{w(b_k+1)\cdots w(b_k+u)} (T^{b_k} - P_k(T))T^u(e_0),$$

$$e_q = \frac{l+1}{w(b_l+1)\cdots w(b_l+v)} (T^{b_l} - P_l(T))T^v(e_0).$$

Thus, for any linear functional $\phi : c_{00} \to \mathbb{K}$, we have

$$|\phi(e_p \cdot e_q)| \leq \frac{(k+1)(l+1)}{2^{u+v}} |\phi(y_{(k,u)(l,v)})|,$$

where

$$y_{(k,u)(l,v)} = (T^{b_k} - P_k(T))(T^{b_l} - P_l(T))T^{u+v}(e_0).$$

So, we have to ensure the convergence of

$$\sum \frac{(k+1)(l+1)}{2^{u+v}}|\phi(y_{(k,u)(l,v)})|.$$

Heuristically speaking, when $u + v$ is large we can expect to take advantage of the smallness of the factor $1/2^{u+v}$, provided that $|\phi(y_{(k,u)(l,v)})|$ is not too big. However, when $u + v$ is not large we will be safe only if the term $\phi(y_{(k,u)(l,v)})$ is small; the most drastic way to ensure this is to impose $\phi(y_{(k,u)(l,v)}) = 0$.

As the next lemma shows, the required properties will be fulfilled if we define ϕ as follows. We put $\phi(e_0) = 1$ and $\phi(T^i e_0) = 0$ if $i \in (0, b_1)$. If $i \in [b_n, b_{n+1})$ for some $n \geq 1$, we set

$$\phi(T^i e_0) = \begin{cases} \phi(P_n(T)T^{i-b_n}e_0) & \text{if } i \in [b_n, 3b_n/2) \cup [2b_n, 5b_n/2), \\ 0 & \text{otherwise.} \end{cases}$$

Notice that $\phi(T^i e_0)$ is indeed well defined if $\phi(T^j e_0)$ is known for all $j < i$, because $\deg(P_n) + i - b_n < i$ and hence $P_n(T)T^{i-b_n}e_0$ is supported on $[0, i)$.

LEMMA 4.23 *Assume that* $\deg P_n < b_n/3$ *for all* n. *Then the following properties hold whenever* $0 \leq k \leq l$.

(1) $\phi(y_{(k,u)(l,v)}) = 0$ *if* $u + v < b_l/6$.
(2) $|\phi(y_{(k,u)(l,v)})| \leq M_l(\mathbf{P}) := \max_{0 \leq j \leq l}(1 + |P_j|_1)^2 \prod_{0 < j \leq l+1} \max(1, |P_j|_1)^2.$

PROOF To prove (1), we observe first that, by the definition of ϕ, we have

$$\phi((T^{b_k} - P_k(T))z) = 0$$

whenever $z \in c_{00}$ is supported on $[0, b_k/2) \cup [b_k, 3b_k/2)$. Now, assume that $u + v < b_l/6$ (so that $l \geq 1$).

When $k = l \geq 1$, we write

$$\begin{aligned} y_{(k,u)(k,v)} &= (T^{b_k} - P_k(T))T^{b_k+u+v}e_0 - (T^{b_k} - P_k(T))P_k(T)T^{u+v}e_0 \\ &= (T^{b_k} - P_k(T))(z_1) - (T^{b_k} - P_k(T))(z_2). \end{aligned}$$

Then z_1 is supported on $[b_k, b_k + u + v) \subset [b_k, 7b_k/6) \subset [b_k, 3b_k/2)$ and z_2 is supported on $[0, \deg(P_k) + u + v) \subset [0, b_k/2)$. Hence $\phi(y_{(k,u)(k,v)}) = 0$.
When $l > k$, we simply write

$$y_{(k,u)(l,v)} = (T^{b_l} - P_l(T))(z),$$

where $z = (T^{b_k} - P_k(T))T^{u+v}e_0$ is supported on $[0, b_k + u + v) \subset [0, b_l/2)$.
To prove (2), we observe first that if R is a polynomial then an application of the triangle inequality readily yields

$$|\phi(R(T)e_0)| \leq |R|_1 \max_{j \leq \deg(R)} |\phi(T^j e_0)|.$$

We also need the following

FACT   It holds that $\max\limits_{i\in[0,b_n)} |\phi(T^i e_0)| \leq \prod\limits_{0<j<n} \max(1,|P_j|_1)^2$ for all $n \in \mathbb{N}$.

PROOF OF THE FACT   Let us set $K_n := \prod_{0<j<n} \max(1,|P_j|_1)^2$. The result is true for $n = 0$ and $n = 1$ if we give the value 1 to an empty product. Let us assume that the inequality holds for some $n \geq 1$, and prove it for $n + 1$.

Setting $\phi_i := |\phi(T^i e_0)|$ we have

$$
\begin{aligned}
\max_{i\in[b_n,2b_n)} \phi_i &= \max_{i\in[b_n,3b_n/2)} |\phi(P_n(T)T^{i-b_n}e_0)| \\
&\leq |P_n|_1 \max_{j<b_n/2+\deg(P_n)} \phi_j \\
&\leq |P_n|_1 K_n,
\end{aligned}
$$

because $b_n/2 + \deg(P_n) < b_n$. Similarly,

$$
\max_{i\in[2b_n,b_{n+1})} \phi_i = \max_{i\in[2b_n,5b_n/2)} \phi_i \leq |P_n|_1 \max_{j<2b_n} \phi_j \leq |P_n|_1^2 K_n.
$$

We conclude that $\max\limits_{i\in[b_n,b_{n+1})} \phi_i \leq K_{n+1}$, and the result follows by induction. $\quad\square$

Now the proof of (2) is easy. Indeed, the vector $y_{(k,u)(l,v)}$ has the form $R(T)e_0$, where R is a polynomial satisfying $\deg(R) \leq b_k + b_l + u + v < 2b_{l+1} < b_{l+2}$ and $|R|_1 \leq (1+|P_l|_1)(1+|P_k|_1)$. Hence, the result follows from the above fact. $\quad\square$

We can now prove that the linear functional ϕ has the required property if the admissible sequence $\mathbf{P}$ is suitably chosen.

PROPOSITION 4.24   *There exists a control sequence* (v_n) *such that the following holds: if the enumeration* $\mathbf{P}$ *is controlled by* (v_n) *then the map* $(x,y) \mapsto \phi(x \cdot y)$ *is continuous on* $c_{00} \times c_{00}$.

PROOF   Put $\Lambda := \{(m,w) \in \mathbb{N} \times \mathbb{N};\ w < b_{m+1} - b_m\}$. Then

$$
\sum_{p,q} |\phi(e_p \cdot e_q)| \leq \sum_{((k,u)(l,v))\in\Lambda\times\Lambda} \frac{(k+1)(l+1)}{2^{u+v}} |\phi(y_{(k,u)(l,v)})|.
$$

Using Lemma 4.23, we get

$$
\begin{aligned}
\sum_{p,q} |\phi(e_p \cdot e_q)| &\leq 2\sum_{k=0}^{+\infty}\sum_{l\geq k}(l+1)^2 M_l(\mathbf{P}) \sum_{u+v\geq b_l/6} \frac{1}{2^{u+v}} \\
&\leq \sum_{k=0}^{+\infty}\sum_{l\geq k}(l+1)^2 M_l(\mathbf{P}) \sum_{i\geq b_l/6} \frac{i+1}{2^i}.
\end{aligned}
$$

Now let (A_n) be a sequence of positive numbers tending to infinity, with $A_n \geq 2$ for all n, such that

$$
\sum_{k=0}^{+\infty}\sum_{l\geq k}(l+1)^2 A_l \sum_{i\geq b_l/6} \frac{i+1}{2^i} < \infty.
$$

From the definition of $M_n(\mathbf{P})$, it is clear that one can find a control sequence (v_n) such that $M_n(\mathbf{P}) \leq A_n$ for all n if $\mathbf{P}$ is controlled by (v_n). By Lemma 4.22 this concludes the proof of Proposition 4.24. □

Putting together the results of Propositions 4.21 and 4.24, the proof of Theorem 4.14 is now complete.

REMARK 4.25   In the above construction, the space $\ell^1(\mathbb{N})$ plays an extremal role. Let (δ_i) be the canonical basis of $\ell^1(\mathbb{N})$, and let T_1 be the operator defined on $\ell^1(\mathbb{N})$ by the same formulae as T, except that everywhere we replace e_i by δ_i. Since the sequence (e_i) is bounded, there is a well-defined operator $J : \ell^1(\mathbb{N}) \to X$ sending δ_i to e_i. Then $TJ = JT_1$, so that T is a *quasi-factor* of T_1 (see Chapter 1). Thus, if T_1 satisfies the Hypercyclicity Criterion, then so does T. This shows that in some sense, the easiest case is $\ell^1(\mathbb{N})$. This is also true for technical reasons. Indeed, to ensure the continuity of T_1 (resp. of the map $(x,y) \mapsto \phi(x \cdot y)$), one just needs to have $\sup_n \|f_n\|_1 < \infty$ (resp. $\sup_{p,q} |\phi(e_p \cdot e_q)| < \infty$).

4.3 Comments and exercises

The results of Sections 4.1.2 and 4.1.3 are due to quite a lot of authors; see [113], [120], [195], [129], [42], [165].

The weak mixing property shares some similarity with the notion of *disjointness*, which was introduced by H. Furstenberg in [112]. Two topological dynamical systems (X_1, T_1) and (X_2, T_2) are said to be **disjoint** if the only closed $T_1 \times T_2$-invariant set $\Delta \subset X_1 \times X_2$ such that $\pi_{X_1}(\Delta) = X_1$ and $\pi_{X_2}(\Delta) = X_2$ is $\Delta = X_1 \times X_2$. In particular, if two hypercyclic operators T_1, T_2 acting on the same space X are disjoint then $T_1 \oplus T_2$ is **diagonally hypercyclic**, which means that $T_1 \oplus T_2$ is hypercyclic and, moreover, one can find $x \in X$ such that $x \oplus x \in HC(T_1 \oplus T_2)$. Disjointness properties in hypercyclicity were studied in the papers [51] by J. Bès and A. Peris and [41] by L. Bernal-González.

The counter-example to the Hypercyclicity Criterion problem presented here is a simplified version of that given in [30], using ideas from [31]. The construction can be modified to produce a hypercyclic non-weakly-mixing operator on the space of entire functions $H(\mathbb{C})$; see [30] or Exercise 4.9. For other applications of this machinery (with, essentially, a better control on the rate of growth of (b_n)), we refer to [31]. Finally, we note that the spectrum of T is the closed disk $4\overline{\mathbb{D}}$ (see Exercise 4.8). This was shown to us by S. Grivaux.

The work in [91] and [30] leaves open the problem of characterizing those F-spaces which support hypercyclic non-weakly mixing operators. In particular, it is unknown whether there exists any Banach space on which every hypercyclic operator is weakly mixing. The so-called *hereditarily indecomposable* Banach spaces (see Definition 6.33) are of course the first candidates that come to mind, since they have very few operators and, moreover, the proof of Theorem 4.14 relies heavily on unconditionality. The sequence space $\mathbb{K}^{\mathbb{N}}$ (endowed with the product topology) is a natural example of a *Fréchet* space with this property: see Exercise 4.10. In fact, it is actually true that every hypercyclic operator on $\mathbb{K}^{\mathbb{N}}$ is mixing [132].

Another open question is the corresponding problem for C_0-semigroups of operators: if $(T(t))_{t\geq 0}$ is a hypercyclic C_0-semigroup, is $(T(t) \oplus T(t))_{t\geq 0}$ hypercyclic?

EXERCISE 4.1   ([184]) Let T be a hypercyclic operator on some topological vector space X, and let $x \in HC(T)$. Show that there exists an increasing sequence of integers (n_k) such that $n_{k+1} - n_k \leq 2$ for all k and the set $\{T^{n_k}(x);\ k \geq 0\}$ is not dense X. (*Hint*: Choose a non-empty open set $U \subset X$ such that $T(U) \cap U = \varnothing$ and consider $\mathbf{N} := \{n \in \mathbb{N}; T^n(x) \notin U\}$.)

EXERCISE 4.2 ([129]) Let X be a separable F-space and let $T \in \mathcal{L}(X)$. Assume that there exists some fixed positive integer p such that for any $U, V \subset X$ non-empty open, one can find an $n \in \mathbb{N}$ such that n and $n + p$ both belong to $\mathbf{N}(U, V)$. Show that T is weakly mixing. (*Hint:* Show first that each difference set $\mathbf{N}(U, V) - \mathbf{N}(U, V)$ contains $p\mathbb{N}$; then use Ansari's theorem 3.1 and Theorem 4.6 to show that T^p is weakly mixing.)

EXERCISE 4.3 **Mixing operators and syndetic sequences** ([87])

1. Let K be a positive integer. A set $\mathbf{F} \subset \mathbb{N}$ is said to be K-**syndetic** if there exists $a \in \mathbb{N}$ such that $\mathbf{F}$ intersects any interval I of length K with $\min(I) \geq a$. Show that if $\mathbf{F}$ is K-syndetic then $\bigcup_{k=0}^{K}(\mathbf{F} + k)$ is a cofinite subset of $\mathbb{N}$.
2. Let $T : X \to X$ be a continuous map acting on some topological space X. Assume that there exists some filter $\mathcal{F} \subset \mathcal{P}(\mathbb{N})$ with the following properties: $\mathbf{N}(U, V) \in \mathcal{F}$ for any non-empty open sets $U, V \subset X$, and each set $\mathbf{F} \in \mathcal{F}$ is K-syndetic for some fixed $K \in \mathbb{N}$. Show that T is mixing. (*Hint:* Start with non-empty open sets U, V, find open sets U_k such that $T^k(U_k) \subset V$ for all $k \in \{0, \ldots, K\}$ and consider $\mathbf{F} := \bigcap_0^K \mathbf{N}(U, U_k)$.)
3. Let X be a topological vector space, and let $T \in \mathcal{L}(X)$. Assume that T satisfies the Hypercyclicity Criterion with respect to some syndetic sequence (n_k). Show that T is mixing.

EXERCISE 4.4 **Cyclic direct sums 1** ([129])
Let X be a topological vector space, and let $T \in \mathcal{L}(X)$. Assume that T is hypercyclic and that the operator $T \oplus T$ is *cyclic*.

1. Prove that, for any non-empty open sets $U_1, V_1, U_2, V_2 \subset X$, there exists an operator $A \in \mathcal{L}(X)$ such that $AT = TA$ and $A(U_1) \cap V_1$, $A(U_2) \cap V_2$ are non-empty.
2. Show that T is weakly mixing. (*Hint:* Proceed as in the proof of Corollary 4.11.)

EXERCISE 4.5 **Cyclic direct sums 2** ([29])
Let X be a topological vector space, and let $T \in \mathcal{L}(X)$ be a supercyclic operator, with supercyclic vector x. Also, let p be a positive integer and let $\omega_1, \ldots, \omega_p$ be pairwise distinct rational numbers of $\mathbb{T}$. Finally, let $S := \omega_1 T \oplus \cdots \oplus \omega_p T$ acting on $Y := X \oplus \cdots \oplus X$.

1. Let $v = v_1 \oplus \cdots \oplus v_p \in Y$, and let $q \in \mathbb{N}$. Show that one can find scalars $\lambda_1, \ldots, \lambda_p \in \mathbb{K}$ and natural numbers $n_1, \ldots, n_p$ with $n_j \equiv j \pmod{q}$, such that $\lambda_1 T^{n_1} x \oplus \cdots \oplus \lambda_p T^{n_p} x$ is close to v.
2. With the notation of part 1 of the exercise, choose q such that $\omega_i^q = 1$ for $i = 1, \ldots, p$. Let P be the polynomial defined by $P(t) = \sum_j \lambda_j T^{n_j}$. Show that

$$P(S)(x \oplus \cdots \oplus x) = M_\omega \cdot (\lambda_1 T^{n_1} x \oplus \cdots \oplus \lambda_p T^{n_p} x),$$

where M_ω is an invertible $p \times p$ matrix depending only on $\omega_1, \ldots, \omega_p$.
3. Show that $S = \omega_1 T \oplus \cdots \oplus \omega_p T$ is cyclic, with cyclic vector $x \oplus \cdots \oplus x$.

EXERCISE 4.6 **Cyclic direct sums 3** ([187])

1. Let X be a Banach space, and let $T \in \mathcal{L}(X)$. Show that $T \oplus T^*$ is not cyclic on $X \oplus X^*$. (*Hint:* Given $x \oplus x^* \in X \oplus X^*$, consider the linear functional ϕ on $X \oplus X^*$ defined by $\phi(z \oplus z^*) = \langle x^*, z \rangle - \langle z^*, x \rangle$. Compute $\phi(T^n(x) \oplus T^{*n}(x^*))$, $n \in \mathbb{N}$.)
2. Let $T \in \mathcal{L}(X)$, and assume that T^* is a linear quasi-factor of T, i.e. there exists a continuous linear operator $J : X \to X^*$ with dense range such that $T^* J = JT$. Show that $T \oplus T$ is not cyclic.
3. Let V be the **Volterra operator** acting on $L^2([0, 1])$, $Vf(x) = \int_0^x f(t)\, dt$. Show that $V \oplus V$ is not cyclic.

EXERCISE 4.7 Let X be a topological vector space, and let $T \in \mathcal{L}(X)$ be cyclic with cyclic vector e_0.

1. Assume that T is weakly mixing. Show that for any $a, b, a', b' \in X$, one can find $p_1, p_2 \in \mathbb{K}[T]e_0$ and $p \in O(e_0, T)$ such that (p_1, p_2) is close to (a, b) and $(p \cdot p_1, p \cdot p_2)$ is close to (a', b').
2. Assume that there exists a non-constant map $\phi : \mathbb{K}[T]e_0 \to Z$, where Z is a Hausdorff topological space, such that the map $(x, y) \mapsto \phi(x \cdot y)$ is continuous on $\mathbb{K}[T]e_0 \times \mathbb{K}[T]e_0$. Show that T is not weakly mixing. (*Hint:* with the notation of 1, write pp_1p_2 in two different ways to show that $\phi(ab') = \phi(ba')$ for any $a, b, a', b' \in X$).

EXERCISE 4.8 **The spectrum of the operator T**
The symbol $\sigma_e(R)$ denotes the essential spectrum of an operator R; see Appendix D.

1. *Weyl's theorem.* Let X be a Banach space, and let $S, L \in \mathcal{L}(X)$ with L compact. Prove that $\sigma(S + L) \backslash \sigma(S)$ consists only of eigenvalues of $S + L$. (*Hint:* Write $S + L - \lambda I = (S - \lambda I)(I + K)$, where K is a compact operator.)
2. Let $T = R + K$ be defined on $\ell^2(\mathbb{N})$ as in Section 4.2.
   (a) Show that $\sigma_e(R) = \overline{4\mathbb{D}} = \sigma(R)$ (see Appendix D).
   (b) Conclude that $\sigma(T) = \sigma(R) = \overline{4\mathbb{D}}$. (*Hint:* $\sigma_p(T^*) = \varnothing$.)

EXERCISE 4.9 ([30]) **Hypercyclic non-weakly-mixing operators on $H(\mathbb{C})$**
We outline here the construction of a hypercyclic non-weakly-mixing operator on $X = H(\mathbb{C})$. Set $b_0 = 0$ and $b_n = 3^n$ for $n \geq 1$. As in Section 4.2, let (P_n) be an enumeration of the polynomials with coefficients in $\mathbf{Q}$. Finally, let (ρ_n) be an increasing sequence of positive numbers tending to infinity. We set $e_i := z^i \in H(\mathbb{C})$ and define T on $c_{00} := \text{span}\{e_i; i \geq 0\}$ by

$$T(e_i) := (i+1)e_{i+1} \qquad \text{if } i \in [b_{n-1}, b_n - 1) \text{ for some } n,$$
$$T^{b_n}(e_0) := P_n(T)e_0 + \rho_n^{-b_n} e_{b_n} \quad \text{for all } n \geq 1.$$

1. Prove the following facts:
   - if $K : c_{00} \to c_{00}$ is a linear map such that $K(e_i)$ is supported on $[0, i]$ for all $i \in \mathbb{N}$ and the sequence $(\|K(e_i)\|_1)$ is bounded then K is continuous on c_{00};
   - if $\phi : c_{00} \to \mathbb{C}$ is a linear functional such that $\sup_{p,q} |\phi(e_p \cdot e_q)| < \infty$ then the map $(x, y) \mapsto \phi(x \cdot y)$ is continuous on $c_{00} \times c_{00}$.
2. (a) With the notation of Section 4.2, compute the numbers ε_n and the vectors f_n.
   (b) Show that there exists a sequence $(u_n)_{n\geq 1}$ tending to ∞ such that if $\deg(P_n) \leq u_n$, $|P_n|_1 \leq u_n$ and $\rho_n \leq u_n$ for all $n \geq 1$ then $\|f_n\|_1 \leq 1$ for all n.
3. (a) Let $k, u, l, v \in \mathbb{N}$ with $k < l$ and $u < b_{k+1} - b_k$. Show that if $u + v \geq b_l/6$, then either $v \geq b_l/12$ or $k \geq l - 2$ and $u \geq b_l/12$.
   (b) Let $\phi : c_{00} \to \mathbb{C}$ be defined as in Section 4.2.
   (c) Compute $\phi(e_p \cdot e_q)$ when $p = b_k + u \in [b_k, b_{k+1})$ and $q = b_l + v \in [b_l, b_{l+1})$. Show that there exists a sequence $(v_n)_{n\geq 1}$ tending to ∞ such that if $\deg(P_n) \leq v_n$, $|P_n|_1 \leq v_n$ and $\rho_n \leq v_n$ for all n then the sequence $(\phi(e_p \cdot e_q))$ is bounded.
4. Conclude.

EXERCISE 4.10 **Hypercyclic operators on $\mathbb{K}^{\mathbb{N}}$** ([30])
The aim of this exercise is to show that every hypercyclic operator on $\mathbb{K}^{\mathbb{N}}$ is weakly mixing. So let $T \in \mathcal{L}(\mathbb{K}^{\mathbb{N}})$ be hypercyclic. Let U_1, U_2, V_1, V_2 be non-empty open sets in $\mathbb{K}^{\mathbb{N}}$ of the form $U_i = J_i^1 \times \cdots \times J_i^q \times \mathbb{K} \times \mathbb{K} \times \cdots$ and $V_i = I_i^1 \times \cdots \times I_i^q \times \mathbb{K} \times \mathbb{K} \times \ldots$, where J_i^j and I_i^j are open subsets of $\mathbb{K}$.

1. Prove that there exists a non-zero polynomial P such that $P(T)$ is represented by a matrix of the form $\begin{pmatrix} 0 & B \\ C & D \end{pmatrix}$ where B has q rows and C has q columns.
2. Show that the rows of B are independent.
3. Deduce that $P(T)(U_i) \cap V_i \neq \varnothing$.
4. Using the result of Exercise 4.4, conclude that T is weakly mixing.

5

Ergodic theory and linear dynamics

Introduction

So far, we have obtained hypercyclic vectors either by a direct construction or by a Baire category argument. The aim of this chapter is to provide another way of doing so, using *ergodic theory*. This will link linear dynamics with measurable dynamics. We first recall some basic definitions from ergodic theory. The classical book of P. Walters [235] is a very readable introduction to that area.

The first important concept is that of invariant measure.

DEFINITION 5.1 *Let* $(X, \mathcal{B}, \mu)$ *be a probability space. We say that a measurable map* $T : (X, \mathcal{B}, \mu) \to (X, \mathcal{B}, \mu)$ *is a* **measure-preserving transformation***, or that* μ *is* T-**invariant***, if* $\mu(T^{-1}(A)) = \mu(A)$ *for all* $A \in \mathcal{B}$.

Measure-preserving transformations already have some important dynamical properties. In particular, the famous **Poincaré recurrence theorem** asserts that if $T : (X, \mu) \to (X, \mu)$ is measure-preserving then, for any measurable set A such that $\mu(A) > 0$, almost every point $x \in A$ is T-**recurrent** with respect to A, which means that $T^n(x) \in A$ for infinitely many $n \in \mathbb{N}$.

Now the central concept in linear dynamics is not recurrence but transitivity. Topological transitivity may be defined in the following two equivalent ways (see Exercise 5.2): a continuous map $T : X \to X$ on some topological space X is topologically transitive iff, for any sets $A, B \subset X$ with non-empty interior, one can find $n \in \mathbb{N}$ such that $T^n(A) \cap B \neq \varnothing$ iff any T-invariant set $A \subset X$ is either dense in X or nowhere dense. The measure-theoretic analogue is the notion of ergodicity.

DEFINITION 5.2 *Let* $(X, \mathcal{B}, \mu)$ *be a probability space. We say that a measurable map* $T : (X, \mathcal{B}, \mu) \to (X, \mathcal{B}, \mu)$ *is* **ergodic** *if it is measure-preserving and satisfies one of the following equivalent conditions:*

(i) *Given any measurable sets* A, B *with positive measures, one can find an integer* $n \geq 0$ *such that* $T^n(A) \cap B \neq \varnothing$;

(ii) *if* $A \in \mathcal{B}$ *satisfies* $T(A) \subset A$ *then* $\mu(A) = 0$ *or* 1.

We note that $T(A) \subset A$ is equivalent to $A \subset T^{-1}(A)$, so that $T^{-1}(A) = A$ up to a set of measure 0 if T is measure-preserving. Conversely, if $\mu(A \triangle T^{-1}(A)) = 0$ then $\tilde{A} := \bigcap_{n \geq 0} T^{-n}(A)$ is equal to A up to a measure-0 set and satisfies $T(\tilde{A}) \subset \tilde{A}$. Thus, assuming that T is measure-preserving, condition (ii) is equivalent to the perhaps more familiar

(ii') *if* $A \in \mathcal{B}$ *satisfies* $T^{-1}(A) = A$ *up to a measure-0 set then* $\mu(A) = 0$ *or* 1.

The starting point in the study of ergodic transformations is **Birkhoff's ergodic theorem**.

THEOREM 5.3 (BIRKHOFF'S ERGODIC THEOREM) *Let* $(X, \mathcal{B}, \mu)$ *be a probability space, and let* $T : (X, \mathcal{B}, \mu) \rightarrow (X, \mathcal{B}, \mu)$ *be a measure-preserving ergodic transformation. For any* $f \in L^1(X, \mu)$,

$$\frac{1}{N} \sum_{n=0}^{N-1} f(T^n x) \xrightarrow{N \to \infty} \int_X f d\mu \qquad \mu\text{-a.e.}$$

From that, it is not difficult to deduce

COROLLARY 5.4 *Let* $T : (X, \mathcal{B}, \mu) \rightarrow (X, \mathcal{B}, \mu)$ *be a measure-preserving transformation. The following are equivalent.*

(i) T *is ergodic;*
(ii) *for any* $f, g \in L^2(X, \mu)$:

$$\lim_{N \to \infty} \frac{1}{N} \sum_{n=0}^{N-1} \int_X f(T^n x) g(x) \, d\mu(x) = \int_X f \, d\mu \int_X g \, d\mu;$$

(iii) *for any* $A, B \in \mathcal{B}$,

$$\lim_{N \to \infty} \frac{1}{N} \sum_{n=0}^{N-1} \mu(A \cap T^{-n}(B)) = \mu(A)\mu(B).$$

Now we come back to the linear setting. Let X be a separable Banach space, with Borel σ-algebra $\mathcal{B}$. Let $T \in \mathcal{L}(X)$, and assume that we have been able to construct a probability measure μ on $(X, \mathcal{B})$ such that T is a measure-preserving ergodic transformation with respect to μ. Moreover, assume that μ has **full support**, which means that $\mu(O) > 0$ for every non-empty open set $O \subset X$. It follows from the definition of ergodicity that T is topologically transitive, hence hypercyclic. Birkhoff's theorem gives us for no extra effort an additional result.

COROLLARY 5.5 *Let* $T \in \mathcal{L}(X)$. *Assume that one can find some T-invariant Borel probability measure μ on X with full support, with respect to which T is an ergodic transformation. Then T is hypercyclic and the set of hypercyclic vectors for T has full μ-measure. More precisely, almost every point $x \in X$ has the following property: for every non-empty open set $V \subset X$, one has*

$$\liminf_{N \to \infty} \frac{\operatorname{card} \{n \in [0, N); \ T^n(x) \in V\}}{N} > 0. \qquad (5.1)$$

PROOF Let $(V_j)_{j \in \mathbb{N}}$ be a countable basis of open sets for X. Applying Birkhoff's theorem to the characteristic function $\mathbf{1}_{V_j}$ of each set V_j, we obtain a sequence of sets (A_j) with $\mu(A_j) = 1$ for each j, such that

$$\frac{1}{N} \sum_{n=0}^{N-1} \mathbf{1}_{V_j}(T^n x) \xrightarrow{N \to \infty} \mu(V_j) > 0 \quad \text{for every } x \in A_j. \qquad (5.2)$$

Then the set $A := \bigcap_{j \geq 0} A_j$ has full measure, and since the left-hand side of (5.2) is exactly N^{-1} card $\{n < N;\ T^n(x) \in V_j\}$, we see immediately that any $x \in A$ is a hypercyclic vector for T and has the property stated above. Hypercyclic operators satisfying the stronger condition (5.1) will be studied in some detail in Chapter 6. They are called *frequently hypercyclic*. □

REMARK   We already know that the set of hypercyclic vectors of a hypercyclic operator is large both in a topological and in an algebraic sense. Under the assumptions of Corollary 5.5, it is also large in a probabilistic sense.

The aim of this chapter should now be clear: under suitable assumptions, we are going to describe a method for associating with an operator $T \in \mathfrak{L}(X)$ some T-invariant Borel probability measure μ on X with full support, in such a way that T is ergodic with respect to μ. This provides an alternate strategy for producing hypercyclic vectors.

From now on, we assume that X is a separable *complex* Banach space, with Borel σ-algebra $\mathcal{B}$. All measures under consideration will be finite Borel measures, unless otherwise specified. The result of the action of a linear functional $x^* \in X$ on a vector $x \in X$ is denoted by $\langle x^*, x \rangle$, or sometimes by $\langle x, x^* \rangle$. Finally, we adopt the following convention: *if $\mathcal{H}$ is a complex Hilbert space then the scalar product $\langle u, v \rangle_{\mathcal{H}}$ is linear with respect to v and conjugate-linear with respect to u.* (The reader should note that in some other chapters of this book the opposite convention is used, e.g. for the scalar product of $H^2(\mathbb{D})$.)

5.1 Gaussian measures and covariance operators

The first problem we have to face is the lack of a distinguished "canonical" measure on an infinite-dimensional space. However, there exist a number of natural infinite-dimensional measures. We will restrict ourselves to those best studied, the so-called *Gaussian* measures, which we introduce in this section. For the sake of completeness, we have tried to give a self-contained exposition of the basic theory of Gaussian measures. As far as linear dynamics is concerned, the details of the proofs can be skipped safely.

Since we are working in a complex Banach space, we first have to introduce complex Gaussian distributions. For any $\sigma > 0$, let us denote by γ_σ the centred Gaussian measure on $\mathbb{R}$ with variance σ^2, that is,

$$ d\gamma_\sigma = \frac{1}{\sigma \sqrt{2\pi}}\, e^{-t^2/2\sigma^2}\, dt. $$

We recall that if ξ_1 and ξ_2 are two independent random variables with respective distributions γ_{σ_1} and γ_{σ_2} then $\xi_1 + \xi_2$ has the distribution γ_σ, where $\sigma^2 = \sigma_1^2 + \sigma_2^2$.

DEFINITION 5.6   *A complex-valued random variable $\xi : \Omega \to \mathbb{C}$ defined on some probability space $(\Omega, \mathcal{F}, \mathbb{P})$ is said to have a **complex symmetric Gaussian distribution** if either ξ is almost surely 0 or the real and imaginary parts of ξ are* *independent and have centred Gaussian distributions with the same variance. If* $\mathbb{E}|\xi|^2 = 1$ *then* ξ *is said to be* **standard**.

Thus, a random variable ξ has complex symmetric Gaussian distribution iff its distribution is either the point mass δ_0 or the product measure $\gamma_\sigma \otimes \gamma_\sigma$, for some $\sigma > 0$. In the latter case, we have $\mathbb{E}|\xi|^2 = 2\sigma^2$.

It is worth noting that if ξ has complex symmetric Gaussian distribution then so does $\lambda\xi$ for any complex number λ. This follows from the *rotational invariance* of Gaussian vectors in $\mathbb{R}^2$. Indeed, writing $\lambda = |\lambda|(c + is)$ and $\xi = \xi_1 + i\xi_2$, we have $\lambda\xi = |\lambda|(\xi_1' + i\xi_2')$, where $\xi_1' = c\xi_1 - s\xi_2$ and $\xi_2' = c\xi_2 + s\xi_1$. Since ξ_1 and ξ_2 have the same real Gaussian distribution, the Gaussian vectors (ξ_1, ξ_2) and (ξ_1', ξ_2') have the same distribution, by rotational invariance. Hence, $\lambda\xi$ does have a symmetric complex Gaussian distribution and in fact the same distribution as ξ if $|\lambda| = 1$.

REMARK   In this chapter, we will often encounter random series of the form $\sum g_n x_n$, where the vectors x_n live in some complex Banach space Z and $(g_n)_{n\in\mathbb{N}}$ is a sequence of independent standard complex Gaussian variables defined on the same probability space $(\Omega, \mathcal{F}, \mathbb{P})$. Such a sequence (g_n) will be called a **standard Gaussian sequence**. It is well-known (but non-trivial) that for Gaussian series $\sum g_n x_n$ all natural notions of convergence are equivalent. In particular, almost sure convergence is equivalent to convergence in $L^2(\Omega, Z)$ or in any $L^p(\Omega, Z), p < \infty$, and, when Z is a Hilbert space (including the scalar field $\mathbb{C}$), this amounts to the convergence of the series $\sum \|x_n\|^2$ by the orthogonality of the Gaussian variables g_n. In what follows we will use these crucial facts without explicit mention. Proofs can be found in many sources, including [164], [77] and [173] (see also Exercise 5.6).

It is now time to define Gaussian measures.

DEFINITION 5.7   *A* **Gaussian measure** *on* X *is a probability measure* μ *on* X *such that each continuous linear functional* $x^* \in X^*$ *has symmetric complex Gaussian distribution, when considered as a random variable on* $(X, \mathcal{B}, \mu)$.

Observe that in our terminology, we only consider *centred* Gaussian measures. Our main references concerning Gaussian measures are [77] and [55].

To aid understanding of the definition, let us give a very important class of examples. It will turn out later that in fact, *any* Gaussian measure falls into that class; see Corollary 5.14. Recall that if $\xi : (\Omega, \mathcal{F}, \mathbb{P}) \to X$ is a random variable with values in X then the *distribution* of ξ is the probability measure μ on X defined by

$$\mu(A) = \mathbb{P}(\xi \in A).$$

EXAMPLE 5.8   Let $(x_n)_{n\in\mathbb{N}}$ be a sequence of vectors in X, and let (g_n) be a standard Gaussian sequence. Assume that the series $\sum g_n x_n$ is almost surely convergent. Then the distribution of $\xi := \sum_0^\infty g_n x_n$ is a Gaussian measure.

PROOF   Let us fix $x^* \in X^*$. Then $\langle x^*, \xi \rangle = \sum_0^\infty \langle x^*, x_n \rangle g_n$, where the series is a.s. convergent. If $\langle x^*, x_n \rangle = 0$ for all n, there is nothing to prove. Otherwise, we may assume that $\langle x^*, x_n \rangle \neq 0$ for all n.

For each $n \in \mathbb{N}$, the random variable $\langle x^*, x_n \rangle g_n$ has a complex symmetric Gaussian distribution; more precisely, its real and imaginary parts have distribution γ_{σ_n}, where $\sigma_n^2 = |\langle x^*, x_n \rangle|^2 / 2$. Let ν be the distribution of $\mathrm{Re}(\langle x^*, \xi \rangle)$ and, for each $N \in \mathbb{N}$, let ν_N be the distribution of the finite sum $\mathrm{Re}\left(\sum_0^N \langle x^*, x_n \rangle g_n \right)$. The Fourier transform of ν_N is given by $\widehat{\nu}_N(t) = e^{-\sum_0^N \sigma_n^2 t^2 / 2}$. By Lebesgue's theorem, $\widehat{\nu}_N$ converges pointwise to $\widehat{\nu}$; hence, we get $\widehat{\nu}(t) = e^{-\sum_0^\infty \sigma_n^2 t^2 / 2}$ (the series is indeed convergent because the Gaussian series $\sum_n \langle x^*, x_n \rangle g_n$ converges in L^2). Treating the imaginary part in the same way, we conclude that $\mathrm{Re}\langle x^*, \xi \rangle$ and $\mathrm{Im}\langle x^*, \xi \rangle$ both have distribution γ_σ, where $\sigma^2 = \sum_0^\infty \sigma_n^2 = \sum_0^\infty |\langle x^*, x_n \rangle|^2 / 2$. $\square$

REMARK   In the above proof we have used Fourier transforms, and we will do that again in what follows; so we should recall the definition! The **Fourier transform** of a probability measure μ on X is the function $\widehat{\mu} : X^* \to \mathbb{C}$ defined by

$$\widehat{\mu}(x^*) = \int_X e^{i \mathrm{Re}\, \langle x^*, x \rangle} \, d\mu(x).$$

It is well known (but not completely obvious) that the Fourier transform uniquely determines the measure. This follows from the finite-dimensional case, together with the fact that a measure μ on X is completely determined by its finite-dimensional "marginals", i.e. its images under all finite-rank operators defined on X (see [77] or Exercise 5.4).

A basic tool for studying a Gaussian measure μ is the so-called *covariance operator* associated with μ. Before stating the next theorem, we observe that if μ is a Gaussian measure on X then each continuous linear functional $x^* : X \to \mathbb{C}$ is square-summable with respect to μ, thus X^* can be considered as a linear subspace of $L^2(\mu)$. In fact, considerably more is true: it follows from a fundamental result of X. Fernique, sometimes referred to as **Fernique's integrability theorem**, that a Gaussian measure μ has finite moments of all orders, in particular,

$$\int_X \|x\|^2 \, d\mu(x) < \infty.$$

See e.g. [55, Theorem 2.8.5] or Exercise 5.5 for a proof of this result.

THEOREM 5.9   *Let μ be a Gaussian measure on X.*

(a) *One can define a continuous conjugate-linear operator $R_\mu : X^* \to X$ such that, for every $(x^*, y^*) \in X^* \times X^*$,*

$$\langle R_\mu(x^*), y^* \rangle = \int_X \overline{\langle x^*, z \rangle} \, \langle y^*, z \rangle \, d\mu(z) = \langle x^*, y^* \rangle_{L^2(\mu)}.$$

*The operator R_μ is called the **covariance operator** of μ.*

(b) *For each $x^* \in X^*$, one has $\widehat{\mu}(x^*) = e^{-\langle R_\mu(x^*), x^* \rangle / 4}$.*

(c) *If μ is the distribution of an almost surely convergent Gaussian series $\sum g_n x_n$ then R_μ is given by*

$$\langle R_\mu(x^*), y^* \rangle = \sum_{n=0}^{\infty} \overline{\langle x^*, x_n \rangle} \langle y^*, x_n \rangle,$$

where the series is absolutely convergent. The Fourier transform of μ is given by

$$\hat{\mu}(x^*) = \exp\left(-\frac{1}{4} \sum_{n=0}^{\infty} |\langle x^*, x_n \rangle|^2\right).$$

PROOF   A simple application of the closed-graph theorem (or an appeal to Fernique's integrability theorem) shows that the inclusion $X^* \subset L^2(\mu)$ is continuous. It follows that, for each $x^* \in X^*$, the formula

$$\Phi_{x^*}(y^*) := \langle x^*, y^* \rangle_{L^2(\mu)}$$

defines a continuous linear functional on X^*. Moreover, the map $x^* \mapsto \Phi_{x^*}$ is conjugate-linear and continuous. What has to be shown is that each Φ_{x^*} is in fact given by an element of X, in other words, that the linear functional Φ_{x^*} is w^*-continuous on X^*. By Grothendieck's completeness theorem (see [4, Chapter 1]), it is enough to check that Φ_{x^*} is w^*-continuous on any bounded subset of X^*; since the bounded subsets of X^* are w^*-metrizable (by the separability of X), this amounts to showing that if (y_n^*) is a (bounded) sequence in X^* w^*-converging to $y^* \in X^*$ then

$$\int_X \overline{\langle x^*, z \rangle} \langle y_n^*, z \rangle \, d\mu(z) \rightarrow \int_X \overline{\langle x^*, z \rangle} \langle y^*, z \rangle \, d\mu(z).$$

This, in turn, follows from Fernique's integrability theorem and Lebesgue's theorem. Thus, we have proved (a).

Part (b) follows from a simple computation, using the corresponding property of one-dimensional Gaussian distributions. Indeed, one has $\hat{\mu}(x^*) = \widehat{\nu_{x^*}}(1)$, where ν_{x^*} is the distribution of the real Gaussian variable $\text{Re}(x^*)$ defined on the probability space (X, μ); hence $\hat{\mu}(x^*) = e^{-\sigma^2/2}$, where $\sigma^2 = \mathbb{E}_\mu |\text{Re}(x^*)|^2$. Now, $\text{Re}(x^*)$ and $\text{Im}(x^*)$ have the same distribution, so $\mathbb{E}_\mu |\text{Re}(x^*)|^2 = \mathbb{E}_\mu |x^*|^2/2 = \langle R_\mu(x^*), x^* \rangle/2$. This explains the unusual coefficient $1/4$.

To prove (c), assume now that μ is the distribution of an a.s. convergent Gaussian series $\sum g_n x_n$. The Fourier transform of μ has already been computed in Example 5.8. Applying (b), we get

$$\langle R_\mu(x^*), x^* \rangle = \sum_{n=0}^{\infty} |\langle x^*, x_n \rangle|^2$$

for every $x^* \in X^*$. By polarization, we conclude that R_μ is given by

$$\langle R_\mu(x^*), y^* \rangle = \sum_{n=0}^{\infty} \overline{\langle x^*, x_n \rangle} \langle y^*, x_n \rangle,$$

where the series is absolutely convergent by the square-summability of the series $\sum \langle x^*, x_n \rangle$ and $\sum \langle y^*, x_n \rangle$. This proves (c). $\qquad\square$

Since the Fourier transform of a measure determines the measure, we get the following corollary.

COROLLARY 5.10 *Two centred Gaussian measures coincide iff they have the same covariance operator.*

We will say that a conjugate-linear operator $R : X^* \to X$ is a **Gaussian covariance operator** if it is the covariance operator of some Gaussian measure μ on X. The covariance operator is very important for us; we shall not construct a Gaussian measure directly, but rather its covariance operator. Unfortunately there is no fully satisfactory characterization of Gaussian covariance operators in a general Banach space. However, we will prove below a somewhat formal characterization which will be sufficient for our purpose.

One obvious necessary condition for a conjugate-linear operator $R : X^* \to X$ to be a covariance operator is that R should be symmetric and positive: $\langle R(x^*), y^* \rangle = \overline{\langle x^*, R(y^*) \rangle}$ and $\langle R(x^*), x^* \rangle \geq 0$. It can be shown that any symmetric positive operator $R : X^* \to X$ admits a **square root**: there exist some separable Hilbert space $\mathcal{H}$ and an operator $K : \mathcal{H} \to X$ such that $R = KK^*$. We will not prove this here, since we do not need the result in full generality. However, we point out two important facts.

- The operator $K^* : X^* \to \mathcal{H}$ is written more properly as IK^*, where $K^* : X^* \to \mathcal{H}^*$ is the "true" (Banach space) adjoint of K and $I : \mathcal{H}^* \to \mathcal{H}$ is the canonical (conjugate-linear) identification operator. We choose to write simply K^* to avoid cumbersome notation. Thus, it should not be forgotten that K^* is a *conjugate-linear* operator. If $x^* \in X^*$ and $h \in \mathcal{H}$ then $\langle K^*(x^*), h \rangle_{\mathcal{H}} = \langle x^*, K(h) \rangle$ and if $x^*, y^* \in X^*$ then $\langle KK^*(x^*), y^* \rangle = \langle K^*(y^*), K^*(x^*) \rangle_{\mathcal{H}}$. In particular, $\langle KK^*(x^*), x^* \rangle = \|K^*(x^*)\|^2$ for every $x^* \in X^*$.
- Replacing $\mathcal{H}$ by $\mathrm{Ker}(K)^\perp$, one can always require K to be one-to-one. Moreover, the square root is essentially unique: *if $K_1 : \mathcal{H}_1 \to X$ is a one-to-one square root of R and $K_2 : \mathcal{H}_2 \to X$ is any other square root then there exists an isometry $V : \mathcal{H}_1 \to \mathcal{H}_2$ such that $K_2^* = VK_1^*$. Indeed, one has $\|K_1^*(x^*)\|^2 = \|K_2^*(x^*)\|^2$ for every $x^* \in X^*$, so one can define V unambiguously on $\mathrm{Ran}(K_1^*)$; then V extends to an isometry defined on the whole $\mathcal{H}_1$ because K_1^* has dense range.*

The next definition is the key to the promised characterization of Gaussian covariance operators. Let us fix once and for all a standard Gaussian sequence $(g_n)_{n \in \mathbb{N}}$ defined on some probability space $(\Omega, \mathcal{F}, \mathbb{P})$.

DEFINITION 5.11 *Let $\mathcal{H}$ be a separable Hilbert space. An operator $K \in \mathfrak{L}(\mathcal{H}, X)$ is said to be γ-**radonifying** if for some (or, equivalently, for any) orthonormal basis (e_n) of $\mathcal{H}$ the Gaussian series $\sum g_n(\omega) K(e_n)$ converges almost surely.*

The fact that the choice of the orthonormal basis (e_n) is irrelevant in the above definition can be seen as follows. Let (e_n) be an orthonormal basis of $\mathcal{H}$ and, for each $N \in \mathbb{N}$, let us denote by μ_N the distribution of the partial sum $\sum_0^N g_n K(e_n)$. Then the series $\sum g_n K(e_n)$ is a.s. convergent iff the Fourier transforms $\widehat{\mu}_N$ converge pointwise to the Fourier transform of some probability measure μ (see [77, Theorem V.2.4]). Now, it follows from Theorem 5.9 that $\widehat{\mu}_N$ is given by

$$\widehat{\mu}_N(x^*) = e^{-\sum_0^N |\langle x^*, K(e_n) \rangle|^2/4}$$
$$= e^{-\sum_0^N |\langle K^*(x^*), e_n \rangle|^2/4},$$

which converges to $\phi(x^*) := e^{-\|K^*(x^*)\|^2/4}$. Thus, the series $\sum g_n K(e_n)$ is a.s. convergent iff the function ϕ is the Fourier transform of some probability measure μ. But this is independent of the orthonormal basis (e_n). We state what we have proved as

REMARK 5.12   An operator $K : \mathcal{H} \to X$ is γ-radonifying if and only if the function $\phi : X^* \to \mathbb{R}$ defined by

$$\phi(x^*) := \exp\left(-\frac{1}{4}\|K^*(x^*)\|^2\right)$$

is the Fourier transform of some probability measure μ on X.

We also note that when X is a Hilbert space an operator $K : \mathcal{H} \to X$ is γ-radonifying iff it is a *Hilbert–Schmidt* operator, since a Gaussian series $\sum g_n x_n$ with terms in a Hilbert space is L^2-convergent iff $\sum_n \|x_n\|^2 < \infty$.

We can now state

THEOREM 5.13   *For a conjugate-linear operator $R : X^* \to X$, the following are equivalent:*

 (i) *R is a Gaussian covariance operator;*
 (ii) *R has a γ-radonifying square root. In other words, one can factorize R as KK^*, where $K : \mathcal{H} \to X$ is a γ-radonifying operator defined on some separable Hilbert space $\mathcal{H}$;*
 (iii) *one can write $R = \sum_{n=0}^{\infty} \langle \cdot, x_n \rangle x_n$, where the series is SOT-convergent and the sequence $(x_n) \subset X$ is such that the Gaussian series $\sum g_n x_n$ is a.s. convergent;*
 (iv) *R is the covariance operator of the distribution of some a.s. convergent Gaussian series $\sum_n g_n x_n$.*

PROOF   (i) $\implies$ (ii)   Assume that R is the covariance operator of some Gaussian measure μ. Then $X^* \subset L^2(\mu)$ and

$$\langle R(x^*), y^* \rangle = \langle x^*, y^* \rangle_{L^2(\mu)}$$

for all $x^*, y^* \in X^*$. Let us denote by $\mathcal{H}$ the closed subspace of $L^2(\mu)$ generated by all functions of the form $\overline{\langle x^*, \cdot \rangle}$, where $x^* \in X^*$. The space $\mathcal{H}$ is separable because $L^2(\mu)$ is.

We first define $K^* : X^* \to \mathcal{H}$ by $K^*(x^*) = \overline{\langle x^*, \cdot \rangle}$. As in the proof of Theorem 5.9, a joint application of Lebesgue's theorem and Fernique's integrability theorem shows that the operator K^* is $w^* - w^*$ continuous. Therefore, K^* is indeed the adjoint of some operator $K : \mathcal{H} \to X$, and since $\langle R(x^*), y^* \rangle = \langle K^*(y^*), K^*(x^*) \rangle_{\mathcal{H}}$ by the definition of K^*, we have $R = KK^*$. Then $\langle R(x^*), x^* \rangle = \|K^*(x^*)\|^2$ for every $x^* \in X^*$, and it follows from Theorem 5.9 that

$$\widehat{\mu}(x^*) = \exp\left(-\frac{1}{4}\|K^*(x^*)\|^2\right).$$

By Remark 5.12 this shows that K is γ-radonifying.

The implication (ii) $\Longrightarrow$ (iii) is easy: if $R = KK^*$ and if $(e_n)_{n \in \mathbb{N}}$ is any orthonormal basis of $\mathcal{H}$ then one can write

$$R(x^*) = K\left(\sum_{n=0}^{\infty} \langle e_n, K^*(x^*) \rangle_{\mathcal{H}} e_n\right)$$

$$= \sum_{n=0}^{\infty} \overline{\langle x^*, K(e_n) \rangle} K(e_n),$$

so one may take $x_n := K(e_n)$.

The implication (iii) $\Longrightarrow$ (iv) is also easy: if R is represented as in (iii) then the distribution of the random variable $\sum_0^{\infty} g_n x_n$ is a Gaussian measure μ (Example 5.8), and $R_\mu = R$ by Theorem 5.9. Finally, (iv) $\Longrightarrow$ (i) is trivial. $\square$

COROLLARY 5.14 *Any Gaussian measure μ on X is the distribution of some a.s. convergent Gaussian series $\sum g_n x_n$.*

PROOF This follows from the previous theorem, since two Gaussian measures are equal iff they have the same covariance operator. $\square$

COROLLARY 5.15 *Assume that X is a Hilbert space, and identify X with X^* in the usual way. Then an operator $R \in \mathfrak{L}(X)$ is a Gaussian covariance operator iff R is a positive trace-class operator.*

PROOF An operator $R \in \mathfrak{L}(X)$ is a positive trace-class operator iff it has a Hilbert–Schmidt square root $K : X \to X$. $\square$

REMARK 5.16 An operator $K : \mathcal{H} \to X$ is γ-radonifying iff the restriction of K to $\mathrm{Ker}(K)^\perp$ is also γ-radonifying. Therefore, an operator $R : X^* \to X$ is a Gaussian covariance operator iff it has a γ-radonifying one-to-one square root. Moreover, *any* square root K of R is then γ-radonifying: this follows from Remark 5.12, Theorem 5.9 and the identity $\|K^*(x^*)\|^2 = \langle R(x^*), x^* \rangle$.

REMARK 5.17 It follows from the proof of Theorem 5.13 that, in the representation (iii) of the Gaussian covariance operator R, one may take $x_n = K(e_n)$ where $K : \mathcal{H} \to X$ is a square root of R and (e_n) is any orthonormal basis of $\mathcal{H}$.

The next proposition describes the support of a Gaussian measure in terms of its covariance operator.

PROPOSITION 5.18 *Let μ be a Gaussian measure on X, and let $R = KK^*$ be its covariance operator, with $K : \mathcal{H} \to X$. Then $\mathrm{supp}(\mu) = \overline{\mathrm{Ran}(K)} = \mathrm{Ker}(R)^\perp$. In particular, μ has full support iff R is one-to-one and iff K has dense range.*

PROOF Since $\langle R(x^*), x^* \rangle = \|K^*(x^*)\|^2$ for all $x^* \in X^*$, we have $\mathrm{Ker}(R) = \mathrm{Ker}(K^*)$, so that $\overline{\mathrm{Ran}(K)} = \mathrm{Ker}(R)^\perp$. We show that $\mathrm{supp}(\mu) = \overline{\mathrm{Ran}(K)}$.

Let (e_n) be an orthonormal basis of $\mathcal{H}$, and set $x_n := K(e_n)$. By Remark 5.17, the measure μ is the distribution of the random variable $\sum_0^\infty g_n x_n$. We have to show that the support of μ is the closed linear span of the vectors x_n.

Clearly, $\mathrm{supp}(\mu) \subset \overline{\mathrm{span}}(x_n; \, n \in \mathbb{N})$, so we just need to check that if O is an open subset of X such that $O \cap \mathrm{span}(x_n; \, n \in \mathbb{N}) \neq \varnothing$ then $\mu(O) > 0$. Let us fix O together with some vector $a \in O \cap \mathrm{span}(x_n; \, n \in \mathbb{N})$. Let us also fix $\varepsilon > 0$ such that $B(a, \varepsilon) \subset O$.

Since the series $\sum g_n x_n$ is a.s. convergent it is also convergent in probability, so that $\mathbb{P}(\|\sum_{n>N} g_n x_n\| \geq \varepsilon/2) \to 0$ as $N \to \infty$. Therefore, one can find an $N \in \mathbb{N}$ such that $\mathbb{P}\left(\sum_{n>N} g_n x_n \in B(0, \varepsilon/2)\right) > 0$ and $a \in \mathrm{span}(x_0, \ldots, x_N)$. Moreover, since the distribution of the finite-dimensional Gaussian vector $(g_0, \ldots, g_N)$ has full support, we also have $\mathbb{P}\left(\sum_{n\leq N} g_n x_n \in B(a, \varepsilon/2)\right) > 0$. By independence, it follows that

$$\mathbb{P}\left(\textstyle\sum_{n\leq N} g_n x_n \in B(a, \varepsilon/2) \text{ and } \sum_{n>N} g_n x_n \in B(0, \varepsilon/2)\right) > 0.$$

Therefore, we obtain $\mu(B(a, \varepsilon)) > 0$, which concludes the proof. □

To conclude this section, we point out two results relating γ-radonifying and *absolutely summing* operators. The relevant definitions will be given below, as well as a sketch of a proof for Proposition 5.19.

It is perhaps worth pointing out that the proof of Proposition 5.19 is rather long and has nothing to do with linear dynamics. Therefore, the impatient reader could simply read the statement (and the definitions if necessary) and then proceed directly to the next section.

PROPOSITION 5.19 *Let $\mathcal{H}$ be a separable Hilbert space, and let $K \in \mathfrak{L}(\mathcal{H}, X)$.*

(a) *Assume that the Banach space X has type 2. Then K is γ-radonifying iff K^* is absolutely 2-summing.*

(b) *Assume that X has cotype 2. Then K is γ-radonifying iff it is absolutely 2-summing.*

COROLLARY 5.20 *Assume that X has type 2, and let $\mathcal{H}$ be a separable Hilbert space. If $K \in \mathfrak{L}(\mathcal{H}, X)$ and if K^* is absolutely 2-summing then $R = KK^*$ is a Gaussian covariance operator.*

Let us now recall the relevant definitions.

- An operator u between two Banach spaces X and Y is said to be **absolutely p-summing** ($p \geq 1$) if it turns weakly p-summable sequences into norm p-summable sequences or, equivalently, if there exists some finite constant C such that

$$\left(\sum_n \|u(x_n)\|^p \right)^{1/p} \leq C \sup_{x^* \in B_{X^*}} \left(\sum_n |\langle x^*, x_n \rangle|^p \right)^{1/p}$$

for every finite sequence $(x_n) \subset X$. For a huge amount of information concerning absolutely p-summing operators, we refer to the book of J. Diestel, H. Jarchow and A. Tonge [97].

- The Banach space X is said to have (Gaussian) **type** $p \in [1, 2]$ if

$$\left\| \sum_n g_n x_n \right\|_{L^2(\Omega, X)} \leq C \left(\sum_n \|x_n\|^p \right)^{1/p},$$

for some finite constant C and every finite sequence $(x_n) \subset X$. Similarly X has (Gaussian) **cotype** $q \in [2, \infty]$ if

$$\left(\sum_n \|x_n\|^q \right)^{1/q} \leq C \left\| \sum_n g_n x_n \right\|_{L^2(\Omega, X)},$$

for some finite constant C and every finite sequence $(x_n) \subset X$.

If the Gaussian variables are replaced by independent *Rademacher* variables then one obtains the classical definitions of type and cotype. However, it is well known (but non-trivial) that X has Gaussian type p (resp. Gaussian cotype q) if and only if it has type p (resp. cotype q). Every Banach space has type 1 and cotype ∞. Any (scalar) Lebesgue space $L^p(\nu)$ has type $\min(p, 2)$ and cotype $\max(p, 2)$ (for $1 \leq p < \infty$) whereas $c_0(\mathbb{N})$ has nothing better than type 1 and cotype ∞. Every Hilbert space has both type 2 and cotype 2, and a famous result of S. Kwapień asserts that, up to isomorphism, this property characterizes Hilbert spaces. Finally, if X has type 2 or cotype 2 then so does $L^2(\Omega, X)$. For more information on these notions, we refer to [97], [173] and B. Maurey's survey paper [180].

Let us now sketch the main steps in the proof of Proposition 5.19. As already said, the reader may safely skip the proof and proceed to the next section.

For convenience, we introduce the following notation: if (z_n) is a finite or infinite sequence in some Banach space Z, we set

$$\|(z_n)\|_{2, weak} := \sup_{z^* \in B_{Z^*}} \left(\sum_n |\langle z^*, z_n \rangle|^2 \right)^{1/2}.$$

Thus, an operator u defined on Z is absolutely 2-summing iff

$$\sum_n \|u(z_n)\|^2 \leq C \|(z_n)\|_{2, weak}^2$$

for some constant $C < \infty$ and any finite sequence $(z_n) \subset Z$.

The first observation is that in neither case does the space X contain $c_0(\mathbb{N})$. By a well-known and useful result of S. Kwapień ([160], see [77, Theorem V.6.1]), it follows that a Gaussian series $\sum g_n x_n$ is convergent in $L^2(\Omega, X)$ iff its partial sums are bounded in $L^2(\Omega, X)$.

Therefore, the operator $K : \mathcal{H} \to X$ is γ-radonifying iff, for any orthonormal basis (e_n) of $\mathcal{H}$, the partial sums of the series $\sum g_n K(e_n)$ are bounded in $L^2(\Omega, X)$. In that case, using the rotational invariance of Gaussian distributions one can prove that this holds in fact for any sequence $(h_n) \in \ell^2_{weak}(\mathcal{H})$. Equivalently, there exists some finite constant C such that

$$\left\| \sum_n g_n K(h_n) \right\|_{L^2(\Omega, X)} \leq C \, \|(h_n)\|_{2, weak}$$

for every finite sequence $(h_n) \subset \mathcal{H}$ (see [97, 12.12 and 12.17]).

Assume that K is γ-radonifying. Then $R := KK^*$ is the covariance operator of some Gaussian measure μ on X. If (x_n^*) is a finite sequence in X^* then

$$\sum_n \|K^*(x_n^*)\|^2 = \int_X \sum_n |\langle x_n^*, z\rangle|^2 \, d\mu(z)$$

$$\leq \int_X \|z\|^2 \, d\mu(z) \times \sup\left\{ \sum_n |\langle x_n^*, z\rangle|^2; \ \|z\| \leq 1 \right\}.$$

By Fernique's integrability theorem, $C := \int_X \|z\|^2 \, d\mu(z)$ is finite, and hence K^* is absolutely 2-summing. If in addition X has cotype 2 then, for any finite sequence $(h_n) \subset \mathcal{H}$, we get

$$\sum_n \|K(h_n)\|^2 \leq C \left\| \sum_n g_n K(h_n) \right\|^2_{L^2(\Omega, X)}$$

$$\leq C \, \|(h_n)\|^2_{2, weak},$$

so that that K is absolutely 2-summing.

Now, assume that K is absolutely 2-summing. By Pietsch's factorization theorem (see [97]), there exists a Borel probability measure ν on $(B_{\mathcal{H}^*}, w^*)$ such that

$$\|K(h)\|^2 \leq C \int_{B_{\mathcal{H}^*}} |\langle \xi, h\rangle|^2 d\nu(\xi)$$

for all $h \in \mathcal{H}$. If (h_n) is any finite sequence in $\mathcal{H}$, one can then write

$$\left\| \sum_n g_n K(h_n) \right\|^2_{L^2(\Omega, X)} = \mathbb{E}_\omega \left\| K\left(\sum_n g_n(\omega) h_n \right) \right\|^2$$

$$\leq C \int_{B_{\mathcal{H}^*}} \mathbb{E}_\omega \left| \sum_n g_n(\omega)\langle \xi, h_n\rangle \right|^2 d\nu(\xi)$$

$$= C \int_{B_{\mathcal{H}^*}} \sum_n |\langle \xi, h_n\rangle|^2 d\nu(\xi)$$

$$\leq C \, \|(h_n)\|^2_{2, weak},$$

where we have used the orthogonality of the Gaussian variables g_n. Hence, K is γ-radonifying.

Finally, assume that X has type 2 and that K^* is absolutely 2-summing. Let (e_n) be an orthonormal basis of $\mathcal{H}$. We have to show that the series $\sum g_n K(e_n)$ is convergent in $L^2(\Omega, X)$. Since X has type 2, the space $L^2(\Omega, X)$ has type 2 as well; in particular, $L^2(\Omega, X)$

does not contain $c_0(\mathbb{N})$. Therefore, by a classical result of C. Bessaga and A. Pełczyński (see [173] or [174]), it is enough to show that the series $\sum g_n K(e_n)$ is weakly unconditionally Cauchy in $L^2(\Omega, X)$, that is, $\sum_0^\infty |\Phi(g_n K(e_n))| < \infty$ for any continuous linear functional $\Phi \in L^2(\Omega, X)^*$.

At this point, we make a short digression. Let us denote by $\mathfrak{G}(X)$ the subspace of $L^2(\Omega, X)$ consisting of all random variables ξ of the form $\xi = \sum_0^\infty g_n z_n$, where $(z_n) \in X^{\mathbb{N}}$ and the series is L^2-convergent. It is not difficult to show that $\mathfrak{G}(X)$ is a closed subspace of $L^2(\Omega, X)$. Moreover, it is proved in [76] that the dual space of $\mathfrak{G}(X)$ can be identified with $\mathfrak{G}(X^*)$ in the following way: any functional $\Phi \in \mathfrak{G}(X)^*$ has the form

$$\Phi(\xi) = \mathbb{E}_\omega \langle \xi^*(\omega), \xi(\omega) \rangle$$

for some $\xi^* \in \mathfrak{G}(X^*)$.

The relevance of that remark should be clear: since the summands of the series $\sum g_n K(e_n)$ belong to $\mathfrak{G}(X)$, we now just have to check that if $\xi^* \in \mathfrak{G}(X^*)$ then

$$\sum_{n=0}^\infty |\mathbb{E}_\omega \langle \xi^*(\omega), g_n(\omega) K(e_n) \rangle| < \infty.$$

Writing $\xi^* = \sum_m g_m x_m^*$ and using the orthogonality of the g_n, we get

$$\mathbb{E}_\omega \langle \xi^*(\omega), g_n(\omega) K(e_n) \rangle = \langle x_n^*, K(e_n) \rangle$$

for all $n \in \mathbb{N}$. Morover, since $\xi^* \in \mathfrak{G}(X^*)$ and X^* has cotype 2, we know that $\sum_0^\infty \|x_n^*\|^2 < \infty$ so one can define a continuous linear operator $A : \mathcal{H} \to X^*$ such that $A(e_n) = x_n^*$. This operator is γ-radonifying by its very definition, hence absolutely 2-summing by the first part of the proof since X^* has cotype 2. Thus we get

$$\sum_{n=0}^\infty |\mathbb{E}_\omega \langle \xi^*(\omega), g_n(\omega) K(e_n) \rangle| = \sum_{n=0}^\infty |\langle K^* A(e_n), e_n \rangle| < \infty,$$

since the operator $K^* A : \mathcal{H} \to \mathcal{H}$ is nuclear, being the product of two absolutely 2-summing operators (see [97]).

REMARK 5.21   It is proved in [76] that the above results in fact give a characterization of type-2 or cotype-2 spaces. This can be expressed in terms of covariance operators as follows: X has type 2 iff any operator $R : X \to X^*$ having a square root with absolutely 2-summing adjoint is a Gaussian covariance operator and X has cotype 2 iff the square roots of any Gaussian covariance operator $R : X^* \to X$ are absolutely 2-summing

5.2 Ergodic Gaussian measures for an operator

Suppose that we have at hand a Gaussian measure μ on X and an operator $T \in \mathfrak{L}(X)$. In order to follow the plan outlined in the introduction, we need some simple way of checking that the measure μ is T-invariant and that T is ergodic with respect to μ. Working with a Gaussian measure μ, we can obtain two efficient criteria involving the covariance operator R_μ.

The criterion for T-invariance is very simple and quite easy to prove.

PROPOSITION 5.22   *Let μ be a Gaussian measure on X with covariance operator R and let $T \in \mathfrak{L}(X, Y)$, where Y is a Banach space. Then the image measure $\mu \circ T^{-1}$ is a Gaussian measure on Y, with covariance operator $T R T^*$. In particular, when $T \in \mathfrak{L}(X)$ the measure μ is T-invariant if and only if $T R T^* = R$.*

PROOF   The measure μ_T is Gaussian by the linearity of T. Its covariance operator R_T is given by

$$
\begin{aligned}
\langle R_T(x^*), y^* \rangle &= \int_Y \overline{\langle x^*, z \rangle} \langle y^*, z \rangle \, d\mu_T(z) \\
&= \int_X \overline{\langle x^*, T(z) \rangle} \langle y^*, T(z) \rangle \, d\mu(z) \\
&= \langle TRT^*(x^*), y^* \rangle.
\end{aligned}
$$

Thus, $R_T = TRT^*$. Since a (centred) Gaussian measure is completely determined by its covariance operator, this concludes the proof.   □

We now come to ergodicity. In fact, we will characterize the stronger properties of *weak mixing* and *strong mixing*. Let us first recall their definitions.

DEFINITION 5.23   *Let $T : (X, \mathcal{B}, \mu) \to (X, \mathcal{B}, \mu)$ be a measure-preserving transformation. Then T is said to be **weakly mixing** with respect to μ if either of the following two equivalent conditions is satisfied:*

(i) $\displaystyle \lim_{N \to \infty} \frac{1}{N} \sum_{n=0}^{N-1} |\mu(A \cap T^{-n}(B)) - \mu(A)\mu(B)| = 0$ *for every $A, B \in \mathcal{B}$.*

(ii) $\displaystyle \lim_{N \to \infty} \frac{1}{N} \sum_{n=0}^{N-1} \left| \int_X f(T^n z) g(z) \, d\mu(z) - \int_X f \, d\mu \int_X g \, d\mu \right| = 0$
for every $f, g \in L^2(X, \mu)$.

*Similarly, T is said to be **strongly mixing** with respect to μ if either of the following two equivalent conditions is satisfied:*

(i) $\displaystyle \lim_{n \to \infty} \mu(A \cap T^{-n}(B)) = \mu(A)\mu(B)$ $(A, B \in \mathcal{B})$;

(ii) $\displaystyle \lim_{n \to \infty} \int_X f(T^n z) g(z) \, d\mu(z) = \int_X f \, d\mu \int_X g \, d\mu$ $(f, g \in L^2(X, \mu))$.

By definition, strong mixing implies weak mixing and weak mixing implies ergodicity. Moreover, the terminology "weakly mixing" is consistent with that used in topological dynamics, since it can be shown that a measure-preserving transformation T is weakly mixing if and only if $T \times T$ is ergodic on $(X \times X, \mathcal{B} \otimes \mathcal{B}, \mu \otimes \mu)$; see [235] or Exercise 5.7. Observe also that mixing in the measure sense is stronger than its topological analogue: if a continuous map $T : X \to X$ (on a topological space X) happens to be (weakly) mixing with respect to some measure μ on X with full support, then T is topologically (weakly) mixing.

THEOREM 5.24   *Let μ be a Gaussian measure on X with full support and covariance operator R. Let $T \in \mathfrak{L}(X)$, and assume that μ is T-invariant. The following are equivalent:*

(i) *T is weakly mixing (resp. strongly mixing) with respect to μ;*
(ii) *for all $x^*, y^* \in X^*$, we have*

$$\lim_{N\to\infty} \frac{1}{N} \sum_{n=0}^{N-1} |\langle RT^{*n}(x^*), y^* \rangle| = 0$$

$(resp. \lim_{n\to\infty} \langle RT^{*n}(x^*), y^* \rangle = 0).$

PROOF   We observe that, for any $x^*, y^* \in X^*$,

$$\langle RT^{*n}(x^*), y^* \rangle = \int_X \overline{\langle x^*, T^n(z) \rangle} \langle y^*, z \rangle \, d\mu(z) = \int_X f(T^n z) g(z) \, d\mu(z)$$

with $f := \overline{\langle x^*, \cdot \rangle}$ and $g := \langle y^*, \cdot \rangle$. Thus, the implication (i) $\implies$ (ii) follows directly from the definition of weak mixing (resp. strong mixing) since $\int_X \langle x^*, z \rangle d\mu(z) = 0$ by the symmetry of μ.

For the converse, we know from (ii) that

$$\lim_{N\to\infty} \frac{1}{N} \sum_{n=0}^{N-1} \left| \int_X f(T^n z) g(z) \, d\mu(z) - \int_X f \, d\mu \int_X g \, d\mu \right| = 0$$

(resp. $\lim_{n\to\infty} \int_X f(T^n z) g(z) \, d\mu(z) = \int_X f \, d\mu \int_X g \, d\mu$) when $f = \overline{\langle x^*, \cdot \rangle}$ and $g = \langle y^*, \cdot \rangle$. We want to show that this holds true for *any* functions $f, g \in L^2(X, \mu)$ or, equivalently, for any indicator functions $\mathbf{1}_A, \mathbf{1}_B$. The following fact will be useful.

FACT   Let (ν_n) be a sequence of Gaussian measures on some finite-dimensional Banach space E, and let ν be a Gaussian measure on E with full support. Assume that $R_{\nu_n} \to R_\nu$ as $n \to \infty$. Then $\nu_n(Q) \to \nu(Q)$ for every Borel set $Q \subset E$.

PROOF OF THE FACT   We may assume that $E = \mathbb{C}^N$ for some $N \geq 1$. Since ν has full support, its covariance matrix is positive definite and ν has density f with respect to the Lebesgue measure on E. Therefore, the covariance matrix of ν_n is positive definite if n is large enough, and ν_n has density f_n. Since the density of a Gaussian measure can be computed using the coefficients of the covariance matrix, convergence of the covariance matrices implies the pointwise convergence of f_n to f. Since f_n and f are probability densities, it is not hard to see that this forces the L^1-convergence of f_n to f (see Exercise 5.8), which concludes the proof since $\nu_n(Q) = \int f_n \mathbf{1}_Q \to \int f \mathbf{1}_Q = \nu(Q)$. $\square$

Returning to the implication (ii) $\implies$ (i), we first assume that the strong property $\langle RT^{*n} (x^*) y^* \rangle \to 0$ holds for every $x^*, y^* \in X^*$. We want to show that T is strongly mixing, i.e. $\mu(A \cap T^{-n}(B)) \xrightarrow{n\to\infty} \mu(A)\mu(B)$ for every $A, B \in \mathcal{B}$. As is well known (see [235, Theorem 1.17]), it is enough to do this for all sets A, B belonging to some algebra $\mathcal{B}_0$ generating the σ-algebra $\mathcal{B}$. In our setting, it is natural to use the algebra of **cylinder sets**. A set $C \subset X$ is a cylinder set if it has the form $C = \pi^{-1}(\widetilde{C})$, where $\pi : X \to F$ is a continuous linear map onto some finite-dimensional space F and $\widetilde{C}$ is a Borel subset of F. Since X is separable, the cylinder sets generate the Borel σ-algebra of X (see [77, Chapter 1] or Exercise 5.4).

So, let us fix two cylinder sets $A = \pi_1^{-1}(\widetilde{A})$ and $B = \pi_2^{-1}(\widetilde{B})$, where $\pi_1 : X \to F_1$ and $\pi_2 : X \to F_2$ are linear (continuous) and onto. Then we have

$$A \cap T^{-n}(B) = L_n^{-1}(\widetilde{A} \times \widetilde{B})$$

for all $n \in \mathbb{N}$, where $L_n : X \to F_1 \times F_2$ is the operator defined by

$$L_n(x) := (\pi_1(x), \pi_2 T^n(x)).$$

Thus, denoting by $\nu_n := \mu \circ L_n^{-1}$ the image measure of μ under L_n and putting $\mu_1 := \mu \circ \pi_1^{-1}$, $\mu_2 := \mu \circ \pi_2^{-1}$, we need to show that

$$\nu_n(\widetilde{A} \times \widetilde{B}) \to (\mu_1 \otimes \mu_2)(\widetilde{A} \times \widetilde{B}).$$

Now, each ν_n is a Gaussian measure on $F_1 \times F_2$ and $\mu_1 \otimes \mu_2$ is a Gaussian measure with full support because π_1, π_2 are onto. By the above fact, it is enough to show that $\langle R_{\nu_n}(a^*), b^* \rangle \to \langle R_{\mu_1 \otimes \mu_2}(a^*), b^* \rangle$ for all $a^*, b^* \in (F_1 \times F_2)^*$. Let us fix a^*, b^* and write $a^* = (a_1^*, a_2^*)$, $b^* = (b_1^*, b_2^*)$, having identified $(F_1 \times F_2)^*$ with $F_1^* \times F_2^*$ in the usual way:

$$\langle u^*, (x_1, x_2) \rangle = \langle u_1^*, x_1 \rangle + \langle u_2^*, x_2 \rangle.$$

By Proposition 5.22, the covariance operator of $\nu_n = \mu \circ L_n^{-1}$ is given by

$$R_{\nu_n} = L_n R_\mu L_n^*.$$

This means that

$$\langle R_{\nu_n}(a^*), b^* \rangle = \langle R_\mu(x_1^* + T^{*n}(x_2^*)), y_1^* + T^{*n}(y_2^*) \rangle,$$

where $x_i^* = \pi_i^*(a_i^*)$ and $y_i^* = \pi_i^*(b_i^*)$.

Since μ is T-invariant, we have $T^n R_\mu T^{*n} = R_\mu$ for all $n \in \mathbb{N}$, and hence

$$\langle R_\mu T^{*n}(x_2^*), T^{*n}(y_2^*) \rangle = \langle R_\mu(x_2^*), y_2^* \rangle.$$

Moreover $\langle R_\mu(x_1^*), T^{*n}(y_2^*) \rangle = \overline{\langle x_1^*, R_\mu T^{*n}(y_2^*) \rangle} \to 0$ and $\langle R_\mu T^{*n}(x_2^*), y_1^* \rangle \to 0$, by assumption. It follows that

$$\langle R_{\nu_n}(a^*), b^* \rangle \to \langle R_\mu(x_1^*), y_1^* \rangle + \langle R_\mu(x_2^*), y_2^* \rangle = \langle R_{\mu_1 \otimes \mu_2}(a^*), b^* \rangle,$$

which concludes the proof for the case of strong mixing.

For weak mixing, we use the following classical lemma (see e.g. [235, Theorem 1.20], or Exercise 5.9). Recall that a set $A \subset \mathbb{N}$ is said to have **density 1** if

$$\lim_{N \to \infty} \frac{\text{card}(A \cap [1, N])}{N} = 1.$$

LEMMA 5.25 *Let* (a_n) *be a sequence of non-negative real numbers. If the Cesàro means* $N^{-1} \sum_{n=0}^{N-1} a_n$ *tend to 0 as* $N \to \infty$ *then there exists a set* $\mathbf{N} \subset \mathbb{N}$ *with density 1 such that* $a_n \to 0$ *as* $n \to \infty$, $n \in \mathbf{N}$. *The converse is true if the sequence* (a_n) *is bounded.*

Now, assume that $N^{-1} \sum_{n=0}^{N-1} |\langle R_\mu T^{*n}(x^*), y^* \rangle| \to 0$ for all $x^*, y^* \in X^*$. With the same notation as above, it is enough to show that if $A = \pi_1^{-1}(\widetilde{A})$ and $B = \pi_2^{-1}(\widetilde{B})$ are cylinder sets in X then

$$\frac{1}{N} \sum_{n=0}^{N-1} \left| \nu_n(\widetilde{A} \times \widetilde{B}) - (\mu_1 \otimes \mu_2)(\widetilde{A} \times \widetilde{B}) \right| \to 0$$

as $N \to \infty$.

For each $(x^*, y^*) \in X^* \times X^*$, the above lemma gives a set $\mathbf{N}_{x^*, y^*} \subset \mathbb{N}$ with density 1 such that $\langle R_\mu T^{*n}(x^*), y^* \rangle \to 0$ as $n \to \infty$ along $\mathbf{N}_{x^*, y^*}$. Since the intersection of finitely many sets with density 1 still has density 1 it follows that, for any finite-dimensional subspace $Z \subset X^*$, one can find a single set $\mathbf{N} \subset \mathbb{N}$ with density 1 such that $\langle R_\mu T^{*n}(x^*), y^* \rangle \to 0$ along $\mathbf{N}$ for all $x^*, y^* \in Z$. Arguing as above, we find that $|\nu_n(\widetilde{A} \times \widetilde{B}) - (\mu_1 \otimes \mu_2)(\widetilde{A} \times \widetilde{B})| \to 0$ along $\mathbf{N}$. Since we are dealing with a bounded sequence, we can now use the "converse" part of the lemma to get the desired conclusion. $\qquad \square$

5.3 How to find an ergodic measure

After this preparatory work, we can now concentrate on the most interesting part of our plan: to find some natural and easily verifiable conditions, given a bounded linear operator $T \in \mathfrak{L}(X)$, that ensure the existence of an invariant Gaussian measure with respect to which T is an ergodic transformation.

As we shall see, the important thing is that T should have sufficiently many *unimodular eigenvalues*. The following terminology will be used constantly: we will say that a vector $x \in X$ is a $\mathbb{T}$-**eigenvector** for T if $T(x) = \lambda x$ for some $\lambda \in \mathbb{T}$. This terminology is perhaps a bit misleading, since a $\mathbb{T}$-eigenvector x is a true eigenvector for T only if $x \neq 0$. However, this additional freedom will be useful.

5.3.1 The unimodular eigenvalues should play a role

We start our discussion with two simple results showing that the unimodular point spectrum is likely to play a crucial role in our matter. The first was observed in [135].

PROPOSITION 5.26 *If $T \in \mathfrak{L}(X)$ admits an invariant measure with full support then all eigenvalues of T^* are unimodular.*

PROOF   Assume that T^* has an eigenvalue λ such that $|\lambda| \neq 1$, and let x^* be an eigenvector of T^* associated with this eigenvalue λ. Set $U := \{x \in X;\ |\langle x^*, x \rangle| > 1\}$ if $|\lambda| < 1$ and $U := \{x \in X;\ 0 < |\langle x^*, x \rangle| < 1\}$ if $|\lambda| > 1$. Since

$$\langle x^*, T^n(x) \rangle = \lambda^n \langle x^*, x \rangle$$

for any $x \in X$ and all $n \in \mathbb{N}$, we see that no T-orbit originating from U can go back to U infinitely many times. By the Poincaré recurrence theorem, it follows that T does not admit any invariant measure with full support. $\qquad \square$

In particular, it follows that if X is finite-dimensional and $T \in \mathfrak{L}(X)$ has an invariant measure with full support then all its eigenvalues have to be unimodular.

Conversely, let us show that if an operator T has a sufficient number of eigenvectors associated with unimodular eigenvalues then it admits an invariant Gaussian measure.

PROPOSITION 5.27   *Let $T \in \mathfrak{L}(X)$, and assume that the $\mathbb{T}$-eigenvectors of T span a dense linear subspace of X. Then T admits an invariant Gaussian measure with full support.*

PROOF   It is enough to show that T is a linear quasi-factor of some operator $S : \ell^2(\mathbb{N}) \to \ell^2(\mathbb{N})$ having an invariant Gaussian measure ν with full support (see Chapter 1). Indeed, if $J : \ell^2(\mathbb{N}) \to X$ is a continuous linear operator with dense range such that the diagram

$$
\begin{array}{ccc}
\ell^2 & \xrightarrow{\ S\ } & \ell^2 \\
{\scriptstyle J}\downarrow & & \downarrow{\scriptstyle J} \\
X & \xrightarrow{\ T\ } & X
\end{array}
$$

commutes then it is straightforward to check that the image measure $\mu := \nu \circ J^{-1}$ has the required properties.

Let $(x_n)_{n\in\mathbb{N}}$ be a sequence of $\mathbb{T}$-eigenvectors for T such that $\overline{\mathrm{span}}(x_n; \; n \in \mathbb{N}) = X$. For each $n \in \mathbb{N}$, we choose $\lambda_n \in \mathbb{T}$ such that $T(x_n) = \lambda_n x_n$. Let $S : \ell^2(\mathbb{N}) \to \ell^2(\mathbb{N})$ be the diagonal operator defined by $S((t_n)) = (\lambda_n t_n)$, and let R be a diagonal operator on $\ell^2(\mathbb{N})$ associated with some summable sequence of positive numbers (r_n). Then R is the covariance operator of a Gaussian measure ν with full support, since it is a one-to-one trace-class positive operator. The equation $SRS^* = R$ is satisfied because $|\lambda_n| = 1$ for all n, so that ν is S-invariant by Proposition 5.22.

Now, if (a_n) is a sufficiently fast decreasing sequence of positive numbers then the formula

$$ J((t_n)) := \sum_{n=0}^{\infty} a_n t_n x_n $$

defines a continuous operator $J : \ell^2(\mathbb{N}) \to X$ with dense range, and we have $TJ = JS$ by definition. Thus T is indeed a linear quasi-factor of S, which concludes the proof. $\square$

5.3.2 Perfectly spanning $\mathbb{T}$-eigenvector fields

Proposition 5.27 says that an operator T having a sufficient number of $\mathbb{T}$-eigenvectors admits an invariant measure with full support. But, of course, we want more: we also expect the operator to be ergodic with respect to that measure. The above condition on eigenvectors is not sufficient; for example, the identity operator satisfies it! As will soon be apparent, what is really needed is that the eigenvalues are somehow "continuously distributed", and that if one removes "not too many" eigenvalues, then T still has a sufficient number of eigenvectors associated with the remaining eigenvalues. Before giving a precise meaning to all this, we first need to introduce some terminology.

DEFINITION 5.28 *Let $T \in \mathfrak{L}(X)$. By a $\mathbb{T}$-**eigenvector field** for T, we mean any bounded map $E : \mathbb{T} \to X$ such that $E(\lambda) \in \mathrm{Ker}(T - \lambda)$ for all $\lambda \in \mathbb{T}$.*

It is important to note that an eigenvector field will always be assumed to be bounded. Of course, $\mathbb{T}$-eigenvector fields are likely to be useful only if they have some regularity. The following lemma shows that one can always parametrize the $\mathbb{T}$-eigenvectors of a given operator T in a reasonable way. Recall that a map $\Phi : Y \to (Z, \mathcal{B})$ from a Polish space (i.e. a separable and completely metrizable topological space) Y into a measurable space $(Z, \mathcal{B})$ is said to be **universally measurable** if the inverse image of any set $B \in \mathcal{B}$ is universally measurable, i.e. measurable with respect to every Borel measure on Y.

LEMMA 5.29 *Let $T \in \mathfrak{L}(X)$; one can find a countable family of universally measurable $\mathbb{T}$-eigenvector fields $(E_i)_{i \in I}$ such that $\mathrm{Ker}(T - \lambda) = \overline{\mathrm{span}}(E_i(\lambda); \; i \in I)$ for every $\lambda \in \mathbb{T}$.*

Of course, in all *concrete* examples that we shall be considering, we will get the eigenvector fields E_i directly in a regular way (for instance, E_i will often be continuous), but Lemma 5.29 will help us to give useful *theoretical* statements. The proof makes use of some classical results from descriptive set theory and can be omitted at first reading.

PROOF We first give a proof for the case when X is a Hilbert space. Then we actually get *Borel* eigenvector fields E_i.

Let $(x_i)_{i \in \mathbb{N}}$ be a sequence in the unit sphere of X whose linear span is dense in X. For each $i \in \mathbb{N}$, let us define $E_i : \mathbb{T} \to X$ by $E_i(\lambda) = \pi_\lambda(x_i)$, where π_λ is the orthogonal projection onto $\mathrm{Ker}(T - \lambda)$. Then each E_i is a $\mathbb{T}$-eigenvector field for T, and it is clear that

$$\overline{\mathrm{span}}(E_i(\lambda); \; i \in \mathbb{N}) = \mathrm{Ran}(\pi_\lambda) = \mathrm{Ker}(T - \lambda)$$

for each $\lambda \in \mathbb{T}$. It remains to show that each eigenvector field E_i is Borel.

By Souslin's theorem (see [156, Theorem 14.12]), it is enough to check that the graph of E_i is a Borel subset of $\mathbb{T} \times X$. Now, if $(\lambda, x) \in \mathbb{T} \times X$ then

$$\pi_\lambda(x_i) = x \Longleftrightarrow T(x) = \lambda x \text{ and } \forall h \in B_X : \; (T(h) \neq \lambda h \text{ or } \langle x - x_i, h \rangle = 0)$$
$$\Longleftrightarrow T(x) = \lambda x \text{ and } \forall h \in B_X : \; R(h, \lambda, x).$$

The relation $R(h, \lambda, x)$ is readily seen to be G_δ in $(B_X, w) \times (\mathbb{T} \times X)$, where (B_X, w) is the closed unit ball of X endowed with the weak topology. Since (B_X, w) is compact (X is a Hilbert space!), this implies that the relation $\forall h \, R$ is G_δ in $\mathbb{T} \times X$ (its complement is the projection of an F_σ-set along a compact factor). Thus the graph of E_i is indeed Borel, actually G_δ in $\mathbb{T} \times X$.

Now we will sketch the proof for a general Banach space X. Let us denote by $\mathcal{F}(X)$ the space of all closed subsets of X, equipped with the **Effros Borel structure**, that is, the σ-algebra generated by all sets of the form $\{F \in \mathcal{F}(X); \ F \cap U \neq \varnothing\}$ where $U \subset X$ is open.

It is well known that one can select in a Borel way a dense sequence in each non-empty closed set $F \subset X$. In other words, there exists a sequence $(d_i)_{i\in\mathbb{N}}$ of Borel maps $d_i : \mathcal{F}(X) \to X$ such that whenever $F \in \mathcal{F}(X)$ is non-empty the sequence $(d_i(F))_{i\in\mathbb{N}}$ is dense in F. This follows from the Kuratowski–Ryll–Nardzewski selection theorem (see [156, Theorem 12.13]).

Putting $E_i(\lambda) := d_i(\mathrm{Ker}(T - \lambda) \cap B_X)$, we get the required vector fields E_i. To conclude the proof, it is enough to check that the map $\Phi : \mathbb{T} \to \mathcal{F}(X)$ defined by $\Phi(\lambda) = \mathrm{Ker}(T - \lambda) \cap B_X$ is universally measurable. Now, for each open set $U \subset X$ the set $\Phi^{-1}(U) := \{\lambda; \ \Phi(\lambda) \cap U \neq \varnothing\}$ is the projection of the Borel set $\{(\lambda, x); \ x \in U \cap B_X \ \text{and} \ T(x) = \lambda x\} \subset \mathbb{T} \times X$ along the first coordinate. As such, $\Phi^{-1}(U)$ is an *analytic* set and hence is universally measurable (see [156]). $\square$

We now recall some basic facts about continuous measures on $\mathbb{T}$.

DEFINITION 5.30 *A complex measure σ on $\mathbb{T}$ is said to be **continuous** if it has no discrete part, i.e. $\sigma(\{\lambda\}) = 0$ for every $\lambda \in \mathbb{T}$. The measure σ is said to be a **Rajchman measure** if $\widehat{\sigma}(n) \to 0$ as $|n| \to \infty$.*

The following classical result of N. Wiener characterizes continuous measures by means of their Fourier coefficients. We state it in a form that will be convenient for us. For a proof, see e.g. [155, I.7.11] or Exercise 5.10.

THEOREM 5.31 (WIENER'S THEOREM) *A complex measure σ on $\mathbb{T}$ is continuous if and only if*

$$\lim_{N\to\infty} \frac{1}{N} \sum_{n=0}^{N-1} |\widehat{\sigma}(n)| = 0.$$

In particular, any Rajchman measure is continuous.

REMARK 5.32 Wiener's theorem is usually stated in terms of $|\widehat{\sigma}(n)|^2$ rather than $|\widehat{\sigma}(n)|$ and with the symmetric Cesàro mean $(2N + 1)^{-1} \sum_{-N}^{N}$ instead of the one-sided mean $N^{-1} \sum_{n=0}^{N-1}$. Since the sequence $(\widehat{\sigma}(n))$ is bounded, the statements with $|\widehat{\sigma}(n)|$ and with $|\widehat{\sigma}(n)|^2$ are equivalent by Lemma 5.25. That one can replace the symmetric Cesàro mean by the one-sided mean follows from any proof of Wiener's theorem.

The next definition is crucial. It gives a precise meaning to the vague sentences formulated in the beginning of this section.

DEFINITION 5.33   *Let $T \in \mathfrak{L}(X)$.*

(a) *Given a probability measure σ on $\mathbb{T}$, we say that $T \in \mathfrak{L}(X)$ has a σ-**spanning set** of $\mathbb{T}$-**eigenvectors** if for every σ-measurable subset A of $\mathbb{T}$ with $\sigma(A) = 1$ the eigenspaces $\mathrm{Ker}(T - \lambda)$, $\lambda \in A$, span a dense subspace of X. If T has a σ-spanning set of $\mathbb{T}$-eigenvectors for some continuous measure σ then we say that T has a **perfectly spanning** set of $\mathbb{T}$-eigenvectors.*

(b) *We define in the same way σ-spanning families of $\mathbb{T}$-eigenvector fields for T: such a family $(E_i)_{i \in I}$ is said to be σ-spanning if, whenever $A \subset \mathbb{T}$ satisfies $\sigma(A) = 1$, it follows that $\overline{\mathrm{span}}(E_i(\lambda);\ i \in I,\ \lambda \in A) = X$.*

To help digestion of the definition, let us note the following simple fact, which will be very useful when looking at concrete examples.

REMARK 5.34   Let $E : \mathbb{T} \to X$ be a continuous $\mathbb{T}$-eigenvector field for T. Assume that $\overline{\mathrm{span}}(E(\lambda);\ \lambda \in \mathbb{T}) = X$. Then E is σ-spanning for any probability measure σ on $\mathbb{T}$ with full support.

PROOF   The result is clear since E is continuous and any σ-measurable set $A \subset \mathbb{T}$ with full measure is dense in $\mathbb{T}$. $\qquad\qquad\qquad\qquad\qquad\qquad\qquad\qquad\qquad\quad\square$

5.3.3 Construction of covariance operators

We are now ready to define a large class of potential covariance operators R. Let σ be a probability measure on $\mathbb{T}$. With each bounded σ-measurable map $E : \mathbb{T} \to X$ one can associate the operator $K_E : L^2(\mathbb{T}, \sigma) \to X$ defined by

$$K_E(f) = \int_{\mathbb{T}} f(\lambda) E(\lambda)\, d\sigma(\lambda).$$

In fact, it is enough to assume that $E \in L^2(\mathbb{T}, \sigma, X)$. We note that the operator K_E is always *compact*. Indeed, one can approximate E by step functions in the norm of $L^2(\mathbb{T}, \sigma, X)$, so that K_E is a limit of finite-rank operators.

The (conjugate-linear) adjoint operator $K_E^* : X^* \to L^2(\mathbb{T}, \sigma)$ is given by

$$K_E^*(x^*) = \overline{\langle x^*, E(\,\cdot\,)\rangle}.$$

The following lemma collects all we need concerning operators of the form K_E.

LEMMA 5.35   *Let $T \in \mathfrak{L}(X)$, and let σ be a probability measure on $\mathbb{T}$. Also, let $(E_i)_{i \in I}$ be a finite or countably infinite family of σ-measurable $\mathbb{T}$-eigenvector fields for T. Let us denote by $\mathcal{H}$ the Hilbert space $\oplus_{i \in I} L^2(\mathbb{T}, \sigma)$, and let $K : \mathcal{H} \to X$ be the operator defined by*

$$K(\oplus_i f_i) = \sum_i \alpha_i K_{E_i}(f_i),$$

where $(\alpha_i)_{i \in I}$ is a family of positive numbers such that $\sum_i \alpha_i^2 \|E_i\|_{L^2(\mathbb{T}, \sigma, X)}^2 < \infty$. Finally, set $R := K K^*$. Then the following facts hold.

(1) The **intertwining equation** $TK = KM$ is satisfied, where $M = \oplus_i M_i : \mathcal{H} \to \mathcal{H}$ is the direct sum of copies M_i of the operator corresponding to the variable λ acting on $L^2(\mathbb{T}, \sigma)$.

(2) $TRT^* = R$.

(3) $\mathrm{Ker}(R) = \{x^* \in X^*; \; \langle x^*, E_i(\lambda) \rangle = 0 \; \sigma\text{-a.e. for each } i \in I\}$.

(4) If $x^*, y^* \in X^*$ then

$$\langle RT^{*n}(x^*), y^* \rangle = \widehat{\sigma_{x^*, y^*}}(n)$$

for all $n \in \mathbb{N}$, where σ_{x^*, y^*} is the complex measure on $\mathbb{T}$ defined by

$$d\sigma_{x^*, y^*}(\lambda) = \left(\sum_i \alpha_i^2 \, \overline{\langle x^*, E_i(\lambda) \rangle} \langle y^*, E_i(\lambda) \rangle \right) d\sigma(\lambda).$$

PROOF   The operator K is well defined by our assumption on (α_i).

(1) Since each E_i is a $\mathbb{T}$-eigenvector field for T, one has

$$TK_{E_i}(f_i) = \int_{\mathbb{T}} f_i(\lambda) \, \lambda E_i(\lambda) \, d\sigma(\lambda)$$

for any $f_i \in L^2(\mathbb{T})$. This gives at once the equation $TK = KM$.

(2) This is clear since M is unitary: $TRT^* = (TK)(K^*T^*) = KMM^*K^* = R$.

(3) We have $K^*(x^*) = \oplus_i \alpha_i K_{E_i}^*(x^*)$ for all $x^* \in X^*$, so that

$$R = \sum_i \alpha_i^2 R_{E_i},$$

where $R_{E_i} = K_{E_i} K_{E_i}^*$. Now, each R_{E_i} is positive. It follows that $\mathrm{Ker}(R) = \bigcap_i \mathrm{Ker}(R_{E_i})$, which gives the result since $K_{E_i}^*(x^*) = \overline{\langle x^*, E_i(\cdot) \rangle}$.

(4) This is a direct computation:

$$\begin{aligned}
\langle RT^{*n}(x^*), y^* \rangle &= \langle K(K^*T^{*n})(x^*), y^* \rangle \\
&= \langle K^*(y^*), M^{*n}K^*(x^*) \rangle_{\mathcal{H}} \\
&= \sum_{i \in I} \alpha_i^2 \langle K_{E_i}^*(y^*), M_i^{*n} K_{E_i}^*(x^*) \rangle_{L^2(\mathbb{T}, \sigma)} \\
&= \sum_{i \in I} \alpha_i^2 \int_{\mathbb{T}} \lambda^{-n} \overline{\langle x^*, E_i(\lambda) \rangle} \langle y^*, E_i(\lambda) \rangle \, d\sigma(\lambda). \qquad \square
\end{aligned}$$

With this lemma at hand it is now easy to prove the following proposition, which is the kind of result we have been looking for from the beginning of the chapter.

PROPOSITION 5.36   *Let $T \in \mathfrak{L}(X)$, and let σ be a probability measure on $\mathbb{T}$. Assume that T admits a σ-spanning countable family of σ-measurable eigenvector fields $(E_i)_{i \in I}$, where each operator $K_{E_i} : L^2(\mathbb{T}, \sigma) \to X$ is γ-radonifying. Then there exists a Gaussian T-invariant measure μ with full support with respect to which*

(a) *T is weakly mixing if σ is a continuous measure;*

(b) *T is strongly mixing if σ is a Rajchman measure.*

PROOF We will keep to the notation of Lemma 5.35. The operator K is well defined provided that $\sum_i \alpha_i^2 \|E_i\|^2_{L^2(\mathbb{T},\sigma,X)} < \infty$. We have $TRT^* = R$ by part (2) of the above lemma; and the operator R is one-to-one by part (3), because (E_i) is countable and σ-spanning. Thus R will be the covariance operator of some Gaussian T-invariant measure μ with full support provided that the operator K is γ-radonifying.

Assume that this is indeed the case. Since the measures σ_{x^*,y^*} defined above are absolutely continuous with respect to σ, they are either continuous or Rajchman if σ is continuous or Rajchman (for the Rajchman case, see [157] or Exercise 5.11). Thus, by Wiener's theorem and Theorem 5.24, T is weakly mixing with respect to μ in the first case and strongly mixing in the second case.

Therefore, the only thing that remains to be done is to show that the operator K will be γ-radonifying if the positive real numbers α_i are suitably chosen. Let (e_n) be an orthonormal basis of $\mathcal{H}$ obtained by putting together orthonormal bases of each copy of the space $L^2(\mathbb{T},\sigma)$. We write $e_n = \oplus_i e_{i,n}$. Then $e_{i,n}$ is either 0 or a normalized vector in $\mathcal{H}_i$ and, for each $i \in I$, the family $\{e_{i,n}; \ e_{i,n} \neq 0\}$ is an orthonormal basis of $L^2(\mathbb{T},\sigma)$.

Since each operator K_{E_i} is γ-radonifying, we know that

$$\beta_{i,N} := \sup_{M>N} \left\| \sum_{n=N}^{M} g_n K_{E_i}(e_{i,n}) \right\|_{L^2(\Omega,X)} \xrightarrow{N\to\infty} 0$$

for each fixed $i \in I$. Setting $\beta_i := \sup_N \beta_i^N$, we now choose the coefficients α_i in such a way that $\sum_i \alpha_i \beta_i < \infty$. Then, for each finite set $J \subset I$ and all $N \in \mathbb{N}$, we have

$$\sup_{M>N} \left\| \sum_{n=N}^{M} g_n K(e_n) \right\|_{L^2(\Omega,X)} = \sup_{M>N} \left\| \sum_{n=N}^{M} g_n \sum_{i\in I} \alpha_i K_{E_i}(e_{i,n}) \right\|_{L^2(\Omega,X)}$$

$$\leq \sum_{i\in J} \alpha_i \beta_{i,N} + \sum_{i\notin J} \alpha_i \beta_i.$$

Choosing first J such that the second term on the right-hand side of the inequality is small and then N large enough to ensure that the first term is also small, we see that the series $\sum g_n K(e_n)$ is convergent in $L^2(\Omega,X)$ and hence that K is γ-radonifying. □

5.4 The results

In view of Proposition 5.36, it remains to find simple conditions ensuring that an operator $K_E : L^2(\mathbb{T},\sigma) \to X$ is γ-radonifying. Here, the geometry of the Banach space X and the regularity of the vector field $E : \mathbb{T} \to X$ come into play. Recall that, for $E \in L^2(\mathbb{T},X)$, the operator $K_E : L^2(\mathbb{T},\sigma) \to X$ is defined by

$$K_E(f) = \int_{\mathbb{T}} f(\lambda)E(\lambda)\,d\sigma(\lambda).$$

The simplest case is when X has type 2.

LEMMA 5.37 *Let σ be a probability measure on $\mathbb{T}$. If $E \in L^2(\mathbb{T}, X)$ then the operator $K_E^* : X^* \to L^2(\mathbb{T}, \sigma)$ is absolutely 2-summing. Therefore K_E is γ-radonifying if X has type 2.*

PROOF The second assertion follows from Proposition 5.19. The first part is well known and easily checked (see [97, Example 2.11]): if (x_n^*) is a finite sequence in X^* then

$$\left(\sum_n \|K_E^* x_n^*\|_2^2 \right)^{1/2} = \left(\sum_n \int_{\mathbb{T}} |\langle x_n^*, E(\lambda)\rangle|^2 d\sigma(\lambda) \right)^{1/2}$$

$$\leq \|E\|_{L^2(\mathbb{T},\sigma)} \sup_{\xi \in B_{X^{**}}} \left(\sum_n |\xi(x_n^*)|^2 \right)^{1/2}. \qquad \square$$

From this lemma, Lemma 5.29 and Proposition 5.36, we get the following result.

THEOREM 5.38 *Let $T \in \mathfrak{L}(X)$, and assume that T has a perfectly spanning set of $\mathbb{T}$-eigenvectors. Suppose moreover that the Banach space X has type 2. Then there exists a Gaussian measure μ on X with full support, with respect to which T is a weakly mixing measure-preserving transformation. If the $\mathbb{T}$-eigenvectors are σ-spanning for some Rajchman measure σ then there exists a Gaussian measure μ on X with full support, with respect to which T is a strongly mixing measure-preserving transformation.*

When X does not have type 2, the situation becomes more involved and we need to impose some regularity on the eigenvector fields E_i. Recall that a map $E : \mathbb{T} \to X$ is said to be α-**Hölderian** ($\alpha \in (0, 1]$) if

$$\|E(\lambda_2) - E(\lambda_1)\| \leq C|\lambda_2 - \lambda_1|^\alpha,$$

for some finite constant C and any $\lambda_1, \lambda_2 \in \mathbb{T}$.

LEMMA 5.39 *Assume that the Banach space X has type $p \in [1, 2]$, and let $E : \mathbb{T} \to X$ be α-Hölderian for some $\alpha > 1/p - 1/2$. Then the operator $K_E : L^2(\mathbb{T}) \to X$ is γ-radonifying.*

PROOF We need to find some orthonormal basis (e_n) of $L^2(\mathbb{T})$ such that the Gaussian series $\sum g_n K_E(e_n)$ is convergent in $L^2(\Omega, X)$. Moreover, since X has type p, it is enough to show that $\sum_n \|K_E(e_n)\|^p < \infty$.

On the one hand, the first, natural, idea is to take $e_n(t) = e^{int}$, $n \in \mathbb{Z}$. Then we get $K_E(e_n) = \widehat{E}(-n)$, the nth Fourier coefficient of the vector-valued function E; and, of course, everybody knows that the Fourier coefficients of a function are related to its regularity. For instance, if $f : \mathbb{T} \to \mathbb{C}$ is a *scalar-valued* α-Hölderian function then a theorem of O. Szász (which improves the classical result of S. Bernstein) ensures that $\sum_{n \in \mathbb{Z}} |\hat{f}(n)|^{2/(2\alpha+1)+\varepsilon} < \infty$ for any $\varepsilon > 0$. If this were true for vector-valued functions then we would get the result, since $2/(2\alpha + 1) < p$ iff $\alpha > 1/p - 1/2$. Unfortunately this is far from being true. For example, there exist continuously differentiable functions with values in ℓ^1 for which $\sum_{n \in \mathbb{Z}} \|\hat{f}(n)\| = \infty$ (see [159]).

On the other hand, it is also well known that the best information on the regularity of a function is usually not given by the Fourier transform but, rather, by suitable *wavelet* transforms. In less pedantic terms, there are orthonormal bases of $L^2(\mathbb{T})$ that are more convenient than the Fourier basis for these matters. Here, we will just need the simplest wavelet basis, namely the classical **Haar basis**.

Let us denote by $\mathcal{J}$ the family of all dyadic intervals of $[0, 2\pi)$, i.e. all intervals of the form $J = [2\pi j/2^n, 2\pi(j+1)/2^n)$, where $n \in \mathbb{N}$ and $0 \le j < 2^n$. For any interval $J \in \mathcal{J}$, let us set

$$h_J := \frac{1}{\sqrt{|J|}} \left(\mathbf{1}_{J^+} - \mathbf{1}_{J^-} \right),$$

where $|J|$ is the length of J and $J = J^- \cup J^+$ is the usual decomposition of J into dyadic intervals of length $|J|/2$. Then $\{h_J;\ J \in \mathcal{J}\} \cup \{\mathbf{1}\}$ is an orthonormal basis of $L^2([0, 2\pi)) \simeq L^2(\mathbb{T})$. Moreover, one can estimate $\|K_E(h_J)\|$ as follows: if the interval J has length 2^{-n} and centre c_J then

$$\|K_E(h_J)\| = \left\| \int_0^{2\pi} E(e^{it}) h_J(t) \frac{dt}{2\pi} \right\|$$

$$= \left\| \int_J \left(E(e^{it}) - E(e^{ic_J}) \right) h_J(t) \frac{dt}{2\pi} \right\|$$

$$\le C\sqrt{2^n} \int_J |t - c_J|^\alpha \frac{dt}{2\pi}$$

$$\le C\sqrt{2^n} \int_0^{2^{-n}} t^\alpha dt$$

$$\le \frac{C}{2^{n(\alpha+1/2)}},$$

where C is a constant which may change from line to line. It follows that

$$\sum_{J \in \mathcal{J}} \|K_E(h_J)\|^p = \sum_{n \ge 0} \sum_{|J| = 2^{-n}} \|K_E(h_J)\|^p$$

$$\le C \sum_{n \ge 0} \frac{1}{2^{n(p\alpha + p/2 - 1)}} < \infty,$$

since by assumption $p\alpha + p/2 - 1 > 0$. $\qquad\square$

Putting together Proposition 5.36 and Lemma 5.39, we finally obtain the following theorem, which improves all the results contained in [27].

THEOREM 5.40 *Assume that the Banach space X has type $p \in [1, 2]$, and let $T \in \mathcal{L}(X)$. Assume that one can find a finite or countably infinite family (E_i) of*

$\mathbb{T}$-*eigenvector fields for T which is spanning with respect to the Lebesgue measure on $\mathbb{T}$ and is such that each E_i is α_i-Hölderian, for some $\alpha_i > 1/p - 1/2$. Then there exists an invariant Gaussian measure on X with full support, with respect to which T is a strongly mixing measure-preserving transformation.*

5.5 Examples

5.5.1 Adjoints of multipliers

Let $\phi \in H^\infty(\mathbb{D})$ and let $M_\phi(f) = \phi f$ be the associated multiplier defined on $H^2(\mathbb{D})$ (see Chapter 1). We recall that M_ϕ^* is hypercyclic on $H^2(\mathbb{D})$ iff if ϕ is non-constant and $\phi(\mathbb{D}) \cap \mathbb{T} \neq \varnothing$. Under this assumption, let us prove the existence of an invariant Gaussian measure with full support on $H^2(\mathbb{D})$ with respect to which M_ϕ^* is strongly mixing.

We have already observed that if $k_z \in H^2(\mathbb{D})$ is the reproducing kernel at $z \in \mathbb{D}$ then $M_\phi^*(k_z) = \overline{\phi(z)}k_z$. Let us fix some $\lambda_0 \in \mathbb{D}$ with $\phi(\lambda_0) \in \mathbb{T}$ and $\phi'(\lambda_0) \neq 0$. Then one can find an open arc $I \subset \mathbb{T}$ containing $\phi(\lambda_0)$ and a curve $\Gamma \subset \mathbb{D}$ containing λ_0 such that ϕ induces a diffeomorphism from Γ onto I. Let ψ be the inverse of ϕ defined on I. Thus, ψ is C^1-smooth on I and, taking I smaller if necessary, we may assume that its derivative ψ' is bounded on I.

Let I^* be the arc made up of all complex conjugates of elements of I, and let us fix a smooth function r defined on $\mathbb{T}$ which is positive on I^* and zero outside $\overline{I^*}$. Now, let us define a $\mathbb{T}$-eigenvector field for M_ϕ^* by setting $E(e^{i\theta}) := r(e^{i\theta})k_{\psi(e^{-i\theta})}$ if $e^{i\theta} \in I^*$ and $E(e^{i\theta}) = 0$ otherwise. The map E is clearly Lipschitz, because $\|k_{z'} - k_z\|_2 \leq C_K|z' - z|$ when z and z' live in a given compact set $K \subset \mathbb{D}$.

It remains to show that the $\mathbb{T}$-eigenvector field E is spanning with respect to the Lebesgue measure on $\mathbb{T}$. This is easy. Indeed, if $A \subset \mathbb{T}$ has Lebesgue measure 1 and if $f \in H^2(\mathbb{D})$ is orthogonal to $E(\lambda)$, $\lambda \in A$, then $f(z) = 0$ for every $z = \psi(\bar{\lambda})$, $\lambda \in A \cap I^*$. Since the set of all such points z has an accumulation point in $\mathbb{D}$, this forces f to vanish identically.

5.5.2 Weighted shifts

Let $B_{\mathbf{w}}$ be a weighted backward shift on $\ell^p(\mathbb{N})$, $1 \leq p < \infty$, with associated weight sequence $\mathbf{w} = (w_n)_{n \geq 1}$. Solving the equation $B_{\mathbf{w}}(x) = \lambda x$, it is easy to find that $B_{\mathbf{w}}$ admits unimodular eigenvalues if and only if

$$\sum_{n \geq 1} \frac{1}{(w_1 \cdots w_n)^p} < \infty.$$

In this case, one defines a $\mathbb{T}$-eigenvector field $E : \mathbb{T} \to \ell^p(\mathbb{N})$ by setting

$$E(\lambda) := \sum_{n \geq 0} \frac{\lambda^n}{w_1 \cdots w_n} e_n,$$

where (e_n) is the canonical basis of $\ell^p(\mathbb{N})$.

It is not hard to show that E is spanning with respect to the Lebesgue measure σ. Indeed, take $A \subset \mathbb{T}$ with $\sigma(A) = 1$, and assume that $y \in \ell^{p^*}(\mathbb{N}) = (\ell^p(\mathbb{N}))^*$ satisfies $\langle y, E(\lambda) \rangle = 0$ for every $\lambda \in A$. Here, p^* is the conjugate exponent of p. Then the function

$$\phi(\lambda) := \sum_{n \geq 0} \frac{y_n}{w_1 \cdots w_n} \lambda^n$$

is almost everywhere 0 on $\mathbb{T}$, hence its Fourier coefficients are all 0; since, however, the series defining ϕ is uniformly convergent on $\mathbb{T}$, the Fourier coefficients are also given by

$$\widehat{\phi}(n) = \frac{y_n}{w_1 \cdots w_n},$$

$n \in \mathbb{N}$ (and $\widehat{\phi}(n) = 0$, $n < 0$), so that $y = 0$. This shows that $\operatorname{span}(E(\lambda); \ \lambda \in A)$ is dense in $\ell^p(\mathbb{N})$.

When $p \geq 2$, the space $\ell^p(\mathbb{N})$ has type 2, so there exists an invariant Gaussian measure μ on $\ell^p(\mathbb{N})$ with full support, with respect to which T is a strongly mixing transformation.

When $1 \leq p < 2$, we need some regularity on the eigenvector field E in order to apply Lemma 5.39; unfortunately, it is possible to choose the weights in such a way that E is not α-Hölderian for any $\alpha > 0$!

However, a more careful look at the operator K_E enables to show directly that K_E is γ-radonifying for any $p \in [1, \infty)$. Indeed, using the Fourier basis of $L^2(\mathbb{T})$ we get

$$\sum_{k \in \mathbb{Z}} g_k(\omega) K_E(e^{-ikt}) = \sum_{k \in \mathbb{Z}} g_k(\omega) \int_{\mathbb{T}} E(\lambda) \bar{\lambda}^k d\sigma(\lambda)$$

$$= \sum_{n \geq 0} \frac{g_n(\omega)}{w_1 \cdots w_n} e_n.$$

Now, the series $\sum (g_n/(w_1 \cdots w_n)) e_n$ is convergent in $L^p(\Omega, \ell^p(\mathbb{N}))$, since

$$\left\| \sum_{n \in I} \frac{g_n(\omega)}{w_1 \cdots w_n} e_n \right\|_{L^p(X)}^p = \sum_{n \in I} \frac{\mathbb{E}_\omega(|g_n(\omega)|^p)}{(w_1 \cdots w_n)^p}$$

for any finite set $I \subset \mathbb{N}$ and the Gaussian variables g_n have the same finite moment of order p. Thus K_E is indeed γ-radonifying, and one can apply Proposition 5.36 directly to get the required result for any $p \in [1, \infty)$.

REMARK The interest of this proof is that the Gaussian measure is explicitly exhibited: it is the distribution of the a.s. convergent Gaussian series

$$\sum_{n \geq 1} \frac{g_n}{w_1 \cdots w_n} e_n.$$

In particular, we see that the random vector

$$\sum_{n\geq 1} \frac{g_n(\omega)}{w_1 \cdots w_n} e_n$$

is almost surely hypercyclic for the operator $B_\mathbf{w}$.

5.5.3 Kalisch-type operators

Let T be the operator defined on $L^p([0, 2\pi])$, $1 \leq p < \infty$, by the formula

$$Tf(\theta) = e^{i\theta} f(\theta) - \int_0^\theta i e^{it} f(t) \, dt.$$

For any $\alpha \in [0, 2\pi)$, the function $E(e^{i\alpha}) := \mathbf{1}_{(\alpha, 2\pi)}$ is an eigenvector of T with associated eigenvalue $e^{i\alpha}$. Moreover,

$$\|E(e^{i\beta}) - E(e^{i\alpha})\|_p = (\beta - \alpha)^{1/p}$$

whenever $0 \leq \alpha < \beta < 2\pi$, which shows that the $\mathbb{T}$-eigenvector field E is $1/p$-Hölderian. Finally, E is spanning with respect to the Lebesgue measure, since any function $g \in L^{p^*}([0, 2\pi])$ which is orthogonal to $E(e^{i\alpha})$ for every α in a subset of $[0, 2\pi]$ of full measure satisfies $\int_\alpha^{2\pi} g = 0$ for almost every $\alpha \in [0, 2\pi]$ and hence is almost everywhere zero. Applying Theorem 5.38 when $p \geq 2$ and Theorem 5.40 when $p < 2$, we conclude that there exists an invariant Gaussian measure on $L^p([0, 2\pi])$ with full support with respect to which T is strongly mixing.

5.5.4 Kalisch-type operators on $\mathcal{C}_0([0, 2\pi])$

The above formula,

$$Tf(\theta) = e^{i\theta} f(\theta) - \int_0^\theta i e^{it} f(t) \, dt,$$

also defines a bounded operator on $\mathcal{C}_0([0, 2\pi])$, the space of all continuous functions on $[0, 2\pi]$ vanishing at 0. However, the $\mathbb{T}$-eigenvectors $E(e^{i\alpha}) = \mathbf{1}_{(\alpha, 2\pi)}$ do not belong to $\mathcal{C}_0([0, 2\pi])$. In fact it is easy to see that T has no $\mathbb{T}$-eigenvector at all in $\mathcal{C}_0([0, 2\pi])$. Indeed, if $Tf = e^{i\alpha} f$ then it follows immediately that f is $\mathcal{C}^1$-smooth on $(0, \alpha) \cup (\alpha, 2\pi)$ and that f' vanishes identically on $(0, \alpha) \cup (\alpha, 2\pi)$. Since f belongs to $\mathcal{C}_0([0, 2\pi])$, it must be identically zero.

Thus, when working on $\mathcal{C}_0([0, 2\pi])$ the eigenvectors disappear. Yet the Gaussian measure remains! Indeed, even if $E(e^{i\alpha})$ does not belong to $\mathcal{C}_0([0, 2\pi])$, the formula

$$Kf(\theta) = \int_0^{2\pi} f(t) E(e^{it})(\theta) \, dt = \int_0^\theta f(e^{it}) \, dt$$

defines an element of $\mathcal{C}_0([0, 2\pi])$ for any $f \in L^2(\mathbb{T})$. So, we have an operator $K : L^2(\mathbb{T}) \to \mathcal{C}_0([0, 2\pi])$ and we can consider $R = KK^*$. Then, the following properties (1)–(3) are satisfied.

(1) *R is the covariance operator of a Gaussian measure μ on $C_0([0, 2\pi])$ with full support.*

We first observe that K has dense range, since $\text{Ran}(K)$ contains all C^1 functions vanishing at 0. The fact that the series $\sum_n g_n(\omega)K(e_n)$ converges a.s. in $C_0([0, 2\pi])$ for any orthonormal basis of $L^2(\mathbb{T})$ is well known to specialists; indeed, the operator K and the Gaussian series $\sum g_n K(e_n)$ provide a way of defining the *Wiener measure* on $C_0([0, 2\pi])$ (see e.g. [55, p. 115]). Let us sketch an elementary proof, which is inspired by a classical inequality of R. Salem and A. Zygmund (see [150]). We take $e_n(t) = e^{int}$, $n \in \mathbb{Z}$, so that $K(e_n) = (in)^{-1}(e_n - 1)$ when $n \neq 0$.

Let $p \geq 0$ and $N, M \in \mathbb{N}$ satisfy $2^p + 1 \leq N < M \leq 2^{p+1}$. Using Bernstein's inequality for trigonometric polynomials together with the mean-value theorem, it is not hard to show that

$$\left\| \sum_{|n|=N}^{M} g_n(\omega)K(e_n) \right\|_\infty \leq 5 \sup_{\gamma \in \Gamma} \left| \sum_{|n|=N}^{M} g_n(\omega) \frac{e^{in\gamma} - 1}{in} \right| \tag{5.3}$$

for any $\omega \in \Omega$, where $\Gamma = \{k\pi/2M; \ k = 0, 1, \dots, 4M - 1\}$ is the set of all $4M$th roots of unity. Indeed, let $P_\omega(t)$ be the trigonometric polynomial $\sum_{|n|=N}^{M} c_n(\omega)(e^{int} - 1)$, where $c_n(\omega) = g_n(\omega)/(in)$. For every $t \in [0, 2\pi]$, one can find a $\gamma \in \Gamma$ such that $|t - \gamma| \leq \pi/(4M)$, and then $|P_\omega(t) - P_\omega(\gamma)| \leq (\pi/4)\|P_\omega\|_\infty$, by Bernstein's inequality. Thus, we get $\|P_\omega\|_\infty \leq (\pi/4)\|P_\omega\|_\infty + \sup_{\gamma \in \Gamma} |P_\omega(\gamma)|$ and (5.3) follows.

By the Orlicz–Jensen inequality (see [150] or [173]) it follows that

$$\mathbb{E} \left\| \sum_{|n|=N}^{M} g_n K e_n \right\|_\infty \leq C\sqrt{\log(M)} \sup_{\gamma \in \Gamma} \left\| \sum_{|n|=N}^{M} g_n \frac{e^{in\gamma} - 1}{in} \right\|_{L^2(\Omega)}$$

$$\leq C\sqrt{\log(M)} \left(\sum_{n=N}^{M} \frac{1}{n^2} \right)^{1/2}$$

$$\leq C \frac{\sqrt{p}}{2^{p/2}}.$$

Thus, for any $p_0 \geq 0$ and $2^{p_0} \leq N < M$, the triangle inequality yields

$$\mathbb{E} \left\| \sum_{|n|=N}^{M} g_n K(e_n) \right\|_\infty \leq C \sum_{p \geq p_0} \frac{\sqrt{p}}{2^p}.$$

This shows that the series $\sum g_n K(e_n)$ is convergent in $L^1(\Omega, C_0([0, 2\pi]))$, hence almost surely.

(2) *The measure μ is T-invariant.*

Indeed, for any $f \in L^2(\mathbb{T})$, integration by parts gives

$$TKf(\theta) = e^{i\theta} \int_0^\theta f(t)\, dt - \int_0^\theta ie^{it} \int_0^t f(u)\, dudt$$

$$= \int_0^\theta e^{it} f(t)\, dt$$

$$= KMf(\theta),$$

where M is the unitary operator of multiplication by the variable λ on $L^2(\mathbb{T})$. In particular, we get $TRT^* = TKK^*T^* = KMM^*K^* = R$.

(3) *The operator T is strongly mixing with respect to μ.*

Indeed, if $x^*, y^* \in (\mathcal{C}_0([0, 2\pi]))^*$ then

$$\langle RT^{*n}(x^*), y^* \rangle = \langle K^*(y^*), K^*T^{*n}(x^*) \rangle_{L^2}$$

$$= \langle K^*(y^*), M^{*n}K^*(x^*) \rangle_{L^2}$$

$$= \widehat{f}(n),$$

where f is the L^1-function $\lambda \mapsto \overline{K^*y^*(\lambda)}K^*x^*(\lambda)$.

In particular, this example shows that there exist Banach space operators which are ergodic with respect to some Gaussian measure with full support and yet do not have any unimodular eigenvalue. Theorem 5.46 below will show that this can never happen if the underlying space has cotype 2.

5.6 Further results

5.6.1 Hypercyclicity and unimodular eigenvalues

The results in Section 5.4 show that if an operator $T \in \mathfrak{L}(X)$ admits a perfectly spanning countable family of $\mathbb{T}$-eigenvector fields then, under some additional assumption on the space X or on the regularity of the eigenvector fields, T turns out to be ergodic with respect to some Gaussian measure with full support and hence hypercyclic. We do not know whether the additional assumptions are really necessary to get an ergodic measure. However, the hypercyclic part remains true in full generality, as shown by the next theorem.

THEOREM 5.41   *Let $T \in \mathfrak{L}(X)$, and assume that T has a σ-spanning set of $\mathbb{T}$-eigenvectors, for some probability measure σ.*

(a) *If the measure σ is continuous then T satisfies the Hypercyclicity Criterion.*

(b) *If σ is a Rajchman measure then T satisfies Kitai's Criterion and hence is topologically mixing.*

PROOF   By Lemma 5.29, T admits a σ-spanning countable family of σ-measurable $\mathbb{T}$-eigenvector fields $(E_i)_{i \in I}$. Set $\mathcal{H} := \oplus_{i \in I} L^2(\mathbb{T}, \sigma)$, and let $K : \mathcal{H} \to X$ be defined as in Lemma 5.35,

$$K(\oplus_i f_i) = \sum_i \alpha_i K_{E_i}(f_i).$$

Then K is a compact operator since each K_{E_i} is compact. Moreover, K has dense range because (E_i) is σ-spanning (see part (3) of Lemma 5.35) and

$$TK = KM,$$

where $M = \oplus_i M_i : \mathcal{H} \to \mathcal{H}$ is the direct sum of copies of the multiplication-by-the-variable operator acting on $L^2(T, \sigma)$. That is, $M_i f_i(\lambda) = \lambda f_i(\lambda)$.

If $f = \oplus_i f_i$ and $g = \oplus_i g_i$ are any two elements of $\mathcal{H}$ then

$$\langle M^n(f), g \rangle_{\mathcal{H}} = \widehat{\nu_{f,g}}(n)$$

for all $n \in \mathbb{Z}$, where $\nu_{f,g}$ is the complex measure on $\mathbb{T}$ defined by $d\nu = (\sum_i \overline{f_i} g_i)\, d\sigma$.

Assume first that σ is a Rajchman measure. Then all measures $\nu_{f,g}$ are Rajchman, so that $M^{\pm n}(f) \xrightarrow{n \to \infty} 0$ *weakly* for any $f \in \mathcal{H}$. Since the operator K is compact, it follows that

$$\|KM^{\pm n}(f)\| \xrightarrow{n \to \infty} 0$$

for all $f \in \mathcal{H}$.

It is now easy to show that T satisfies Kitai's Criterion. Indeed, let $\mathcal{D}_0$ be a countable dense subset of $\mathcal{H}$. Since K has dense range, the set $\mathcal{D} := K(\mathcal{D}_0)$ is dense in X. For each $u \in \mathcal{D}$, we *choose* some $f \in \mathcal{D}_0$ such that $u = Kf$. Then, we define maps $S_n : \mathcal{D} \to X$ by

$$S_n(Kf) = KM^{-n}(f).$$

Since $T^n K = KM^n$, we have $T^n S_n = I$ on $\mathcal{D}$ for all $n \in \mathbb{N}$ and, by what we already observed, we also have $\lim_{n \to \infty} T^n(u) = 0 = \lim_{n \to \infty} S_n(u)$ for each $u \in \mathcal{D}$. This proves (b).

Now, assume only that the measure σ is continuous. Then each measure $\nu_{f,g}$ is continuous. Hence, by Wiener's theorem and Lemma 5.25, there exists a set of integers $\mathbf{N}_{f,g}^+$ with density 1 such that $\langle M^n(f), g \rangle \to 0$ as $n \to \infty, n \in \mathbf{N}_{f,g}^+$. Similarly, we can find $\mathbf{N}_{f,g}^- \subset \mathbb{N}$ with density 1 such that $\langle M^{-n}(f), g \rangle \to 0$ along $\mathbf{N}_{f,g}^-$.

Let us fix a countable dense set $\mathcal{D}_0 \subset \mathcal{H}$. Since the intersection of two subsets of $\mathbb{N}$ with density 1 still has density 1, a diagonal argument provides an increasing sequence $(n_k) \subset \mathbb{N}$ such that $\lim_{k \to \infty} \langle M^{\pm n_k}(f), g \rangle = 0$ for all $f, g \in \mathcal{D}_0$. Since M is unitary, it follows that $M^{\pm n_k}(f) \to 0$ weakly for each $f \in \mathcal{D}_0$; since the operator K is compact, this shows that $\|KM^{\pm n_k}(f)\| \to 0$ for any $f \in \mathcal{D}_0$. Then one may prove exactly as above that T satisfies the Hypercyclicity Criterion, by putting $\mathcal{D} := K(\mathcal{D}_0)$ and using the maps $S_{n_k} : \mathcal{D} \to X$ defined by $S_{n_k}(Kf) = KM^{-n_k}(f)$. $\square$

REMARK The above proof consists essentially in observing that $\langle M^{\pm n}(f), g \rangle \to 0$ for all $f, g \in \mathcal{H}$, at least in the Cesàro sense. Since $K^* T^{*n} = M^{-n} K^*$, this implies that $\langle RT^{*n}(x^*), y^* \rangle \to 0$ for all $x^*, y^* \in X^*$, where as usual $R = KK^*$. When R is a covariance operator, the latter property guarantees ergodicity. Thus, even if we have not been able to find an ergodic measure, a kind of ergodicity remains!

5.6.2 Semigroups of operators

Using the same methods, one can produce invariant Gaussian measures for C_0-semi-groups of operators. Recall that a semigroup of measurable transformation $(T_t)_{t>0}$ defined on some probability space $(X, \mathcal{B}, \mu)$ is said to be **measure-preserving** if $\mu(T_t^{-1}(B)) = \mu(B)$ for any $B \in \mathcal{B}$ and all $t > 0$. The semigroup is said to be **strongly mixing** if it is measure-preserving and if

$$\lim_{t \to \infty} \int_X f(T_t z) g(z) \, d\mu(z) = \int_X f \, d\mu \int_X g \, d\mu$$

for any $f, g \in L^2(X, \mu)$.

Suppose now that μ is a Gaussian measure on the separable Banach space X and that $(T_t)_{t>0}$ is a semigroup of operators on X. Then analogues of Proposition 5.22 and Theorem 5.24 hold, with exactly the same proofs: the measure μ is (T_t)-invariant iff $T_t R T_t^* = R$ for any $t > 0$ and the semigroup (T_t) is strongly mixing with respect to μ iff $\langle R T_t^*(x^*), y^* \rangle \xrightarrow{t \to \infty} 0$ for any $x^*, y^* \in X^*$.

To apply these results, the unimodular eigenvalues of a single operator T have to be replaced by the purely imaginary eigenvalues of the infinitesimal generator of the semigroup. Recall that the **infinitesimal generator** of a C_0-semigroup (T_t) is the (possibly) unbounded operator A densely defined by the formula

$$A(x) = \lim_{t \to 0} \frac{T_t(x) - x}{t} \, .$$

One can show that if x_λ is an eigenvector of A corresponding to the eigenvalue λ then $T_t(x_\lambda) = e^{\lambda t} x_\lambda$ for any $t > 0$ (see e.g. [192]). Thus we get the following result.

THEOREM 5.42   *Let $(T_t)_{t>0}$ be a C_0-semigroup in $\mathfrak{L}(X)$ with infinitesimal gen-erator A. Assume that there exists a countable family $(F_j)_{j \in J}$ of locally bounded measurable maps $F_j : \mathbb{R} \to X$ such that $AF_j(t) = it F_j(t)$ for each j and all $t \in \mathbb{R}$ and $\overline{\mathrm{span}}(F_j(t); t \notin B, j \in J) = X$ for any measure-0 set $B \subset \mathbb{R}$. Then there exists a Gaussian measure on X with full support with respect to which (T_t) is a strongly mixing measure-preserving semigroup provided that either X has type 2, or X has type $p \in [1, 2]$ and each F_j is α_j-Hölderian for some $\alpha_j > 1/p - 1/2$.*

PROOF   For $k \geq 0$ and $\theta \in [0, 2\pi)$, set $E_{j,k}(e^{i\theta}) := F_j(\theta + 2k\pi)$ and define $K_{E_{j,k}} : L^2(\mathbb{T}) \to X$ as usual. Define $K : \oplus_{j,k} L^2(\mathbb{T}) \to X$ by $K(\oplus_{j,k} f_{j,k}) = \sum_{j,k} \alpha_{j,k} K_{E_{j,k}}$, where $\alpha_{j,k} > 0$ and $\sum_{j,k} \alpha_{j,k}^2 \|E_{j,k}\|_2^2 < \infty$. Finally, set $R := KK^*$. Then one shows exactly as in the proof of Proposition 5.36 that the opera-tor R is one-to-one, that $TRT^* = R$ and that $\lim_{n \to \infty} \langle R T^{*n}(x^*), y^* \rangle = 0$ for any $x^*, y^* \in X^*$. The last two properties follow from the identity $T_t(E_{j,k}(e^{i\theta})) = e^{it\theta} E_{j,k}(e^{i\theta})$. So the first point is proved as in Theorem 5.38 and the second as in Theorem 5.40 (see the discussions before the statements of these theorems). There is only one small difference: the maps $E_{j,k}$ have no reason for being continuous at 1. However, a look at the proof of Theorem 5.40 shows that this is unimportant. $\square$

5.6.3 Nuclear Fréchet spaces

Gaussian measures are not restricted to live on Banach spaces. It makes perfect sense to define them on a separable Fréchet space and the results proved in Section 5.1 remain true in that setting, with essentially the same proofs. Moreover, there is an important class of Fréchet spaces where Gaussian covariance operators are completely characterized: the so-called **nuclear** Fréchet spaces.

We shall say nothing about these spaces except that the space of entire functions $H(\mathbb{C})$ is nuclear and that Gaussian covariance operators on nuclear Fréchet spaces are characterized by a theorem of R. A. Minlos (see [55, p.155] or [77, Theorem VI.4.2]). Using this characterization, one can obtain the following theorem.

THEOREM 5.43 *Let X be a nuclear Fréchet space, and let $T \in \mathfrak{L}(X)$. Assume that T admits a countable family of measurable $\mathbb{T}$-eigenvector fields (E_i) which is spanning with respect to the Lebesgue measure on $\mathbb{T}$ and uniformly bounded, i.e. there exists some bounded set $B \subset X$ such that $\mathrm{Ran}(E_i) \subset B$ for all i. Then there exists a Gaussian measure on X with full support with respect to which T is a strongly mixing measure-preserving transformation.*

We do not give the proof of this theorem, which can be found in K.-G. Grosse-Erdmann's paper [135]. Nevertheless, we will illustrate it with the following

EXAMPLE Let T be a linear operator on $H(\mathbb{C})$. Assume that T commutes with the translation operators and is not a scalar multiple of the identity. Then there exists a Gaussian measure μ on $H(\mathbb{C})$ with respect to which T is a strongly mixing measure-preserving transformation.

PROOF By Lemma 1.43, there exists a non-constant entire function of exponential type $\phi(z) = \sum_{k=0}^{\infty} a_k z^k$ such that $T = \phi(D) = \sum_{k \geq 0} a_k D^k$. Then, for any $\lambda \in \mathbb{C}$, the function $e_\lambda(z) = e^{\lambda z}$ is an eigenvector of T associated with the eigenvalue $\phi(\lambda)$. Fix some $\lambda_0 \in \mathbb{C}$ such that $\phi(\lambda_0) \in \mathbb{T}$ and $\phi'(\lambda_0) \neq 0$. There exists an open arc $J \subset \mathbb{T}$ containing $\phi(\lambda_0)$ and a bounded curve $\Gamma \subset \mathbb{C}$ containing λ_0 such that ϕ maps homeomorphically Γ onto J. Let $E : \mathbb{T} \to H(\mathbb{C})$ be defined by $E(\lambda) = e_{\phi^{-1}(\lambda)}$ if $\lambda \in J$ and $E(\lambda) = 0$ otherwise. It is clear that the map E is bounded and that $TE(\lambda) = \lambda E(\lambda)$ for any $\lambda \in \mathbb{T}$. Moreover, if $A \subset \mathbb{T}$ has full measure then A is dense in $\mathbb{T}$, hence $\phi^{-1}(A)$ has an accumulation point in $\mathbb{C}$ because ϕ is an open map. By Lemma 1.44, it follows that $\overline{\mathrm{span}}(E(\lambda); \lambda \in A) = H(\mathbb{C})$. This concludes the proof. □

5.6.4 When unimodular eigenvalues are necessary

In this subsection, we return to the case of a separable Banach space X. The results in Section 5.4 say that, for an operator $T \in \mathfrak{L}(X)$, a large supply of unimodular eigenvectors ensures the existence of an invariant ergodic measure with full support. It is natural to ask whether the converse holds. The example in subsection 5.5.4 shows that it is not true in full generality. However, when the underlying Banach space X has *cotype 2*, it does turn out to be true.

To prove this, we will need the following two lemmas. The first has already been proved (see Proposition 5.19).

LEMMA 5.44   *Assume that the Banach space X has cotype 2. If $R : X^* \to X$ is a Gaussian covariance operator then any square root of R is absolutely 2-summing.*

A proof of the lemma below can be found in [77, Theorem VI.5.3, Corollary 2].

LEMMA 5.45   *Let $(\Omega, \mathcal{F}, \nu)$ be a probability space, and let $L : L^2(\Omega, \nu) \to X$ be an absolutely 2-summing operator. Then there exists a measurable map $E : \Omega \to X$ in $L^2(\Omega, \nu, X)$ such that $L^*(x^*) = \langle x^*, E(\,\cdot\,)\rangle$ for every $x^* \in X^*$.*

We can now prove the following theorem.

THEOREM 5.46   *Assume that the Banach space X has cotype 2, and let $T \in \mathfrak{L}(X)$. Assume that T admits a Gaussian invariant measure μ with full support. Then the $\mathbb{T}$-eigenvectors of T span a dense subspace of X. If T is weakly mixing with respect to μ then the $\mathbb{T}$-eigenvectors of T are perfectly spanning.*

PROOF   Let R be the covariance operator of μ. We first show that there exist a square root $K : \mathcal{H} \to X$ and a unitary operator $U : \mathcal{H} \to \mathcal{H}$ such that $TK = KU$. Indeed, let $K_0 : \mathcal{H}_0 \to X$ be a one-to-one square root of R. Since μ is T-invariant we have $(TK_0)(TK_0)^* = K_0 K_0^*$, so that TK_0 is also a square root of R. By the essential uniqueness of the square root, one can find an isometry $V_0 : \mathcal{H}_0 \to \mathcal{H}_0$ such that $K_0^* T^* = V_0 K_0^*$. Now let $V : \mathcal{H} \to \mathcal{H}$ be a unitary operator defined on some larger Hilbert space $\mathcal{H} \supset \mathcal{H}_0$, such that $V \equiv V_0$ on $\mathcal{H}_0$ or, more accurately, $V \pi^* = \pi^* V_0$ where $\pi : \mathcal{H} \to \mathcal{H}_0$ is the orthogonal projection from $\mathcal{H}$ onto $\mathcal{H}_0$. It is straightforward to check that $K := K_0 \pi$ is a square root of R and that $K^* T^* = V K^*$. Thus one may take $U := V^*$.

Now, we apply the spectral theorem to the unitary operator U. This gives a sequence of Hilbert spaces $(\mathcal{H}_i)_{i \geq 1}$, where each $\mathcal{H}_i$ is a space $L^2(\mathbb{T}, \sigma_i)$ for some probability measure σ_i on $\mathbb{T}$, such that U is unitarily equivalent to the direct sum of the multiplication-by-the-variable operators on each of the spaces $\mathcal{H}_i$. We denote this operator by $M : \oplus_{i \geq 1} \mathcal{H}_i \to \oplus_{i \geq 1} \mathcal{H}_i$, and we choose a unitary operator $J : \oplus_{i \geq 1} \mathcal{H}_i \to \mathcal{H}$ such that $U = JMJ^*$.

Let us denote by $L : \oplus_i \mathcal{H}_i \to X$ the operator K viewed as acting on $\oplus_i \mathcal{H}_i$, that is, $L = KJ$. Then L is a square root of R and, since $TK = KU$, we have $TL = LM$. We also note that L has dense range, since the measure μ has full support (see Proposition 5.18).

Since X has cotype 2, the operator L is absolutely 2-summing by Lemma 5.44. Then, denoting by L_i the restriction of L to $\mathcal{H}_i$, each operator L_i is 2-summing as well. By Lemma 5.45, we get a σ_i-measurable map $E_i : \mathbb{T} \to X$ which is in $L^2(\mathbb{T}, X, \sigma_i)$ such that

$$L_i^*(x^*) = \overline{\langle x^*, E_i(\,\cdot\,)\rangle} \tag{5.4}$$

for every $x^* \in X^*$. Equivalently,

$$L_i(f) = \int_{\mathbb{T}} f(\lambda) E_i(\lambda) \, d\sigma_i(\lambda) \qquad (5.5)$$

for every $f \in \mathcal{H}_i$.

From (5.4) and the equation $TL = LM$ it is easily inferred that, for each fixed $x^* \in X^*$, one has $\langle x^*, TE_i(\lambda) \rangle = \langle x^*, \lambda E_i(\lambda) \rangle$ σ_i-almost everywhere. Since X is separable (so that one can find a countable separating set in X^*), it follows that $TE_i(\lambda) = \lambda E_i(\lambda)$ σ_i-a.e. Therefore, each E_i becomes a $\mathbb{T}$-eigenvector field after having been modified to take the value 0 on some set of σ_i-measure 0.

For each $i \geq 1$, set $\Lambda_i := \{\lambda \in \mathbb{T}; \, \sigma_i(\{\lambda\}) > 0 \text{ and } E_i(\lambda) = 0\}$. If $L_i \neq 0$ then $\sigma_i(\mathbb{T} \setminus \Lambda_i) > 0$ and we may denote by $\widetilde{\sigma}_i$ the normalized restriction of σ_i to $\mathbb{T} \setminus \Lambda_i$; otherwise, we put $\widetilde{\sigma}_i = 0$. Finally, let $\sigma := \sum_1^\infty 2^{-i} \widetilde{\sigma}_i$. We show that the family $(E_i)_{i \geq 1}$ is σ-spanning and that σ is a continuous measure if T is weakly mixing with respect to μ.

Let $A \subset \mathbb{T}$ be a σ-measurable set with $\sigma(A) = 1$. Looking at the definitions of σ and of the sets Λ_i, we see that if $x^* \in X^*$ satisfies $\langle x^*, E_i(\lambda) \rangle = 0$ for each i and every $\lambda \in A$ then $\langle x^*, E_i(\lambda) \rangle = 0$ almost everywhere with respect to σ_i, for each $i \geq 1$. By (5.5), it follows that $x^*(\text{Ran}(L)) = 0$, hence $x^* = 0$ since the operator L has dense range. This shows that the family (E_i) is σ-spanning.

Now, assume that the measure σ is not continuous. By the definition of σ, one can find $\lambda_0 \in \mathbb{T}$ and some $i_0 \geq 1$ such that $\sigma_{i_0}(\{\lambda_0\}) > 0$ and $E_{i_0}(\lambda_0) \neq 0$. Choose $x_0^* \in X^*$ such that $\langle x_0^*, E_{i_0}(\lambda_0) \rangle \neq 0$. Then

$$\begin{aligned}
\langle RT^{*n}(x_0^*), x_0^* \rangle &= \langle K^*(x_0^*), U^{*n} K^*(x_0^*) \rangle_{\mathcal{H}} \\
&= \langle L^*(x_0^*), M^{*n} L^*(x_0^*) \rangle_{\oplus_i \mathcal{H}_i} \\
&= \sum_{i \geq 1} \lambda^{-n} \int_{\mathbb{T}} |\langle x_0^*, E_i(\lambda) \rangle|^2 \, d\sigma_i(\lambda) \\
&= \widehat{\nu}(n)
\end{aligned}$$

for all $n \in \mathbb{N}$, where ν is the positive measure defined by $d\nu = \sum_i |\langle x_0^*, E_i(\cdot) \rangle|^2 \, d\sigma_i$. The measure ν is not continuous since $\nu(\{\lambda_0\}) > 0$. By Wiener's theorem, the Cesàro means $N^{-1} \sum_0^{N-1} |\widehat{\nu}(n)|$ do not tend to 0 as $N \to \infty$, and hence T is not weakly mixing with respect to μ. $\square$

5.7 Comments and exercises

The problem of the existence of an invariant Gaussian measure for a linear operator was investigated by E. Flytzanis in [111], where Proposition 5.27 is stated in a Hilbert space setting. One should also mention the paper [67] by P. Brunovsky and J. Komornik, where it is proved that the shift semigroup on $\mathcal{C}([0, \infty))$ is ergodic with respect to some non-degenerate Gaussian measure. The connection between ergodic theory and linear dynamics was made by the first author and S. Grivaux in [26].

Ergodic theory and linear dynamics

Theorem 5.38 is proved in [26] in a Hilbert space setting. The proof given in [26] uses the theory of *Fock spaces*. The more elementary proof given in this chapter was pointed out by K.-G. Grosse-Erdmann in [135], following the work of R. Rudnicki ([211]).

The extension to a Banach space setting was carried out in [27]. However, one can find there only weaker versions of Theorem 5.40, in which either $\alpha > 1/2$ or some assumption is made on the *Fourier type* of the Banach space X. The missing tool was Lemma 5.39, which appeared in [152] after the publication of [27]. The proof of Lemma 5.39 given in [152] uses the language of Besov spaces (of course, this is tantamount to using wavelet decompositions). The extension to nuclear Fréchet spaces comes from [135].

Lemma 5.39 is optimal with respect to α and p (see [152]). However, we do not know whether Theorem 5.40 is likewise optimal: as already mentioned, we do not have any example of an operator with a perfectly spanning set of $\mathbb{T}$-eigenvectors for which one cannot find any Gaussian measure turning it into a weakly mixing transformation. A plausible conjecture would be that Theorem 5.38 characterizes spaces of type 2, whereas Theorem 5.46 characterizes spaces of cotype 2.

One can find other examples of operators admitting an invariant Gaussian measure in [26] or [25]. In particular, we point out that if ϕ is a hyperbolic or a parabolic automorphism of the disk then the composition operator C_ϕ acting on $H^p(\mathbb{D})$ is strongly mixing with respect to some Gaussian measure with full support. The examples in subsections 5.5.3 and 5.5.4 were inspired by a paper of G. Kalisch ([151]), in which operators with an arbitrary F_σ point spectrum are constructed.

EXERCISE 5.1 Recurrence theorems

1. *Poincaré* Let $(X, \mathcal{B}, \mu)$ be a probability space, and let $T : X \to X$ be a measure-preserving transformation.

   (a) Let $A \in \mathcal{B}$, and let $E := A \setminus \bigcup_{n \geq 1} T^{-n}(A)$. Show that $E, T^{-1}(E), T^{-2}(E), \dots$ are pairwise disjoint. Deduce that $\mu(T^{-k}(E)) = 0$ for all $k \in \mathbb{N}$.

   (b) Let $A \in \mathcal{B}$, and assume that $\mu(A) > 0$. Show that almost every point $x \in A$ is T-recurrent with respect to A, i.e. $T^n(x) \in A$ for infinitely many $n \in \mathbb{N}$.

2. *Birkhoff* Let X be a compact metric space, and let $T : X \to X$ be a continuous map.

   (a) A non-empty closed set $C \subset X$ is said to be T-**minimal** if $T(C) \subset C$ and no proper closed non-empty subset of C is T-invariant. Show that any T-invariant closed set $E \neq \varnothing$ contains a T-minimal set.

   (b) Show that a non-empty closed set $C \subset X$ is T-minimal iff $\overline{O(z, T)} = C$ for any $z \in C$.

   (c) Show that T admits a **recurrent point**, i.e. one can find $x \in X$ such that $T^{n_k}(x) \to x$ for some sequence $(n_k) \subset \mathbb{N}$ tending to ∞. (*Hint:* Take any point in some T-minimal set.)

EXERCISE 5.2 Let X be an abstract set, and let $T : X \to X$.

1. Let $\mathcal{F}$ be a family of subsets of X which is closed under supersets, i.e. $(A \in \mathcal{F}$ and $A \subset A')$ implies $(A' \in \mathcal{F})$. Show that the following are equivalent:

   (i) for any $A, B \in \mathcal{F}$, there exists $n \in \mathbb{N}$ such that $T^n(A) \cap B \neq \varnothing$;
   (ii) if $A \subset X$ satisfies $T(A) \subset A$ then either $A \notin \mathcal{F}$ or $X \setminus A \notin \mathcal{F}$.

   (*Hint:* To prove that (i) $\Longrightarrow$ (ii), consider $B := X \setminus A$. For the converse, consider $\widetilde{A} := \bigcup_{n \geq 0} T^n(A)$.)

2. How do the above statements read when "$A \in \mathcal{F}$" means "A has non-empty interior" (in a topological space X) or "A contains a measurable set with positive measure" (in a measure space (X, μ))?

EXERCISE 5.3 Extreme points and ergodic measures

Let X be a compact metric space, and let $T : X \to X$ be a continuous map. We denote by $\mathbf{I}_T$ the set of all T-invariant Borel probability measures on X.

1. Let $\mathbf{P}(X)$ be the space of all Borel probability measures on X equipped with the w^*-topology. Show that $\mathbf{I}_T$ is a compact convex subset of $\mathbf{P}(X)$.
2. Show that $\mathbf{I}_T$ is non-empty. (*Hint:* Apply your favourite fixed point theorem. Alternatively, pick $a \in X$ and consider any w^*-cluster point of the sequence (μ_n), where $\mu_n = n^{-1} \sum_{i<n} \delta_{T^i(a)}$.)
3. Let $\mu \in \mathbf{I}_T$. Show that μ is an ergodic measure for T if and only if μ is an extreme point of $\mathbf{I}_T$. (*Hint:* For the "if" part observe that if $A \subset X$ is a Borel set such that $T^{-1}(A) = A$ up to a set of μ-measure 0 then the measure $\mu_{|A}$ is T-invariant. For the converse, use Birkhoff's theorem to show that if μ is T-ergodic then the only measure in $\mathbf{I}_T$ that is absolutely continuous with respect to μ is μ itself.)
4. Show that T admits at least one invariant ergodic measure.
5. Let a be an irrational number lying in $\mathbb{T}$, i.e. $a \in \mathbb{T} \setminus e^{2i\pi\mathbb{Q}}$, and let $\tau_a : \mathbb{T} \to \mathbb{T}$ be the associated translation, $\tau_a(\zeta) = a\zeta$. Show that τ_a is ergodic with respect to the Lebesgue measure σ. (*Hint:* Find all the τ_a-invariant probability measures on $\mathbb{T}$.)

EXERCISE 5.4 **Cylinder sets** (see e.g. [77])
Let X be a separable Fréchet space. We recall that a cylinder set in X is any Borel set $C \subset X$ of the form $C = \pi^{-1}(\widetilde{C})$, where $\pi : X \to F$ is a continuous linear map onto some finite-dimensional space F and $\widetilde{C}$ is a Borel subset of F.

1. Let p be a continuous seminorm on X. Show that, for any $a \in X$ and each $r > 0$, the closed ball $\overline{B}_p(a,r) := \{x \in X; \; p(x-a) \le r\}$ is a countable intersection of closed half-spaces. (*Hint:* For "countable", use the Lindelöf property of $X \setminus \overline{B}_p(a,r)$.)
2. Show that the Borel σ-algebra of X is generated by the cylinder sets.
3. Show that any Borel probability measure on X is determined by its finite-dimensional marginals, i.e., if μ and ν are Borel probability measures on X such that $\mu \circ A^{-1} = \nu \circ A^{-1}$ for every finite-rank operator A defined on X then $\mu = \nu$.
4. Show that the Fourier transform of a measure on X determines the measure.

EXERCISE 5.5 **Fernique's integrability theorem** (see e.g. [55])
Let E be a separable Fréchet space, and let ξ be a Gaussian random variable with values in E. Also, let $q : E \to [0,\infty]$ be a Borel seminorm. Assume that $q(\xi(\omega)) < \infty$ almost surely. In what follows, we put $M(\omega) := q(\xi(\omega))$.

1. Let ξ_1 and ξ_2 be two independent copies of ξ.
   (a) Show that the pairs (ξ_1, ξ_2) and $(\xi_1', \xi_2') := \left(\frac{\xi_1+\xi_2}{\sqrt{2}}, \frac{\xi_2-\xi_1}{\sqrt{2}}\right)$ have the same distribution and that
   $$\{q(\xi_1) \le s, q(\xi_2) > t\} \subset \left\{q(\xi_1') > \frac{t-s}{\sqrt{2}}, q(\xi_2') > \frac{t-s}{\sqrt{2}}\right\}$$
   whenever $0 < s < t$.
   (b) Deduce that if $0 < s < t$ then
   $$\mathbb{P}(M \le s)\,\mathbb{P}(M > t) \le \mathbb{P}\left(M > \frac{t-s}{\sqrt{2}}\right)^2.$$
2. Let s_0 be any positive number such that $\mathbb{P}(M > s_0) \le 1/4$.
   (a) For any $t > 0$, set
   $$r(t) := \frac{\mathbb{P}(M > t)}{\mathbb{P}(M \le s_0)}.$$
   Let (t_n) be the sequence of positive numbers defined by $t_0 = s_0$ and $t_{n+1} = s_0 + \sqrt{2}\,t_n$. Show that $t_n \le 4(\sqrt{2})^n s_0$ and $r(t_n) \le (1/3)^{2^n}$.
   (b) Show that one can find absolute constants $a, b > 0$ such that
   $$\mathbb{P}(M > t) \le ae^{-b(t/s_0)^2}$$

for all $t > 0$.

(c) *Fernique's integrability theorem* Deduce that one can find some absolute constant C such that

$$\mathbb{E}\left(e^{(M/Cs_0)^2}\right) \leq C < \infty.$$

In particular, $M \in L^p$ for all $p \in [1, \infty)$. (*Hint:* Use the integration by parts formula $\mathbb{E}(f(\xi)) = f(0) + \int_0^\infty f'(t)\mathbb{P}(\xi > t)\, dt$.)

3. Show that $\|M\|_p \leq C_p \|M\|_1$ for all $p \in [1, \infty[$, where C_p is a finite constant depending only on p. (*Hint:* One may take $s_0 = 4\|M\|_1$.)

4. Prove the Gaussian version of the Khinchine–Kahane inequalities: for any finite sequence of vectors (x_n) in a Banach space X,

$$\left\|\sum_n g_n x_n\right\|_{L^p(\Omega, X)} \leq C_p \left\|\sum_n g_n x_n\right\|_{L^1(\Omega, X)}.$$

5. Let X be a separable Banach space, and let $\sum g_n x_n$ be a Gaussian series with terms in X. Show that if $\sum g_n x_n$ is almost surely convergent then it is convergent in $L^p(\Omega, X)$ for every $p \in [1, \infty)$. (*Hint:* Consider $E := X^{\mathbb{N}}$, the E-valued random variable $\xi := (g_n x_n)_{n \in \mathbb{N}}$ and the semi-norm $q((z_n)) := \sup_{N \in \mathbb{N}} \|\sum_0^N z_n\|$. Use Lebesgue's theorem.)

EXERCISE 5.6 **Series of independent random variables** (see e.g. [173])

1. Let $\xi_1, \ldots, \xi_N$ be independent symmetric random variables with values in some Banach space X. For each $n \in [1, N]$, put $S_n := \xi_1 + \cdots + \xi_n$. Finally, let $t > 0$.

(a) Let $T : \Omega \to [0, \infty]$ be defined by

$$T(\omega) = \inf\{n \in [1, N];\ \|S_n(\omega)\| > t\},$$

with the convention that $\inf \varnothing = \infty$. Show that, for all $n \in [1, N]$,

$$\{T = n\} \subset \{\|S_N\| > t\} \cup \{\|2S_n - S_N\| > t\}$$

and

$$\mathbb{P}(\|S_N\| > t, T = n) = \mathbb{P}(\|2S_n - S_N\| > t, T = n).$$

(*Hint:* For the second statement, put $\xi_i' = \xi_i$ if $i \leq n$ and $\xi_i' = -\xi_i$ if $i > n$. Then $(S_1', \ldots, S_N')$ and $(S_1, \ldots, S_N)$ have the same distribution.) Deduce that

$$\mathbb{P}(T = n) \leq 2\,\mathbb{P}(\|S_N\| > t, T = n).$$

(b) Show that

$$\mathbb{P}\left(\max_{1 \leq n \leq N} \|S_n\| > t\right) \leq 2\,\mathbb{P}(\|S_N\| > t).$$

2. Let X be a separable Banach space, and let (ξ_n) be a sequence of independent symmetric random variables with values in X. Show that the series $\sum \xi_n$ is a.s. convergent if and only if it is convergent in probability.

3. Using Exercise 5.5, conclude that a Gaussian series $\sum g_n x_n$ with terms in a Banach space X converges in probability iff it is a.s. convergent and iff it is convergent in $L^p(\Omega, X)$ for some $p \in [1, \infty)$ and iff it is convergent in L^p for all $p \in [1, \infty)$.

EXERCISE 5.7 **Characterizations of weak mixing** (see e.g. [235])

Let $(X, \mathcal{B}, \mu)$ be a probability space, and let $T : (X, \mathcal{B}, \mu) \to (X, \mathcal{B}, \mu)$ be an invertible measure-preserving transformation. Then T induces a unitary operator $U_T : L^2(X, \mu) \to L^2(X, \mu)$, namely $U_T(f) = f \circ T$. The aim of this exercise is to prove the equivalence of the following three properties:

(i) T is weakly mixing;

(ii) $T \times T$ is ergodic on $(X \times X, \mu \otimes \mu)$;
(iii) the only eigenvectors of U_T are constant functions.

1. Show that (i) implies (ii). (*Hint:* Compute $(\mu \otimes \mu) \left((T \times T)^{-n} (A \times C) \cap (B \times D) \right)$ for $n \in \mathbb{N}$ and $A, B, C, D \in \mathcal{B}$.)
2. Show that (ii) implies (iii). (*Hint:* If $f \in L^2(X, \mu)$ is an eigenvector of U_T, compute $g \circ (T \times T)$ where $g(x, y) = f(x)\overline{f(y)}$.)
3. Let E be the spectral measure of U_T and, for any $f, g \in L^2(X, \mu)$, let $\sigma_{f,g}$ be the complex measure on $\mathbb{T}$ defined by $\sigma_{f,g}(A) = \langle E(A)f, g \rangle$. Finally, let V be the closed subspace of $L^2(X, \mu)$ generated by the eigenvectors of U_T.

   (a) Show that if $f \in L^2(X, \mu)$ then $U_T E(\{\lambda\})f = \lambda E(\{\lambda\})f$ for any $\lambda \in \mathbb{T}$.
   (b) Deduce that if $f \in V^\perp$ and $g \in L^2(X, \mu)$ is arbitrary then $\sigma_{f,g}$ is a continuous measure.
   (c) Show that if $f \in V^\perp$ then

   $$\lim_{N \to \infty} \frac{1}{N} \sum_{n=0}^{N-1} |\langle U_T^n(f), g \rangle| = 0$$

   for any $g \in L^2(X, \mu)$.
   (d) Show that (iii) implies (i). (*Hint:* If (iii) holds then $f - \langle 1, f \rangle 1 \in V^\perp$ for any $f \in L^2(X, \mu)$.)

EXERCISE 5.8 Let (X, μ) be a measure space, and let (f_n) be a sequence in $L^1(\mu)$. Assume that (f_n) converges a.e. to some function $f \in L^1(\mu)$ and that $\|f_n\|_1 \to \|f\|_1$. Show that $f_n \xrightarrow{\|\cdot\|_1} f$. (*Hint:* Apply Fatou's lemma with $g_n := |f_n| + |f| - |f_n - f|$.)

EXERCISE 5.9 **Cesàro convergence and density** 1 (see e.g. [235])
In this exercise, $(a_n)_{n \in \mathbb{N}}$ is a sequence of nonnegative real numbers.

1. Assume that $N^{-1} \sum_0^{N-1} a_n \to 0$ as $N \to \infty$.
   (a) Show that for each $\varepsilon > 0$ the set $A_\varepsilon := \{n \in \mathbb{N}; \ a_n < \varepsilon\}$ has density 1.
   (b) Show that one can construct an increasing sequence of integers $(N_k)_{k \geq 0}$ with $N_0 = 0$ and a sequence $(I_k)_{k \geq 1}$ of finite subsets of $\mathbb{N}$ with $I_k \subset (N_{k-1}, N_k]$ such that $a_n < 2^{-k}$ for all $n \in I_k$ and $\mathrm{card}(I_k) \geq (1 - 2^{-k})N_k$.
   (c) Show that one can find a set $\mathbf{N} \subset \mathbb{N}$ with density 1 such that $a_n \to 0$ as $n \to \infty$ along $\mathbf{N}$.
2. Assume that the sequence (a_n) is bounded and that $a_n \to 0$ along a set of density 1. Show that $N^{-1} \sum_0^{N-1} a_n \to 0$ as $N \to \infty$.

EXERCISE 5.10 **Wiener's theorem** (see e.g. [155])
In this exercise, σ is a complex Borel measure on $\mathbb{T}$.

1. Let $n \in \mathbb{Z}$. Show that $|\hat{\sigma}(n)|^2 = \int_{\mathbb{T} \times \mathbb{T}} (\bar{z}w)^n d\sigma(z) d\sigma^*(w)$, where σ^* is the measure defined by $\sigma^*(A) = \overline{\sigma(A)}$.
2. Let $\Delta = \{(z, w) \in \mathbb{T} \times \mathbb{T}; \ z = w\}$. Compute $(\sigma \otimes \sigma^*)(\Delta)$.
3. Show that

$$\lim_{N \to \infty} \frac{1}{N} \sum_{n=0}^{N} |\hat{\sigma}(n)|^2 = \sum_{a \in \mathbb{T}} |\sigma(\{a\})|^2 = \lim_{N \to \infty} \frac{1}{2N+1} \sum_{n=-N}^{N} |\hat{\sigma}(n)|^2.$$

4. Conclude that σ is continuous iff $N^{-1} \sum_0^N |\hat{\sigma}(n)| \to 0$ as $N \to \infty$.

EXERCISE 5.11 Let σ be a Rajchman measure on $\mathbb{T}$. Show that every measure absolutely continuous with respect to σ is Rajchman. (*Hint:* This is clear for measures of the form $P\sigma$, where P is a trigonometric polynomial.)

6

Beyond hypercyclicity

Introduction

In this chapter we study some variants of hypercyclicity. First, we show that a Banach space operator T is hypercyclic provided that every point of the underlying space stays at a bounded distance (not depending on the point) from some fixed T-orbit. Then, we consider two qualitative strengthenings of hypercyclicity, namely *chaoticity* and *frequent hypercyclicity*. We point out several interesting similarities and differences between hypercyclicity and these two variants. In particular, any rotation and any power of a chaotic or frequently hypercyclic operator has the same property; however, chaotic or frequently hypercyclic operators cannot be found in every separable Banach space. Moreover we show that frequently hypercyclic operators need not be chaotic, and we construct an operator which is both chaotic and frequently hypercyclic but not topologically mixing.

6.1 Operators with d-dense orbits

Given a Banach space X and $d \in (0, \infty)$, we say that a set $A \subset X$ is d-**dense** in X if, for each $x \in X$, one can find $z \in A$ such that $\|z - x\| < d$. The following interesting theorem is due to N. S. Feldman [107].

THEOREM 6.1 *Let X be a separable infinite-dimensional Banach space, and let $T \in \mathfrak{L}(X)$. Assume that T has a d-dense orbit for some $d \in (0, \infty)$. Then T is hypercyclic.*

PROOF We first observe that if T has a d-dense orbit for some d then in fact it has an ε-dense orbit for any $\varepsilon > 0$. Indeed, if $O(x, T)$ is d-dense in X, let us consider $x_\varepsilon := (\varepsilon/d)x$. For any $y \in X$, one may find an $n \in \mathbb{N}$ such that $\|T^n(x) - (d/\varepsilon)y\| < d$. Multiplying by ε/d, we get $\|T^n(x_\varepsilon) - y\| < \varepsilon$. Thus, x_ε has an ε-dense orbit.

Next, we recall that, since X is infinite-dimensional, any non-empty open set $O \subset X$ contains a sequence of disjoint open balls with the same (positive) radius. Indeed, if we fix $a \in O$ and $r > 0$ such that $\overline{B}(a, 2r) \subset O$ then we can find $\varepsilon > 0$ such that $\overline{B}(a, r)$ cannot be covered by finitely many balls with radius 2ε. It follows at once that one can construct inductively a sequence $(a_i) \subset \overline{B}(a, r)$ such that $\|a_i - a_j\| \geq 2\varepsilon$ whenever $i \neq j$. Since $\varepsilon < r$ and $\overline{B}(a, 2r) \subset O$, the balls $B(a_i, \varepsilon)$ have the required properties.

Now we show that T is topologically transitive. Let U, V be two non-empty open subsets of X. Let us choose $\varepsilon > 0$ and two sequences of balls (B_i), (B_i'), with radius ε, such that $B_i \subset U$, $B_i' \subset V$ and $B_i \cap B_j = \varnothing = B_i' \cap B_j'$ whenever $i \neq j$.

Let $x_\varepsilon \in X$ be a vector with an ε-dense orbit. Then $O(x_\varepsilon, T)$ intersects all balls B_i, B_i' and hence visits the open sets U and V infinitely many times. In particular, one can find $p, q \in \mathbb{N}$ with $p < q$ such that $T^p(x_\varepsilon) \in U$ and $T^q(x_\varepsilon) \in V$. Then $T^{q-p}(U) \cap V \neq \emptyset$. $\quad\square$

Using the language of the Bourdon–Feldman theorem (see Chapter 3), Theorem 6.1 does *not* say that d-dense orbits are everywhere dense. In fact this is usually not the case, as shown by the forthcoming theorem and example.

THEOREM 6.2 *Let X be a separable Banach space, and let $T \in \mathfrak{L}(X)$. Assume there exist a set $\mathcal{D} \subset X$ which is d-dense for some $d \in (0, \infty)$, a vector $u \in X$, a real number $\delta > 0$ and a map $S : \mathcal{D} \to \mathcal{D}$ such that the following properties hold true:*

(1) $\|T^n(x) - u\| \geq \delta$ *and* $\|S^n(x) - u\| \geq \delta$ *for each* $x \in \mathcal{D}$ *and all* $n \in \mathbb{N}$;
(2) $T^n(x) \to 0$ *and* $S^n(x) \to 0$ *for any* $x \in \mathcal{D}$;
(3) $TS = I$ *on* $\mathcal{D}$.

Then for any $\varepsilon > 0$ there exists a vector $x_\varepsilon \in X$ whose T-orbit is ε-dense but not dense in X.

PROOF It is enough to construct a single vector $x \notin HC(T)$ whose orbit is c-dense for some $c \in (0, \infty)$ (then we may take $x_\varepsilon := (\varepsilon/c)x$; see the proof of Theorem 6.1).

We may and do assume that $\mathcal{D}$ is countable, and we enumerate it as a sequence $(x_k)_{k \geq 1}$. Also, let $(\varepsilon_k)_{k \geq 1}$ be a sequence of positive numbers to be chosen later, with $\sum_1^\infty \varepsilon_k < \infty$.

For any $k \geq 1$ we may choose $m_k \in \mathbb{N}$ such that if $n \geq m_k$ then

$$\|S^n(x_k)\| \leq \varepsilon_k \qquad \text{and} \qquad \|T^n(x_i)\| \leq \varepsilon_k \text{ for all } i \leq k.$$

We define a sequence of integers (n_k) by setting $n_1 := m_1$ and $n_{k+1} := n_k + m_k + m_{k+1}$. Finally, we put $x := \sum_1^\infty S^{n_k}(x_k)$; the series is obviously convergent since $n_k \geq m_k$ for all k.

For any $k \geq 1$ we can write

$$T^{n_k}(x) = x_k + \sum_{j<k} T^{n_k - n_j}(x_j) + \sum_{j>k} S^{n_j - n_k}(x_j).$$

Since $n_q - n_p \geq m_p$ whenever $q > p$, this yields

$$\|T^{n_k}(x) - x_k\| \leq k\varepsilon_k + \sum_{j>k} \varepsilon_j := \alpha_k, \tag{6.1}$$

and since the sequence (x_k) is d-dense in X, it follows that $O(x, T)$ is $(d+1)$-dense provided that the ε_k are small enough.

However, we can show that the vector u will not be in the closure of the set $\{T^p(x); \ p \geq n_1\}$ if (ε_k) is suitably chosen. Let us fix $p \geq n_1$, and let k be the unique integer such that $p \in [n_k, n_{k+1})$.

Suppose first that $p \in [n_k, n_k + m_k]$. Then

$$T^p(x) - u = (T^{p-n_k}(x_k) - u) + \sum_{j<k} T^{p-n_j}(x_j)$$
$$+ \sum_{j>k+1} S^{n_j-p}(x_j) + S^{n_{k+1}-p}(x_{k+1}).$$

Estimating the two sums by an argument like that leading to (6.1) and using the inequality $n_{k+1} - p \geq m_{k+1}$, we deduce that

$$\|T^p(x) - u\| \geq \delta - (\alpha_k + \varepsilon_{k+1}).$$

When $p \in [n_k + m_k, n_{k+1})$ we write

$$T^p(x) - u = (S^{n_{k+1}-p}(x_{k+1}) - u) + \sum_{j<k} T^{p-n_j}(x_j)$$
$$+ \sum_{j>k+1} S^{n_j-p}(x_j) + T^{p-n_k}(x_k),$$

which yields $\|T^p(x) - u\| \geq \delta - (\alpha_k + \varepsilon_k)$. This concludes the proof. □

EXAMPLE 6.3 ([107])   Let $T = 2B$, where B is the canonical backward shift on $\ell^2(\mathbb{N})$. Then one can find a vector $x \in \ell^2(\mathbb{N})$ whose T-orbit is d-dense for some $d \in (0, \infty)$ but not dense in $\ell^2(\mathbb{N})$.

The proof relies on the following elementary lemma (where $\| \cdot \|$ is the euclidean norm on $\mathbb{C}^{n+1}$).

LEMMA 6.4   *Given $n \in \mathbb{N}$ and $y = (y_0, \ldots, y_n) \in \mathbb{C}^{n+1}$, one can find $z \in \mathbb{C}^{n+1}$ such that $\|z - y\| \leq 8/\sqrt{3}$ and z has the form $z = (z_0, z_1/2, \ldots, z_n/2^n)$, where $|z_k| \geq 2$ for all $k \in \{0, \ldots, n\}$.*

PROOF   It is easy to check that, for any $k \in \{0, \ldots, n\}$, one may find a complex number z_k with $|z_k| \geq 2$ and $|z_k/2^k - y_k| \leq 1/2^{k-1}$ (consider the cases $|y_k| \geq 1/2^{k-1}$ and $|y_k| < 1/2^{k-1}$). If z is defined as above, then $\|z-y\|^2 \leq \sum_0^\infty 1/4^{k-2} = 64/3$. □

PROOF OF EXAMPLE 6.3   We set

$$\mathcal{D} := \left\{ (z_0, z_1/2, \ldots, z_n/2^n, 0, \ldots); \ n \geq 0, \ |z_k| \geq 2 \right\} \subset \ell^2(\mathbb{N}).$$

By the lemma, $\mathcal{D}$ is $(8/\sqrt{3}+\varepsilon)$-dense in $\ell^2(\mathbb{N})$, for any $\varepsilon > 0$. Of course $T^n(x) \to 0$ and $S^n(x) \to 0$ for any $x \in \mathcal{D}$, where S is half the forward shift. Finally, since the first coordinate of $T^n(x)$ (resp. of $S^n(x)$) is either 0 or a complex number of modulus at least 2, we see that $\|T^n(x) - e_0\| \geq 1$ (resp. $\|S^n(x) - e_0\| \geq 1$) for every $n \in \mathbb{N}$, where e_0 is the first vector of the canonical basis of $\ell^2(\mathbb{N})$. By Theorem 6.2 the result follows. □

6.2 Chaotic operators

In common language, the word *chaotic* usually refers to phenomena with "unpredictable behaviour". In his classical book [94], R. L. Devaney proposed the following precise mathematical definition.

DEFINITION 6.5   *Let $f : E \to E$ be a continuous map acting on some metric space (E, d). The map f is said to be **chaotic** if*

(1) *f is topologically transitive;*
(2) *f has a dense set of periodic points ($x \in E$ is a **periodic** point of f if $f^k(x) = x$ for some $k \geq 1$);*
(3) *f has a **sensitive dependence on initial conditions**: there exists $\delta > 0$ such that, for any $x \in E$ and every neighbourhood U of x, one can find $y \in U$ and an integer $n \geq 0$ such that $d(f^n(x), f^n(y)) \geq \delta$.*

At first sight, property (3) seems to be the most important, since it captures rather well the intuitive idea of unpredictable behaviour. However, it turns out that if a continuous map $f : E \to E$ is topologically transitive and has a dense set of periodic points then f *automatically* has a sensitive dependence on initial conditions, provided that E is infinite ([18]; see Exercise 6.1). In other words, property (3) is redundant in the above definition. We note that in particular, the chaoticity of a continuous map $f : E \to E$ does not depend on the choice of a compatible metric d on E.

In the linear setting, one can say more: hypercyclicity alone implies a very strong form of sensitive dependence on initial conditions. This observation was made in [123].

PROPOSITION 6.6   *Let X be a separable F-space, and let $T \in \mathfrak{L}(X)$. Assume that T is hypercyclic. Then, for every $x \in X$, there is a dense G_δ set $G(x) \subset X$ such that the set $\{T^n(y) - T^n(x); \ n \geq 0\}$ is dense in X for any $y \in G(x)$.*

PROOF   We are looking for a dense G_δ set such that $y - x \in HC(T)$ for any $y \in G(x)$. Just put $G(x) := x + HC(T)$!   □

From now on, we assume that X is a complex (separable, infinite-dimensional) F-space. Then an operator $T \in \mathfrak{L}(X)$ is chaotic iff T is hypercyclic and

$$\mathrm{Per}(T) := \{x \in X; \exists k \geq 1 : \ T^k(x) = x\}$$

is dense in X. We first note the following simple fact.

REMARK 6.7   Per(T) is the linear span of $\{\mathrm{Ker}(T - \lambda); \ \lambda \in e^{2\pi i \mathbb{Q}}\}$.

PROOF   Using the Bezout identity, we see that $\mathrm{Ker}(T^k - I) = \oplus_{\omega \in \Gamma_k} \mathrm{Ker}(T - \omega)$ for any $k \geq 1$, where $\Gamma_k = \{\omega \in \mathbb{T}; \ \omega^k = 1\}$. Since Per$(T)$ is easily seen to be a vector space and $\mathrm{Per}(T) = \bigcup_{k \geq 1} \mathrm{Ker}(T^k - I)$, the result follows.   □

Thus, a hypercyclic operator is chaotic iff it has "sufficiently many" eigenvectors associated with rational unimodular eigenvalues. This remark itself enables us to give non-trivial examples.

EXAMPLE 6.8   Let $\phi \in H^{\infty}(\mathbb{D})$ and let $M_{\phi} : H^2(\mathbb{D}) \to H^2(\mathbb{D})$ be the associated multiplication operator, $M_{\phi}(f) = \phi f$. Then M_{ϕ}^* is chaotic if and only if it is hypercyclic, i.e. ϕ is non-constant and $\phi(\mathbb{D}) \cap \mathbb{T} \neq \varnothing$.

PROOF   Assume that M_{ϕ}^* is hypercyclic. To prove that it is chaotic, we note that $A = \phi^{-1}(e^{2i\pi\mathbb{Q}})$ has no isolated point (this is easy to check using the fact that ϕ is an open map). In particular A has an accumulation point in $\mathbb{D}$, so the reproducing kernels k_z, $z \in A$ span a dense linear subspace of $H^2(\mathbb{D})$. Since k_z is a periodic point of M_{ϕ}^* if $z \in A$ (because $M_{\phi}^*(k_z) = \overline{\phi(z)}k_z$), we conclude that $\mathrm{Per}(M_{\phi}^*)$ is dense in $H^2(\mathbb{D})$.   □

EXAMPLE 6.9   Let T be a continuous linear operator on $H(\mathbb{C})$. Assume that T commutes with every translation operator and is not a scalar multiple of the identity. Then T is chaotic.

The proof is similar to that of the previous example, using the arguments given for the proof of Theorem 1.42. The details are left to the reader. In particular, the non-trivial translation operators and the derivative operator are chaotic.

To proceed further we need the following criterion for chaoticity, which is a very natural strengthening of the Hypercyclicity Criterion. It was stated by A. Bonilla and K.-G. Grosse-Erdmann in [61], using ideas from M. Taniguchi [231].

We recall that a series $\sum x_n$ with terms in X is said to be **unconditionally convergent** if, for every neighbourhood O of 0 in X, one can find $N \geq 1$ such that $\sum_{k \in F} x_k \in O$ for any finite set $F \subset \mathbb{N}$ with $F \cap [0, N] = \varnothing$. Equivalently, $\sum x_n$ is unconditionally convergent iff all subseries of $\sum x_n$ are convergent in X and iff $\sum x_{\sigma(n)}$ is convergent for any permutation σ of the index set. When X is a Banach space the convergence of $\sum \|x_n\|$ implies the unconditional convergence of $\sum x_n$, but the converse is not true unless X is finite-dimensional (this is the so-called *Dvoretzky–Rogers theorem*; see e.g. [96, Chapter VI] or [97, Chapter 1]).

THEOREM 6.10 (CHAOTICITY CRITERION)   *Let $T \in \mathfrak{L}(X)$. Assume that there exist a dense set $\mathcal{D} \subset X$ and a map $S : \mathcal{D} \to \mathcal{D}$ such that*

(1) $\sum T^n(x)$ *and* $\sum S^n(x)$ *are unconditionally convergent, for each* $x \in \mathcal{D}$;
(2) $TS = I$ *on* $\mathcal{D}$.

Then T is chaotic.

PROOF   It follows from (1) that $T^n(x) \to 0$ and $S^n(x) \to 0$ for any $x \in \mathcal{D}$. Hence T is hypercyclic because it satisfies the Hypercyclicity Criterion (it is even topologically mixing, see Chapter 2). Now, let $x \in \mathcal{D}$ and, for each $k \geq 1$, set

$$x_k := \sum_{n=1}^{\infty} S^{nk}(x) + x + \sum_{n=1}^{\infty} T^{nk}(x).$$

By (1), the series are indeed convergent and $x_k \to x$ as $k \to \infty$. Moreover, it follows from (2) that $T^k(x_k) = x_k$ for all $k \geq 1$. Thus, any point $x \in \mathcal{D}$ may be approximated by periodic points of T, so that $\mathrm{Per}(T)$ is dense in X. $\qquad\square$

The above criterion is only a sufficient condition for chaoticity. However, it is worth noting that any chaotic operator satisfies the Hypercyclicity Criterion.

PROPOSITION 6.11   *Chaotic operators are weakly mixing.*

PROOF   This follows immediately from Corollary 4.13, since a chaotic operator has a dense set of points with bounded orbit. $\qquad\square$

We now illustrate Theorem 6.10 with our two other favourite examples: weighted shifts and composition operators.

THEOREM 6.12   *Let $B_{\mathbf{w}}$ be a weighted backward shift on $\ell^p(\mathbb{N})$, $1 \leq p < \infty$. The following are equivalent:*

(i) *$B_{\mathbf{w}}$ is chaotic;*
(ii) *$B_{\mathbf{w}}$ admits a non-zero periodic point;*
(iii) *the series $\sum_{n \geq 1} (w_1 \cdots w_n)^{-p}$ is convergent.*

PROOF   Of course, (i) $\implies$ (ii) is trivial.

(ii) $\implies$ (iii): Assume that (ii) holds and let $x = (x_n) \in \ell^p(\mathbb{N})$ be a non-zero periodic point for $B_{\mathbf{w}}$. Let us choose $N \geq 1$ such that $B_{\mathbf{w}}^N(x) = x$, and let us also fix $a \in \mathbb{N}$ with $x_a \neq 0$. If we compare the entries of $x = B_{\mathbf{w}}^{kN}(x)$ at positions a and $a + kN$, we find

$$x_a = w_{a+1} \cdots w_{a+kN}\, x_{a+kN}$$

for all $k \geq 1$. Since $x \in \ell^p$ it follows that $\sum_{k=1}^{\infty} (w_{a+1} \cdots w_{a+kN})^{-p} < \infty$. Moreover, if $k \geq 1$ and $n \in [a + (k-1)N, a + kN)$ then

$$(w_1 \cdots w_n)^{-p} = \left(\frac{w_{n+1} \cdots w_{a+kN}}{w_1 \cdots w_a} \right)^p (w_{a+1} \cdots w_{a+kN})^{-p}$$
$$\leq \frac{\|\mathbf{w}\|_\infty^{Np}}{(w_1 \cdots w_a)^p} (w_{a+1} \cdots w_{a+kN})^{-p}.$$

Thus, we get

$$\sum_{n \geq a} \frac{1}{(w_1 \cdots w_n)^p} \leq \frac{N\|\mathbf{w}\|_\infty^{Np}}{(w_1 \cdots w_a)^p} \sum_{k \geq 1} (w_{a+1} \cdots w_{a+kN})^{-p} < \infty.$$

(iii) $\implies$ (i): Assume that (iii) holds. Let $\mathcal{D} \subset \ell^p(\mathbb{N})$ be the set of all finitely supported vectors, and let $S_{\mathbf{w}}$ be the forward shift defined on $\mathcal{D}$ by $S_{\mathbf{w}}(e_i) = w_{i+1}^{-1} e_{i+1}$. Then $TS = I$ on $\mathcal{D}$. Moreover, it follows from (iii) that the series $\sum S^n(x)$ is unconditionally convergent for any $x \in \mathcal{D}$. Since $B_{\mathbf{w}}^n(x) = 0$ for large enough n if $x \in \mathcal{D}$, we conclude that the assumptions of Theorem 6.10 are fulfilled. $\qquad\square$

THEOREM 6.13   *If ϕ is an automorphism of the unit disk $\mathbb{D}$ without fixed points in $\mathbb{D}$ then the composition operator C_ϕ is chaotic on $H^2(\mathbb{D})$.*

PROOF   The difficulty here is that we cannot simply reproduce the proof given in Chapter 1 for the hypercyclicity of C_ϕ, since the latter was based on Lebesgue's theorem with no quantitative estimate. Now we have to choose more carefully the dense set $\mathcal{D}$, in order to control the decay of $C_\phi^n(f)$ for $f \in \mathcal{D}$.

As usual, it is easier to work on the upper half-plane $\mathbb{P}_+ = \{\text{Im}(s) > 0\}$, which is biholomorphic to $\mathbb{D}$ via the Cayley map $\omega(z) = i(1+z)/(1-z)$. A straightforward computation gives $\omega^{-1}(s) = (s-i)/(s+i)$. Recall that the automorphisms of the disk without fixed points in $\mathbb{D}$ are either *parabolic* or *hyperbolic* (see Chapter 1). A hyperbolic automorphism (having $\alpha = +1$ as attractive fixed point) is associated with a dilation $h_{\lambda,b}(s) = \lambda(s-b) + b \, (\lambda > 1, b \in \mathbb{R})$ on the half-plane $\mathbb{P}_+$, whereas a parabolic automorphism is associated with a translation $T_a(s) = s + a \, (a \in \mathbb{R}^*)$. As a matter of notation, we will still denote by $h_{\lambda,b}$ or T_a the associated composition operator acting on $\mathcal{H}^2 := \{f \circ \omega^{-1}; \, f \in H^2(\mathbb{D})\}$. We recall that the norm of the space $\mathcal{H}^2$ is given by $\|F\|_2^2 = \pi^{-1} \int_{\mathbb{R}} |F(t)|^2 dt/(1+t^2)$. Let us now see how one can apply Theorem 6.10.

We will start with the hyperbolic case: thus let us assume that ϕ is a hyperbolic automorphism of $\mathbb{D}$. Upon conjugating by a suitable automorphism, we may assume that the attractive fixed point of ϕ is $\alpha = +1$ and that its repulsive fixed point is $\beta = -1$. Thus C_ϕ is unitarily equivalent to some operator $T = h_{\lambda,0}$ defined on $\mathcal{H}^2$, and we have to prove that T is chaotic. We set $S := T^{-1} = h_{1/\lambda,0}$ and try to apply Theorem 6.10.

We need some dense set $\mathcal{D} \subset \mathcal{H}^2$ on which both T^n and S^n tend to zero sufficiently rapidly. In view of the proof of Theorem 1.47, it is natural to consider the set $\mathcal{P}$ of all holomorphic polynomials satisfying $P(1) = 0 = P(-1)$ and to put $\mathcal{D} := \{P \circ \omega^{-1}; \, P \in \mathcal{P}\}$. An easy modification of Lemma 1.48 shows that $\mathcal{P}$ is dense in $H^2(\mathbb{D})$. Hence, $\mathcal{D}$ is dense in $\mathcal{H}^2$.

Let us now fix $Q = P \circ \omega^{-1} \in \mathcal{D}$. Then $|P(z)| \le C|z-1|$ and $|P(z)| \le C|z+1|$ on $\mathbb{T}$, by the definition of $\mathcal{P}$. (Here, as usual, C is a finite constant which may change from line to line). This gives $|Q(t)| \le C/\sqrt{1+t^2}$ and $|Q(t)| \le C|t|$, $t \in \mathbb{R}$. We estimate $\|T^n(Q)\|$ as follows:

$$\|T^n(Q)\|^2 = \frac{1}{\pi} \int_{|t| \le \lambda^{-n/2}} |Q(\lambda^n t)|^2 \frac{dt}{1+t^2} + \frac{1}{\pi} \int_{|t| > \lambda^{-n/2}} |Q(\lambda^n t)|^2 \frac{dt}{1+t^2}$$

$$\le C\lambda^{-n/2} + \frac{C}{1+\lambda^n} \int_{|t| > \lambda^{-n/2}} \frac{dt}{1+t^2} \le C\lambda^{-n/2},$$

where we have used the boundedness of Q on $\mathbb{R}$ and the inequality $|Q(x)|^2 \le C/(1+x^2)$. This shows that the series $\sum \|T^n(Q)\|$ is convergent. One may check in the same way that $\sum_n \|S^n(Q)\|$ is convergent, which concludes the proof in the hyperbolic case.

We now turn to the parabolic case: assume that ϕ is a parabolic automorphism of $\mathbb{D}$, with attractive fixed point $\alpha = +1$. Then C_ϕ is unitarily equivalent to a translation operator $T_a \in \mathfrak{L}(\mathcal{H}^2)$, and we may assume that $a > 0$. A new difficulty arises here: we *never* have $\sum_n \|T_a^n(Q)\| < \infty$ for $Q \in \mathcal{H}^2 \setminus \{0\}$ (see Exercise 6.3). However, it is still possible to prove the unconditional convergence of $\sum_n T_a^n(Q)$ for suitable functions Q.

Here we consider the set $\mathcal{P}$ of all holomorphic polynomials satisfying $P(1) = 0 = P'(1)$. It is not hard to check that $\mathcal{P}$ is again dense in $H^2(\mathbb{D})$. As before, we set $\mathcal{D} := \{P \circ w^{-1}; \ P \in \mathcal{P}\}$. Let us fix $Q = P \circ w^{-1} \in \mathcal{D}$, some positive integer N and some finite set $F \subset \mathbb{N}$ with $F \cap [0, N] = \varnothing$. By the definition of $\mathcal{P}$, we have $|P(z)| \le C|z - 1|^2$ on $\mathbb{T}$, so that $|Q(x)| \le C/(1 + x^2)$ on $\mathbb{R}$. This yields

$$\left| \sum_{n \in F} T_a^n Q(t) \right| \le C \sum_{n \ge N} \frac{1}{1 + (t + na)^2}$$

for all $t \in \mathbb{R}$, so that

$$\left\| \sum_{n \in F} T_a^n(Q) \right\|^2 \le C \int_{\mathbb{R}} \left(\sum_{n \ge N} \frac{1}{1 + (t + na)^2} \right)^2 \frac{dt}{1 + t^2}.$$

The integrand is dominated by

$$\left(\sum_{n \in \mathbb{Z}} \frac{1}{1 + (t + na)^2} \right)^2,$$

which is continuous and a-periodic, hence bounded on $\mathbb{R}$. Therefore, we may let N tend to infinity and apply Lebesgue's convergence theorem to conclude that the series $\sum_n T_a^n(Q)$ is unconditionally convergent. The same holds true for $\sum_n S^n(Q)$, where $S = T_a^{-1} = T_{-a}$. Thus, the operator T_a is chaotic. $\square$

REMARK 6.14 It follows from Theorem 6.13, but not directly from the above proof, that the "rational $\mathbb{T}$-eigenvectors" of an invertible parabolic or hyperbolic composition operator span a dense subspace of $H^2(\mathbb{D})$. This can also be proved directly (see e.g. [26, Example 3.6] for the hyperbolic case and [114] for the parabolic case). However, to do this is not particularly easy, and it is in fact rather less elementary than the above calculations. Moreover, we prefer to deduce the result from the Chaoticity Criterion, Theorem 6.10, because we will use this criterion again in the next section.

6.3 Frequently hypercyclic operators

6.3.1 Definition and examples

By definition, an operator $T \in \mathfrak{L}(X)$ is hypercyclic iff there is some vector $x \in X$ whose T-orbit visits each non-empty open set $V \subset X$. Using the notation of Chapter 4, this means that the sets

$$\mathbf{N}(x, V) := \{n \in \mathbb{N};\ T^n(x) \in V\}$$

are all non-empty. It is natural to ask how often $O(x,T)$ visits each open set V or, equivalently, how large can the sets $\mathbf{N}(x, V)$ be. This question was already in the air in Chapter 4. Indeed, we saw there that an operator T is weakly mixing iff all sets $\mathbf{N}(U, V) := \{n \in \mathbb{N};\ T^n(U) \cap V \neq \varnothing\}$ are *thick*.

We recall that the **lower density** of a set of natural numbers A is defined by

$$\underline{\mathrm{dens}}(A) := \liminf_{N \to \infty} \frac{\mathrm{card}(A \cap [1, N])}{N}.$$

Similarly, the **upper density** of A is

$$\overline{\mathrm{dens}}(A) := \limsup_{N \to \infty} \frac{\mathrm{card}(A \cap [1, N])}{N}.$$

If we enumerate an infinite set $A \subset \mathbb{N}$ as an increasing sequence $(n_k)_{k \in \mathbb{N}}$, it is easy to check that A has positive lower density if and only if $n_k = O(k)$, i.e. $n_k \leq Ck$ for some finite constant C and all $k \in \mathbb{N}$. In particular, we have the following lemma, which we quote for future reference.

LEMMA 6.15 *Let $A = \{n_k;\ k \in \mathbb{N}\}$ and I be two subsets of $\mathbb{N}$ with positive lower density. Then the set $\{n_k;\ k \in I\}$ has positive lower density.*

The following class of operators was introduced in [26].

DEFINITION 6.16 *Let X be a topological vector space, and let $T \in \mathcal{L}(X)$. The operator T is said to be **frequently hypercyclic** if there exists some vector $x \in X$ such that $\mathbf{N}(x, V)$ has positive lower density for every non-empty open set $V \subset X$. Such a vector x is said to be frequently hypercyclic for T, and the set of all frequently hypercyclic vectors for T is denoted by $FHC(T)$.*

By the above remark concerning sets with positive density, a vector $x \in X$ is frequently hypercyclic for T iff for each non-empty open set V there exist an increasing sequence of integers (n_k) and some constant C such that

$$T^{n_k}(x) \in V \quad \text{and} \quad n_k \leq Ck$$

for all $k \in \mathbb{N}$. Thus, in some sense, the orbit of a frequently hypercyclic vector has the most extreme possible behavior.

Having introduced a definition, we must take care of two obvious objections: at this stage of the story it is not clear that there exists any frequently hypercyclic operator, neither is it clear that there exist hypercyclic operators which are not frequently hypercyclic!

The second objection is easy to answer. Indeed, let $B_{\mathbf{w}}$ be any weighted backward shift on $\ell^2(\mathbb{N})$. If V is the open ball $B(2e_0, 1) \subset \ell^2(\mathbb{N})$ and if $x = (x_n)$ is any vector in $\ell^2(\mathbb{N})$ then

$$|w_1 \cdots w_n x_n| \geq 1$$

for all $n \in \mathbf{N}(x, V)$. It follows that

$$\sum_{n\in\mathbf{N}(x,V)} \frac{1}{(w_1 \cdots w_n)^2} < \infty,$$

for any vector $x \in \ell^2(\mathbb{N})$. This leads to the following example.

EXAMPLE 6.17 Let $B_{\mathbf{w}}$ be the weighted backward shift on $\ell^2(\mathbb{N})$ with weight sequence $w_n = \sqrt{(n+1)/n}$. Then $B_{\mathbf{w}}$ is hypercyclic but not frequently hypercyclic.

PROOF On the one hand, since $w_1 \cdots w_n = \sqrt{n+1} \to \infty$ as $n \to \infty$, it follows from Salas' theorem that $B_{\mathbf{w}}$ is hypercyclic. On the other hand,

$$\sum_{n\in\mathbf{N}(x,V)} \frac{1}{n+1} = \sum_{n\in\mathbf{N}(x,V)} \frac{1}{(w_1 \cdots w_n)^2} < \infty$$

for any $x \in \ell^2(\mathbb{N})$, where $V = B(2e_0, 1)$. In particular, $\mathbf{N}(x, V)$ never has a positive lower density. □

To answer the second objection, i.e. to exhibit frequently hypercyclic operators, we need to work a little harder. Fortunately (and perhaps surprisingly), there is a very simple criterion for detecting frequent hypercyclicity. It should look familiar, since it is exactly the same as the Chaoticity Criterion stated in the previous section.

THEOREM 6.18 (FREQUENT HYPERCYCLICITY CRITERION) *Let X be a separable F-space, and let $T \in \mathfrak{L}(X)$. Assume that there exist a dense set $\mathcal{D} \subset X$ and a map $S : \mathcal{D} \to \mathcal{D}$ such that:*

(1) $\sum T^n(x)$ *and* $\sum S^n(x)$ *are unconditionally convergent for each $x \in \mathcal{D}$;*
(2) $TS = I$ *on* $\mathcal{D}$.

Then T is frequently hypercyclic.

The proof of Theorem 6.18 is divided into two parts. We first prove a simple combinatorial lemma saying that it is possible to find an infinite sequence of sets with positive density such that each set in the sequence is "well separated" from the others. These sets will then be used to construct a frequently hypercyclic vector.

LEMMA 6.19 *Let $(N_p)_{p\geq 1}$ be any sequence of positive real numbers. Then one can find a sequence $(\mathbf{N}_p)$ of pairwise disjoint subsets of $\mathbb{N}$ such that*

(1) *Each set $\mathbf{N}_p$ has positive lower density;*
(2) $\min \mathbf{N}_p \geq N_p$, *and* $|n-m| \geq N_p + N_q$ *whenever $n \neq m$ and $(n, m) \in \mathbf{N}_p \times \mathbf{N}_q$.*

PROOF Let (a_p) be an increasing sequence of integers with $a_1 = 1$, such that a_{p+1} is a multiple of a_p for each p and $\sum_1^\infty (N_p/a_p) < \infty$. For each $p \geq 1$, we set $I_p := a_p \mathbb{N} \setminus a_{p+1}\mathbb{N}$. Clearly, $(I_p)_{p\geq 1}$ is a partition of $\mathbb{N}$. Moreover, each set I_p has positive lower density since $I_p = a_p \big(\mathbb{N} \setminus (a_{p+1}/a_p)\mathbb{N}\big)$ and $\mathbb{N} \setminus a\mathbb{N}$ has positive lower density for any $a \geq 2$.

For any $k \in \mathbb{N}$, let us denote by $\phi(k)$ the unique natural number such that $k \in I_{\phi(k)}$. We define an increasing sequence of integers $(n_k)_{k \in \mathbb{N}}$ by setting

$$n_k := N_{\phi(k)} + 2 \sum_{j < k} N_{\phi(j)}.$$

This definition ensures that $n_k \geq N_p$ whenever $k \in I_p$ and that $n_k - n_j \geq N_p + N_q$ whenever $k > j$ and $(k, j) \in I_p \times I_q$. Thus, if we set

$$\mathbf{N}_p := \{n_k; \ k \in I_p\}$$

then property (2) above is satisfied. In view of Lemma 6.15, the proof will be complete if we are able to show that the set $\{n_k; \ k \in \mathbb{N}\}$ has a positive lower density.

For any $k, p \geq 1$, the set $[1, k] \cap I_p \subset [1, k] \cap a_p\mathbb{N}$ has cardinality at most k/a_p. Therefore, the equation $\phi(j) = p$ admits at most k/a_p solutions $j \in \{1, \ldots, k\}$. It follows that

$$n_k \leq 2k \sum_{p=1}^{\infty} \frac{N_p}{a_p}$$

for all $k \geq 1$, which concludes the proof. $\blacksquare$

REMARK   It should be clear that something like Lemma 6.19 is needed to prove the existence of frequently hypercyclic operators. Indeed, disjoint open sets V give rise to disjoint sets of integers $\mathbf{N}(x, V)$. So, if we are aiming to construct a frequently hypercyclic vector for some operator, we should at least be able to produce infinitely many pairwise disjoint subsets of $\mathbb{N}$ with positive density.

PROOF OF THEOREM 6.18   Let ρ be a (complete) translation-invariant metric generating the topology of X. For convenience, we shall write $\|x\|$ instead of $\rho(x, 0)$.

Since X is separable we may assume that the dense set $\mathcal{D}$ is countable, and we enumerate it as a sequence $(x_p)_{p \geq 1}$. Also, let $(\varepsilon_p)_{p \geq 1}$ be a sequence of positive numbers to be chosen later, with $\sum_1^{\infty} \varepsilon_p < \infty$. Our assumptions may be formulated as follows: for any integer $p \geq 1$, one can find a positive integer N_p such that, for any finite set $F \subset \mathbb{N} \cap [N_p, \infty)$, one has

$$\left\| \sum_{n \in F} T^n(x_i) \right\| + \left\| \sum_{n \in F} S^n(x_i) \right\| < \varepsilon_p \quad \text{for all } i \leq p.$$

Let $(\mathbf{N}_p)_{p \geq 1}$ be the sequence of subsets of $\mathbb{N}$ obtained by applying Lemma 6.19 to the sequence (N_p). The frequently hypercyclic vector we are looking for is defined by

$$x := \sum_{p=1}^{\infty} \sum_{n \in \mathbf{N}_p} S^n(x_p).$$

First, we note that x is well defined. Indeed, each series $\sum_{n \in \mathbf{N}_p} S^n(x_p)$ is convergent by unconditionality and

$$\sum_{p \geq 1} \left\| \sum_{n \in \mathbf{N}_p} S^n(x_p) \right\| \leq \sum_{p \geq 1} \varepsilon_p < \infty .$$

Let us fix $p \geq 1$ and $n \in \mathbf{N}_p$. Then

$$\|T^n(x) - x_p\| \leq \sum_{q=1}^{\infty} \left\| \sum_{m \in \mathbf{N}_q, m > n} S^{m-n}(x_q) \right\| + \sum_{q=1}^{\infty} \left\| \sum_{m \in \mathbf{N}_q, m < n} T^{n-m}(x_q) \right\| .$$

We evaluate the first sum by decomposing it as

$$\sum_{q=1}^{p} \left\| \sum_{m \in \mathbf{N}_q, m > n} S^{m-n}(x_q) \right\| + \sum_{q=p+1}^{\infty} \left\| \sum_{m \in \mathbf{N}_q, m > n} S^{m-n}(x_q) \right\| .$$

Since $n \in \mathbf{N}_p$, we know that $m - n > \max(N_p, N_q)$ whenever $m \in \mathbf{N}_q$ and $m > n$. By the choice of the sequence (N_q), this gives immediately

$$\sum_{q=1}^{\infty} \left\| \sum_{m \in \mathbf{N}_q, m > n} S^{m-n}(x_q) \right\| \leq p \varepsilon_p + \sum_{q=p+1}^{\infty} \varepsilon_q := \alpha_p .$$

Evaluating the second sum in the same way, we conclude that $\|T^n(x) - x_p\| \leq 2\alpha_p$, for each $p \geq 1$ and all $n \in \mathbf{N}_p$. Thus, we have proved that

$$\mathbf{N}_p \subset \mathbf{N}(x, B(x_p, 3\alpha_p))$$

for each $p \geq 1$. In particular, each set $\mathbf{N}(x, B(x_p, 3\alpha_p))$ has positive lower density. Since the sequence (x_p) is dense in X, this shows that x is a frequently hypercyclic vector for T provided that $\alpha_p \to 0$ (i.e. $\varepsilon_p = o(1/p)$). $\qquad \square$

COROLLARY 6.20 *The following operators are frequently hypercyclic:*

(a) *weighted backward shifts $B_{\mathbf{w}}$ on $\ell^p(\mathbb{N})$ for which $\sum_1^{\infty} (w_1 \cdots w_n)^{-p} < \infty$;*
(b) *composition operators on $H^2(\mathbb{D})$ associated with automorphisms of the disk without fixed points;*
(c) *non-trivial translation operators on $H(\mathbb{C})$.*

PROOF   To prove (a) and (b), there is nothing to do. Indeed, we have already observed in Section 6.2 that in both cases the assumptions given in Theorem 6.18 are fulfilled.

To prove (c), we use a method due to R. M. Gethner and J. H. Shapiro (see [119]). Let T be a non-trivial translation operator on $H(\mathbb{C})$. We may assume that T is given by $Tf(z) = f(z+1)$. For $m \in \mathbb{N}$ and $k \geq 1$, we consider the entire function

$$f_{m,k}(z) := z^m \left(\frac{\sin(z/k)}{z/k} \right)^{m+2} = k^{m+2} \frac{\sin^{m+2}(z/k)}{z^2} .$$

Since $f_{m,k}(z) \to z^m$ in $H(\mathbb{C})$ as $k \to \infty$, the linear span of $\{f_{m,k};\ m \in \mathbb{N}, k \geq 1\}$ is dense in $H(\mathbb{C})$. Moreover, since the sine function is bounded in any horizontal strip $\{a \leq \mathrm{Im}(z) \leq b\}$, each series

$$\sum_{n \geq 1} T^{\pm n} f_{m,k}(z) = k^{m+2} \sum_{n \geq 1} \frac{\sin^{m+2}((z \pm n)/k)}{(z \pm n)^2}$$

is uniformly absolutely convergent on compact subsets of $\mathbb{C}$, hence unconditionally convergent in $H(\mathbb{C})$. Thus, we may apply Theorem 6.18 with $S := T^{-1}$.   □

REMARK 6.21   It is proved in [60] that in fact any operator on $H(\mathbb{C})$ which commutes with translations and is not a multiple of the identity is frequently hypercyclic.

REMARK 6.22   In view of Corollary 6.20, it is natural to "conjecture" that a weighted backward shift $B_{\mathbf{w}}$ is frequently hypercyclic on $\ell^p(\mathbb{N})$ iff $\sum_{n \geq 1} (w_1 \cdots w_n)^{-p} < \infty$. The problem is still open on ℓ^p. However, the corresponding conjecture is false on c_0: there exist frequently hypercyclic weighted shifts $B_{\mathbf{w}}$ on $c_0(\mathbb{N})$ such that $(w_1 \cdots w_n)^{-1}$ does not define an element of c_0; in fact, one may require that $w_1 \cdots w_n = 1$ for infinitely many n. An example can be found in [27].

6.3.2 Frequent hypercyclicity and ergodic theory

The reader has probably noticed that frequent hypercyclicity lies at the heart of Chapter 5, although it is not explicitly mentioned there.

PROPOSITION 6.23   *Let X be a separable F-space, and let $T \in \mathfrak{L}(X)$. Assume that one can find some T-invariant Borel probability measure μ on X with full support, with respect to which T is an ergodic transformation. Then T is frequently hypercyclic. Moreover, the set of frequently hypercyclic vectors for T has full measure.*

PROOF   The proof is exactly that of Corollary 5.5, using the fact that the hypercyclic vectors provided by Birkhoff's ergodic theorem are in fact frequently hypercyclic.   □

Thus, all the work done in Chapter 5 is relevant to frequent hypercyclicity. In particular, we may state the following results.

COROLLARY 6.24   *Let X be a complex separable and infinite-dimensional Banach space, and let $T \in \mathfrak{L}(X)$. In each of the following two cases, the operator T is frequently hypercyclic.*

(a) *X has type 2 and T has a perfectly spanning set of $\mathbb{T}$-eigenvectors;*
(b) *X has type $p \in [1,2]$, and one can find a countable family (E_i) of $\mathbb{T}$-eigenvector fields for T which is spanning with respect to the Lebesgue measure on $\mathbb{T}$ and such that each E_i is α_i-Hölderian, for some $\alpha_i > 1/p - 1/2$.*

6.3.3 Frequent hypercyclicity vs. hypercyclicity

In this subsection we show that some general properties of hypercyclic operators have their counterparts in the frequent hypercyclicity setting but others do not.

We first observe that the Frequent Hypercyclicity Criterion has been proved without using the Baire category theorem. The reason is that it is impossible to use Baire's theorem: the assumptions of Theorem 6.18 imply that $FHC(T)$ *cannot* be a residual subset of X.

THEOREM 6.25 *Let X be a separable F-space, and let $T \in \mathcal{L}(X)$. Assume that there exists an open set $O \subset X$ with $\overline{O} \neq X$ such that, for every $a \in (0,1)$, the set*

$$\{x \in X; \ \overline{\mathrm{dens}}(\mathbf{N}(x,O)) \geq a\}$$

is dense in X. Then $FHC(T)$ is a set of the first Baire category in X. In particular, this is the case if $T^n(x) \to 0$ for every x in some dense set $\mathcal{D} \subset X$.

PROOF For $a \in (0,1)$ we set

$$\begin{aligned}
G_a &:= \{x \in X; \ \overline{\mathrm{dens}}(\mathbf{N}(x,O)) \geq a\} \\
&= \bigcap_{m \geq 1} \bigcap_{K \geq 1} \bigcup_{N \geq K} \{x \in X; \ \mathrm{card}\{n \leq N; \ T^n(x) \in O\} > (1-1/m)aN\} \\
&= \bigcap_{m \geq 1} \bigcap_{K \geq 1} \bigcup_{N \geq K} O_{m,N} \, .
\end{aligned}$$

Since each set $O_{m,N}$ is clearly open, G_a is a dense G_δ subset of X. Hence, so is $G := \bigcap_{a \in (0,1)} G_a = \bigcap_{n \in \mathbb{N}} G_{1-2^{-n}}$.

By the definition of G, any point $x \in G$ satisfies $\overline{\mathrm{dens}}(\mathbf{N}(x,O)) = 1$. Since $\overline{\mathrm{dens}}(A) = 1 - \underline{\mathrm{dens}}(\mathbb{N} \backslash A)$ for any set $A \subset \mathbb{N}$, it follows that $\underline{\mathrm{dens}}(\mathbf{N}(x, X \backslash \overline{O})) = 0$ for all $x \in G$. Since by assumption $X \backslash \overline{O}$ is a non-empty open set, we conclude that G contains no frequently hypercyclic vector for T. Thus, $FHC(T) \subset X \backslash G$ is a set of the first Baire category. If $T^n(x) \to 0$ for all x in some dense set $\mathcal{D} \subset X$, we may take as O any (non-dense) neighbourhood of 0. Indeed, if $x \in \mathcal{D}$ then $\mathbf{N}(x,O)$ contains all but finitely many $n \in \mathbb{N}$. $\qquad\square$

COROLLARY 6.26 *Let $T \in \mathcal{L}(X)$. Assume that T satisfies either the assumptions of the Frequent Hypercyclicity Criterion or one of the assumptions of Corollary 6.24, where in (a) the T-eigenvectors are σ-spanning with respect to some Rajchman measure. Then $FHC(T)$ is a set of the first Baire category in X.*

PROOF This follows directly from Theorem 6.25, recalling that an operator satisfying one of the assumptions of Corollary 6.26 also satisfies the Hypercyclicity Criterion (see Theorem 5.41). $\qquad\square$

REMARK It follows that under the assumptions of Corollary 6.24, $FHC(T)$ is both large in a probabilistic sense and small in the Baire category sense. This is noteworthy but should not be too surprising: after all, there are dense G_δ sets in $\mathbb{R}$ which have Lebesgue measure 0! What is more amusing is that, in a way, the largeness of $FHC(T)$ in a probabilistic sense *implies* its smallness in the category sense.

In spite of the above theorem, some properties of hypercyclic operators do have "frequent" counterparts. In particular, as a rule all properties which rely on a connectedness argument remain true. For example, here is a version of the Bourdon–Feldman theorem (see Chapter 3).

THEOREM 6.27 *Let X be a topological vector space, and let $T \in \mathfrak{L}(X)$. Also, let $x \in X$. Assume that there exists a non-empty open set $O \subset X$ such that $\underline{\mathrm{dens}}(\mathbf{N}(x, W)) > 0$ for every non-empty open set $W \subset O$. Then x is a frequently hypercyclic vector for T.*

PROOF Our assumption implies that $O(x, T)$ is somewhere dense. By the Bourdon–Feldman theorem T is hypercyclic and hence topologically transitive. In particular, for any non-empty open set $V \subset X$, one can find $p \in \mathbb{N}$ and a non-empty open set $W \subset O$ such that $T^p(W) \subset V$. Then $\mathbf{N}(x, W) + p \subset \mathbf{N}(x, V)$, hence $\mathbf{N}(x, V)$ has positive lower density. □

Frequent hypercyclicity is also invariant under rotations.

THEOREM 6.28 *Let X be a complex topological vector space, and let $T \in \mathfrak{L}(X)$ be frequently hypercyclic. Then λT is frequently hypercyclic for any $\lambda \in \mathbb{T}$, with $FHC(\lambda T) = FHC(T)$.*

To prove this result, we first need an elementary lemma.

LEMMA 6.29 *Let $A \subset \mathbb{N}$ have positive lower density. Also, let $I_1, \ldots, I_q \subset \mathbb{N}$ with $\bigcup_{j=1}^q I_j = \mathbb{N}$ and let $n_1, \ldots, n_q \in \mathbb{N}$. Then $B := \bigcup_{j=1}^q (n_j + A \cap I_j)$ has positive lower density.*

PROOF If $N \geq \max(n_1, \ldots, n_q)$ then each set $(n_j + A \cap I_j) \cap [0, 2N]$ contains $n_j + (A \cap I_j \cap [0, N])$. Therefore, we get

$$\mathrm{card}(B \cap [0, 2N]) \geq \frac{1}{q} \sum_{j=1}^q \mathrm{card}\left((n_j + A \cap I_j) \cap [0, 2N]\right)$$

$$\geq \frac{1}{q} \sum_{j=1}^q \mathrm{card}(A \cap I_j \cap [0, N])$$

$$\geq \frac{1}{q} \mathrm{card}(A \cap [0, N])$$

for all $N \geq \max(n_1, \ldots, n_q)$. This proves the lemma. □

PROOF OF THEOREM 6.28 Let $x \in FHC(T)$, and let V be any non-empty open subset of X. We have to show that the set

$$B := \{n \in \mathbb{N};\ (\lambda T)^n(x) \in V\}$$

has positive lower density.

Let us choose a non-empty open set $V' \subset V$ and $\varepsilon > 0$ such that $zV' \subset V$ for any complex number z satisfying $|z - 1| < \varepsilon$. The key observation is that one may cover $\mathbb{N}$ with finitely many translates of the set

$$I := \{n \in \mathbb{N}; \ |\lambda^n - 1| < \varepsilon\}.$$

Indeed, this is clear if $\lambda \in e^{2\pi i \mathbb{Q}}$ since in that case I contains $p\mathbb{N}$ for some positive integer p. Otherwise the set $\{\lambda^n; \ n \in \mathbb{N}\}$ is dense in $\mathbb{T}$. By compactness, one can find $N \geq 1$ such that $\mathbb{T} = \bigcup_{j=0}^{N}\{z \in \mathbb{T}; \ |z - \lambda^j| < \varepsilon\}$ and then we may write $\mathbb{N} = \bigcup_{j=0}^{N}\{m; \ |\lambda^m - \lambda^j| < \varepsilon\} = \bigcup_0^N (j + I)$. Let us fix $m_1, \ldots, m_q \in \mathbb{N}$ such that

$$\mathbb{N} = \bigcup_{j=1}^{q}(m_j + I) := \bigcup_{j=1}^{q} I_j.$$

By the León–Müller theorem (Theorem 3.2), we know that for each $j \in \{1, \ldots, q\}$ the vector $\lambda^{m_j} x$ is a hypercyclic vector for λT. Thus, we may find an integer n_j such that $\lambda^{m_j}\lambda^{n_j}T^{n_j}(x) \in V'$. In particular, we get a neighbourhood W of x such that $\lambda^{m_j}\lambda^{n_j}T^{n_j}(W) \subset V'$ for all $j \in \{1, \ldots, q\}$. Let us set

$$A := \mathbf{N}(x, W) = \{k \in \mathbb{N}; \ T^k(x) \in W\}.$$

Since $x \in FHC(T)$, the set A has positive lower density. Moreover, if $j \in \{1, \ldots, q\}$ and $k \in A \cap I_j$ then

$$\lambda^{n_j+k}T^{n_j+k}(x) = \lambda^{k-m_j}\lambda^{m_j}\lambda^{n_j}T^{n_j}(T^k(x))$$
$$\in D(1, \varepsilon) \cdot \lambda^{m_j}\lambda^{n_j}T^{n_j}(W) \subset V.$$

Thus, we see that B contains the set $\bigcup_{j=1}^{q}(n_j + A \cap I_j)$. By Lemma 6.29, this concludes the proof. $\qquad\square$

Similar (but simpler) considerations lead to the analogue of Ansari's theorem.

THEOREM 6.30 *If T is a frequently hypercyclic operator then T^p is frequently hypercyclic for every positive integer p, with $FHC(T) = FHC(T^p)$.*

PROOF Let $x \in FHC(T)$, and let V be any non-empty open subset of X. By Ansari's theorem, we know that $T^{p-j}(x)$ is a hypercyclic vector for T^p for each $j \in \{1, \cdots, p\}$, so that we may find an integer k_j such that $T^{k_j p - j}(x) \in V$. It follow that there exists a neighbourhood W of x such that $T^{k_j p - j}(W) \subset V$ for all $j \in \{1, \cdots, p\}$.

We now set $I_j = \{n \in \mathbb{N}; n \equiv j \pmod{p}\}$ and $B = \bigcup_{j=1}^{p}((k_j p - j) + A \cap I_j)$, where $A = N(x, W)$. Then B has positive lower density by Lemma 6.29, and $B \subset p\mathbb{N}$. Moreover, for any $m = k_j p - j + n \in B$, one has $T^m(x) \in T^{k_j p - j}(W) \subset V$. This shows that x is a frequently hypercyclic vector for T^p. $\qquad\square$

We conclude this section by showing that frequently hypercyclic operators satisfy the Hypercyclicity Criterion. Recall that on separable F-spaces the Hypercyclicity Criterion is equivalent to the weak mixing property, which in turn is related to the size of the sets $\mathbf{N}(U,V)$ for U, V non-empty and open in X. Since frequent hypercyclicity is defined in terms of sets $\mathbf{N}(x,V)$, the following theorem is quite natural. It is due to K.-G. Grosse-Erdmann and A. Peris [136].

THEOREM 6.31 *Frequently hypercyclic operators are weakly mixing.*

For the proof, we need the following well-known lemma on sets with positive density.

LEMMA 6.32 *If $A \subset \mathbb{N}$ has positive lower density then the difference set $A - A$ has bounded gaps.*

In fact, the same conclusion holds if A is only assumed to have positive *upper* density. A proof can be found in [212], where the result is attributed to P. Erdös and A. Sárközy.

PROOF OF THEOREM 6.31 Let $T \in \mathfrak{L}(X)$ be frequently hypercyclic, where X is an arbitrary topological vector space. We show that T satisfies the three open sets condition (see Theorem 4.10).

Let U, V be non-empty open subsets of X, and let W be a neighbourhood of 0. One can find $m \in \mathbb{N}$ and a non-empty open set $U' \subset U$ such that $T^m(U') \subset W$. Let x be a frequently hypercyclic vector for T. We see from Lemma 4.5 that

$$\mathbf{N}(U', U') = \mathbf{N}(x, U') - \mathbf{N}(x, U').$$

By Lemma 6.32, it follows that $\mathbf{N}(U', U')$ has bounded gaps. Then $\mathbf{N}(U, W)$ has bounded gaps as well since $\mathbf{N}(U, W) \supset \mathbf{N}(U', U') + m$. Since $\mathbf{N}(W, V)$ is thick, by Lemma 4.9, we conclude that $\mathbf{N}(U, W) \cap \mathbf{N}(W, V) \neq \varnothing$. □

REMARK We have in fact proved that a linear operator T is weakly mixing provided that all sets $\mathbf{N}(U, V)$ have bounded gaps. In other words, if all the sets $\mathbf{N}(U, V)$ have bounded gaps then they are all thick (!) This is no longer true in a nonlinear setting, however; see [121].

6.4 Spaces without chaotic or frequently hypercyclic operators

In this section we show that, unlike hypercyclic operators, chaotic and frequently hypercyclic operators cannot live in every Banach space. As may be expected, one gets into trouble if the underlying Banach space has "few" operators. Such exotic spaces were first exhibited by W. T. Gowers and B. Maurey in [126]. For a comprehensive survey, we recommend Maurey's paper [179].

DEFINITION 6.33 *An infinite-dimensional Banach space is said to be **indecomposable** if it cannot be written as the direct sum of two infinite-dimensional closed subspaces. An infinite-dimensional Banach space X is said to be **hereditarily indecomposable (HI)** if every closed infinite-dimensional subspace of X is indecomposable.*

Obviously, no "classical" Banach space is HI. In fact, the mere existence of HI spaces is a deep result due to Gowers and Maurey [126].

As indicated above, the main property of HI spaces that we will use is the fact that they have few operators. Recall that a Banach space operator $T \in \mathfrak{L}(X)$ is said to be **strictly singular** if there is no closed infinite-dimensional subspace $X_0 \subset X$ such that $T_{|X_0}$ is one-to-one with closed range; in other words, for any infinite-dimensional subspace $X_0 \subset X$ and every $\varepsilon > 0$, there exists $x \in X_0$ such that $\|T(x)\| < \varepsilon \|x\|$.

PROPOSITION 6.34 *Any continuous linear operator on a complex HI space has the form $T = \lambda I + S$, where $\lambda \in \mathbb{C}$ and S is strictly singular.*

PROOF   Let X be a complex HI space. We use the assumption on X in the following way: if Y and Z are any two closed infinite-dimensional subspaces of X then the unit spheres of Y and Z *almost meet*; that is, for every $\varepsilon > 0$ one can find $(y, z) \in Y \times Z$ such that $\|y\| = 1 = \|z\|$ and $\|y - z\| < \varepsilon$. Indeed, otherwise the map $(y, z) \mapsto y - z$ is an isomorphism from $Y \times Z$ onto $Y - Z = Y + Z$; hence $Y \cap Z = \{0\}$ and the subspace $Y \oplus Z$ is closed in X, which contradicts the HI property.

Now, let $T \in \mathfrak{L}(X)$ be an arbitrary operator. Let us choose some complex number λ such that $S := T - \lambda$ is not *left-Fredholm* (see Proposition D.3.6 in Appendix D). We will show that S is strictly singular. Accordingly, let us fix a closed infinite-dimensional subspace $X_0 \subset X$ and $\varepsilon > 0$. Since the operator S is not left-Fredholm, one can find an infinite-dimensional closed subspace $E \subset X$ such that $\|S_{|E}\| < \varepsilon$. Since the unit spheres of X_0 and E almost meet, one can find $(x, e) \in X_0 \times E$ such that $\|x\| = 1 = \|e\|$ and $\|x - e\| < \varepsilon$. Then $\|S(x)\| \leq \|S(x - e)\| + \|S(e)\| < 2\varepsilon \|S\|$. Since ε is arbitrary, this concludes the proof. $\Box$

REMARK   Very recently, S. A. Argyros and R. Haydon [9] were able to construct a (separable infinite-dimensional) Banach space on which any operator has the form $T = \lambda I + K$, where K is a *compact* operator.

It is well known that strictly singular operators have the same spectral properties as compact operators (see e.g. [174, Vol. 1, 2.c]). In particular, if $S \in \mathfrak{L}(X)$ is strictly singular and $\lambda \in \mathbb{C}$ then the spectrum of $\lambda I + S$ is either finite or has the form $\{\lambda\} \cup \{\lambda_n; \; n \in \mathbb{N}\}$, where $\lambda_n \to \lambda$. Thus, we get

COROLLARY 6.35 *If X is a complex HI space then the spectrum of any operator $T \in \mathfrak{L}(X)$ is countable, with at most one limit point.*

We are now ready to prove the following theorem. The chaotic part is essentially due to J. Bonet, F. Martínez-Giménez and A. Peris [57], and the frequently hypercyclic part was proved by S. Shkarin [226].

THEOREM 6.36 *A complex hereditarily indecomposable Banach space carries no chaotic or frequently hypercyclic operator.*

The key point is to observe that the spectrum of a chaotic or frequently hypercyclic operator cannot be too small, whereas operators on HI spaces always have a small spectrum. More precisely, Theorem 6.36 follows at once from Corollary 6.35 and the next proposition. We recall that a non-empty compact subset of $\mathbb{C}$ is said to be **perfect** if it has no isolated point.

PROPOSITION 6.37 *Let X be a Banach space, and let $T \in \mathfrak{L}(X)$ be chaotic or frequently hypercyclic. Then $\sigma(T)$ is a perfect set.*

PROOF Assume that $\sigma(T)$ has an isolated point λ. By the Riesz decomposition theorem, we may write $X = X_1 \oplus X_2$ and $T = T_1 \oplus T_2$, where X_i is T-invariant, $\sigma(T_1) = \{\lambda\}$ and $\sigma(T_2) = \sigma(T)\setminus\{\lambda\}$. It is straightforward to check that $T_1^n = \pi_1(T^n)_{|X_1}$ for all $n \in \mathbb{N}$, where $\pi_1 : X \to X_1$ is the associated projection. In particular, π_1 turns periodic point into periodic points, and $\mathbf{N}_T(x, V) \subset \mathbf{N}_{T_1}(\pi_1(x), \pi_1(V))$ for any $x \in X$ and every non-empty open set $V \subset X$. It follows that the operator T_1 is chaotic or frequently hypercyclic whenever T is. Therefore, we may in fact assume that $\sigma(T)$ consists of the single point $\{\lambda\}$. Then $|\lambda| = 1$, because the spectrum of any hypercyclic operator must intersect the unit circle (see Theorem 1.18).

Assume first that T is chaotic. Since $\sigma(T) = \{\lambda\}$ and $\mathrm{Per}(T) \neq \{0\}$, we then have $\lambda \in e^{2i\pi\mathbb{Q}}$ and $\mathrm{Per}(T) = \bigcup_{k \geq 1} \mathrm{Ker}(T - \lambda)^k$. If we choose a positive integer N such that $\lambda^N = 1$, it follows that $T^N(x) = x$ for all $x \in \mathrm{Per}(T)$. Since $\mathrm{Per}(T)$ is dense in X, we conclude that $T^N = I$. In particular the T-orbit of any vector is finite, which makes it hard for T to be hypercyclic.

The proof is more involved in the frequently hypercyclic case; it is also rather elegant. First, we need some information on the distribution of the zeros of entire functions. Recall that an entire function $F : \mathbb{C} \to \mathbb{C}$ is said to be of **exponential type 0** if, for every $\varepsilon > 0$, there exists some finite constant $C = C_\varepsilon$ such that

$$|F(z)| \leq Ce^{\varepsilon|z|}$$

for all $z \in \mathbb{C}$. The result we need is the following. Its proof is a typical application of Jensen's formula; see e.g. B. Ya. Levin's book [172] or Appendix A.

FACT Let F be a (non-zero) entire function of exponential type 0 and, for each $r > 0$, let us denote by $n_F(r)$ the number of zeros of F in the disk $D(0, r)$. Then $n_F(r) = o(r)$ as $r \to \infty$.

Now, assume that $T \in \mathfrak{L}(X)$ is frequently hypercyclic, with $\sigma(T) = \{\lambda\}$. Then $\lambda^{-1}T$ is frequently hypercyclic as well, by Theorem 6.28. Thus, we may in fact assume that $\sigma(T) = \{1\}$.

In what follows, we fix two non-empty open sets $U, V \subset X$ such that $T(V) \subset U$ and U, V are strictly separated by some linear functional x^*, i.e. $\mathrm{Re}\langle x^*, u \rangle < 0 < \mathrm{Re}\langle x^*, v \rangle$ for any $(u, v) \in U \times V$. To obtain these sets U and V, start with any point $a \in X$ such that $T(a) \neq a$, pick two balls $V \ni a$ and $U \ni T(a)$ with $\overline{U} \cap \overline{V} = \varnothing$ and $T(V) \subset U$ and separate these balls using the Hahn–Banach theorem.

Let us write $T = I + S$. Then $\sigma(S) = \{0\}$, so that $\|S^n\|^{1/n} \to 0$ as $n \to \infty$, by the spectral radius formula. Thus, we may define an operator $\log(T)$ by the formula

$$\log(T) = \log(I + S) := \sum_{m \geq 1} (-1)^{m-1} \frac{S^m}{m}$$

and set $T^t := \exp(t \log(T))$, $t \in \mathbb{R}$. Then

$$T^t = \sum_{n=0}^{\infty} A_n \frac{t^n}{n!},$$

where $A_n = (\log(T))^n$. Moreover, writing

$$A_n = \left(\sum_{m \geq 1} \frac{(-1)^{m-1} S^{m-1}}{m} \right)^n S^n,$$

we see that $\|A^n\|^{1/n} \to 0$ as $n \to \infty$.

Now let us fix an arbitrary vector $x \in X$. We define a function $F : \mathbb{R} \to \mathbb{R}$ by

$$F(t) := \operatorname{Re}\langle x^*, T^t(x) \rangle .$$

The function F has the power series expansion

$$F(t) = \sum_{n=0}^{\infty} a_n \frac{t^n}{n!},$$

where $a_n = \operatorname{Re}\langle x^*, A_n(x) \rangle$. Clearly, $|a_n|^{1/n} \to 0$ as $n \to \infty$. Hence, the formula $F(z) = \sum_0^\infty a_n (z^n/n!)$ makes sense for any $z \in \mathbb{C}$ and defines an entire function of exponential type 0.

We now enumerate the set $\mathbf{N}(x, V)$ as an increasing sequence $(r_k)_{k \in \mathbb{N}}$. Since $T^{r_k}(x) \in V$ and $T^{1+r_k}(x) \in U$, we have $F(r_k) > 0$ and $F(1 + r_k) < 0$ for all $k \in \mathbb{N}$. Since the function F is real on $\mathbb{R}$, it follows that F must have a zero in each interval $(r_k, 1 + r_k)$. By the above fact, this yields $\lim_{k \to \infty} r_k/k = \infty$, i.e. $\overline{\mathrm{dens}}(\mathbf{N}(x, V)) = 0$. In particular, x is not a frequently hypercyclic vector for T. $\square$

REMARK   The following result can be extracted from the proof: *if $T \in \mathfrak{L}(X)$ satisfies $\sigma(T) = \{1\}$ and $T \neq I$ then one can find a non-empty open set $V \subset X$ such that $\overline{\mathrm{dens}}(\mathbf{N}(x, V)) = 0$ for all $x \in X$.*

6.5 Almost closing the circle

So far, we have introduced several strong forms of hypercyclicity: weak mixing, mixing, chaoticity, frequent hypercyclicity. The following diagram summarizes the known implications between these properties.

mixing   frequently hypercyclic   chaotic

weakly mixing

We also know that there exist mixing operators which are neither chaotic nor frequently hypercyclic, for example, any mixing operator on a hereditarily indecomposable Banach space. In this section, we study the remaining implications. More precisely, we shall prove that frequently hypercyclic operators need not be chaotic and that chaotic or frequently hypercyclic operators need not be mixing. This will almost complete the above diagram.

Before proceeding further, let us specify some terminology. By a **Cantor set** we mean any topological space homeomorphic to the abstract Cantor space $\{0, 1\}^{\mathbb{N}}$. It is well known (but non-trivial, see e.g. [156, Theorem 7.4]) that a topological space is a Cantor set iff it is compact, metrizable, perfect (i.e. with no isolated point) and zero dimensional. In particular, the Cantor sets in $\mathbb{T}$ are exactly the nowhere dense perfect compact subsets of $\mathbb{T}$. We shall need the following facts. Recall that a probability measure is said to be **continuous** if every point has measure 0 and that a topological space is said to be **Polish** if it is separable and completely metrizable.

- Any Cantor set $\mathcal{C}$ carries a continuous probability measure: just consider the image of the Haar measure of $\{0, 1\}^{\mathbb{N}}$ under some homeomorphism $\phi : \{0, 1\}^{\mathbb{N}} \to \mathcal{C}$.
- Any uncountable Polish space E contains a Cantor set (see [156, Corollary 6.5]).

The operators that we are going to discuss are frequently hypercyclic, thanks to a large supply of eigenvectors associated with unimodular eigenvalues. The next lemma summarizes all what we need to know regarding the role of the unimodular point spectrum.

We recall that a $\mathbb{T}$-**eigenvector field** for an operator $T \in \mathfrak{L}(X)$ is a map $E : \mathbb{T} \to X$ such that $TE(\lambda) = \lambda E(\lambda)$ for every $\lambda \in \mathbb{T}$. More generally, we allow the domain of E to be some compact set $K \subset \mathbb{T}$. The $\mathbb{T}$-eigenvector field $E : K \to H$ is said to be **spanning** if $\overline{\text{span}}(E(\lambda); \; \lambda \in K) = H$.

LEMMA 6.38   *Let H be a separable Hilbert space, and let $T \in \mathfrak{L}(H)$. Let also K be a perfect compact subset of $\mathbb{T}$. Assume that T admits a continuous and spanning $\mathbb{T}$-eigenvector field $E : K \to H$. Then*

(1) *T is frequently hypercyclic;*
(2) *T is chaotic provided $e^{2i\pi\mathbb{Q}} \cap K$ is dense in K.*

PROOF   Since K is perfect, one can find some Borel continuous probability measure σ on K with support exactly K. To produce such a measure σ, one may argue as follows. Let $(V_n)_{n \geq 1}$ be a countable basis of open sets for K, and choose non-empty open sets W_n such that $\overline{W}_n \subset V_n$. Then each $\overline{W}_n$ is a perfect compact set because K is perfect; hence $\overline{W}_n$ contains a Cantor set. It follows that, for each $n \geq 1$, one can find a continuous Borel probability measure σ_n on K with $\mathrm{supp}(\sigma_n) \subset V_n$. Set $\sigma := \sum_1^\infty 2^{-n}\sigma_n$.

Since σ has support K, any measurable set $A \subset K$ with $\sigma(A) = 1$ is dense in K. Since E is spanning and continuous, it follows that $\mathrm{span}(E(\lambda); \lambda \in A)$ is dense in H, for any such set A. In the terminology of Chapter 5, this means that the $\mathbb{T}$-eigenvector field is σ-*spanning*. By Corollary 6.24, it follows that T is frequently hypercyclic.

If $e^{2i\pi\mathbb{Q}} \cap K$ is dense in K then $\overline{\mathrm{span}}(E(\lambda); \lambda \in e^{2i\pi\mathbb{Q}}) = H$ by the continuity of E. In particular, the periodic points of T are dense in H, so that T is chaotic since it is already known to be hypercyclic. $\qquad \square$

To exhibit a frequently hypercyclic operator which is not chaotic, we need to find an operator with many $\mathbb{T}$-eigenvectors (to ensure frequent hypercyclicity) but not too many (to avoid chaoticity). That this can be done will follow from the next two lemmas. Recall that we denote by $\sigma_p(T)$ the point spectrum of an operator T, i.e. the set of all eigenvalues of T.

LEMMA 6.39 *Let K be an arbitrary compact subset of $\mathbb{T}$. Then there exist a separable Hilbert space H and an operator $T \in \mathfrak{L}(H)$ such that*

(1) $\sigma_p(T) = K$;

(2) *T admits a continuous and spanning $\mathbb{T}$-eigenvector field $E : K \to H$.*

PROOF We start with the Kalisch operator, defined on $L^2([0, 2\pi])$ by

$$Sf(\theta) := e^{i\theta}f(\theta) - \int_0^\theta ie^{it}f(t)\,dt.$$

We saw in subsection 5.5.3 that $\sigma_p(S) = \mathbb{T}$ and that for any $\alpha \in [0, 2\pi)$ the eigenspace of S associated with the eigenvalue $e^{i\alpha}$ is the one-dimensional space spanned by $\mathbf{1}_{(\alpha, 2\pi)}$. We set $H := \overline{\mathrm{span}}(\mathbf{1}_{(\alpha, 2\pi)}; e^{i\alpha} \in K)$ and denote by T the restriction of S to the invariant subspace H. We claim that T has the required properties.

We first note that if $0 \le \alpha < \beta < 2\pi$ then

$$\|\mathbf{1}_{(\alpha, 2\pi)} - \mathbf{1}_{(\beta, 2\pi)}\|_2 \le (\beta - \alpha)^{1/2}.$$

It follows that the map $E : K \to H$ defined by $E(e^{i\theta}) = \mathbf{1}_{(\theta, 2\pi)}$ is a continuous $\mathbb{T}$-eigenvector field for T, which is spanning by the definition of H.

Clearly $K \subset \sigma_p(T)$. For the converse inclusion, it suffices to check that $\mathbf{1}_{(\alpha, 2\pi)} \notin H$ if $\alpha \in [0, 2\pi)$ and $e^{i\alpha} \notin K$. Assuming that $\alpha > 0$, let us choose some open interval $J = (\alpha - \varepsilon, \alpha + \varepsilon) \subset (0, 2\pi)$ such that $e^{iJ} \cap K = \emptyset$. Then J is either contained in or disjoint from $(\beta, 2\pi)$ when $e^{i\beta} \in K$; hence, every function $f \in H$ is constant on J. It follows that any $f \in H$ is far from $\mathbf{1}_{(\alpha, 2\pi)}$. Indeed, if $f = c$ on J then

$$\|f - \mathbf{1}_{(\alpha,2\pi)}\|_2^2 \geq \|\mathbf{1}_J(f - \mathbf{1}_{(\alpha,2\pi)})\|_2^2$$
$$= \|c\mathbf{1}_{(\alpha-\varepsilon,\alpha+\varepsilon)} - \mathbf{1}_{[\alpha,\alpha+\varepsilon)}\|_2^2$$
$$= \varepsilon(c^2 + (c-1)^2) \geq \varepsilon/2 \,.$$

If $\alpha = 0$ the proof is the same; one observes that all functions $f \in H$ are 0 on some fixed interval $[0, \varepsilon)$. $\qquad\square$

The second lemma is quite well known.

LEMMA 6.40 *If D is a countable subset of $\mathbb{T}$ then one can find a Cantor set $C \subset \mathbb{T}$ such that $C \cap D = \varnothing$.*

PROOF The set $G := \mathbb{T} \setminus D$ is a G_δ subset of $\mathbb{T}$ and hence a Polish space (see [156, Theorem 3.11]). Moreover, G is perfect since it is dense in $\mathbb{T}$. Hence, G contains a Cantor set.

Alternatively (and in a more elementary way), one can build the Cantor set C by imitating the construction of the usual Cantor ternary set: the set C will appear as a countable intersection $\bigcap_{n \in \mathbb{N}} C_n$ where, at stage n, the set C_n avoids the first n points of the countable set D. $\qquad\square$

We can now prove

THEOREM 6.41 *There exists a Hilbert space operator which is frequently hypercyclic but not chaotic.*

PROOF By Lemma 6.40 one can choose a Cantor set $C \subset \mathbb{T}$ with $C \cap e^{2\pi i \mathbb{Q}} = \varnothing$. Let $T \in \mathfrak{L}(H)$ be the operator given by Lemma 6.39 applied for $K = C$. Then T is frequently hypercyclic by Lemma 6.38 and not chaotic since it has no rational unimodular eigenvalue. $\qquad\square$

REMARK 6.42 It is possible to construct a frequently hypercyclic backward shift on $c_0(\mathbb{N})$ which does not have any unimodular eigenvalue (see [27]). This is another example of a frequently hypercyclic but not chaotic operator, which is quite different from the one we have just given.

We now turn to the construction of chaotic or frequently hypercyclic operators which are not mixing. Our approach relies on the following easy remark: if X is a Banach space and if $T \in \mathfrak{L}(X)$ is mixing then $\|T^n\| \to \infty$ as $n \to \infty$. Indeed, let U be the open ball $B(0,1) \subset X$ and, for any $C > 0$, let $V_C := \{z \in X; \|z\| > C\}$. By the mixing property, the set $\{n \in \mathbb{N}; T^n(U) \cap V_C \neq \varnothing\}$ contains all but finitely many n, which means that $\|T^n\| > C$ if n is large enough.

Thus, it will be enough to construct an operator T having many unimodular eigenvalues but such that $\|T^n\|$ remains bounded along some increasing sequence of integers (n_k).

The strategy is the following. We start with some operator S acting on a separable Hilbert space H_0 and admitting a continuous $\mathbb{T}$-eigenvector field $E : \mathbb{T} \to H_0 \setminus \{0\}$; for example, S could be twice the backward shift on $\ell^2(\mathbb{N})$. We also fix an increasing sequence of integers $(n_k)_{k \in \mathbb{N}}$. The idea is to define a new "norm" $| \cdot | : H_0 \to [0, \infty]$ such that $\mathcal{H} := \{x \in H_0; |x| < \infty\}$ is invariant under S and $|S^{n_k}(x)| \leq |x|$ for all $x \in H_0$. We also hope that $E(\lambda) \in \mathcal{H}$ for sufficiently many $\lambda \in \mathbb{T}$, more precisely for all λ in some perfect set $K \subset \mathbb{T}$. Then, we will consider $H := \overline{\mathrm{span}(E(\lambda); \, \lambda \in K)}^{|\cdot|}$. Clearly, $T := S_{|H}$ is a good candidate for being a frequently hypercyclic or chaotic non-mixing operator.

To be sure that $|S^{n_k} x| \leq |x|$ for all $x \in H_0$, perhaps the most natural choice for $| \cdot |$ would be to set

$$|x|^2 = \|x\|^2 + \sum_{j=1}^{\infty} \sum_{k_1, \ldots, k_j \in \mathbb{N}} \left\| S^{n_{k_1} + \cdots + n_{k_j}} x \right\|^2 .$$

It should be more or less clear that $|S^{n_k} x| \leq |x|$ and even that $\sum_k |S^{n_k} x|^2 \leq |x|^2$ for all $x \in H_0$. Indeed,

$$\sum_{k \in \mathbb{N}} |S^{n_k} x|^2 = \sum_{k \in \mathbb{N}} \|S^{n_k} x\|^2 + \sum_{k \in \mathbb{N}} \sum_{j=1}^{\infty} \sum_{k_1, \ldots, k_j \in \mathbb{N}} \left\| S^{n_k + n_{k_1} + \cdots + n_{k_j}} x \right\|^2$$

$$= \sum_{k \in \mathbb{N}} \|S^{n_k} x\|^2 + \sum_{j=2}^{\infty} \sum_{k_1, \ldots, k_j \in \mathbb{N}} \left\| S^{n_{k_1} + \cdots + n_{k_j}} x \right\|^2$$

$$\leq |x|^2 .$$

However, since $\|S^{n_k} E(\lambda)\| = \|\lambda^{n_k} E(\lambda)\| = \|E(\lambda)\|$ for all $k \in \mathbb{N}$, we have $|E(\lambda)| = \infty$ for any $\lambda \in \mathbb{T}$, which is not very interesting for our purpose. Thus, we have to modify slightly the above definition.

The main observation is that we may replace each S^{n_k} by the operator $S^{n_k} - I$. This will yield the boundedness of the sequence $(S^{n_k} - I)$, which is equivalent to that of (S^{n_k}). The great advantage is that $\|(S^{n_k} - I)E(\lambda)\| = |\lambda^{n_k} - 1| \|E(\lambda)\|$, and there are good reasons for expecting $|\lambda^{n_k} - 1|$ to be very small for many $\lambda \in \mathbb{T}$ if the sequence (n_k) is suitably chosen.

There is no extra complication in considering operators of the form $P_k(S)$, where P_k is an arbitrary polynomial; in fact, this makes the "formal" part of the proof perhaps neater. Accordingly, we consider a sequence of polynomials $\mathbf{P} = (P_k)_{k \in \mathbb{N}}$. For any $x \in H_0$, we set

$$|x|_{\mathbf{P}}^2 := \|x\|^2 + \sum_{j=1}^{\infty} \sum_{k_1, \ldots, k_j \in \mathbb{N}} \left\| P_{k_1}(S) \cdots P_{k_j}(S) x \right\|^2 ,$$

and we put $\mathcal{H}_{\mathbf{P}} := \{x \in H_0; |x|_{\mathbf{P}} < \infty\}$. The main properties of $\mathcal{H}_{\mathbf{P}}$ are summarized in the next lemma.

LEMMA 6.43 *The following properties hold.*

(1) *The normed space $(\mathcal{H}_{\mathbf{P}}, | \cdot |_{\mathbf{P}})$ is a Hilbert space invariant under the operators $P_k(S)$, and each $P_k(S)$ acts as a bounded operator on $(\mathcal{H}_{\mathbf{P}}, | \cdot |_{\mathbf{P}})$, with $|P_k(S)| \leq 1$.*

(2) *Let K be a compact subset of $\mathbb{T}$, and assume that the series $\sum |P_k(\lambda)|^2$ is uniformly convergent on K, with $\sum_{k\in\mathbb{N}} |P_k(\lambda)|^2 < 1$ for every $\lambda \in K$. Then $E(\lambda) \in \mathcal{H}_{\mathbf{P}}$ for each $\lambda \in K$ and $E_{|K}$ is continuous from K into $(\mathcal{H}_{\mathbf{P}}, |\cdot|_{\mathbf{P}})$.*

PROOF   That $\mathcal{H}_{\mathbf{P}}$ is a Hilbert space is left as an exercise (the inner product is easy to find, and completeness follows from the fact that $|\cdot|_{\mathbf{P}} \geq \|\cdot\|$). To prove the second part of (1), we proceed exactly as we did for the norm $|\cdot|$; the computation gives

$$\sum_{k\in\mathbb{N}} |P_k(S)x|_{\mathbf{P}}^2 \leq |x|_{\mathbf{P}}^2$$

for all $x \in \mathcal{H}$.

In the proof of (2), we shall use the following notation. For each finite sequence of natural numbers $\bar{k} = (k_1, \ldots, k_j)$, we put $P_{\bar{k}} := P_{k_1} \cdots P_{k_j}$. Similarly, for any sequence of scalars $a = (a_k)_{k\in\mathbb{N}} \in \mathbb{C}^{\mathbb{N}}$ and every finite sequence $\bar{k} \in \mathbb{N}^j$, we put $a_{\bar{k}} := a_{k_1} \cdots a_{k_j}$. Finally, we introduce the Hilbert space

$$\mathcal{H} := H_0 \oplus \bigoplus_{j\geq 1} \ell^2(\mathbb{N}^j, H_0),$$

where the symbols $\oplus$ denote an ℓ^2-direct sum. We note that, by the definition of $\mathcal{H}_{\mathbf{P}}$, the map $J : \mathcal{H}_{\mathbf{P}} \to \mathcal{H}$ defined by

$$J(x) := x \oplus \bigoplus_{j\geq 1} (P_{\bar{k}}(S)x)_{\bar{k}\in\mathbb{N}^j}$$

is an isometric embedding.

For each sequence of scalars $a = (a_k)_{k\in\mathbb{N}} \in \mathbb{C}^{\mathbb{N}}$ and every $j \geq 1$, we denote by $\phi_j(a)$ the sequence $(a_{\bar{k}})_{\bar{k}\in\mathbb{N}^j}$. We note that if $a \in \ell^2(\mathbb{N})$ then $\phi_j(a) \in \ell^2(\mathbb{N}^j)$ for each $j \geq 1$, with $\|\phi_j(a)\| = \|a\|^j$. Indeed,

$$\|\phi_j(a)\|_{\ell^2(\mathbb{N}^j)}^2 = \sum_{\bar{k}\in\mathbb{N}^j} |a_{k_1} \cdots a_{k_j}|^2 = \left(\sum_{k\in\mathbb{N}} |a_k|^2\right)^j.$$

It follows that if $(x, a) \in H_0 \times \ell^2(\mathbb{N})$ and $\|a\| < 1$ then

$$\Phi(x, a) := x \oplus \bigoplus_{j\geq 1} (a_{\bar{k}}x)_{\bar{k}\in\mathbb{N}^j}$$

is a well-defined element of $\mathcal{H}$. Moreover, the map Φ is continuous from $H_0 \times B_{\ell^2}(0, 1)$ into $\mathcal{H}$. Indeed, let us set

$$\Phi_n(x, a) := x \oplus \bigoplus_{j=1}^{n} (\phi_j(a) \otimes x) \oplus 0 \oplus 0 \oplus \cdots,$$

where $\phi_j(a) \otimes x = (a_{\bar{k}}x)_{\bar{k}\in\mathbb{N}^j} \in \ell^2(\mathbb{N}^j, H_0)$. The maps Φ_n are continuous because each map ϕ_j is a j-linear continuous map, and they converge to Φ uniformly on every set of the form $B \times B_{\ell^2}(0, \delta)$, where B is a bounded subset of H_0 and $\delta < 1$, because $\|\phi_j(a) \otimes x\| \leq \delta^j \|x\|$ for all j if $\|a\| \leq \delta$.

Now, we observe that if $\lambda \in K$ then $P_{\bar{k}}(S)E(\lambda) = P_{\bar{k}}(\lambda)E(\lambda)$ for any sequence of natural numbers $\bar{k} = (k_1, \ldots, k_j)$. It follows at once that $E(\lambda) \in \mathcal{H}_\mathbf{P}$, with

$$J(E(\lambda)) = \Phi\left(E(\lambda), (P_k(\lambda))_{k \in \mathbb{N}}\right).$$

Since E is continuous from K into H_0 and since the map $\lambda \mapsto (P_k(\lambda))_{k \in \mathbb{N}}$ is continuous from K into $B_{\ell^2}(0, 1)$ by assumption, this shows that

$$E(\lambda) = J^{-1} \circ \Phi\left(E(\lambda), (P_k(\lambda))_{k \in \mathbb{N}}\right)$$

is indeed a continuous map from K into $\mathcal{H}_\mathbf{P}$. $\qquad\square$

We now show that, as suggested above, $|\lambda^{n_k} - 1|$ is small for sufficiently many $\lambda \in \mathbb{T}$ if the sequence (n_k) is suitably chosen.

LEMMA 6.44 *Let $(n_k)_{k \in \mathbb{N}}$ be an increasing sequence of integers, and let us assume that $\sum_0^\infty (n_k/n_{k+1})^2 < \infty$. Then one can find a Cantor set $C \subset \mathbb{T}$ such that the series $\sum |\lambda^{n_k} - 1|^2$ is uniformly convergent on K. Moreover, if n_k divides n_{k+1} for each $k \in \mathbb{N}$, one may require that $C \cap e^{2i\pi\mathbb{Q}}$ is dense in C.*

PROOF It is not hard to show that the assumption $\sum_0^\infty (n_k/n_{k+1})^2 < \infty$ is equivalent to

$$\sum_{k \in \mathbb{N}} n_k^2 \left(\sum_{j > k} \frac{1}{n_j}\right)^2 < \infty;$$

see Exercise 6.7. In particular, one can find $k_0 \in \mathbb{N}$ such that

$$\sum_{k \geq k_0} \frac{1}{n_k} < \frac{1}{2} \quad \text{and} \quad \sum_{j > k} \frac{1}{n_j} < \frac{1}{2n_k}$$

for all $k \geq k_0$. For notational convenience, we will assume that $k_0 = 0$.

It is enough to find some homeomorphic embedding $\theta : \{0, 1\}^\mathbb{N} \to (-1/2, 1/2)$ such that the series $\sum |e^{2i\pi n_k \theta(\varepsilon)} - 1|^2 = 2 \sum |\sin(\pi n_k \theta(\varepsilon))|^2$ is uniformly convergent with respect to $\varepsilon \in \{0, 1\}^\mathbb{N}$; the required Cantor set will then be $C := \{e^{2i\pi\theta(\varepsilon)}; \varepsilon \in \{0, 1\}^\mathbb{N}\}$. The embedding θ can be constructed by successive approximations, as follows.

First, we set $\theta_0(\varepsilon) := \varepsilon_0/n_0$. The map θ_0 is continuous on $\{0, 1\}^\mathbb{N}$, since $\theta_0(\varepsilon)$ depends only on the first coordinate of ε, and $n_0\theta_0(\varepsilon) \in \mathbb{N}$ for any $\varepsilon \in \{0, 1\}^\mathbb{N}$.

Assume that we have already constructed continuous maps $\theta_0(\varepsilon), \ldots, \theta_k(\varepsilon)$ with $n_k\theta_k(\varepsilon) \in \mathbb{N}$ for every $\varepsilon \in \{0, 1\}^\mathbb{N}$. We would like to define a continuous map $\theta_{k+1}(\varepsilon)$ with $n_{k+1}\theta_{k+1}(\varepsilon) \in \mathbb{N}$ and $\theta_{k+1}(\varepsilon)$ as close as possible to $\theta_k(\varepsilon)$. One reasonable way of doing this is to consider $q_{k+1}(\varepsilon)$, the largest integer not exceeding $n_{k+1}\theta_k(\varepsilon)$, and to put

$$n_{k+1}\theta_{k+1}(\varepsilon) := q_{k+1}(\varepsilon) + \varepsilon_{k+1}.$$

We note that $n_{k+1}|\theta_k(\varepsilon) - \theta_{k+1}(\varepsilon)| \leq 1$, so that

$$|\theta_{k+1}(\varepsilon) - \theta_k(\varepsilon)| \leq \frac{1}{n_{k+1}}. \qquad (6.2)$$

Moreover $n_{k+1}\theta_{k+1}(\varepsilon) \in \mathbb{N}$. Finally, we also note that if n_k divides n_{k+1} then $n_{k+1}\theta_k(\varepsilon) \in \mathbb{N}$ by the induction hypothesis, so that $\theta_{k+1}(\varepsilon) = \theta_k(\varepsilon)$ if $\varepsilon_{k+1} = 0$.

By (6.2) and the assumption on $k_0 = 0$, the sequence of continuous maps (θ_k) is uniformly convergent to some continuous map $\theta : \{0,1\}^{\mathbb{N}} \to \mathbb{R}$ with values in $[-\sum_{k\geq 1} n_k^{-1}, \sum_{k\geq 1} n_k^{-1}] \subset (-1/2, 1/2)$. We will show that θ has the required properties.

The map θ is one-to-one. Indeed, if $\varepsilon \neq \varepsilon'$, let k be the least integer with $\varepsilon_k \neq \varepsilon'_k$. Then

$$|\theta(\varepsilon) - \theta(\varepsilon')| \geq |\theta_k(\varepsilon) - \theta_k(\varepsilon')| - \sum_{j>k} |\theta_j(\varepsilon) - \theta_{j-1}(\varepsilon)| - \sum_{j>k} |\theta_j(\varepsilon') - \theta_{j-1}(\varepsilon')|$$

so that $|\theta(\varepsilon) - \theta(\varepsilon')| \geq n_k^{-1} - 2\sum_{j>k} n_j^{-1} > 0$. Since $\{0,1\}^{\mathbb{N}}$ is compact, it follows that θ is a homeomorphic embedding.

Since $n_k\theta_k(\varepsilon) \in \mathbb{N}$ and $n_k|\theta(\varepsilon) - \theta_k(\varepsilon)| \leq n_k \sum_{j>k} n_j^{-1}$, one has

$$\left|e^{2i\pi n_k\theta(\varepsilon)} - 1\right|^2 = 4\left|\sin(\pi n_k\theta(\varepsilon))\right|^2$$
$$= 4\left|\sin(\pi n_k(\theta(\varepsilon) - \theta_k(\varepsilon)))\right|^2$$
$$\leq 4\pi^2 n_k^2 \left(\sum_{j>k} \frac{1}{n_j}\right)^2$$

for every $\varepsilon \in \{0,1\}^{\mathbb{N}}$ and all $k \in \mathbb{N}$. Hence, the series $\sum |e^{2i\pi n_k\theta(\varepsilon)} - 1|^2$ is uniformly convergent on $\{0,1\}^{\mathbb{N}}$.

Finally, assume that n_k divides n_{k+1} for all $k \in \mathbb{N}$. Let us denote by $\mathbf{Q}$ the set of all $\varepsilon \in \{0,1\}^{\mathbb{N}}$ having only finitely many non-zero coordinates. Our inductive construction shows that if $\varepsilon \in \mathbf{Q}$ then the sequence $(\theta_k(\varepsilon))$ is eventually constant. Since $\theta_k(\varepsilon) \in \mathbb{Q}$ for all k, it follows that $\theta(\mathbf{Q}) \subset \mathbb{Q}$ and, since $\mathbf{Q}$ is dense in $\{0,1\}^{\mathbb{N}}$, we conclude that $\mathcal{C} \cap e^{2i\pi\mathbb{Q}}$ is dense in $\mathcal{C}$. $\qquad\square$

Let us now summarize our discussion. Let $(n_k)_{k\geq 1}$ be an increasing sequence of integers such that $\sum_1^{\infty}(n_k/n_{k+1})^2 < \infty$, and set $n_0 := 1$. By Lemma 6.44, we have at our disposal a Cantor set $\mathcal{C}$ such that the series $\sum |\lambda^{n_k} - 1|^2$ is uniformly convergent on $\mathcal{C}$. Choosing $M < \infty$ such that $\sum_{k\geq 0} |\lambda^{n_k} - 1|^2 < M^2$ for every $\lambda \in \mathcal{C}$, we may consider the Hilbert space $\mathcal{H}_{\mathbf{P}}$, where $P_k(\lambda) = M^{-1}(\lambda^{n_k} - 1)$. By Lemma 6.43, we know that the $\mathbb{T}$-eigenvector field E sends $\mathcal{C}$ continuously into $\mathcal{H}_{\mathbf{P}}$. Since $\mathcal{C}$ is separable, it follows that $H := \overline{\operatorname{span}(E(\lambda), \lambda \in \mathcal{C})}^{\mathcal{H}_{\mathbf{P}}}$ is a *separable* Hilbert space. The space H is invariant under S, and $T := S_{|H}$ acts as a bounded operator on H (because $n_0 = 1$) with $\sup_{k\in\mathbb{N}} \|M^{-1}(T^{n_k} - I)\| < \infty$, i.e. $\sup_k \|T^{n_k}\| < \infty$. Thus, T is not mixing. However, T admits a continuous and spanning $\mathbb{T}$-eigenvector field $E : \mathcal{C} \to H$, hence T is frequently hypercyclic by Lemma 6.38. Finally, if n_k divides n_{k+1} for each $k \geq 1$ then we may require that $e^{2i\pi\mathbb{Q}} \cap \mathcal{C}$ is dense in $\mathcal{C}$ so that T is chaotic, again by Lemma 6.38. This means that we have proved the following beautiful theorem due to C. Badea and S. Grivaux [14].

THEOREM 6.45 *Let* $(n_k)_{k\geq 1}$ *be an increasing sequence of integers, and let us assume that* $\sum_{k\geq 1}(n_k/n_{k+1})^2 < \infty$. *Then there exist a separable Hilbert space* H *and an operator* $T \in \mathfrak{L}(H)$ *such that:*

(1) $\sigma_p(T) \cap \mathbb{T}$ *is uncountable;*
(2) $\sup_{k\geq 1} \|T^{n_k}\| < \infty$;
(3) T *is frequently hypercyclic.*

Moreover, if n_k *divides* n_{k+1} *for each* $k \geq 1$ *then one may also require that* T *is chaotic.*

COROLLARY 6.46 *There exist Hilbert space operators which are both frequently hypercyclic and chaotic but not topologically mixing.*

REMARK The operator T satisfies a much stronger property than the mere boundedness of $\|T^{n_k}\|$, namely, $\sum_{k\geq 1} \|(T^{n_k} - I)x\|^2 \leq C\|x\|^2$ for some finite constant C and all $x \in H$ (see the proof of Lemma 6.43).

We end this chapter with an open question.

OPEN QUESTION Do there exist any chaotic operators which are not frequently hypercyclic?

An example could be obtained by showing that Corollary 6.24(a) is not valid for an arbitrary Banach space X. However, one may observe that if $T \in \mathfrak{L}(X)$ is a chaotic operator then for any non-empty open set $V \subset X$ one can find an $x \in X$ such that $\underline{\mathrm{dens}}(\mathbf{N}(x, V)) > 0$; consider any periodic point $x \in V$.

6.6 Comments and exercises

Theorem 6.1 and Example 6.3 are due to N. S. Feldman [107]. Theorem 6.2 formalizes some ideas from [107].
The study of chaotic linear operators was initiated in [123]. Theorem 6.12 is due to K.-G. Grosse-Erdmann [134]. Theorem 6.13 was obtained in [231] (see also [146]). Chaotic *semigroups* of operators also deserve some attention; see e.g. [93] and the references given therein.
Frequently hypercyclic operators were introduced in [26], and further studied in [27], [61], [60] and [136]. The Frequent Hypercyclicity Criterion was proved in [26], under stronger assumptions. In the form stated in Theorem 6.18, it appears in [60]. The proof of Theorem 6.28 is an adaptation of the ideas of [79], where it is shown that any operator in a frequently hypercyclic semigroup is already frequently hypercyclic. A probabilistic version of the Frequent Hypercyclicity Criterion, based on the ideas of Chapter 5, can be found in [130].
One can relax the definition of a frequently hypercyclic operator, by requiring that, for some vector $x \in X$, each set $\mathbf{N}(x, V)$ can be enumerated as an increasing sequence (n_k) with $n_k = O(m_k)$, where (m_k) is a given sequence of integers. Operators with this property are called (m_k)-*hypercyclic* in [31] and are studied in some detail. In particular, one can find there an (m_k)-Hypercyclicity Criterion and it is shown that when the sequence (m_k) satisfies $\lim_{k\to\infty} m_k/k = \infty$ it is possible to construct (m_k)-hypercyclic operators which are not weakly mixing.

That some Banach spaces fail to support any chaotic operator was observed by J. Bonet, F. Martínez-Giménez and A. Peris in [57]. Their result was formulated for a Banach space with a hereditarily indecomposable dual space, using essentially the same proof as the one we have given. As already mentioned, the frequently hypercyclic part of Theorem 6.36 is due to S. Shkarin [226].

Theorem 6.45 is just one of the many interesting results contained in [14]. The first example of an operator T such that $(\|T^n\|)$ is bounded along some increasing sequence of integers (n_k) and yet $\sigma_p(T) \cap \mathbb{T}$ is uncountable appears in the paper [200] of T. Ransford and M. Roginskaya. Their work was motivated by a classical result of B. Jamison [147], according to which the unimodular point spectrum of any power-bounded operator is countable. Exercise 6.8 outlines a proof of Jamison's theorem.

EXERCISE 6.1   **Transitivity plus periodicity implies chaos** ([18])
Let (E, d) be an infinite metric space, and let $f : E \to E$ be a continuous map. Assume that f is topologically transitive and has a dense set of periodic points.

1. Show that one can find $\delta > 0$ with the following property: for any $x \in E$, there exists a $p \in E$ such that $\mathrm{dist}(x, O(p, f)) \geq \delta$. (*Hint*: First show that one can find two periodic points q_1 and q_2 with disjoint orbits.)
2. Let $\delta > 0$, $x \in E$ and p be as in part 1 of the exercise. Let U be an open neighbourhood of x, and let $q \in B(x, \delta/4)$ be a periodic point with period m. Finally, set $V := \bigcap_{i=1}^m f^{-i}(B(f^i(p), \delta/4))$.
   (a) Why is it possible to find a $y \in B(x, \delta/4) \cap U$ such that $f^k(y) \in V$ for some $k \in \mathbb{N}$?
   (b) Write $k = am - r$, where $a \in \mathbb{N}$ and $r \in \{1, \ldots, m\}$. Show that either $d(f^{am}(x), q) \geq \delta/4$ or $d(f^{am}(x), f^{am}(y)) \geq \delta/4$. (*Hint*: Observe that $f^{am}(y) \in B(f^r(p), \delta/4)$.)
3. Conclude that f has a sensitive dependence on the initial conditions.

EXERCISE 6.2   Let $\alpha_1, \ldots, \alpha_p \in \mathbb{C} \backslash \mathbb{D}$ and $m_1, \ldots, m_p \in \mathbb{N}$. Let $\mathcal{P}$ be the set of all holomorphic polynomials satisfying $P^{(j)}(\alpha_i) = 0$ for each i and all $j \in \{0, \ldots, m_i\}$. Show that $\mathcal{P}$ is dense in $H^2(\mathbb{D})$.

EXERCISE 6.3 ([29])   Let $\mathcal{H}^2$ be the image of the Hardy space $H^2(\mathbb{D})$ under the Cayley map $\omega(z) = i(1 + z)/(1 - z)$. For any $a \in \mathbb{R}$, let T_a be the translation operator defined on $\mathcal{H}^2$ by $T_a F(s) = F(s + a)$. Show that if $F \in \mathcal{H}^2$ is arbitrary then $\|T_a^n(F)\| \geq c \|F\|/n$ for some constant $c > 0$ and all $n \geq 1$.

EXERCISE 6.4   **Semigroups of chaotic operators**
Given a semigroup of operators $\mathcal{T} \subset \mathfrak{L}(X)$, a point $x \in X$ is said to be a **periodic** point for $\mathcal{T}$ if one can find $T \in \mathcal{T}$, $T \neq I$, such that $T(x) = x$. The semigroup $\mathcal{T}$ is said to be **chaotic** if it is hypercyclic and has a dense set of periodic points.

1. Let D_1, D_2 be countable subsets of $\mathbb{R}$ such that D_2 is dense in $\mathbb{R}$ and $D_1 \cap D_2 = \varnothing$. Show that one can find a Cantor set $\mathcal{C} \subset [0, 1]$ such that $\mathcal{C} \cap D_1 = \varnothing$ and $\mathcal{C} \cap D_2$ is dense in $\mathcal{C}$.
2. Let $t_0 \in \mathbb{R} \setminus \mathbb{Q}$, and let $\lambda := e^{2i\pi t_0}$. Show that there exists an invertible chaotic Hilbert space operator T such that λT is not chaotic.
3. Let $t_0 \in \mathbb{R} \setminus \mathbb{Q}$. Show that there exists a chaotic C_0-semigroup $(T_t)_{t \geq 0}$ such that T_{t_0} is not chaotic. (*Hint*: take a logarithm of the above operator T.)

Remark: With the help of a more sophisticated Cantor set $\mathcal{C}$, one can obtain a chaotic C_0-semigroup $(T_t)_{t \geq 0}$ such that no operator T_t is chaotic. See [24] for details.

EXERCISE 6.5   Let D be the derivative operator on $H(\mathbb{C})$. Show that D is frequently hypercyclic.

EXERCISE 6.6   Let X be a topological vector space, and let $T \in \mathfrak{L}(X)$. For any $x \in X$, denote by E_x the set of all $y \in X$ satisfying the following property: there exists some set $\mathbf{N}_y \subset \mathbb{N}$ with positive lower density such that $T^n(x) \to y$ as $n \to \infty$ along $\mathbf{N}_y$. Show that each set E_x is countable. (*Hint*: If $y \neq y'$ then $\mathbf{N}_y \cap \mathbf{N}_{y'}$ is finite.)

EXERCISE 6.7   Let (a_k) be an increasing sequence of positive real numbers. Show that the convergence of $\sum_{k\geq 1} a_k^2 \left(\sum_{j>k}(1/a_j)\right)^2$ is equivalent to that of $\sum_{k\geq 1} (a_k/a_{k+1})^2$. (*Hint:* The inequality $(x_1 + \cdots + x_N)^2 \leq \sum_{j=1}^{N} 2^j x_j^2$ may be useful.)

EXERCISE 6.8   **Jamison's theorem** ([147])
Let X be a separable Banach space, and let $T \in \mathfrak{L}(X)$. Assume that $\sup_{k\geq 1} \|T^k\| < \infty$.

1. Let $\lambda, \mu \in \mathbb{T}$. Assume that λ, μ are eigenvalues of T, and let e_λ, e_μ be associated eigenvectors, with $\|e_\lambda\| = 1 = \|e_\mu\|$. Show that $|\mu^k - \lambda^k| \leq 2\|T^k\| \, \|e_\mu - e_\lambda\|$ for all $k \geq 1$. (*Hint:* Write $T^k(e_\mu) - T^k(e_\lambda) = (\mu^k - \lambda^k)e_\mu + \lambda^k(e_\mu - e_\lambda)$ and observe that $\|T^k\| \geq 1$.) Deduce that if $\lambda \neq \mu$ then $\|e_\lambda - e_\mu\| \geq 1/(2M)$, where $M = \sup_{k\geq 1} \|T^k\|$.
2. Show that $\sigma_p(T) \cap \mathbb{T}$ is countable.

EXERCISE 6.9   **Jamison sequences** ([13], [14])
Let $(n_k)_{k\geq 1}$ be an increasing sequence of positive integers. The sequence (n_k) is said to be a **Jamison sequence** if, whenever a bounded operator T acting on a separable Banach space X satisfies $\sup_{k\geq 1} \|T^{n_k}\| < \infty$, it follows that $\sigma_p(T) \cap \mathbb{T}$ is countable. By the previous exercise, the sequence $(n_k) = (k)$ is a Jamison sequence.

1. Assume that, for some $\varepsilon > 0$, the set $\Lambda_\varepsilon := \left\{\zeta \in \mathbb{T}; \sup_{k\geq 1} |\zeta^{n_k} - 1| < \varepsilon\right\}$ is countable. Show that (n_k) is a Jamison sequence. (*Hint:* Adapt the method of the previous exercise.)
2. Assume that $n_{k+1}/n_k \to \infty$ as $k \to \infty$.
   (a) Show that, for any given $\delta \in (0,1)$, one can find an integer $k_0 \geq 1$ such that the set $\{\lambda \in \mathbb{T}; |\lambda^{n_k} - 1| < \delta \text{ for all } k \geq k_0\}$ contains a Cantor set.
   (b) Let S be twice the backward shift acting on $X_0 := \ell^2(\mathbb{N})$. For any $x \in X_0$, set
   $$|x| := \max\left(\|x\|, \max_{j\geq 1} \max_{k_1,\ldots,k_j\geq 1} \left\|\prod_{l=1}^{j}(S^{n_{k_l}} - I)x\right\|\right).$$
   Show that $|S^{n_k}x| \leq 2|x|$ for each k and all $x \in X_0$.
   (c) Show that (n_k) is not a Jamison sequence. (*Hint:* Argue as in the proof of Theorem 6.45.)

7

Common hypercyclic vectors

Introduction

Let $(T_\lambda)_{\lambda \in \Lambda}$ be a family of hypercyclic operators acting on the same F-space X. When the family (T_λ) is countable, it follows from the Baire category theorem that $\bigcap_{\lambda \in \Lambda} HC(T_\lambda)$ is a residual subset of X, since we already know that $HC(T_\lambda)$ is residual for each $\lambda \in \Lambda$. Thus, there are (many) vectors $x \in X$ which are hypercyclic for *all* operators T_λ. Any such vector x will be called a **common hypercyclic vector** for the family (T_λ).

However, the following two examples show that one cannot hope to get common hypercyclic vectors for an arbitrary uncountable family of hypercyclic operators (T_λ). The second is due to A. Borichev (quoted in [1]), and we will refer to it as Borichev's example.

EXAMPLE 7.1   The hypercyclic weighted shifts on $\ell^2(\mathbb{N})$ have no common hypercyclic vectors.

PROOF   Let $x = (x_n) \in \ell^2(\mathbb{N})$ be arbitrary. We have to find some weight sequence $\mathbf{w} = (w_n)_{n \geq 1}$ such that the associated weighted shift $B_\mathbf{w}$ is hypercyclic but $x \notin HC(B_\mathbf{w})$. Since $x_n \to 0$, we may find an increasing sequence of integers (n_k) such that $|x_n| \leq 2^{-k}$ whenever $n \geq n_k$. We define $\mathbf{w}$ by setting $w_n := 2$ if $n = n_k$ for some k and $w_n := 1$ otherwise. Then $B_\mathbf{w}$ is hypercyclic, since $w_1 \cdots w_n \to \infty$. However, $\|B_\mathbf{w}^n(x) - 2e_0\| \geq |2 - w_1 \cdots w_n x_n| \geq 1$ for any $n \geq n_1$, so that x is not a hypercyclic vector for $B_\mathbf{w}$. □

EXAMPLE 7.2 (BORICHEV'S EXAMPLE)   Let $\Lambda_0 = (1, \infty) \times (1, \infty)$. For $\lambda = (s, t) \in \Lambda_0$, define $T_\lambda := sB \oplus tB$ acting on $\ell^2(\mathbb{N}) \oplus \ell^2(\mathbb{N})$, where B is the canonical backward shift on $\ell^2(\mathbb{N})$. If $\Lambda \subset \Lambda_0$ is such that $\bigcap_{\lambda \in \Lambda} HC(T_\lambda) \neq \varnothing$ then Λ has Lebesgue measure 0.

PROOF   First we note that each operator T_λ is hypercyclic, being a direct sum of two mixing operators. Let us fix $\Lambda \subset \Lambda_0$ such that $\bigcap_{\lambda \in \Lambda} HC(T_\lambda) \neq \varnothing$ and choose a common hypercyclic vector $z = x \oplus y \in \ell^2(\mathbb{N}) \oplus \ell^2(\mathbb{N})$. Then, for each $\lambda = (s, t) \in \Lambda$, one can approximate the vector $e_0 \oplus e_0$ by vectors of the form $s^n B^n(x) \oplus t^n B^n(y)$. Here, of course, e_0 is the first vector of the canonical basis of $\ell^2(\mathbb{N})$. Looking at the first coordinates of $s^n B^n(x)$ and $t^n B^n(y)$, it follows that one can find $n \in \mathbb{N}^*$ such that $|s^n x_n - 1|$ and $|t^n y_n - 1|$ are arbitrarily small. Then x_n and y_n have positive real parts and, putting $a_n := -n^{-1} \log(\text{Re}(x_n))$, $b_n := -n^{-1} \log(\text{Re}(y_n))$, we see that $n\,|\log(s) - a_n|$ and $n\,|\log(t) - b_n|$ are arbitrarily small. Denoting by $C_{n,\varepsilon}$ the square with center (a_n, b_n) and side ε/n, it follows that the set $\log(\Lambda) := \{(\log(s), \log(t)); (s, t) \in \Lambda\}$ is contained in

$\bigcup_{n \geq 1} C_{n,\varepsilon}$ for every $\varepsilon > 0$. Since the series $\sum_{n \geq 1} n^{-2}$ is convergent, we conclude that $\log(\Lambda)$ has Lebesgue measure 0, hence that Λ has measure 0 as well. $\qquad\square$

Nevertheless, several positive results have been discovered quite recently concerning the existence of common hypercyclic vectors for uncountable families of operators. The first non-trivial example is a theorem due to E. Abakumov and J. Gordon [1] according to which the operators λB, $\lambda > 1$, have a residual set of common hypercyclic vectors, where B is the usual backward shift acting on $\ell^2(\mathbb{N})$; see Example 7.17. This result certainly motivated to a large extent the interest in common hypercyclicity.

Some other interesting examples were met in Chapter 3. Indeed, the León–Müller and Conejero–Müller–Peris theorems proved there can be viewed as very strong common hypercyclicity results, since they show that some (uncountable) families of hypercyclic operators share the *same* set of hypercyclic vectors. Here is an application. Recall that we denote by $Aut(\mathbb{D})$ the set of all automorphisms of the unit disk $\mathbb{D}$ and that the composition operator C_ϕ induced by any automorphism $\phi \in Aut(\mathbb{D})$ is hypercyclic on $H^2(\mathbb{D})$ (see Chapter 1).

EXAMPLE 7.3   Let Λ be the set of all $\phi \in Aut(\mathbb{D})$ having $\alpha = +1$ as attractive fixed point. Then $\bigcap_{\phi \in \Lambda} HC(C_\phi)$ is a residual subset of $H^2(\mathbb{D})$.

PROOF   As usual, we move to the upper half-plane $\mathbb{P}_+ = \{\operatorname{Im}(s) > 0\}$, thanks to the Cayley map $\omega(z) = i(1+z)/(1-z)$. We denote by $\mathcal{H}^2 = \{f \circ \omega^{-1}; f \in H^2(\mathbb{D})\}$ the image of $H^2(\mathbb{D})$ under the Cayley map, endowed with the norm

$$\|F\|_2 = \left(\int_{\mathbb{R}} |F(t)|^2 \frac{dt}{1+t^2} \right)^{1/2}.$$

We recall that a hyperbolic automorphism of $\mathbb{D}$ (having $\alpha = +1$ as attractive fixed point) is associated via ω with a dilation $h_{\lambda,b}(s) = \lambda(s-b)+b$ on $\mathbb{P}_+$ ($\lambda > 1, b \in \mathbb{R}$), whereas a parabolic automorphism is associated with a translation $T_a(s) = s+a$ ($a \in \mathbb{R}^*$). We will denote by $h_{\lambda,b}$ and T_a the corresponding composition operators acting on $\mathcal{H}^2$.

We first show that $HC(h_{\lambda,0}) \subset HC(h_{\lambda,b})$ for each $\lambda > 1$ and every $b \in \mathbb{R}$. Let us fix λ, b, and let $f \in HC(h_{\lambda,0})$. If $g \in \mathcal{H}^2$ then

$$\|h_{\lambda,b}^n(f) - g\|_2^2 = \int_{\mathbb{R}} |f(\lambda^n(t-b)+b) - g(t)|^2 \frac{dt}{1+t^2}$$

$$= \int_{\mathbb{R}} |f(\lambda^n \tau) - g(\tau + b - \lambda^{-n}b)|^2 \frac{d\tau}{1 + (\tau + b - \lambda^{-n}b)^2}$$

$$\leq C \|h_{\lambda,0}^n(f) - T_{b-\lambda^{-n}b}(g)\|_2^2,$$

for all $n \in \mathbb{N}$ and some finite constant C. It follows that if n is a large integer such that $h_{\lambda,0}^n(f)$ is close to $T_b(g)$ then $h_{\lambda,b}^n(f)$ is close to g. Thus, f is indeed a hypercyclic vector for $h_{\lambda,b}$.

We now come to the semigroup argument. The family $(T_a)_{a>0}$ is a hypercyclic C_0-semigroup on $\mathcal{H}^2$. By the Conejero–Müller–Peris theorem 3.5, $HC(T_a)$ does not depend on $a > 0$; in particular, $\bigcap_{a>0} HC(T_a)$ is a residual subset of $\mathcal{H}^2$. Likewise the family $(\widetilde{T}_t)_{t>0}$ is a hypercyclic C_0-semigroup, where $\widetilde{T}_t = h_{e^t,0}$; hence $\bigcap_{\lambda>1} HC(h_{\lambda,0})$ is also residual. By the Baire category theorem, $\bigcap_{a>0} HC(T_a) \cap \bigcap_{\lambda>1} HC(h_{\lambda,0})$ is residual as well. $\square$

REMARK   It is not hard to see that the family of all hypercyclic composition operators on $H^2(\mathbb{D})$ does not have common hypercyclic vectors; see Exercise 7.1.

In this chapter, our aim is to study the problem of common hypercyclicity in some detail, for families of operators which are not semigroups. We shall formulate several criteria giving sufficient conditions for the existence of a common hypercyclic vector, which apply in various situations. These criteria may be viewed as "uncountable Baire category theorems" (applying to very special families of open sets), since in fact they provide residual sets of common hypercyclic vectors. Of course, our main tool will be the Baire category theorem itself.

In all cases under consideration, we shall deal with a family of operators $(T_\lambda)_{\lambda \in \Lambda} \subset \mathfrak{L}(X)$ parametrized by some topological space Λ. We will always make the following assumptions:

(A1) the map $(\lambda, x) \mapsto T_\lambda(x)$ is continuous from $\Lambda \times X$ into X;
(A2) the parameter space Λ is a K_σ set, i.e. a countable union of compact sets.

The continuity assumption (A1) is quite natural, and (A2) will be needed in order to use some Baire category arguments. This is still too general a setting, however. Typically, the criteria we will encounter in this chapter apply to *one-dimensional* families of operators, i.e. for which the parameter set Λ is an interval of the real line. Borichev's example given above indicates rather clearly that some troubles may occur when considering higher-dimensional families. It is rather amusing that the León–Müller and Conejero–Müller–Peris theorems also apply typically to one-dimensional families. Sometimes these two kinds of result may be combined, and then one obtains two-dimensional positive results!

7.1 Common hypercyclic vectors and transitivity

In this section, we put ourselves in the most general situation. Thus, X is a separable F-space and $(T_\lambda)_{\lambda \in \Lambda} \subset \mathfrak{L}(X)$ is a family of operators satisfying assumptions (A1), (A2) above.

7.1.1 General facts

If V is an open subset of X and $K \subset \Lambda$, we set

$$V_K := \left\{ x \in X; \; \forall \lambda \in \Lambda \; \exists n \in \mathbb{N} : T_\lambda^n(x) \in V \right\}.$$

Clearly, if $\mathcal{O}$ is a countable basis of open sets for X then

$$\bigcap_{\lambda \in K} HC(T_\lambda) = \bigcap_{V \in \mathcal{O}} V_K.$$

This very simple observation will be crucial in what follows, since it allows us to replace an uncountable intersection by a countable one. In fact, all the results given in this chapter will depend on it.

Another simple fact will be used repeatedly: if K is a compact subset of Λ then each set V_K is *open* in X. Indeed, $X \setminus V_K$ is closed in X because it is the projection of the closed set $\{(\lambda, x) \in \Lambda \times X; \ \forall n \in \mathbb{N} \ : T_\lambda^n(x) \in X \setminus V\}$ along the compact set K (see [100, Chapter XI, Theorem 2.5]). In particular, we get the following result.

PROPOSITION 7.4 *Under the above assumptions, $\bigcap_{\lambda \in \Lambda} HC(T_\lambda)$ is a G_δ set. It is dense in X if and only if all sets V_K with $K \subset \Lambda$ compact are dense.*

PROOF Write $\Lambda = \bigcup_i K_i$, where (K_i) is a countable family of compact sets. Then

$$\bigcap_{\lambda \in \Lambda} HC(T_\lambda) = \bigcap_i \bigcap_{\lambda \in K_i} HC(T_\lambda),$$

so the result follows from the above observations and the Baire category theorem. $\square$

This result suggests that if we are able to find *one* vector in $\bigcap_{\lambda \in \Lambda} HC(T_\lambda)$ then in fact there will be a residual set of common hypercyclic vectors. For example, we have the following:

COROLLARY 7.5 *If the family (T_λ) is commuting then $\bigcap_{\lambda \in \Lambda} HC(T_\lambda)$ is either empty or residual in X.*

PROOF Assume $\bigcap_{\lambda \in \Lambda} HC(T_\lambda) \neq \varnothing$, and pick $x \in \bigcap_{\lambda \in \Lambda} HC(T_\lambda)$. Fix also $\lambda_0 \in \Lambda$. Using the commutativity assumption, it is straightforward to check that the dense set $\{T_{\lambda_0}^n(x); \ n \geq 1\}$ is contained in $\bigcap_{\lambda \in \Lambda} HC(T_\lambda)$. Since $\bigcap_{\lambda \in \Lambda} HC(T_\lambda)$ is a G_δ set, this concludes the proof. $\square$

We now show that, just as in the case of a single operator, the existence of common hypercyclic vectors can be rephrased in terms of transitivity. Recall that an operator $T \in \mathfrak{L}(X)$ is hypercyclic iff, for each pair of non-empty open sets (U, V), one can find an integer n such that $T^n(U) \cap V \neq \varnothing$ or, equivalently, iff for each pair of non-empty open sets $(U, V) \subset X$ one can find an $n \in \mathbb{N}$ and a non-empty open set $W \subset U$ such that $T^n(W) \subset V$. The parametrized version reads as follows.

PROPOSITION 7.6 *The following are equivalent.*

(i) $\bigcap_{\lambda \in \Lambda} HC(T_\lambda)$ *is a dense G_δ subset of X.*

(ii) *For each pair of non-empty open sets (U, V) and for each compact set $K \subset \Lambda$, one can find sets of parameters $\Lambda_1, \ldots, \Lambda_q \subset \Lambda$, integers $n_1, \ldots, n_q$ and a non-empty open set $W \subset U$ such that*

(1) $\bigcup_{i=1}^{q} \Lambda_i \supset K$;

(2) $T_\lambda^{n_i}(W) \subset V$ for each i and every $\lambda \in \Lambda_i$.

PROOF   We already know that $\bigcap_{\lambda \in \Lambda} HC(T_\lambda)$ is a G_δ set.

Assume that (i) holds, and let us prove (ii). Let U and V be two non-empty open subsets of X, and let $K \subset \Lambda$ be compact. Pick $x \in \bigcap_{\lambda \in \Lambda} HC(T_\lambda) \cap U$. For each $\lambda \in \Lambda$, one can find an integer n_λ such that $T_\lambda^{n_\lambda}(x) \in V$. By continuity of the map $(\mu, y) \mapsto T_\mu^{n_\lambda}(y)$ at (λ, x), one can find an open neighbourhood Λ_λ of λ and an open neighbourhood W_λ of x such that $T_\mu^{n_\lambda}(W_\lambda) \subset V$ for all $\mu \in \Lambda_\lambda$. By compactness, one can cover K by finitely many Λ_λ, say $K \subset \Lambda_{\lambda_1} \cup \cdots \cup \Lambda_{\lambda_q}$. Then the integers $n_{\lambda_1}, \ldots, n_{\lambda_q}$ and the open set $W := U \cap \bigcap_{i=1}^{q} W_{\lambda_i}$ have the required properties (W is non-empty, since $x \in W$).

Conversely, conditions (1), (2) above imply that $U \cap V_K \neq \varnothing$ for any compact set $K \subset \Lambda$ and each pair of non-empty open sets (U, V): indeed, the open set W is contained in $U \cap V_K$. Thus, all sets V_K are dense in X. By Proposition 7.4, this concludes the proof.   □

We conclude this subsection by showing that for a reasonable parameter space Λ there is always a large subset allowing common hypercyclicity. Recall that a topological space is said to be **Polish** if it is separable and completely metrizable.

PROPOSITION 7.7   *Assume that the parameter space Λ is Polish. If all operators T_λ are hypercyclic then there is a dense G_δ set $G \subset \Lambda$ such that $\bigcap_{\lambda \in G} HC(T_\lambda) \neq \varnothing$.*

PROOF   The set $\mathcal{G} := \{(\lambda, x) \in \Lambda \times X; \ x \in HC(T_\lambda)\}$ is easily seen to be G_δ in the product space $\Lambda \times X$. Moreover, if all operators T_λ are hypercyclic then each λ-section $\mathcal{G}_\lambda$ is comeager in X. By the Kuratowski–Ulam theorem (the analogue of Fubini's theorem for Baire categories, see [156]), one can find an $x \in X$ such that the x-section $\mathcal{G}^x := G$ is dense in Λ.   □

7.1.2 Translation operators

To illustrate Proposition 7.6, let us see how it applies to translation operators on $H(\mathbb{C})$, the first examples of hypercyclic operators encountered in this book. Recall that, for any $a \in \mathbb{C}$, the translation operator $T_a : H(\mathbb{C}) \to H(\mathbb{C})$ is defined by $T_a f(z) = f(z + a)$. We saw in Chapter 1 that T_a is hypercyclic when $a \neq 0$.

THEOREM 7.8   *The operators T_a, $a \neq 0$, have a residual set of common hyper-cyclic vectors.*

PROOF   We first show that $\bigcap_{\theta \in [0, 2\pi)} HC(T_{e^{i\theta}})$ is residual. Let $K = [a, b]$ be a compact subinterval of $[0, 2\pi)$ and let (U, V) be a pair of non-empty open sets in $H(\mathbb{C})$. One can find $\varepsilon > 0$, $R < \infty$ and $u, v \in H(\mathbb{C})$ such that

$$U \supset \left\{ h \in H(\mathbb{C}); \ \|h - u\|_{\overline{D}(0,R)} < \varepsilon \right\}$$

Fig. 7.1 The shape of the compact set E.

and

$$V \supset \left\{ h \in H(\mathbb{C}); \; \|h - v\|_{\overline{D}(0,R)} < 2\varepsilon \right\}.$$

Since v is continuous, one can choose a $\tau > 0$ such that

$$(|z| \leq R \text{ and } |z - z'| < \tau) \implies |v(z) - v(z')| < \varepsilon.$$

Finally, let N be an integer such that $N > 2R$. Since the series $\sum_{j \geq 1} \tau/(jN)$ is divergent, one may find a subdivision $a = \theta_0 < \theta_1 < \cdots < \theta_q = b$ of the interval $[a, b]$ such that $|\theta_j - \theta_{j-1}| \leq \tau/(jN)$ for all $j > 0$. We set $\Lambda_j := [\theta_{j-1}, \theta_j]$, $n_j := jN$ and $E_j := \overline{D}(0, R) + n_j e^{i\Lambda_j}$. Since $N > 2R$, the sets E_j are pairwise disjoint and do not intersect $\overline{D}(0, R)$. Moreover, $E := \overline{D}(0, R) \cup \bigcup_{j=1}^{q} E_j$ has a connected complement if N is large enough to ensure that no "circular tube" E_j contains the associated circle $\{|z| = n_j\}$ (see Figure 7.1).

We define $W \subset H(\mathbb{C})$ as the open set made up of all $f \in H(\mathbb{C})$ satisfying

$$\|f - u\|_{\overline{D}(0,R)} < \varepsilon$$

and

$$\|f - v(\cdot - n_j e^{i\theta_j})\|_{E_j} < \varepsilon \qquad \text{for any } j \in \{1, \ldots, q\}. \tag{7.1}$$

Clearly $W \subset U$, and W is *non-empty* by Runge's approximation theorem. Moreover, $T_{e^{i\theta}}^{n_j}(W) \subset V$ for each j and every $\theta \in \Lambda_j$, by the definition of W. Indeed, if $f \in W$ then

$$\left| f\left(z + n_j e^{i\theta}\right) - v(z) \right| \leq \left| f\left(z + n_j e^{i\theta}\right) - v\left(\left(z + n_j e^{i\theta}\right) - n_j e^{i\theta_j}\right) \right|$$
$$+ \left| v\left(z + n_j\left(e^{i\theta} - e^{i\theta_j}\right)\right) - v(z) \right|$$
$$< \varepsilon + \varepsilon = 2\varepsilon$$

for all $z \in \overline{D}(0, R)$, where we have used (7.1) and the inequality $|n_j(e^{i\theta} - e^{i\theta_j})|$ $< \tau$. By Proposition 7.6, it follows that $\bigcap_{\theta \in [0, 2\pi)} HC(T_{e^{i\theta}})$ is a dense G_δ subset of $H(\mathbb{C})$.

To conclude the proof, we now observe that $HC(T_{e^{i\theta}}) = HC(T_{re^{i\theta}})$ for each $\theta \in [0, 2\pi)$ and all $r > 0$. Indeed, the family $(T_{re^{i\theta}})_{r>0}$ is a hypercyclic C_0-semigroup on $H(\mathbb{C})$, so that by the Conejero–Müller–Peris theorem $HC(T_{re^{i\theta}})$ does not depend on r.   □

REMARK   The subdivision of $[a, b]$ used in the proof foreshadows a method that will be used several times in Section 7.2.

7.1.3 A parametrized Hypercyclicity Criterion

With Proposition 7.6 at hand, we may now formulate a criterion for common hypercyclicity which looks very much like the Hypercyclicity Criterion. It is due to K. C. Chan and R. Sanders [73].

THEOREM 7.9   *Assume that, for each compact set $K \subset \Lambda$, one can find a dense set $\mathcal{D} \subset X$ such that the following properties hold. For each $v \in \mathcal{D}$ and each open neighbourhood O of 0 in X, there exist a dense set $\mathcal{D}_v \subset X$, a sequence of integers (n_k) and vectors $S_{\lambda,k}(v) \in X$ ($\lambda \in K$) satisfying the following:*

(1) *$S_{\lambda,k}(v) \to 0$ pointwise on K as $k \to \infty$;*
(2) *$T_\lambda^{n_k}(u) \to 0$ uniformly on K, for each $u \in \mathcal{D}_v$;*
(3) *one can find compact sets of parameters $\Lambda_1, \ldots, \Lambda_q \subset \Lambda$ such that*

    (i) *$\bigcup_{i=1}^q \Lambda_i \supset K$,*
    (ii) *$T_\lambda^{n_k} S_{\mu,k}(v) - v \in O$ for all $k \in \mathbb{N}$ whenever $i \in \{1, \ldots, q\}$ and $\lambda, \mu \in \Lambda_i$.*

Then $\bigcap_{\lambda \in \Lambda} HC(T_\lambda)$ is a dense G_δ subset of X.

PROOF   In order to apply Proposition 7.6, let us fix a pair (U, V) of non-empty open sets in X and a compact set $K \subset \Lambda$. Let $\mathcal{D} \subset X$ be the dense set associated with K. Then V has the form $V = v + N$, where $v \in \mathcal{D}$ and N is an open neighbourhood of 0. Let O be a neighbourhood of 0 such $O + O \subset N$, and choose $\mathcal{D}_v$, (n_k), $S_{\lambda,k}(v)$, (Λ_i) satisfying (1)–(3). The key point is the following

FACT   For each $i \in \{1, \ldots, q\}$ and each non-empty open set W_{i-1}, there exist a non-empty open set $W_i \subset W_{i-1}$ and an integer k_i such that $T_\lambda^{n_{k_i}}(W_i) \subset V$ for every $\lambda \in \Lambda_i$.

PROOF OF THE FACT   Pick $u \in \mathcal{D}_v \cap W_{i-1}$, and let Ω be a neighbourhood of 0 such that $u + \Omega \subset W_{i-1}$. Let us also fix $\lambda_i \in \Lambda_i$. By (1) and (2), one can find $k_i \in \mathbb{N}$ such that $S_{\lambda_i, k_i}(v) \in \Omega$ and $T_\lambda^{n_{k_i}}(u) \in O$ for every $\lambda \in \Lambda_i$. By (3), we know that we also have $T_\lambda^{n_{k_i}} S_{\lambda_i, k_i}(v) - v \in O$. Set $w := u + S_{\lambda_i, k_i}(v)$. Then $w \in W_{i-1}$ and, writing

$$T_\lambda^{n_{k_i}}(w) = T_\lambda^{n_{k_i}}(u) + \left(T_\lambda^{n_{k_i}} S_{\lambda_i, k_i}(v) - v\right) + v,$$

we see that $T_\lambda^{n_{k_i}}(w) \in V$ for any $\lambda \in \Lambda_i$. Hence $W_{i-1} \cap \bigcap_{\lambda \in \Lambda_i} (T_\lambda^{n_{k_i}})^{-1}(V) \neq \varnothing$. Now, $\bigcap_{\lambda \in \Lambda_i} (T_\lambda^{n_{k_i}})^{-1}(V)$ is open in X by the compactness of Λ_i and the continuity of the map $(\lambda, x) \mapsto T_\lambda^{n_{k_i}}(x)$. Hence we may take $W_i := W_{i-1} \cap \bigcap_{\lambda \in \Lambda_i} (T_\lambda^{n_{k_i}})^{-1}(V)$. $\qquad\square$

Coming back to the proof of Theorem 7.9, we start with $W_0 = U$. Using the above fact, we define inductively non-empty open sets $W_q \subset \cdots \subset W_0 = U$ and integers $m_1, \ldots, m_q$ such that $T_\lambda^{m_i}(W_i) \subset V$ for any $\lambda \in \Lambda_i$. Then $W := W_q$ satisfies the assumptions of Proposition 7.6. $\qquad\square$

REMARK 7.10   If Λ is an interval of the real line, condition (3) can be replaced by the following.

(3′)   One can find $\delta > 0$ such that $T_\lambda^{n_k} S_{\mu,k}(v) - v \in O$ whenever $\lambda, \mu \in K$ satisfy $|\lambda - \mu| < \delta$.

Indeed, it is enough to consider the case of a compact interval $K = [a, b]$. Let $\lambda_0 = a < \lambda_1 < \cdots < \lambda_q = b$ be a subdivision of $[a, b]$ such that $\lambda_i - \lambda_{i-1} < \delta$. Setting $\Lambda_i := [\lambda_{i-1}, \lambda_i]$ for $i = 1, \ldots, q$, we see at once that (3) is satisfied.

REMARK 7.11   When the family (T_λ) consists of a single operator T, the assumptions of Theorem 7.9 reduce to that of the Hypercyclicity Criterion. They are even formally weaker, since the sequence (n_k) may depend on the point $v \in \mathcal{D}$, but nevertheless are still strong enough to imply the three open sets condition (see Chapter 4) and hence are equivalent to the original criterion. We have stated Theorem 7.9 in this form because it will be applied later in a case where it is indeed useful to allow the sequence (n_k) to depend on v; see the proof of Theorem 7.25 below.

In view of the last remark, Theorem 7.9 is "philosophically" quite satisfactory. But unfortunately, it cannot be applied to any of the concrete examples that we will consider in this chapter! The reason is that conditions (3) and (3′) are very restrictive. Indeed, usually one cannot hope to control $T_\lambda^n S_\mu^n(v) - v$ under the assumption $|\lambda - \mu| < \delta$, without any reference to n. A more natural assumption would be to require that $T_\lambda^n S_\mu^n(v) - v$ is close to 0 when $|\lambda - \mu| < \delta_n$, for some suitable sequence (δ_n). But of course, there will be a price to pay for such a weakening of (3′): conditions (1) and (2) will have to be strengthened. In the next two sections, we show how this can be done.

7.2 Common hypercyclicity criteria

In this section, we consider again a family of operators $(T_\lambda)_{\lambda \in \Lambda}$ acting on a separable F-space X and satisfying conditions (A1), (A2). We make one additional assumption.

(A3)   There exists a dense set $\mathcal{D} \subset X$ such that each operator T_λ has a (partial) right-inverse $S_\lambda : \mathcal{D} \to \mathcal{D}$; that is, $T_\lambda S_\lambda(x) = x$ for all $x \in \mathcal{D}$.

7.2.1 Basic Criterion

The following criterion for common hypercyclicity is a very simple consequence of
Proposition 7.4. Yet all the later results will ultimately rely on it.

LEMMA 7.12 (BASIC CRITERION)   *Assume that the following holds true. For each
compact set* $K \subset \Lambda$, *each pair* $(u, v) \in \mathcal{D} \times \mathcal{D}$ *and each open neighbourhood* O *of* 0
in X, *one can find parameters* $\lambda_1, \dots, \lambda_q \in \Lambda$, *sets of parameters* $\Lambda_1, \dots, \Lambda_q \subset \Lambda$
with $\lambda_i \in \Lambda_i$ *for all* i *and integers* $n_1, \dots, n_q \in \mathbb{N}$, *such that*

(BC1) $\bigcup_i \Lambda_i \supset K$;
(BC2) $\forall i \in \{1, \dots, q\}, \forall \lambda \in \Lambda_i : \sum_{j=1}^q S_{\lambda_j}^{n_j}(v) \in O$ *and* $\sum_{j \neq i} T_\lambda^{n_i} S_{\lambda_j}^{n_j}(v) \in O$;
(BC3) $\forall i \in \{1, \dots, q\}, \forall \lambda \in \Lambda_i : T_\lambda^{n_i}(u) \in O$;
(BC4) $\forall i \in \{1, \dots, q\}, \forall \lambda \in \Lambda_i : T_\lambda^{n_i} S_{\lambda_i}^{n_i}(v) - v \in O$.

Then $\bigcap_{\lambda \in \Lambda} HC(T_\lambda)$ *is a dense* G_δ *subset of* X.

PROOF   By Proposition 7.4, it is enough to show that $U \cap V_K \neq \varnothing$ for any compact
set $K \subset \Lambda$ and each pair (U, V) of non-empty open sets in X. Recall that

$$V_K := \{x \in X;\ \forall \lambda \in \Lambda\ \exists n \in \mathbb{N} : T_\lambda^n(x) \in V\}.$$

Since $\mathcal{D}$ is dense in X, we may assume that $U = u + N$ and $V = v + N$,
where $u, v \in \mathcal{D}$ and N is a neighbourhood of 0. Let $O \subset X$ be an open neighbourhood of 0 such that $O + O + O \subset N$, and let us choose $n_i, \lambda_i, \Lambda_i$ satisfying the
assumptions of the lemma. Finally, set $x := u + \sum_{j=1}^q S_{\lambda_j}^{n_j}(v)$. Since $O \subset N$,
we immediately get from (BC2) that $x \in U$, so we just have to check that
$x \in V_K$.
    Let us fix $\lambda \in \Lambda$. Choose $i \in \{1, \dots, q\}$ such that $\lambda \in \Lambda_i$, and put $n := n_i$.
Writing

$$T_\lambda^n(x) = T_\lambda^n(u) + \sum_{j \neq i} T_\lambda^n S_{\lambda_j}^{n_j}(v) + \left(T_\lambda^{n_i} S_{\lambda_i}^{n_i}(v) - v\right) + v$$

and using (BC2)–(BC4) we see that $T_\lambda^n(x) \in V$. Since $\lambda \in K$ is arbitrary, this shows
that $x \in V_K$.   □

REMARK 7.13   In all the common hypercyclicity criteria to be given below, it will
turn out that every product family $(T_\lambda \times \cdots \times T_\lambda)_{\lambda \in \Lambda}$ satisfies the assumptions
of the Basic Criterion, so that $\bigcap_{\lambda \in \Lambda} HC(T_\lambda \times \cdots \times T_\lambda)$ is a dense G_δ subset of
$X \times \cdots \times X$. This will be useful in Chapter 8.

    We now discuss several applications of the Basic Criterion. From now on, we
restrict the generality further by assuming that

$$\Lambda \text{ is an interval of the real line.}$$

7.2.2 The Costakis–Sambarino Criterion

The following theorem due to G. Costakis and M. Sambarino [86] gives a general and powerful criterion for common hypercyclicity. Let us fix a translation-invariant metric d generating the topology of our underlying F-space X. For notational simplicity, we write $\|x\|$ instead of $d(x, 0)$.

THEOREM 7.14 *Assume that, for each $f \in \mathcal{D}$ and each compact interval $K \subset \Lambda$, the following properties hold true, where all parameters α, λ, μ belong to K.*

(1) *There exist $\kappa \in \mathbb{N}$ and a sequence of positive numbers $(c_k)_{k \geq \kappa}$ such that*

- $\displaystyle\sum_{k=\kappa}^{\infty} c_k < \infty;$
- $\|T_\lambda^{n+k} S_\alpha^n(f)\| \leq c_k$ *whenever $n \in \mathbb{N}$, $k \geq \kappa$ and $\alpha \leq \lambda$;*
- $\|T_\lambda^n S_\alpha^{n+k}(f)\| \leq c_k$ *whenever $n \in \mathbb{N}$, $k \geq \kappa$ and $\lambda \leq \alpha$.*

(2) *Given $\eta > 0$, one can find $\tau > 0$ such that, for all $n \geq 1$,*

$$0 \leq \mu - \lambda < \tau/n \implies \|T_\lambda^n S_\mu^n(f) - f\| < \eta.$$

Then $\bigcap_{\lambda \in \Lambda} HC(T_\lambda)$ is a dense G_δ subset of X.

PROOF We will apply the Basic Criterion (Lemma 7.12, conditions (BC1)–(BC4)). So, let us fix a compact interval $K = [a, b] \subset \Lambda$, a pair $(u, v) \in \mathcal{D} \times \mathcal{D}$ and an open ball $O = B_d(0, \eta)$. Let us choose $\kappa \in \mathbb{N}$ and a summable sequence $(c_k)_{k \geq \kappa}$ satisfying property (1) of the theorem for both $f := u$ and $f := v$. Finally, let us fix $\tau > 0$ such that property (2) holds for our given η and $f := v$.

We choose some large positive integer $N \geq \kappa$. Since the series $\sum (Ni)^{-1}$ is divergent, one can find a subdivision $a = \lambda_0 < \cdots < \lambda_q = b$ of K such that $\lambda_i - \lambda_{i-1} < \tau(Ni)^{-1}$ for all $i \in \{1, \ldots, q\}$. For $i \in \{1, \ldots, q\}$, we put $\Lambda_i := [\lambda_{i-1}, \lambda_i]$ (so that condition (BC1) is satisfied) and $n_i := Ni$.

Writing $\sum_{j \neq i} T_\lambda^{n_i} S_{\lambda_j}^{n_j}(v) = \sum_{j < i} + \sum_{j > i}$, property (1) of the theorem yields

$$\left\| \sum_{j \neq i} T_\lambda^{n_i} S_{\lambda_j}^{n_j}(v) \right\| \leq \sum_{j=1}^{i-1} c_{N(i-j)} + \sum_{j=i+1}^{q} c_{N(j-i)} \leq 2 \sum_{k \geq N} c_k$$

for any $\lambda \in \Lambda_i$. The right-hand side is less than η if N is large enough, in which case the second half of condition (BC2) is satisfied. Similarly, the first half of condition (BC2) and condition (BC3) are satisfied if N is large (apply property (1) with $n = 0$), whereas condition (BC4) follows from property (2) and our choice of the subdivision $(\lambda_0, \ldots, \lambda_q)$. $\square$

REMARK 7.15 In property (2) of Theorem 7.14 one can put the parameters λ, μ in reverse order. That is, one can replace property (2) by the following.

$(\widetilde{2})$ Given $\eta > 0$, one can find $\tau > 0$ such that, for all $n \geq 1$,

$$0 \leq \lambda - \mu < \tau/n \implies \|T_\lambda^n S_\mu^n(f) - f\| < \eta.$$

Indeed, everything works as above provided that we make the following modification: the subdivision is now written as $a = \lambda_1 < \cdots < \lambda_q < \lambda_{q+1} = b$, and we put $\Lambda_i := [\lambda_i, \lambda_{i+1}]$.

One can also state a "multiplicative" version of the Costakis–Sambarino Criterion; namely, for a parameter interval $\Lambda \subset (0, \infty)$ one can replace property (2) of Theorem 7.14 by the following.

(2') Given $\eta > 0$, one can find $\tau \in (0, 1)$ such that

$$\tau^{1/n} \leq \frac{\lambda}{\mu} \leq 1 \implies \|T_\lambda^n S_\mu^n(f) - f\| < \eta.$$

Indeed, if the family $(T_\lambda)_{\lambda \in \Lambda}$ satisfies (1) and (2') and if we set $\widetilde{T}_{\tilde\lambda} := T_{e^{\tilde\lambda}}$ then the family $(\widetilde{T}_{\tilde\lambda})_{\tilde\lambda \in \log(\Lambda)}$ satisfies (1) and (2).

REMARK 7.16   The continuity assumption (2) of Theorem 7.14 can be weakened as follows. Set

$$\delta_n(f, \eta) := \sup\left\{\delta \geq 0; \ (0 \leq \mu - \lambda < \delta) \implies \|T_\lambda^n S_\mu^n(f) - f\| < \eta\right\}.$$

Then it is enough to assume that $\sum_0^\infty \delta_n(f, \eta) = \infty$ for each $\eta > 0$. Indeed, given N as in the above proof, one can find $p \in [0, N)$ such that $\sum_0^\infty \delta_{p+Ni} = \infty$, and hence a subdivision $a = \lambda_0 < \cdots < \lambda_q = b$ such that $\lambda_i - \lambda_{i-1} < \delta_{p+Ni}$ for all i. Then put $n_i := p + Ni$.

To illustrate the Costakis–Sambarino Criterion, let us see how it applies to the Abakumov–Gordon theorem mentioned in the introductory section of this chapter. More general results will be proved later.

EXAMPLE 7.17   Let B be the usual backward shift on $\ell^2(\mathbb{N})$. Then $\bigcap_{|\lambda|>1} HC(\lambda B)$ is a dense G_δ subset of $\ell^2(\mathbb{N})$.

PROOF   For any real number $\lambda > 1$ the sets $HC(T_{\lambda e^{i\theta}})$ do not depend on $\theta \in \mathbb{R}$, by the León–Müller theorem (Theorem 3.2). Therefore, it is enough to show that $\bigcap_{\lambda \in (1,\infty)} HC(\lambda B)$ is a dense G_δ subset of $\ell^2(\mathbb{N})$. Let $\mathcal{D} := c_{00}(\mathbb{N}) \subset \ell^2(\mathbb{N})$ be the set of all finitely supported sequences, and set $S_\lambda := \lambda^{-1} S$, where S is the canonical forward shift.

Observe that $T_\lambda^{n+k} S_\alpha^n = (\lambda^{n+k}/\alpha^n) B^k$ and $T_\lambda^n S_\alpha^{n+k} = (\lambda^n/\alpha^{n+k}) S^k$ for any $\alpha, \lambda > 1$ and $k, n \in \mathbb{N}$. Hence, given a compact interval $K = [a, b] \subset (1, \infty)$ and $f \in \mathcal{D}$, we see that

$$\begin{aligned} T_\lambda^{n+k} S_\alpha(f) &= 0 && \text{whenever } k \geq \max(\operatorname{supp}(f)), \\ \|T_\lambda^n S_\alpha^{n+k}(f)\| &\leq \tfrac{1}{a^k}\|f\| && \text{if } a \leq \lambda \leq \alpha. \end{aligned}$$

Thus, property (1) in Theorem 7.14 holds with $\kappa := \max(\operatorname{supp}(f))$ and $c_k := \|f\|/a^k$. Moreover, the identity $T_\lambda^n S_\mu^n f - f = (\lambda^n/\mu^n - 1) f$ shows that condition (2') from Remark 7.15 is satisfied as well. $\square$

The Costakis–Sambarino Criterion is quite general, and it does apply to a number of natural examples. However, the summability condition $\sum_k c_k < \infty$ is very strong, which is not surprising since no assumption is made on the F-space X. Thus, precisely because of its generality, the criterion reflects neither the geometry of the space X nor the structure of the operators T_λ. We shall see now how the summability condition can be relaxed in two different contexts.

7.2.3 A criterion on Banach lattices

In this section, we prove a common hypercyclicity criterion which applies when the underlying space X is a Banach lattice. We first recall some terminology. The interested reader should consult [174] for more information on the subject.

A **Banach lattice** is a real Banach space X equipped with a partial ordering $\leq$ satisfying the following properties:

 (i) $x \leq y$ implies $x + z \leq y + z$ for all $z \in X$;
 (ii) $\lambda x \geq 0$ whenever $x \geq 0$ in X and $\lambda \in [0, \infty)$;
 (iii) each pair $(x, y) \in X$ admits a least upper bound $x \vee y$ and a greatest lower bound $x \wedge y$;
 (iv) $\|x\| \leq \|y\|$ whenever $|x| \leq |y|$, the absolute value being defined by $|a| = a \vee (-a)$.

The canonical examples of Banach lattices are the spaces $L^p(\Omega, \mu)$ for an arbitrary measure space (Ω, μ). Every Banach space with an unconditional basis is also a Banach lattice, after a suitable renorming.

A Banach lattice X is said to be p-**convex** $(1 \leq p < \infty)$ if

$$\left\| \left(\sum |x_i|^p \right)^{1/p} \right\| \leq M \left(\sum \|x_i\|^p \right)^{1/p}$$

for some $M < \infty$ and every choice of (finitely many) vectors (x_i) in X. Clearly, any space $L^p(\mu)$ is p-convex. It is well known that a p-convex Banach lattice satisfies the following upper estimate:

$$\left\| \sum x_i \right\| \leq M \left(\sum \|x_i\|^p \right)^{1/p}$$

for every choice of disjoint vectors (x_i) in X (two vectors $x, y \in X$ are **disjoint** if $|x| \wedge |y| = 0$).

We can now state the following theorem.

THEOREM 7.18 *Assume that X is a p-convex Banach lattice $(1 \leq p < \infty)$ and that, for each $f \in \mathcal{D}$ and each compact set $K \subset \Lambda$, the following properties hold true (where all parameters belong to K).*

(1) *There exist $\kappa \in \mathbb{N}$ and a sequence of positive numbers $(c_k)_{k \geq \kappa}$ such that*

- $\displaystyle\sum_{k=\kappa}^{\infty} c_k^p < \infty;$

- $\|T_\lambda^{n+k} S_\alpha^n(f)\| \leq c_k$ whenever $n \in \mathbb{N}$, $k \geq \kappa$ and $\alpha \leq \lambda$;
- $\|T_\lambda^n S_\alpha^{n+k}(f)\| \leq c_k$ whenever $n \in \mathbb{N}$, $k \geq \kappa$ and $\lambda \leq \alpha$.

(2) *There exists an $s \in \mathbb{N}$ such that, whenever $m, m' \in \mathbb{N}$ satisfy $m' - m \geq s$ and $\mu, \mu' \in K$ satisfy $\mu' \geq \mu$, the vectors $T_\lambda^n S_\mu^m(f)$ and $T_\lambda^n S_{\mu'}^{m'}(f)$ are disjoint for any choice of $n \in \mathbb{N}$ and $\lambda \in K$.*

(3) *Given $\eta > 0$, one can find $\tau > 0$ such that, for all $n \geq 1$,*

$$0 \leq \mu - \lambda < \frac{\tau}{n} \implies \|T_\lambda^n S_\mu^n(f) - f\| < \eta.$$

Then $\bigcap_{\lambda \in \Lambda} HC(T_\lambda)$ is a dense G_δ subset of X.

PROOF We start exactly as in the proof of Theorem 7.14, except that we require additionally that $N \geq \max(s_u, s_v)$, where s_u and s_v are the natural numbers associated with $f := u$ and $f := v$ in condition (2). By (2) (with $n = 0$), the vectors $S_{\lambda_i}^{Ni}(v)$ are disjoint ($1 \leq i \leq q$). By the p-convexity of X and condition (1), this yields

$$\left\| \sum_{i=1}^q S_{\lambda_i}^{n_i}(v) \right\| \leq M \left(\sum_{i=1}^q \|S_{\lambda_i}^{Ni}(v)\|^p \right)^{1/p}$$

$$\leq M \left(\sum_{k \geq N} c_k^p \right)^{1/p} < \eta$$

provided that N is large enough.

Similarly, for any fixed $i \in \{1, \dots, q\}$, the vectors $T_\lambda^{Ni} S_{\lambda_j}^{Nj}(u)$, $j \neq i$ are disjoint. Hence

$$\left\| \sum_{j \neq i} T_\lambda^{n_i} S_{\lambda_j}^{n_j}(v) \right\| \leq \left\| \sum_{j=1}^{i-1} T_\lambda^{n_i} S_{\lambda_j}^{n_j}(v) \right\| + \left\| \sum_{j=i+1}^q T_\lambda^{n_i} S_{\lambda_j}^{n_j}(v) \right\|$$

$$\leq 2M \left(\sum_{k \geq N} c_k^p \right)^{1/p} < \eta$$

if N is large. Thus, condition (BC2) in the Basic Criterion (Lemma 7.12) is satisfied for large enough N. The other assumptions are checked exactly as in the proof of Theorem 7.14.   □

Typically, the disjointness property (2) in Theorem 7.18 is rather easy to check for weighted shifts or change of variable operators acting on weighted L^p spaces. Here are two illustrations.

EXAMPLE 7.19 Let $p \in [1, \infty)$, and let $(\mathbf{w}(\lambda))_{\lambda \in \Lambda}$ be a family of (bounded, positive) weight sequences parametrized by some interval $\Lambda \subset \mathbb{R}$. Let us denote by $B_{\mathbf{w}(\lambda)}$ the associated weighted shifts acting on $\ell^p(\mathbb{N})$. Suppose that

(a) all functions $\log(w_n)$ are non-decreasing and are Lipschitz on compact sets with uniformly bounded Lipschitz constants;

(b) all series $\displaystyle\sum_n \frac{1}{(w_1(\lambda)\cdots w_n(\lambda))^p}$ are convergent.

Then $\bigcap_{\lambda\in\Lambda} HC(B_{\mathbf{w}(\lambda)})$ is a residual subset of $\ell^p(\mathbb{N})$.

PROOF Let $\mathcal{D} := c_{00}(\mathbb{N}) \subset \ell^p(\mathbb{N})$ be the set of all finitely supported sequences, and let $S_{\mathbf{w}(\lambda)} : \mathcal{D} \to \mathcal{D}$ be the right-inverse of $B_{\mathbf{w}(\lambda)}$, defined by $S_{\mathbf{w}(\lambda)}(e_i) = (w_{i+1}(\lambda))^{-1}e_{i+1}$, where $(e_i)_{i\in\mathbb{N}}$ is the canonical basis of $\ell^p(\mathbb{N})$. In order to apply Theorem 7.18, let us fix a compact interval $K = [a, b] \subset \Lambda$.

We first note that condition (2) in Theorem 7.18 is clearly satisfied. Indeed, if $f \in \mathcal{D}$ is supported on some interval $[0, A]$ then the vectors $B_{\mathbf{w}(\lambda)}^n S_{\mathbf{w}(\mu)}^m(f)$ and $B_{\mathbf{w}(\lambda)}^n S_{\mathbf{w}(\mu')}^{m'}(f)$ are disjointly supported for all $n \in \mathbb{N}$ and any $\mu, \mu' \in \Lambda$, provided that $m' - m > A$.

By linearity, it is enough to check the remaining assumptions (1) and (3) of Theorem 7.18 for $f = e_i$, $i \in \mathbb{N}$; so we fix $i \in \mathbb{N}$.

To prove condition (1) of Theorem 7.18, we first note that on the one hand $B_{\mathbf{w}(\lambda)}^{n+k} S_{\mathbf{w}(\alpha)}^n(e_i) = 0$ for all $n \in \mathbb{N}$ provided that $k > i$. On the other hand,

$$B_{\mathbf{w}(\lambda)}^n S_{\mathbf{w}(\alpha)}^{n+k}(e_i) = \frac{w_{i+k+1}(\lambda)\cdots w_{i+k+n}(\lambda)}{w_{i+1}(\alpha)\cdots w_{i+k+n}(\alpha)} e_i.$$

Since all the functions w_n are non-decreasing by (a), it follows that

$$\|B_{\mathbf{w}(\lambda)}^n S_{\mathbf{w}(\alpha)}^{n+k}(e_i)\| \leq \frac{1}{w_{i+1}(a)\cdots w_{i+k}(a)}$$

for any $n, k \in \mathbb{N}$ and any $\lambda \leq \alpha \in [a, b]$. By (b), this shows that condition (1) of Theorem 7.18 is satisfied.

To prove condition (3), we observe that, for any $\lambda, \mu \in [a, b]$ with $0 \leq \mu - \lambda \leq \tau/n$, the following inequalities hold true:

$$\begin{aligned}
\|B_{\mathbf{w}(\lambda)}^n S_{\mathbf{w}(\mu)}^n(e_i) - e_i\| &= \left|\frac{w_{i+1}(\lambda)\cdots w_{i+n}(\lambda)}{w_{i+1}(\mu)\cdots w_{i+n}(\mu)} - 1\right| \|e_i\| \\
&\leq \left|\exp\left(\sum_{k=i+1}^{i+n} |\log(w_k(\lambda)) - \log(w_k(\mu))|\right) - 1\right| \\
&\leq |\exp(Mn|\lambda - \mu|) - 1| \\
&\leq |\exp(M\tau) - 1|,
\end{aligned}$$

where M is chosen in such a way that all the functions $\log(w_n)$ are M-Lipschitz on $[a, b]$ (which can be done thanks to condition (a)). Thus condition (3) in Theorem 7.18 is satisfied if τ is small enough. $\qquad\square$

REMARK Taking $\mathbf{w}(\lambda) := \lambda\mathbf{w}$, where $\mathbf{w} = (w_n)_{n\geq 1}$ is a fixed weight sequence, one obtains the following natural extension of the Abakumov–Gordon theorem.

Let $B_{\mathbf{w}}$ be a weighted shift on $X = \ell^p(\mathbb{N})$, and put

$$\lambda_{\mathbf{w}} = \inf\left\{\lambda > 0;\ \text{the series}\ \sum \frac{\lambda^{-n}}{w_1 \cdots w_n}\, e_n\ \text{is convergent in } X\right\}.$$

Then $\bigcap_{\lambda > \lambda_{\mathbf{w}}} HC(\lambda B_{\mathbf{w}})$ is a dense G_δ subset of X.

We also note that a minor modification of the above proof yields a similar result for backward weighted shifts acting on a *Fréchet* space X with an unconditional basis $(e_n)_{n\in\mathbb{N}}$: just replace condition (b) of Example 7.19 by

(b′) All series $\sum \dfrac{1}{w_1(\lambda) \cdots w_n(\lambda)}\, e_n$ are convergent in X.

The details can be found in [29]. In particular, this condition may be applied to the family $(\lambda D)_{\lambda \neq 0}$, where D is the derivative operator on the space of entire functions $H(\mathbb{C})$. Indeed, the sequence $(z^n)_{n\in\mathbb{N}}$ is an unconditional basis of $H(\mathbb{C})$ and D acts as the weighted backward shift with weights $w_n = n$. Condition (b′) is clearly satisfied by $w_n(\lambda) = \lambda n$, since $w_1(\lambda) \cdots w_n(\lambda) = n!\lambda^n$. The conclusion is that $\bigcap_{\lambda \neq 0} HC(\lambda D)$ is a dense G_δ subset of $H(\mathbb{C})$.

EXAMPLE 7.20   Let $w : \mathbb{C} \to \mathbb{R}$ be a positive bounded and continuous function such that $w(z - a)/w(z)$ is bounded for each $a \in \mathbb{C}$. Under these assumptions, the translation operators T_a defined by $T_a f(z) = f(z + a)$ are continuous (and invertible) on the weighted Lebesgue space $X_p := L^p(\mathbb{C}, w(z)dA(z))$ for any $p \in [1, \infty)$, where dA is the Lebesgue measure on $\mathbb{C}$.

Assume that $w(z) = O\left(|z|^{-\rho}\right)$ as $|z| \to \infty$, where $\rho > 1$. Then $\bigcap_{a \neq 0} HC(T_a)$ is a dense G_δ subset of X_p, for any $p \in [1, \infty)$.

PROOF   For any fixed $\theta \in \mathbb{R}$, the family $(T_{re^{i\theta}})_{r>0}$ is a C_0-semigroup on X_p, so that $HC(T_{re^{i\theta}})$ does not depend on $r > 0$ by the Conejero–Müller–Peris theorem. Therefore, we just need to prove that $\bigcap_{\theta \in [0, 2\pi]} HC(T_{e^{i\theta}})$ is a dense G_δ subset of X_p. Let $\mathcal{D} \subset X_p$ be the set of all compactly supported $\mathcal{C}^1$-smooth functions. Of course, we define the right-inverses $S_a : \mathcal{D} \to \mathcal{D}$ by $S_a := T_a^{-1}$, i.e. $S_a f(z) = f(z - a)$. In order to check the assumptions of Theorem 7.18, let us fix $f \in \mathcal{D}$ and choose $R > 0$ such that f is supported on the closed disk $\overline{D}(0, R)$.

For any θ, α, n, m, the function $T_{e^{i\theta}}^n S_{e^{i\alpha}}^m (f)$ has support in the closed disk with radius R centred at $-ne^{i\theta} + me^{i\alpha}$. In particular, if $m' - m > 2R$ then the functions $T_{e^{i\theta}}^n S_{e^{i\alpha}}^m (f)$ and $T_{e^{i\theta}}^n S_{e^{i\alpha}}^{m'} (f)$ are disjointly supported, whatever the choice of θ, α, n may be. Thus, the disjointness condition (2) of Theorem 7.18 is satisfied.

Moreover, $|z - (n+k)e^{i\theta} + ne^{i\alpha}| \geq k - R$ for any n, k, θ, α and every $z \in \overline{D}(0, R)$. Keeping in mind that f is bounded and using the assumption on w, it follows that

$$\left\| T_{e^{i\theta}}^{n+k} S_{e^{i\alpha}}^n (f) \right\|^p = \int_{\overline{D}(0,R)} |f(z)|^p\, w(z - (n+k)e^{i\theta} + ne^{i\alpha})\, dA(z)$$

$$= O\left(\frac{1}{k^\rho}\right),$$

where the O-constant is independent of n, θ, α. Since $\rho > 1$, this gives condition (1) of Theorem 7.18.

Finally, it follows from the mean-value theorem that if $\tau > 0$ and $0 \leq \varphi - \theta \leq \tau/n$ then

$$\|T_{e^{i\theta}}^n S_{e^{i\varphi}}^n(f) - f\|^p \leq \int_{\overline{D}(0, R+1)} |f(z + n(e^{i\theta} - e^{i\varphi})) - f(z)|^p \, w(z) \, dA(z)$$
$$\leq M\tau^p$$

for some constant $M := M_{f,w}$. This gives (3). $\qquad\square$

REMARK The reader is invited to check that for $w(z) := (1 + |z|^r)^{-1}$, one can apply the Costakis–Sambarino Criterion only if $r > p$.

7.3 A probabilistic criterion

7.3.1 Why a probabilistic criterion?

Theorem 7.18 suffers from one obvious drawback: it applies only in a Banach lattice! In particular, the disjointness condition makes no sense in a general Banach space, not to mention spaces of holomorphic functions. In this section, we discuss another Banach space criterion which takes into account the geometry of the underlying space X. As in the case of the existence of invariant measures, the relevant parameter turns out to be the *type* of the Banach space X.

Recall that a *Rademacher* variable is a real random variable taking only the values 1 and -1, each with probability $1/2$. We fix once and for all a sequence of independent Rademacher variables $(\varepsilon_n)_{n \in \mathbb{N}}$ defined on some probability space $(\Omega, \mathcal{F}, \mathbb{P})$. The Banach space X is said to have (Rademacher) **type** $p \in [1, 2]$ if

$$\left\| \sum_n \varepsilon_n x_n \right\|_{L^2(\Omega, X)} \leq C \left(\sum_n \|x_n\|^p \right)^{1/p},$$

for some constant $C < \infty$ and every finite sequence $(x_n) \subset X$. This is not the definition given in Chapter 5, which was formulated in terms of Gaussian variables. However, the two definitions are in fact equivalent (see Appendix C).

Let us explain why the type comes into play in this context. The main observation is that one can relax the assumptions in the Basic Criterion by using the Costakis–Peris theorem (Corollary 3.14).

REMARK 7.21 In the Basic Criterion, one can replace condition (BC2) by the following.

(BC2') *One can find* $\varepsilon_1, \ldots, \varepsilon_q \in \{-1, 1\}$ *such that* $\sum_{j=1}^{q} \varepsilon_j S_{\lambda_j}^{n_j}(v) \in O$ *and*
  $\sum_{j \neq i} \varepsilon_j T_\lambda^{n_i} S_{\lambda_j}^{n_j}(v) \in O$ *for each* $i \in \{1, \ldots, q\}$ *and every* $\lambda \in \Lambda_i$.

PROOF We proceed as in the proof of the Basic Criterion, except that x is now defined by $x := u + \sum_{j=1}^{q} \varepsilon_j S_{\lambda_j}^{n_j}(v)$. The same proof shows that $x \in U$ and $\varepsilon_i T_\lambda^{n_i}(x) \in V$ for each i and every $\lambda \in \Lambda_i$. It follows that $U \cap \widetilde{V}_K \neq \varnothing$, where

$$\widetilde{V}_K := \left\{ x \in X;\ \forall \lambda \in K\ \exists n : T_\lambda^n(x) \in V \text{ or } T_\lambda^n(-x) \in V \right\}.$$

Thus, each open set $\widetilde{V}_K$ is dense in X. By the Baire category theorem, this gives a dense G_δ set of vectors $x \in X$ such that $\{T_\lambda^n(x);\ n \in \mathbb{N}\} \cup \{T_\lambda^n(-x);\ n \in \mathbb{N}\}$ is dense in X for every $\lambda \in \Lambda$. By the Costakis–Peris theorem, such a vector x is hypercyclic for all operators T_λ. $\square$

Assume now that our family of operators (T_λ) satisfies the assumptions of Theorem 7.14, except that we replace the summability condition $\sum_k c_k < \infty$ by the weaker assumption $\sum_k c_k^p < \infty$, where $p > 1$. If we try to follow the proof of Theorem 7.14, we need to find a choice of signs $\varepsilon_1, \dots, \varepsilon_q$ such that condition (BC2′) above is satisfied. If ($p \leq 2$ and) the underlying Banach space X has type p, this can be done by a probabilistic argument.

Indeed, the expectation $\mathbb{E} \left\| \sum_{j=1}^{q} \varepsilon_j(\omega) S_{\lambda_j}^{n_j}(u) \right\|^p$ is controlled by $\sum_{k \geq N} c_k^p$, according to the definition of a type-p Banach space; hence, with large probability $\sum_{j=1}^{q} \varepsilon_j(\omega) S_{\lambda_j}^{n_j}(u)$ has a small norm. Likewise, for a *fixed* $\lambda \in \Lambda$, with large probability the sum $\sum_{j \neq i} \varepsilon_j(\omega) T_\lambda^{n_i} S_{\lambda_j}^{n_j}(u)$ also has a small norm. However, we need to manage all the parameters λ at the same time. That is, we would like to control $\mathbb{E}\left(\sup_{i, \lambda \in \Lambda_i} \left\| \sum_{j \neq i} \varepsilon_j(\omega) T_\lambda^{n_i} S_{\lambda_j}^{n_j}(u) \right\| \right)$. Thus, we are facing one of the most basic problems in probability theory: to estimate the supremum of a family of random variables. Fortunately, quite efficient tools are available, one of which we describe in the next subsection.

7.3.2 Dudley's theorem

Dudley's majorization theorem is a very useful result, which gives an estimate for the expectation of the supremum of a subgaussian random process. For the convenience of the reader, we briefly review some well-known facts. For much more information, we refer to [173], [164] or [230].

Let $(\Omega, \mathcal{F}, \mathbb{P})$ be a probability space. If $\psi : [0, \infty] \to [0, \infty]$ is an increasing convex function with $\psi(0) = 0$ and $\psi(\infty) = \infty$, the **Orlicz space** $L^\psi(X) = L^\psi(\Omega, \mathbb{P}, X)$ is the Banach space of all random variables $Z : \Omega \to X$ such that $\mathbb{E}\left[\psi(\|Z\|/c) \right] < \infty$ for some real number $c > 0$ (depending on Z). The norm of the space $L^\psi(X)$ is defined by

$$\|Z\|_\psi = \inf \left\{ c > 0;\ \mathbb{E}\left[\psi\left(\frac{\|Z\|}{c} \right) \right] \leq 1 \right\}.$$

We will be concerned with the case $\psi(x) = \psi_2(x) := e^{x^2} - 1$. Elementary computations using Stirling's formula and the Taylor expansion of the exponential function show that

$$a\|Z\|_{\psi_2} \leq \sup_{p\geq 1} \frac{\|Z\|_p}{\sqrt{p}} \leq b\|Z\|_{\psi_2} \tag{7.2}$$

for some numerical constants a, b and all $Z \in L^{\psi_2}(X)$. This is an interesting exercise.

If (T, d) is a compact metric space, we denote by $N_d(\varepsilon)$, $\varepsilon > 0$, the ε-covering number of (T, d), that is, the minimal number of d-open balls of radius ε which are needed to cover T. The function N_d is the **entropy function** of the metric space (T, d). The **entropy integral** $J(d)$ is defined by

$$J(d) = \int_0^\infty \sqrt{\log(N_d(\varepsilon))}\, d\varepsilon.$$

We can now state Dudley's theorem, in a form due to G. Pisier [196].

THEOREM 7.22 (DUDLEY'S THEOREM) *Let (T, d) be a compact metric space with finite entropy integral $J(d)$, and let $(Z_t)_{t\in T}$ be an X-valued random process such that $Z_t \in L^{\psi_2}(X)$ for all $t \in T$. Assume that the process satisfies the Lipschitz condition*

$$\|Z_s - Z_t\|_{\psi_2} \leq d(s, t) \quad (s, t \in T).$$

Then the following estimate holds (where t_0 is a fixed arbitrary point in T):

$$\mathbb{E}(\sup_{t\in T}\|Z_t\|) \leq \kappa_1 (J(d) + \|Z_{t_0}\|_{\psi_2}),$$

for some numerical constant κ_1.

Processes satisfying a Lipschitz estimate of the above type are called **subgaussian** (see Exercise 7.6 for an explanation of this terminology). For the application we have in mind, we will need a simple variant of Dudley's theorem involving finitely many subgaussian processes.

COROLLARY 7.23 *Let q be a positive integer and, for each $i \in \{1, \ldots, q\}$, let $(Z_{i,\lambda})_{\lambda\in\Lambda_i}$ be a random process with values in X, where Λ_i is a compact interval of $\mathbb{R}$. Assume that $Z_{i,\lambda} \in L^{\psi_2}(X)$ for each i and every $\lambda \in \Lambda_i$ and that each process $(Z_{i,\lambda})_{\lambda\in\Lambda_i}$ satisfies the Lipschitz condition*

$$\|Z_{i,\lambda} - Z_{i,\mu}\|_{\psi_2} \leq c_i |\lambda - \mu| \quad (\lambda, \mu \in \Lambda_i).$$

Then the following estimate holds:

$$\mathbb{E}\left(\sup_{i,\lambda}\|Z_{i,\lambda}\|\right) \leq \kappa_2 \left(\sup_i c_i |\Lambda_i| + \sup_{i,\lambda}\|Z_{i,\lambda}\|_{\psi_2}\right)\sqrt{\log(q+1)},$$

where $|\Lambda_i|$ is the length of the interval Λ_i and κ_2 is a numerical constant.

PROOF We define a compact metric space (T, d) and a random process on it in order to deduce Corollary 7.23 from Dudley's theorem. Put

$$T := \{(i, \lambda);\ i \in \{1, \ldots, q\},\ \lambda \in \Lambda_i\}, \quad M_Z := 2\sup_{i,\lambda}\|Z_{i,\lambda}\|_{\psi_2},$$

and let us define a metric d on T as follows:

$$d\big((i,\lambda),(j,\mu)\big) := \begin{cases} c_i |\lambda - \mu| & \text{if } i = j, \\ M_Z & \text{otherwise.} \end{cases}$$

The entropy function $N_d(\varepsilon)$ is dominated by $2M_\Lambda q/\varepsilon$, and the diameter of (T,d) is not greater than $M_\Lambda + M_Z$, where $M_\Lambda := \sup_i c_i |\Lambda_i|$. Hence, we may estimate the entropy integral as follows:

$$\begin{aligned} J(d) &\le \int_0^{M_\Lambda + M_Z} \sqrt{\log(2M_\Lambda q/\varepsilon)}\, d\varepsilon \\ &\le \left(\sqrt{\log 2} + \sqrt{\log q} + \int_0^1 \sqrt{\log(1/t)}\, dt \right) (M_\Lambda + M_Z). \end{aligned}$$

The result now follows directly from Dudley's theorem. $\qquad\square$

We intend to apply Corollary 7.23 to a Rademacher process $Z(\omega) = \sum_k \varepsilon_k(\omega) x_k$. In this setting, one can take advantage of **Kahane's inequalities** (see e.g. [97]): for any $p \in [1,\infty)$, there exist finite positive constants A_p, B_p such that

$$A_p \left\| \sum_k \varepsilon_k x_k \right\|_2 \le \left\| \sum_k \varepsilon_k x_k \right\|_p \le B_p \left\| \sum_k \varepsilon_k x_k \right\|_2$$

for every Banach space X and all finite sequences $(x_k) \subset X$. Moreover, it was shown by S. Kwapień [161] that $B_p \le C\sqrt{p}$, for some absolute constant C. Comparing this inequality with (7.2), we see that the Orlicz norm $\|Z\|_{\psi_2}$ of a Rademacher process is equivalent to its L^2 norm. In other words, one can simply forget the definition of the Orlicz norm and use the more familiar L^2 norm ...

7.3.3 The criterion

The following criterion was exhibited in [29].

THEOREM 7.24 *Assume that the Banach space X has type $p \in [1,2]$ and that for each $f \in \mathcal{D}$ and each compact set $K \subset \Lambda$ there exists a sequence of positive numbers $(c_k)_{k \in \mathbb{N}}$ such that the following properties hold (where all parameters belong to K):*

(1) *(c_k) is non-increasing, and $\sum_0^\infty c_k^p < \infty$;*

(2a) *$\|T_\lambda^{n+k} S_\alpha^n(f)\| \le c_k$ for any $n, k \in \mathbb{N}$ and $\lambda \ge \alpha$;*

(2b) *$\|T_\lambda^n S_\alpha^{n+k}(f)\| \le c_k$ for any $n, k \in \mathbb{N}$ and $\lambda \le \alpha$;*

(3a) *$\|(T_\lambda^{n+k} - T_\mu^{n+k})(S_\alpha^n f)\| \le (n+k)|\lambda - \mu|\, c_k$, for $n, k \in \mathbb{N}$ and $\lambda, \mu \ge \alpha$;*

(3b) *$\|(T_\lambda^n - T_\mu^n)(S_\alpha^{n+k}(f))\| \le n|\lambda - \mu|\, c_k$, for $n, k \in \mathbb{N}$ and $\lambda, \mu \le \alpha$.*

Then $\bigcap_{\lambda \in \Lambda} HC(T_\lambda)$ is a dense G_δ subset of E.

PROOF   In what follows, the symbol C denotes an "absolute" constant (depending only on X and p) whose value may change from line to line.

We start as in the proof of Theorem 7.14. Let us fix a compact set $K \subset \Lambda$, an open neighbourhood $O = B_d(0, \eta)$ of 0 in X and a pair $(u, v) \in \mathcal{D} \times \mathcal{D}$. Let (c_k) be a sequence of positive numbers satisfying the five conditions above for both u and v. Given a large integer N and a small real number $\tau > 0$, we choose a subdivision $a = \lambda_0 < \cdots < \lambda_q = b$ such that $\lambda_i - \lambda_{i-1} < \tau/(Ni)$ and we put $\Lambda_i = [\lambda_{i-1}, \lambda_i]$, $n_i = Ni$.

By condition (2b) with $n = 0$ and $k = Ni$, we have $\|S_{\lambda_i}^{Ni}(v)\| \le c_{Ni}$ for all $i \in \{1, \ldots, q\}$. Since X has type p, it follows that

$$\mathbb{E}\left(\left\|\sum_{j=1}^q \varepsilon_j(\omega) S_{\lambda_j}^{n_j}(v)\right\|\right) \le C\left(\sum_{j=1}^q c_{Nj}^p\right)^p \le C\left(\sum_{k \ge N} c_k^p\right)^{1/p},$$

so that $\|\sum_{j=1}^q \varepsilon_j(\omega) S_{\lambda_j}^{n_j}(v)\| < \eta$ with large probability if N is large enough.

Next, we use Corollary 7.23 to show that the second half of (BC2$'$) (see Remark 7.21) is satisfied for many choices of sign. For each $i \in \{1, \ldots, q\}$ and $\lambda \in \Lambda_i$, we set $Z_{i,\lambda}(\omega) := \sum_{j \ne i} \varepsilon_j(\omega) T_\lambda^{n_i} S_{\lambda_j}^{n_j}(v)$. Since X has type p, we have

$$\|Z_{i,\lambda} - Z_{i,\mu}\|_{\psi_2} \le C\left(\sum_{j \ne i} \|(T_\lambda^{n_i} - T_\mu^{n_i})(S_{\lambda_j}^{n_j}(v))\|^p\right)^{1/p}$$

for any $\lambda, \mu \in \Lambda_i$. Using (3a) and (3b), it follows that

$$\|Z_{i,\lambda} - Z_{i,\mu}\|_{\psi_2} \le C\left(\sum_{j \ne i} (Ni)^p |\lambda - \mu|^p c_{N|i-j|}^p\right)^{1/p} \le M_i |\lambda - \mu|,$$

where $M_i = CNi\left(\sum_{l \ge 1} c_{Nl}^p\right)^{1/p}$. Similarly, conditions (2a) and (2b) give

$$\|Z_{i,\lambda}\|_{\psi_2} \le C\left(\sum_{j \ne i} \|T_\lambda^{n_i} S_{\lambda_j}^{n_j}(v)\|^p\right)^{1/p} \le C\left(\sum_{l \ge 1} c_{Nl}^p\right)^{1/p}.$$

Observe now that $M_i |\lambda_i - \lambda_{i-1}| \le C\tau\left(\sum_{l \ge 1} c_{Nl}^p\right)^{1/p}$ and that we may choose the subdivision $(\lambda_0, \ldots, \lambda_q)$ in such a way that $\log q \le C N/\tau$. By Corollary 7.23, it follows that

$$\mathbb{E}\left(\sup_{i,\lambda} \|Z_{i,\lambda}\|\right) \le \frac{C}{\tau^{1/2}} \sqrt{N}\left(\sum_{l \ge 1} c_{Nl}^p\right)^{1/p}.$$

Assume that N is an even integer, $N = 2N'$. Since the sequence (c_k) is non-increasing, we have

$$c_{Nl}^p \leq \frac{1}{N'} \sum_{k=(2l-1)N'}^{2lN'-1} c_k^p$$

for each $l \geq 1$. It follows that

$$\mathbb{E}\left(\sup_{i,\lambda} \|Z_{i,\lambda}\| \right) \leq \frac{C}{\tau^{1/2}} N^{1/2-1/p} \left(\sum_{k \geq N'} c_k^p \right)^{1/p}$$

$$\leq \frac{C}{\tau^{1/2}} \left(\sum_{k \geq N'} c_k^p \right)^{1/p},$$

keeping in mind that $p \leq 2$.

We conclude that if $N = N(\tau)$ is large enough then, with large probability, the vector $\sum_{j \neq i} \varepsilon_j(\omega) T_\lambda^{n_i} S_{\lambda_j}^{n_j}(v)$ is close to 0 for each $i \in \{1, \ldots, q\}$ and every $\lambda \in \Lambda_i$. In particular, there is at least one choice of signs $\varepsilon_1, \ldots, \varepsilon_q$ for which condition (BC2') of Remark 7.21 is satisfied.

Moreover, by (2a) with $n = 0$ and $k = n_i$, the deterministic quantity $\|T_\lambda^{n_i}(u)\|$ is small if N is large enough, and, using (3b) with $k = 0$, $n = n_i$, $\alpha = \mu = \lambda_i$, we see that $\|T_\lambda^{n_i} S_{\lambda_i}^{n_i}(v) - v\|$ is small as well, for each $i \in \{1, \ldots, q\}$ and every $\lambda \in \Lambda_i$ provided that τ is small enough, whatever the choice of N may be. In particular, conditions (BC3) and (BC4) of the Basic Criterion are satisfied, which concludes the proof of Theorem 7.24. □

REMARK   Since every Banach space has type 1, the above criterion can be applied with $p = 1$ in an arbitrary Banach space X. However, in this case the hypotheses are stronger than for the Costakis–Sambarino Criterion, so this does not help much! Observe also the different meanings of the conditions in Theorem 7.24: (2a) and (2b) are used to estimate the L^2 norm of the Rademacher process, whereas the continuity conditions (3a) and (3b) give the Lipschitz estimate needed to apply Dudley's theorem.

In [29], Theorem 7.24 was used to show that the family of all parabolic automorphic composition operators with attractive fixed point $\alpha = +1$ has a common hypercyclic vector. This has some "philosophical" interest, since it is not possible to apply the Costakis–Sambarino Criterion in that case (see Exercise 6.3). However, the proof based on the Conejero–Müller–Peris theorem given in the introduction to this chapter (see also Exercise 7.3) is definitely much simpler.

In some sense, this is a rather disappointing situation since we have given no application for Theorem 7.24. But this does not mean that no interesting application will ever be found. More importantly, the idea of mixing Baire-category and probabilistic arguments is rather seducing and we think it may have some usefulness in the near future, perhaps.

7.4 Paths of weighted shifts

In this section, we consider the continuous paths of weighted shifts on the Hilbert space $\ell^2(\mathbb{N})$. As usual, we will denote by $B_{\mathbf{w}}$ the weighted backward shift on $\ell^2(\mathbb{N})$ associated with the (bounded) weight sequence $\mathbf{w} = (w_n)_{n \geq 1}$. By a **continuous path of operators**, we mean any continuous map $\lambda \mapsto T_\lambda$ from some compact interval $[a, b] \subset \mathbb{R}$ into $\mathfrak{L}(\ell^2)$, where $\mathfrak{L}(\ell^2)$ is endowed with the *operator norm* topology. Of course, this is a much stronger continuity property than the rather "minimalist" condition (H1) we have been working with from the beginning of the chapter.

Using Salas' condition, it is not too hard to check that the set of all hypercyclic weighted shifts on $\ell^2(\mathbb{N})$ is path connected (see Exercise 7.7). In other words, given any two hypercyclic weighted shifts $B_{\mathbf{w}}$, $B_{\mathbf{w}'}$, one can find a continuous path of operators $(T_\lambda)_{\lambda \in [a,b]}$ such that $T_a = B_{\mathbf{w}}$, $T_b = B_{\mathbf{w}'}$ and each T_λ is a hypercyclic weighted shift. It is natural to ask about the existence of common hypercyclic vectors along such a path of weighted shifts. We shall prove here two results due to K. C. Chan and R. Sanders [73], which go in two opposite directions: on the one hand, any two hypercyclic weighted shifts can be connected by a continuous path of weighted shifts having a common hypercyclic vector; on the other hand, one can also construct a path of hypercyclic weighted shifts without common hypercyclic vectors.

7.4.1 Paths of weighted shifts with a common hypercyclic vector

In this subsection, we show that any two hypercyclic weighted shifts can be connected by a path of weighted shifts in such a way that all shifts in the path have a common hypercyclic vector (see [73]).

THEOREM 7.25 *Let $B_{\mathbf{w}}$ and $B_{\mathbf{w}'}$ be two hypercyclic weighted shifts on $\ell^2(\mathbb{N})$. Then one can find a path of weighted shifts $(T_\lambda)_{\lambda \in [0,1]}$ such that $T_0 = B_{\mathbf{w}}$, $T_1 = B_{\mathbf{w}'}$ and $\bigcap_{\lambda \in [0,1]} HC(T_\lambda)$ is residual.*

The proof is in two steps. We shall first show that it is possible to construct such a path between any two hypercyclic weighted shifts whose weights are bounded below by 1. This path is very simple, and common hypercyclicity will be obtained thanks to Theorem 7.18. Then we shall construct a path between any hypercyclic weighted shift and some weighted shift whose weights are bounded below by 1. This path will be much more sophisticated, and common hypercyclicity will follow from Theorem 7.9.

In several places we shall use without comment the following elementary fact: the relation "B and B' can be connected by a path of weighted shifts (T_λ) such that $\bigcap_\lambda HC(T_\lambda)$ is residual" is transitive. In other words, if one can connect B to B' and B' to B'' by suitable paths then one can also connect B to B'': just glue the two paths together and apply the Baire category theorem. Let us now proceed to prove Theorem 7.25.

STEP 1 *Let $B_{\mathbf{w}}$ and $B_{\mathbf{w}'}$ be two hypercyclic weighted shifts on $\ell^2(\mathbb{N})$, and assume that $\min(w_n, w_n') \geq 1$ for all $n \geq 1$. Then one can connect $B_{\mathbf{w}}$ to $B_{\mathbf{w}'}$ by a path of weighted shifts (T_λ) such that $\bigcap_\lambda HC(T_\lambda)$ is residual.*

PROOF Set $v_n := \max(2, w_n, w_n')$. By symmetry, it is enough to construct a suitable path between $B_{\mathbf{w}}$ and $B_{\mathbf{v}}$. The most obvious one will do the job: just put $T_\lambda := B_{\mathbf{w}(\lambda)}$, $\lambda \in [0,1]$, where $w_n(\lambda) = (1-\lambda)w_n + \lambda v_n$. The functions $\log(w_n(\lambda))$ are clearly non-decreasing and easily seen to be 1-Lipschitz on $[0,1]$. Moreover, the series $\sum_n (w_1(\lambda) \cdots w_n(\lambda))^{-2}$ is convergent when $\lambda \in (0,1]$. By Example 7.19, it follows that $\bigcap_{\lambda > 0} HC(T_\lambda)$ is residual. Hence $\bigcap_{\lambda \geq 0} HC(T_\lambda)$ is residual as well, by the Baire category theorem. $\square$

STEP 2 *Let $B_{\mathbf{w}}$ be an arbitrary hypercyclic weighted shift. Then one can connect $B_{\mathbf{w}}$ to some weighted shift $B_{\mathbf{w}'}$ with $w'(n) \geq 1$ for all $n \geq 1$ by a path of weighted shifts (T_λ) such that $\bigcap_\lambda HC(T_\lambda)$ is residual.*

As already pointed out, the situation here becomes much more involved. Indeed, since the weights w_n are not bounded below we have no control on the decay of $(w_1 \cdots w_n)^{-1}$, so it is not possible to apply the criteria of Example 7.2. To overcome this difficulty, we shall group the weights into blocks in such a way that the product of the weights inside each block is large. In any block, it will then be possible to find a path along which the product does not change and each weight becomes in the end larger than 1. This is the content of the next two lemmas.

LEMMA 7.26 *Let $\mathbf{w} = (w_n)_{n \geq 1}$ be a bounded weight sequence satisfying Salas' condition $\limsup_n w_1 \cdots w_n = \infty$. Also, let $E \geq \|\mathbf{w}\|_\infty$. Then one can find an increasing sequence of integers $(m_k)_{k \geq 0}$ with $m_0 = 0$ such that*

(1) $w_{m_k+1} \cdots w_{m_{k+1}} \geq E$ *for all $k \geq 0$;*
(2) $w_{m_k - j} \geq E^{-j}$ *for each $k \geq 1$ and all $j \in \{0, \ldots, k-1\}$.*

PROOF Assume that $m_0, \ldots, m_k$ have been defined, and set

$$A_{k+1} := \left\{ n \in \mathbb{N};\ n \geq m_k + (k+1) \text{ and } \prod_{l=m_k+1}^{n} w_l \geq E^{k+1} \right\}.$$

The set A_{k+1} is non-empty, so we can put $m_{k+1} := \min(A_{k+1})$. Then (1) is satisfied. To prove (2) for $k+1$, observe that, for any $j \in \{0, \ldots, k\}$,

$$\prod_{l=m_k+1}^{m_{k+1}-j-1} w_l \times w_{m_{k+1}-j} \times \prod_{l=m_{k+1}-j+1}^{m_{k+1}} w_l \geq E^{k+1}.$$

The first factor in the product is less than E^{k+1} by the choice of m_{k+1} (consider the cases $m_{k+1} - j - 1 \geq m_k + (k+1)$ and $m_{k+1} - j - 1 < m_k + (k+1)$ separately), and the third is not greater than E^j. Hence, the second factor is at least E^{-j}. $\square$

LEMMA 7.27 *Let $d \geq 1$ and $a_1, \ldots, a_d > 0$. Let $M := (a_1 \cdots a_d)^{1/d}$ be the geometric mean of $a_1, \ldots, a_d$, and let $e \geq \max_k(a_k) - \min_k(a_k)$. Then there exists a map $\psi : [0, e] \to \mathbb{R}^d$ such that*

(i) *$\psi(0) = (a_1, \ldots, a_d)$;*
(ii) *$\psi(e) = (M, \ldots, M)$;*
(iii) *$\min(a_i, M) \leq \psi_i(\lambda) \leq \max(a_i, M)$ and $\psi_1(\lambda) \cdots \psi_n(\lambda) = M^d$, for each i and every $\lambda \in [0, e]$;*
(iv) *ψ is 1-Lipschitz from $[0, e]$ into $(\mathbb{R}^d, \| \cdot \|_\infty)$.*

PROOF   The obvious way to define a function satisfying the first three conditions would be to set $\psi_i(\lambda) := a_i^\lambda M^{1-\lambda}$ (assuming $e = 1$). But this cannot work because the Lipschitz constant of this map is very large when a_i is close to 0. So we need a more elaborate construction.

We proceed inductively, since the result is certainly true for $d = 1$. Without loss of generality, we may assume that $a_1 \leq \cdots \leq a_p \leq M \leq a_{p+1} \leq \cdots \leq a_d$ for some $p < d$. Let us set $u(s) := (a_1 + s) \cdots (a_p + s)$ and $v(t) := (a_{p+1} - t) \cdots (a_d - t)$. The equation $u(s)v(t) = M^d$ defines $s \in [0, \infty)$ as an increasing differentiable function of $t \in [0, a_{p+1})$. We shall write $s = f(t)$. Clearly, $f(0) = 0$ and $f(t) \to \infty$ as $t \to a_{p+1}$.

Differentiating the equation $u(f(t))v(t) = M^d$, we get

$$f'(t) = -\frac{u(f(t))}{u'(f(t))} \frac{v'(t)}{v(t)} = \left(\sum_{i=1}^{p} \frac{1}{a_i + f(t)} \right)^{-1} \left(\sum_{j=p+1}^{d} \frac{1}{a_j - t} \right).$$

In particular, we see that $f'(t)$ is an increasing function of t which tends to $+\infty$ as $t \to a_{p+1}$. Thus, we may set $\alpha := \inf\{t \in [0, a_{p+1}); f'(t) \geq 1\}$. Observe that $\alpha = 0$ if $f'(t)$ is always greater than 1. Now, we define a map $\phi : [0, \infty) \to \mathbb{R}^d$ as follows.

- If $t \in [0, \alpha]$ then $\phi_i(t) = \begin{cases} a_i + f(t) & i \leq p, \\ a_i - t & i > p. \end{cases}$

- If $t > \alpha$, then $\phi_i(t) = \begin{cases} a_i + f(\alpha) + t - \alpha & i \leq p, \\ a_i - f^{-1}(t - \alpha + f(\alpha)) & i > p. \end{cases}$

Observe that $\phi_p(0) = a_p \leq M$ and $\phi_p(t) \to \infty$ as $t \to \infty$. Therefore, the number $\beta := \inf\{t \geq 0; \exists i \in \{1, \ldots, d\} : \phi_i(t) = M\}$ is well defined. We put $\psi_{|[0,\beta]} := \phi_{|[0,\beta]}$. Then ψ satisfies conditions (i), (iii), (iv) on $[0, \beta]$, and at least one coordinate of ψ takes the value M at β. We also point out that ψ is in some sense "optimal" on $[0, \beta]$ since, at each point $t \in [0, \beta]$, $|\psi'_j(t)| = 1$ for at least one coordinate.

Let us now enumerate the set $\{\psi_i(\beta); \psi_i(\beta) \neq M\}$ as $\{a'_1, \ldots, a'_k\}$. Then $k < d$. Moreover, we claim that $\max_j(a'_j) - \min_j(a'_j)$ is not greater than $e - \beta$ (so that in particular $\beta \leq e$). Indeed, we first note that $\psi_i(t) \neq M$ for each $i \in \{1, \ldots, d\}$ and all $t \in [0, \beta)$. Since $\psi_i(0) = a_i$ and $a_p \leq M \leq a_{p+1}$, it follows that

$\psi_i(t) < M$ on $[0, \beta)$ if $i \leq p$ whereas $\psi_i(t) > M$ if $i > p$. Looking at the definition of ψ, we deduce that $\psi_1(\beta) \leq \cdots \leq \psi_p(\beta) \leq \psi_{p+1}(\beta) \leq \cdots \leq \psi_d(\beta)$, so that $\max_j(a'_j) - \min_j(a'_j) \leq \psi_d(\beta) - \psi_1(\beta)$. Now,

$$\psi_d(\beta) - \psi_1(\beta) = \begin{cases} (a_d - \beta) - (a_1 + f(\beta)) & \beta \leq \alpha, \\ (a_d - f^{-1}(f(\alpha) + \beta - \alpha)) - (a_1 + f(\alpha) + \beta - \alpha) & \beta > \alpha. \end{cases}$$

Writing $\psi_d(\beta) - \psi_1(\beta) \leq a_d - \beta - a_1$ in the first case and $\psi_d(\beta) - \psi_1(\beta) \leq (a_d - \alpha) - (a_1 + \beta - \alpha)$ in the second case, we see that $\psi_d(\beta) - \psi_1(\beta) \leq e - \beta$. This proves our claim.

Applying the induction hypothesis to the sequence $(a'_1, \ldots, a'_k)$, we get a map $\rho : [0, e - \beta] \to \mathbb{R}^k$. Then we define ψ on $[\beta, e]$ by setting $\psi_i(\lambda) := \rho_j(\lambda + \beta)$ if $\psi_i(\beta) = b_j$ for some j, and $\psi_i(\lambda) := M$ if we already have $\psi_i(\beta) = M$. Clearly, the map $\psi : [0, e] \to \mathbb{R}^d$ has the required properties. $\qquad \square$

With these two lemmas to hand, we can now proceed to connect our given hypercyclic weighted shift $B_{\mathbf{w}}$ to some hypercyclic weighted shift $B_{\mathbf{w}'}$ with $w'_n \geq 1$ for all $n \geq 1$.

We set $E := \|\mathbf{w}\|_\infty$. Let (m_k) be the sequence given by Lemma 7.26. For each $k \geq 1$, let $\psi_k : [0, E] \to \mathbb{R}^d$ be a map given by Lemma 7.27 applied with $d := m_k - m_{k-1}$, $a_1 := w_{m_{k-1}+1}, \ldots, a_d := w_{m_k}$ and $e := E$. We denote by $w_{m_k+j}(\lambda)$ the jth entry of the vector $\psi_k(\lambda)$ $(1 \leq j \leq m_k - m_{k-1})$. Thus we have defined for each $\lambda \in [0, E]$ an infinite weight sequence $\mathbf{w}(\lambda) = (w_n(\lambda))_{n \geq 1}$. We put $T_\lambda := B_{\mathbf{w}(\lambda)}$. Observe that $\mathbf{w}(0) = \mathbf{w}$ and that $w_n(E) \geq 1$ for all $n \geq 1$ since the geometric mean of $w_{m_k+1}, \ldots, w_{m_{k+1}}$ is always greater than 1, by condition (1) in Lemma 7.26. Moreover, each map $\lambda \mapsto w_n(\lambda)$ is 1-Lipschitz, so that $(T_\lambda)_{\lambda \in [0, E]}$ is a continuous path of weighted shifts.

It remains to prove that $\bigcap_{\lambda \in [0, E]} HC(T_\lambda)$ is residual in $\ell^2(\mathbb{N})$, which can be done using Theorem 7.9. Let $\mathcal{D} := c_{00}(\mathbb{N}) \subset \ell^2(\mathbb{N})$ be the set of all finitely supported sequences and, for each $v \in \mathcal{D}$, set $\mathcal{D}_v := \mathcal{D}$. We fix a vector $v = \sum_{p=0}^q a_p e_p \in \mathcal{D}$ and an open ball $O = \{\|x\| < \varepsilon\} \subset \ell^2(\mathbb{N})$.

We put $n_k := m_k - q$ for all $k \geq q$ and define the vectors $S_{\lambda, n_k}(v)$ by $S_{\lambda, n_k}(v) := \sum_p a_p S_{\lambda, n_k}(e_p)$, where

$$S_{\lambda, n_k}(e_p) := \frac{1}{w_{p+1}(\lambda) \cdots w_{p+n_k}(\lambda)} e_{p+n_k}$$
$$= \frac{1}{w_{p+1}(\lambda) \cdots w_{m_k-(q-p)}(\lambda)} e_{m_k-(q-p)}.$$

We note that the sequence (n_k) does depend on the vector v, so we really need Theorem 7.9 as we stated it. By linearity, it is enough to show that, for each $p \in \{0, \ldots, q\}$, properties (1), (2), (3) in Theorem 7.9 are satisfied with e_p in place of v. Accordingly, we fix $p \in \{0, \ldots, q\}$.

By the choice of (m_k) and $\mathbf{w}(\lambda)$, we know that $w_1(\lambda) \cdots w_{m_k}(\lambda) \geq E^k$ for each $k \geq 1$ and every $\lambda \in [0, E]$. Hence

$$\|S_{\lambda, n_k}(e_p)\| = \frac{w_1(\lambda) \cdots w_p(\lambda)}{w_1(\lambda) \cdots w_{m_k}(\lambda)} w_{m_k - (q-p)+1}(\lambda) \cdots w_{m_k}(\lambda)$$
$$\leq \frac{E^p}{E^k} E^{q-p},$$

so that $\|S_{\lambda, n_k}(e_p)\| \to 0$ as $k \to \infty$ (because $E = \|\mathbf{w}\|_\infty > 1$). Moreover, if $u \in \mathcal{D}$ then $T_\lambda^{n_k}(u) = 0$ provided that $n_k \geq \max(\text{supp}(u))$. Thus, properties (1) and (2) of Theorem 7.9 are satisfied. It remains to check (3), or its variant (3') pointed out in Remark 7.10.

If $k \in \mathbb{N}$ and $\lambda, \mu \in [0, E]$ then

$$T_\mu^{n_k} S_{\lambda, n_k}(e_p) = \frac{w_{p+1}(\mu) \cdots w_{p+n_k}(\mu)}{w_{p+1}(\lambda) \cdots w_{p+n_k}(\lambda)} e_p.$$

Since the product $w_1(\cdot) \cdots w_{m_k}(\cdot)$ is constant on $[0, E]$, the quotient on the right-hand side can be written as

$$\frac{w_1(\lambda) \cdots w_p(\lambda)}{w_1(\mu) \cdots w_p(\mu)} \times \frac{w_{m_k - (q-p-1)}(\lambda) \cdots w_{m_k}(\lambda)}{w_{m_k - (q-p-1)}(\mu) \cdots w_{m_k}(\mu)},$$

and it follows that

$$\|T_\mu^{n_k} S_{\lambda, n_k}(e_p) - e_p\| = \left| \frac{w_1(\lambda) \cdots w_p(\lambda)}{w_1(\mu) \cdots w_p(\mu)} \times \frac{w_{m_k - (q-p-1)}(\lambda) \cdots w_{m_k}(\lambda)}{w_{m_k - (q-p-1)}(\mu) \cdots w_{m_k}(\mu)} - 1 \right|.$$

Now, each function w_n is 1-Lipschitz and it follows from property (2) in Lemma 7.26 that all functions $w_{m_k - j}$ ($0 \leq j \leq q - p - 1$) are bounded below on $[0, E]$ by some positive constant $c_{p,q}$ which does not depend on k. The same is of course true for the functions $w_1, \dots, w_p$. Looking at the right-hand side of the last equation, since only q functions are involved it should now be clear that $\|T_\mu^{n_k} S_{\lambda, k}(e_p) - e_p\| < \varepsilon$ for all k if $|\mu - \lambda|$ is small enough. This concludes the proof. $\qquad \square$

7.4.2 A path without common hypercyclic vectors

We conclude this chapter as we started it, namely by giving a counter-example to common hypercyclicity. This example is due to K. C. Chan and R. Sanders [73]. It is particularly interesting since it involves a one-parameter family satisfying a strong continuity assumption. Thus, it shows clearly that one cannot hope for too general a positive statement regarding common hypercyclicity. As in the previous subsection, we work on the Hilbert space $\ell^2(\mathbb{N})$.

THEOREM 7.28 *There exists a continuous path of hypercyclic weighted shifts* (T_λ) *such that* $\bigcap_\lambda HC(T_\lambda) = \varnothing$.

PROOF   The strategy is the following. We will construct a path of hypercyclic weighted shifts $(B_{\mathbf{w}(\lambda)})$ such that for any given $x \in \ell^2(\mathbb{N})$ the following properties hold:

(1) there are only a few natural numbers n for which $B_{\mathbf{w}(\lambda)}^n(x)$ is close to $2e_0$ for some λ;

(2) for any fixed $n \in \mathbb{N}$, there are only a few λ such that $B_{\mathbf{w}(\lambda)}^n(x)$ is close to $2e_0$.

Here, as usual, e_0 is the first vector of the canonical basis of $\ell^2(\mathbb{N})$. Property (1) will be ensured essentially because the products $w_1(\lambda) \cdots w_n(\lambda)$ are often smaller than 1. For property (2), the key point will be that the products $w_1(\lambda) \cdots w_n(\lambda)$ are very sensitive to variations in the parameter λ. Once these two properties are established, a simple measure-theoretic argument (as in Borichev's example 7.2) will show that the weighted shifts $B_{\mathbf{w}(\lambda)}$ have no common hypercyclic vectors.

For the construction of the path, let us fix a fast increasing sequence of integers $(n_k)_{k \geq 0}$ with $n_0 = 0$ and $n_{k+1} > 2n_k$ for all k, and set $m_k := 2n_k$. For every $\lambda \in [2, 3]$, we consider the weight sequence $\mathbf{w}(\lambda)$ defined as follows:

$$
\begin{cases}
w_{n_k+1}(\lambda) = \cdots = w_{m_k}(\lambda) = 4, \\
w_{m_k+1}(\lambda) = \dfrac{\lambda^{n_{k+1}}}{4^{n_{k+1}}}, \\
w_{m_k+2}(\lambda) = \cdots = w_{n_{k+1}}(\lambda) = 1.
\end{cases}
$$

We quote the following two identities, which hold true for any $k \geq 0$:

$$
w_1(\lambda) \cdots w_n(\lambda) = \frac{\lambda^{n_1 + \cdots + n_{k+1}}}{4^{n_{k+1}}} \quad \text{for any } n \in (m_k, n_{k+1}]; \tag{7.3}
$$

$$
w_1(\lambda) \cdots w_n(\lambda) = \frac{\lambda^{n_1 + \cdots + n_k}}{4^{m_k - n}} \quad \text{for any } n \in (n_k, m_k]. \tag{7.4}
$$

It follows from (7.4) applied with $n = m_k$ that $w_1(\lambda) \cdots w_{m_k}(\lambda) \to \infty$ as $k \to \infty$. Hence, each weighted shift $B_{\mathbf{w}(\lambda)}$ is hypercyclic. Moreover, for any $\lambda, \mu \in [2, 3]$, an application of the mean-value theorem yields

$$
\|\mathbf{w}(\lambda) - \mathbf{w}(\mu)\|_\infty = \sup_{k \geq 0} |w_{m_k+1}(\lambda) - w_{m_k+1}(\mu)|
$$

$$
\leq \frac{1}{4}|\lambda - \mu| \sup_{k \geq 0} \left[n_{k+1} \left(\frac{3}{4} \right)^{n_{k+1}-1} \right].
$$

Therefore, the family $(B_{\mathbf{w}(\lambda)})_{\lambda \in [2,3]}$ is a continuous path of operators.

We claim that if the sequence (n_k) is sufficiently fast increasing then the following properties hold true for every $k \geq 0$ and $\lambda, \mu \in [2, 3]$:

(i) if $n \in (m_k, n_{k+1}]$ then $w_1 \cdots w_n(\lambda) < 1$;

(ii) if $k \geq 1$ and $m_{k-1} < p \leq n_k < n + p \leq m_k$ then $\dfrac{w_{1+p}(\lambda) \cdots w_{n+p}(\lambda)}{w_1(\mu) \cdots w_{n+p}(\mu)} > 1$;

(iii) if $n \in (n_k, n_{k+1}]$ and $\mu \geq \lambda$ then $\dfrac{w_1(\mu) \cdots w_n(\mu)}{w_1(\lambda) \cdots w_n(\lambda)} \geq \left(\dfrac{\mu}{\lambda} \right)^{n_k}$.

Indeed, by (7.3), property (i) holds if we assume that $3^{n_1+\cdots+n_{k+1}} < 4^{n_{k+1}}$ for all k. By (7.3) and (7.4), property (ii) amounts to the inequality $4^{n_k}/\mu^{n_1+\cdots+n_k} > 1$, which holds if (n_k) is fast increasing, since $\mu \leq 3$. Finally, property (iii) follows at once from (7.3) and (7.4).

Now, we assume that (i), (ii), (iii) hold and we proceed to show that the $B_{\mathbf{w}(\lambda)}$ have no common hypercyclic vectors, following the strategy outlined at the beginning of the proof. Towards a contradiction, assume that $\bigcap_{\lambda \in [2,3]} HC(B_{\mathbf{w}(\lambda)}) \neq \varnothing$ and let us fix a vector $x = (x_n) \in \bigcap_\lambda HC(B_{\mathbf{w}(\lambda)})$, with $\|x\| \leq 1$.

For each $k \geq 0$, we set

$$A_k := \{n \in (n_k, n_{k+1}]; \ \exists \lambda \in [2,3] \ : \ \|B^n_{\mathbf{w}(\lambda)}(x) - 2e_0\| < 1\}.$$

Since $|x_n| \leq 1$, we see that if $n \in A_k$ then $w_1(\lambda) \cdots w_n(\lambda) \geq 1$ for some $\lambda \in [2,3]$. By (i) above, it follows that A_k is contained in $(n_k, m_k]$.

Put $n := \min A_k$ and $n + p := \max A_k$. Then $p \leq m_k - n_k = n_k$. Choose $\lambda, \mu \in [2,3]$ with $\|B^n_{\mathbf{w}(\lambda)}(x) - 2e_0\| < 1$ and $\|B^{n+p}_{\mathbf{w}(\mu)}(x) - 2e_0\| < 1$. Looking at the first coordinate of $B^{n+p}_{\mathbf{w}(\mu)}(x)$ and at the pth coordinate of $B^n_{\mathbf{w}(\lambda)}(x)$, we get

$$w_1(\mu) \cdots w_{n+p}(\mu)|x_{n+p}| \geq 1 \quad \text{and} \quad w_{p+1}(\lambda) \cdots w_{p+n}(\lambda)|x_{p+n}| \leq 1.$$

Dividing these two inequalities, this yields

$$\frac{w_{1+p}(\lambda) \cdots w_{n+p}(\lambda)}{w_1(\mu) \cdots w_{n+p}(\mu)} \leq 1.$$

By (ii), it follows that $p \leq m_{k-1}$ (if $k \geq 1$). Thus we have shown that, for any $k \geq 1$, the cardinality of A_k is at most m_{k-1} and hence less than $n_k^{1/2}$ under a suitable growth condition on (n_k).

We now come to the second part of the proof. For each $n \in \mathbb{N}$, we set

$$E_n := \{\lambda \in [2,3]; \ \|B^n_{\mathbf{w}(\lambda)}(x) - 2e_0\| < 1\}.$$

Assuming $E_n \neq \varnothing$, we put $a := \inf(E_n)$ and $b := \sup(E_n)$. We also choose $k \geq 0$ such that $n \in (n_k, n_{k+1}]$. Clearly, $1 \leq w_1(a) \cdots w_n(a)|x_n| \leq 3$ and $1 \leq w_1(b) \cdots w_n(b)|x_n| \leq 3$. Dividing these two inequalities and using (iii), we get

$$\left(\frac{b}{a}\right)^{n_k} \leq 3,$$

so that

$$b - a \leq \left(3^{1/n_k} - 1\right) a \leq \frac{C}{n_k}$$

for some finite constant C. This shows that $m(E_n) \leq C/n_k$ for each $k \in \mathbb{N}$ and all $n \in (n_k, n_{k+1}]$, where m is the Lebesgue measure on $\mathbb{R}$.

Finally, here is the measure-theoretic argument. Since $x \in \bigcap_\lambda HC(B_{\mathbf{w}(\lambda)})$, we know that

$$[2,3] = \bigcup_{n \geq n_\kappa} E_n = \bigcup_{k \geq \kappa} \bigcup_{n \in A_k} E_n$$

for any $\kappa \geq 1$. Therefore,

$$1 \leq \sum_{k \geq \kappa} m\left(\bigcup_{n \in A_k} E_n\right) \leq \sum_{k \geq \kappa} n_k^{1/2} \frac{C}{n_k}$$

for all $\kappa \geq 1$. If we assume that the series $\sum n_k^{-1/2}$ is convergent, this is the contradiction for which we were looking. $\square$

7.5 Comments and exercises

The first question concerning the existence of common hypercyclic vectors appears in the work of G. Godefroy and J. H. Shapiro [123]. The authors asked there about the existence of a common hypercyclic vector for all hypercyclic adjoints of multipliers. This question remains unsolved, even for small subfamilies (see Exercise 7.8 for a partial positive result).

Later on, H. N. Salas asked in [216] about the existence of a common hypercyclic vector for the family $(\lambda B)_{\lambda > 1}$, where B is the usual backward shift on $\ell^2(\mathbb{N})$. This was first answered by E. Abakumov and J. Gordon in [1]. Their proof was quite different from the one presented in this book; it uses a rather difficult construction of sequences of natural numbers with nice properties.

The idea of using Baire category arguments even in that kind of problem is due to G. Costakis and M. Sambarino. In particular, Theorem 7.14 comes from their paper [86], as well as a number of applications.

The results of Section 7.2 and most of the subsequent results in this chapter were obtained either in [73] or in [29], sometimes under stronger assumptions. The (difficult) results of Section 7.4 are taken from [73], and we have tried to simplify the proofs as much as we could. Theorem 7.18 is new.

Most results described in this chapter could be formulated in the more general setting of universal sequences of operators. Recall that a sequence $\mathbf{T} = (T_n) \subset \mathfrak{L}(X)$ is said to be universal if there exists a vector x such that the set $\{T_n(x); \ n \in \mathbb{N}\}$ is dense in X. Under reasonable assumptions, the set of universal vectors $Univ(\mathbf{T})$ is residual in X provided that it is non-empty; for example, this holds if the operators T_n have dense range and commute with each other. Exactly as in the hypercyclic case, one can ask for the residuality of $\bigcap_{\lambda \in \Lambda} Univ(\mathbf{T}_\lambda)$ for a given family of universal sequences $(\mathbf{T}_\lambda) = (T_{\lambda,n})_{n \in \mathbb{N}, \lambda \in \Lambda}$. Under the natural continuity assumption for the maps $(\lambda, x) \mapsto T_{n,\lambda}(x)$, most results proved in this chapter remain true after some appropriate (and minor) modification. Essentially, one just has to replace T_λ^n by $T_{\lambda,n}$ everywhere. Details can be found in the references given above. Only Theorem 7.24 is specific to hypercyclic operators. This is due to the use of a connectedness argument (see Remark 7.21), which is not available in the "universal" setting.

EXERCISE 7.1   **Size of the set of attractive fixed points**

1. Let ϕ be a hyperbolic automorphism of $\mathbb{D}$. Show that $(\phi_n(0))$ converges non-tangentially to the attractive fixed point of ϕ. (*Hint*: Look at the sister of ϕ living on $\mathbb{P}_+$.)
2. Let $\Lambda \subset Aut(\mathbb{D})$ be a family of hyperbolic automorphisms of $\mathbb{D}$ and put

$$\mathrm{Fix}(\Lambda) := \{\xi \in \mathbb{T}; \ \exists \phi \in \Lambda : \ \xi \text{ is the attractive fixed point of } \phi\}.$$

Show that if the family $(C_\phi)_{\phi \in \Lambda}$ has a common hypercyclic vector then $\mathrm{Fix}(\Lambda)$ has Lebesgue measure 0. (*Hint*: If $f \in HC(C_\phi)$ then f does not admit a non-tangential limit at ξ, the attractive fixed point of ϕ.)

EXERCISE 7.2   Let $(x_k)_{k \geq 1}$ be a dense sequence in $\ell^2(\mathbb{N})$ made up of finitely supported vectors, with $|x_k(n)| \geq 1/k$ or $x_k(n) = 0$ for every $n \in \mathbb{N}$. Show that if (n_k) is a sufficiently fast increasing sequence of integers then the vector $x := \sum_{k=0}^{+\infty}(2^{-1}S)^{n_k}(x_k)$ is a hyper-cyclic vector for $2B$ which is not hypercyclic for $3B$. Here, B is the usual backward shift and S is the forward shift.

EXERCISE 7.3 We intend to prove here a refinement of the Costakis–Sambarino Criterion. To be precise, the aim of the exercise is to show that one can replace the summability assumption $\sum_k c_k < \infty$ by the following one: *the sequence (c_k) is non-decreasing and $c_k = o(1/k)$.* We keep the notations of the proof of Theorem 7.14 and choose the subdivision in such a way that $\log(q) \le CN/\tau$.

1. Show that $\|\sum_{j \ne i} T_\lambda^{n_i} S_{\lambda_j}^{n_j}(v)\| \le 2\sum_{l \le A(N,\tau)} c_{lN}$, where $A(N,\tau) = \exp(Cn/\tau)$.
2. Prove that $\sum_{l \le A(N,\tau)} c_{lN} \le N^{-1}\sum_{l \le A(N,\tau)} c_l$.
3. Conclude.

EXERCISE 7.4 Let D be the derivative operator on $H(\mathbb{C})$. Use the Costakis–Sambarino Criterion to show that $\bigcap_{\lambda \ne 0} HC(\lambda D)$ is a residual subset of $H(\mathbb{C})$.

EXERCISE 7.5 **Common frequently hypercyclic vectors** ([26])
Let Λ be an uncountable subset of $(1, \infty)$, and assume that the family $(\lambda B)_{\lambda \in \Lambda}$ admits a common *frequently* hypercyclic vector $x \in \ell^2(\mathbb{N})$ (see Chapter 6). For each $\lambda \in \Lambda$, we set

$$E_\lambda := \{n \in \mathbb{N};\ \|(\lambda B)^n(x) - 2e_0\| < 1\} \quad \text{and} \quad \delta_\lambda := \underline{\text{dens}}(E_\lambda) = \liminf_{N \to \infty} \frac{\text{card}(E_\lambda \cap [1, N])}{N}.$$

1. Show that one can find $\lambda_1, \ldots, \lambda_q \in \Lambda$ such that $\sum_{i=1}^q \delta_{\lambda_i} > 1$.
2. Show that there exist $\lambda \ne \mu$ such that $E_\lambda \cap E_\mu$ is infinite.
3. Obtain a contradiction.

Using the same arguments as in Example 7.17 together with the results of [79], one can prove that $\bigcap_{|\lambda|=r} FHC(\lambda B)$ is a residual subset of $\ell^2(\mathbb{N})$ for any $r \in (1, \infty)$.

EXERCISE 7.6 In this exercise, we define the function ψ_2 by $\psi_2(x) = e^{-x^2/2} - 1$.

1. Let M be a non-negative random variable defined on some probability space $(\Omega, \mathcal{F}, \mathbb{P})$. Show that if $\mathbb{P}(M \ge \varepsilon) \le e^{-\varepsilon^2/2}$ for each $\varepsilon > 0$ then $\|M\|_{\psi_2} \le 1$. Conversely, show that if $\|M\|_{\psi_2} \le 1$ then $\mathbb{P}(M \ge \varepsilon) \le 2e^{-\varepsilon^2/2}$ for each $\varepsilon > 0$.
2. Explain the terminology "subgaussian" for stochastic processes satisfying a Lipschitz L^{ψ_2}-estimate.

EXERCISE 7.7 Let $\mathcal{W}_{HC}$ be the set of all hypercyclic weighted shifts on $\ell^2(\mathbb{N})$.

1. Show that if $B_{\mathbf{w}} \in \mathcal{W}_{HC}$ and if $C \ge 1 + \|\mathbf{w}\|_\infty$ then the line segment $[B_{\mathbf{w}}, CB]$ is contained in $\mathcal{W}_{HC}$ (B is the canonical backward shift). Deduce that $\mathcal{W}_{HC}$ is path connected.
2. Show that $\mathcal{W}_{HC}$ is not convex (see [73]).

EXERCISE 7.8 **Adjoints of multipliers** ([29], [118])

1. Let X be a separable Fréchet space, and let $T \in \mathfrak{L}(X)$. Also, let $\lambda_0 \ge 0$. Assume that
   (i) $\mathcal{D} := \bigcup_{n=1}^\infty \text{Ker}(T^n)$ is dense in X and T has a right-inverse $S : \mathcal{D} \to X$;
   (ii) for any $\lambda > \lambda_0$ and each $u \in \mathcal{D}$, the set $\{\lambda^{-n}S^n(u);\ n \in \mathbb{N}\}$ is bounded in X.
   Show that $\bigcap_{\lambda > \lambda_0} HC(\lambda T)$ is a dense G_δ subset of X.
2. *Application to adjoints of multipliers* Let $\phi \in H^\infty(\mathbb{D})$. Assume that ϕ is not an outer function and that it satisfies $1/\phi^* \in L^\infty(\mathbb{T})$, where ϕ^* is the boundary value of ϕ. We denote by $M_\phi : H^2(\mathbb{D}) \to H^2(\mathbb{D})$ the associated multiplication operator, $M_\phi(f) = \phi f$. Finally, let $\phi = u\theta$ be the inner–outer factorization of ϕ, where θ is the (non-constant) inner function.

(a) For each $n \in \mathbb{N}$, set $K_n := (\theta^n H^2(\mathbb{D}))^\perp$. Show that $\bigcup_{n=1}^{\infty} K_n$ is dense in $H^2(\mathbb{D})$.

(b) Let $S := M_{1/u}^* M_\theta$. Show that $M_\phi^* S_\lambda = I$.

(c) Deduce that $\bigcap_{\lambda \in (\|1/\phi^*\|_\infty, \infty)} HC(\lambda M_\phi^*)$ is a residual subset of $H^2(\mathbb{D})$.

EXERCISE 7.9   **Changes of variable**
Let $w : I \to \mathbb{R}$ be a positive bounded and continuous function on some open interval $I \subset \mathbb{R}$ and, for each $p \in [1, \infty)$, put $X_p := L^p(I, w(t)dt)$. Also, let $(\phi_\lambda)_{\lambda \in \Lambda}$ be a family of increasing diffeomorphisms of I parametrized by some interval $\Lambda \subset \mathbb{R}$. Assume that

$$w(t) \leq C(\lambda) |\phi_\lambda'(t)| w(\phi_\lambda(t))$$

for some locally bounded function $C(\lambda)$ and that the map $(\lambda, t) \mapsto \phi_\lambda(t)$ is continuous on $\Lambda \times I$. For each $\lambda \in \Lambda$, let us denote by $T_\lambda : X_p \to X_p$ the change-of-variable operator defined by

$$T_\lambda(f) = f \circ \phi_\lambda.$$

1. Show that T_λ is well defined and bounded on X_p and that the map $(\lambda, f) \mapsto T_\lambda(f)$ is continuous from $\Lambda \times X_p$ into X_p.

2. Assume that $I = \mathbb{R}$ and that the following properties hold, where $\phi_\lambda^n = \phi_\lambda \circ \cdots \circ \phi_\lambda$:

   (ia) the family (ϕ_λ) is monotonic, and the ϕ_λ have no fixed point;

   (ib) $|\phi_\lambda^k(\tau)| \to \infty$ as $k \to +\infty$, for each $\lambda \in \Lambda$ and every $\tau \in \mathbb{R}$;

   (ii) $\sum_{j \in \mathbb{Z}} \int_{|t| \geq |\phi_\lambda^j(\tau)|} w(t) \, dt < \infty$ for any $\lambda \in \Lambda$ and every $\tau \in \mathbb{R}$.

   (iii) all functions ϕ_λ^{-1} are 1-Lipschitz, and $|\phi_\mu^{-1} \circ \phi_\lambda(t) - t| \leq C|\lambda - \mu|$.

   Show that $\bigcap_{\lambda \in \Lambda} HC(T_\lambda)$ is a residual subset of X_p. (*Hint:* Apply Theorem 7.18. Let $\mathcal{D} \subset X_p$ be the set of all compactly supported smooth functions and define $S_\lambda : \mathcal{D} \to \mathcal{D}$ by $S_\lambda(f) = f \circ \phi_\lambda^{-1}$.)

3. Assume now that $I = (0, \infty)$, $w(t) = (1 + t^\rho)^{-1}$ for some $\rho > 1$, and $\phi_\lambda(t) = t + \lambda - (t + \lambda)^{-1}$, $\lambda \in (0, \infty)$. Show that $\bigcap_{\lambda > 0} HC(T_\lambda)$ is a residual subset of X_p.

8

Hypercyclic subspaces

Introduction

In Chapter 1, we observed that the set of hypercyclic vectors of any hypercyclic operator T has a rich algebraic structure: there exists a dense linear subspace of the underlying space X consisting entirely of hypercyclic vectors, except 0.

It may also happen that $HC(T)$ contains a *closed* infinite-dimensional subspace (except 0). This is the topic that we will consider in this chapter. For brevity, we adopt the following terminology: by a **hypercyclic subspace** for an operator $T \in \mathfrak{L}(X)$, we shall always mean a closed infinite-dimensional subspace $Z \subset X$ such that $Z \setminus \{0\} \subset HC(T)$.

Our main goal in this chapter is to prove the two theorems stated below, which show that hypercyclic subspaces exist under quite natural assumptions. For simplicity, we will restrict ourselves to the case of Banach space operators.

The following basic result is due to A. Montes-Rodríguez [183].

THEOREM 8.1 *Let X be a separable Banach space, and let $T \in \mathfrak{L}(X)$. Assume that the following hold for some increasing sequence of integers (n_k).*

(1) *T satisfies the Hypercyclicity Criterion with respect to (n_k).*
(2) *There exists a closed infinite-dimensional subspace $E \subset X$ such that $T^{n_k}(x) \to 0$ for all $x \in E$.*

Then T has a hypercyclic subspace.

It is important to note that properties (1) and (2) in Theorem 8.1 have to be satisfied by the *same* sequence (n_k). However, it turns out that this restriction is in fact not necessary. This follows from the next theorem, which gives a complete characterization of the operators admitting a hypercyclic subspace, provided that they satisfy the Hypercyclicity Criterion. This theorem is due to F. León-Saavedra and A. Montes-Rodríguez when X is a Hilbert space [167], and to M. González, F. León-Saavedra and A. Montes-Rodríguez in the general case [125].

We recall the definition of the **essential spectrum** of an operator. Let X be a (complex) Banach space, and let $\mathcal{K}(X)$ the two-sided ideal of $\mathfrak{L}(X)$ consisting of all compact operators. If $T \in \mathfrak{L}(X)$, we denote by $[T]_{\mathfrak{L}/\mathcal{K}}$ the image of T in the Calkin algebra $\mathfrak{L}(X)/\mathcal{K}(X)$ under the canonical quotient map. The essential spectrum of T is the spectrum of $[T]_{\mathfrak{L}/\mathcal{K}}$ in $\mathfrak{L}(X)/\mathcal{K}(X)$.

THEOREM 8.2 *Let X be a separable complex Banach space, and let $T \in \mathfrak{L}(X)$. Assume that T satisfies the Hypercyclicity Criterion. Then the following are equivalent.*

(i) T *has a hypercyclic subspace.*

(ii) *There exists some closed infinite-dimensional subspace* $E \subset X$ *and an increasing sequence of integers* (n_k) *such that* $T^{n_k}(x) \to 0$ *for all* $x \in E$.

(iii) *There exists some closed infinite-dimensional subspace* $E \subset X$ *and an increasing sequence of integers* (n_k) *such that* $\sup_k \|T_{|E}^{n_k}\| < \infty$.

(iv) *The essential spectrum of* T *intersects the closed unit disk.*

As an immediate corollary, one obtains the following result, which is due to León-Saavedra and Montes-Rodríguez [167].

COROLLARY 8.3 *Let* X *be a separable complex Banach space, and let* $T \in \mathfrak{L}(X)$. *Assume that* T *satisfies the Hypercyclicity Criterion, and that there exists some compact operator* K *such that* $\|T - K\| \leq 1$. *Then* T *has a hypercyclic subspace.*

The reader will probably have noticed that Theorem 8.1 is contained in Theorem 8.2. However, we think it is better to have stated Theorem 8.1 separately, for at least three reasons. First, it is in itself an interesting result. Second, it is a main ingredient in the proof of Theorem 8.2. Finally, two quite different and equally interesting proofs of Theorem 8.1 can be found in the literature.

Accordingly, we start the chapter by giving these two different proofs of Theorem 8.1. The first is a direct construction of a hypercyclic subspace, which uses *basic sequence* techniques. The second proof relies on a clever and useful idea of K. C. Chan (see [70], [74]), which can also be used to prove similar statements concerning the existence of a *common* hypercyclic subspace for countable or uncountable families of operators.

The proof of Theorem 8.2 is given in Section 8.3. In Section 8.4, we illustrate the results with two natural examples, weighted shifts and composition operators. Section 8.5 is completely independent of the rest of the chapter. We show there that one can construct a non-trivial *algebra* of entire functions consisting only of hypercyclic vectors for the derivative operator D (except for the zero vector).

8.1 Hypercyclic subspaces via basic sequences

In this section we prove Theorem 8.1 using basic sequence techniques. Let us first recall some well-known facts concerning basic sequences. For more information, the reader may consult [174], [96], [173] or [3] (see also Appendix C).

A sequence $(e_n)_{n \in \mathbb{N}}$ in a Banach space X is a **basic sequence** if each vector $x \in [(e_n)] := \overline{\mathrm{span}}\{e_n; \ n \geq 0\}$ can be uniquely written as the sum of a convergent series $\sum_0^\infty \alpha_n e_n$. In that case, the projections $\pi_m : [(e_n)] \to [(e_n)]$ defined by $\pi_m(\sum_0^\infty \alpha_n e_n) = \sum_0^m \alpha_n e_n$ are uniformly bounded. The finite constant $C := \sup_m \|\pi_m\|$ is called the **basis constant** of (e_n). Conversely, if (e_n) is linearly independent and if the projections π_m are uniformly bounded on span$\{e_n;\ n \in \mathbb{N}\}$ then (e_n) is a basic sequence.

If (e_n) is a basic sequence in X, we denote by (e_n^*) the associated sequence of coordinate functionals, defined on $[(e_n)]$ by $\langle e_n^*, \sum_k \alpha_k e_k \rangle = \alpha_n$. We will also denote by e_n^* any Hahn–Banach extension of e_n^* to the whole space X; thus, we may consider (e_n^*) as a sequence in X^*.

It is not hard to check that $\|e_n\|\|e_n^*\| \leq 2C$ for all $n \in \mathbb{N}$, where C is the basis constant of (e_n). In particular, it is important to keep in mind that if the basic sequence (e_n) is bounded below, i.e. $\inf_n \|e_n\| > 0$, then the sequence (e_n^*) is bounded.

We shall need two very useful facts concerning basic sequences.

- Every infinite-dimensional Banach space X contains a basic sequence. More precisely, given any sequence (E_n) of infinite-dimensional subspaces of X, one can produce a basic sequence (e_n) such that $e_n \in E_n$ for all $n \in \mathbb{N}$. This is achieved by a classical method due to S. Mazur, which will be referred to as **Mazur's construction** (see e.g. [96, p. 39] or Appendix C).
- Basic sequences are stable under "small perturbations". This is the content of the following lemma (see [96, p. 46] or Appendix C). Recall that two basic sequences (e_n) and (f_n) are said to be **equivalent** if the convergence of any series $\sum_n \alpha_n e_n$ is equivalent to that of $\sum_n \alpha_n f_n$.

LEMMA 8.4 *Let (e_n) be a basic sequence in X, and let (f_n) be a sequence in X satisfying $\sum_0^\infty \|e_n^*\|\|e_n - f_n\| < 1$. Then (f_n) is a basic sequence equivalent to (e_n).*

Finally, we recall that a sequence $(e_n) \subset X$ is said to be **normalized** if $\|e_n\| = 1$ for all n and **semi-normalized** if $0 < \inf_n \|e_n\| \leq \sup_n \|e_n\| < \infty$.

Basic sequence arguments were used in A. Montes-Rodríguez' original proof of Theorem 8.1. This proof was modified in an elegant way by F. León-Saavedra and V. Müller in [169]. We present this modification now.

First, we recall the assumptions of the Hypercyclicity Criterion (Definition 1.5): there exist two dense sets $\mathcal{D}_1, \mathcal{D}_2 \subset X$ and a sequence of maps $S_{n_k} : \mathcal{D}_2 \to X$ such that

(1) $T^{n_k}(x) \to 0$ for any $x \in \mathcal{D}_1$;
(2) $S_{n_k}(y) \to 0$ for any $y \in \mathcal{D}_2$;
(3) $T^{n_k} S_{n_k}(y) \to y$ for each $y \in \mathcal{D}_2$.

Now, we turn to the first proof of Theorem 8.1. In what follows, we put $\mathbf{N} := \{n_k;\ k \in \mathbb{N}\}$. We also fix some well-ordering $\prec$ of $\mathbb{N} \times \mathbb{N}$ induced by any bijection between $\mathbb{N} \times \mathbb{N}$ and $\mathbb{N}$.

Let $(e_i)_{i \in \mathbb{N}}$ be a normalized basic sequence in E. Also, let $(y_j)_{j \geq 1}$ be a dense sequence in X. Finally, let $(\varepsilon_{i,j})_{(i,j) \in \mathbb{N} \times \mathbb{N}}$ be a double sequence of positive numbers such that $\sum_{i,j} \varepsilon_{i,j} \leq 1$. By induction with respect to $\prec$, one can construct two double sequences $(z_{i,j}) \subset X$ and $(n_{i,j}) \subset \mathbf{N}$ such that the following properties hold true for each $(i, j) \in \mathbf{N} \times \mathbf{N}$:

(i) $z_{i,0} = e_i$ and $\|z_{i,j}\| \leq \varepsilon_{i,j}$ if $j \geq 1$;

(ii) $\|T^{n_{i,j}}(z_{i',j'})\| \leq 2^{-j} \varepsilon_{i',j'}$ for all $(i', j') \prec (i, j)$;

(iii) $\|T^{n_{i,j}}(z_{i,j}) - y_j\| \leq 2^{-j}$ if $j \geq 1$.

Indeed, at each step (i, j) with $j \geq 1$ we may choose a vector $z_{i,j}$ and an integer $n_{i,j} \in \mathbf{N}$ as large as we want satisfying (i) and (iii), thanks to conditions (2) and (3) in the Hypercyclicity Criterion. Furthermore, condition (1) in the Hypercyclicity Criterion tells us that we can also assume that $T^n(z_{i,j}) \to 0$ as $n \to \infty$, $n \in \mathbf{N}$, and since $e_i \in E$, the same is true if $j = 0$. This will ensure (ii) for the later steps.

Let us put $z_i = \sum_{j=0}^\infty z_{i,j}$, $i \in \mathbf{N}$. This makes sense by (i). Moreover, we have

$$\|z_i - e_i\| \leq \sum_{j \geq 1} \|z_{i,j}\| \leq \sum_{j \geq 1} \varepsilon_{i,j}$$

for all $i \in \mathbf{N}$, so that (z_i) is a basic sequence with $\|z_i\| \geq 1/2$ provided that the $\varepsilon_{i,j}$ are small enough. We assume that this is indeed the case, and we show that

$$Z := \overline{\mathrm{span}}\{z_i; \ i \in \mathbf{N}\}$$

is a hypercyclic subspace for T if the $\varepsilon_{i,j}$ are suitably chosen.

We note that, since $\inf \|z_i\| > 0$, the sequence of coordinate functionals (z_i^*) is bounded in Z^*, say $\|z_i^*\| \leq C$ for all i.

Let z be any non-zero vector in Z, say $\|z\| = 1$. Write $z = \sum_i \alpha_i z_i$ (so that $|\alpha_i| \leq C$) and let us fix $i_0 \in \mathbf{N}$ such that $\alpha_{i_0} \neq 0$. For each $r \geq 1$, we have

$$\|T^{n_{i_0,r}}(z) - \alpha_{i_0} y_r\| \leq \sum_{i \neq i_0} |\alpha_i| \|T^{n_{i_0,r}}(z_i)\| + |\alpha_{i_0}| \|T^{n_{i_0,r}}(z_{i_0}) - y_r\|$$

$$\leq C \left(\sum_{(i,j) \neq (i_0,r)} \|T^{n_{i_0,r}}(z_{i,j})\| + \|T^{n_{i_0,r}}(z_{i_0,r}) - y_r\| \right)$$

$$\leq C \left(2^{-r} \sum_{(i,j) \prec (i_0,r)} \varepsilon_{i,j} + \|T^{n_{i_0,r}}\| \sum_{(i,j) \succ (i_0,r)} \varepsilon_{i,j} + 2^{-r} \right).$$

Thus, if we assume that $\sum_{(i,j) \succ (i',j')} \varepsilon_{i,j} \leq \|T^{n_{i',j'}}\|^{-1} \times 2^{-j'}$ for all $(i', j') \in \mathbf{N} \times \mathbf{N}$, we get

$$\|T^{n_{i_0,r}}(z) - \alpha_{i_0} y_r\| \leq 3C\|z\| \times 2^{-r}$$

for all $r \geq 1$. This shows that z is indeed a hypercyclic vector for T.

8.2 Hypercyclicity in the operator algebra

In this section, we give another proof of Theorem 8.1. The key idea is to introduce the operator $\mathbf{L}_T : \mathfrak{L}(X) \to \mathfrak{L}(X)$ defined by

$$\mathbf{L}_T(U) := TU.$$

We note that $\mathbf{L}_T$ cannot be hypercyclic because $\mathfrak{L}(X)$ is not separable in the operator norm topology. However, we will see below that if T satisfies the Hypercyclicity Criterion then the operator $\mathbf{L}_T$ is hypercyclic with respect to the *strong operator topology* of $\mathfrak{L}(X)$. Theorem 8.1 follows very easily from this. As already mentioned, the idea of this proof is due to K. C. Chan. We recall the **Bès–Peris theorem** (Theorem 4.2): T satisfies the Hypercyclicity Criterion iff T is hereditarily hypercyclic and iff $T \oplus T$ is hypercyclic.

For the sake of clarity, from the beginning the Banach space X will not be assumed to be separable. Recall that the strong operator topology on $\mathfrak{L}(X)$ is the weakest topology on $\mathfrak{L}(X)$ for which the evaluation maps $T \mapsto T(x)$, $x \in X$, are continuous. We will need the following easily verified facts.

(i) Each $W_0 \in \mathfrak{L}(X)$ has an SOT-neighbourhood basis consisting of sets of the form

$$\mathcal{N}_{\overline{x},\varepsilon}(W_0) := \{W \in \mathfrak{L}(X); \ \|(W - W_0)(x_i)\| < \varepsilon \ \text{for} \ i = 1, \ldots, K\} \quad (8.1)$$

where $\varepsilon > 0$ and $\overline{x} = (x_1, \ldots, x_K)$ is a finite *linearly independent* sequence in X.

(ii) The finite-rank operators are SOT-dense in $\mathfrak{L}(X)$.

If $(x, x^*) \in X \times X^*$, we denote by $x \otimes x^*$ the rank-1 operator defined by

$$x \otimes x^*(u) = \langle x^*, u \rangle \, x.$$

If $D \subset X$ and $D^* \subset X^*$, we denote by $D \otimes D^*$ the set of all operators $W \in \mathfrak{L}(X)$ of the form $W = \sum_i x_i \otimes x_i^*$, where the sum is finite and $(x_i, x_i^*) \in D \times D^*$. In particular, $X \otimes X^*$ is the set of all finite-rank operators.

LEMMA 8.5   *Let $D \subset X$ and $D^* \subset X^*$. Assume that D is dense in X and that D^* is w^*-dense in X^*. Then $D \otimes D^*$ is SOT-dense in $\mathfrak{L}(X)$.*

PROOF   By the definition of the strong operator topology, the bilinear map $(x, x^*) \mapsto x \otimes x^*$ is continuous from $(X, \| \cdot \|) \times (X^*, w^*)$ into $(\mathfrak{L}(X), SOT)$. Therefore, our assumptions imply that each rank-1 operator $x \otimes x^*$ is in the SOT-closure of $D \otimes D^*$. By linearity it follows that the latter contains all finite rank operators, which concludes the proof.   □

In deriving the hypercyclicity of $\mathbf{L}_T$, one cannot apply the Baire category theorem directly to $\mathfrak{L}(X)$, because $(\mathfrak{L}(X), SOT)$ is not a Baire space. We will in fact apply Baire's theorem for the *operator norm* topology, but to a smaller space. In what follows, we denote by $\mathcal{FIN} \subset \mathfrak{L}(X)$ the norm closure of the set of all finite-rank operators. It is plain that $\mathcal{FIN}$ is $\mathbf{L}_T$-invariant. We also note that in many cases (for example, if X has a Schauder basis), $\mathcal{FIN}$ is just the set of all *compact* operators on X.

The following terminology will be useful.

DEFINITION 8.6   *Let Y be a topological vector space, and let $S \in \mathfrak{L}(Y)$. Also, let* $\mathbf{N}$ *be an infinite subset of* $\mathbb{N}$. *We say that S is*

(a) $\mathbf{N}$**-hypercyclic** *if there exists $y \in Y$ such that $\{S^n(y); \ n \in \mathbf{N}\}$ is dense in Y;*
(b) $\mathbf{N}$**-mixing** *if for any non-empty open subsets U, V of Y there exists a cofinite subset A of* $\mathbf{N}$ *such that $T^n(U) \cap V \neq \varnothing$ for every $n \in A$.*

It is plain that if S satisfies the Hypercyclicity Criterion with respect to $\mathbf{N}$ then S is $\mathbf{N}$-mixing. In the same vein, the proof of Birkhoff's transitivity theorem shows that if Y is a separable F-space and if S is $\mathbf{N}$-mixing then S is $\mathbf{N}$-hypercyclic. Moreover, it follows from the Bès–Peris theorem that an operator $T \in \mathfrak{L}(X)$ satisfies the Hypercyclicity Criterion if and only if it is $\mathbf{N}$-mixing for some infinite $\mathbf{N} \subset \mathbb{N}$.

LEMMA 8.7   *Let $T \in \mathfrak{L}(X)$, and let $\mathbf{N}$ be an infinite subset of $\mathbb{N}$. The following are equivalent:*

 (i) *T is $\mathbf{N}$-mixing;*
 (ii) *$\mathbf{L}_T : \mathfrak{L}(X) \to \mathfrak{L}(X)$ is $\mathbf{N}$-mixing for the strong operator topology;*
 (iii) *$(\mathbf{L}_T)_{|\mathcal{FIN}} : \mathcal{FIN} \to \mathcal{FIN}$ is $\mathbf{N}$-mixing for the operator norm topology.*

PROOF   (i) $\implies$ (ii): Assume that (i) holds. Let $(\mathcal{U}, \mathcal{V})$ be a pair of non-empty SOT-open sets in $\mathfrak{L}(X)$ of the form

$$\mathcal{U} = \mathcal{N}_{\overline{x}, \varepsilon}(U_0), \quad \mathcal{V} = \mathcal{N}_{\overline{y}, \varepsilon}(V_0);$$

see (8.1) above. Upon adding some vectors to $\overline{x}$ or $\overline{y}$, we may assume that $\overline{x}$ and $\overline{y}$ have the same length. We write $\overline{x} = (x_1, \ldots, x_K)$ and $\overline{y} = (y_1, \ldots, y_K)$.

Since T is $\mathbf{N}$-mixing, the operator

$$\underbrace{T \times \cdots \times T}_{K \text{ times}}$$

is $\mathbf{N}$-mixing as well. Therefore, one can find an $n_0 \in \mathbf{N}$ such that for each $n \in \mathbf{N}$, $n \geq n_0$, there exist $x_{1,n}, \ldots, x_{K,n} \in X$ satisfying $\|x_{i,n} - U_0(x_i)\| < \varepsilon$ and $\|T^n(x_{i,n}) - V_0(y_i)\| < \varepsilon$ for $i = 1, \ldots, K$.

Let us fix $n \in \mathbf{N}$, $n \geq n_0$. Since $x_1, \ldots, x_K$ are linearly independent, one can find an operator $U_n \in \mathfrak{L}(X)$ such that $U_n(x_i) = x_{i,n}$ for $i = 1, \ldots, K$. Then $U_n \in \mathcal{U}$ and $\mathbf{L}_T^n(U_n) = T^n U_n \in \mathcal{V}$. Thus, we have proved that $\mathbf{L}_T$ is $\mathbf{N}$-mixing for the strong operator topology.

(ii) $\implies$ (iii): Assume that (ii) holds. To prove that $(\mathbf{L}_T)_{|\mathcal{FIN}}$ is $\mathbf{N}$-mixing for the operator norm topology, it is enough to show that if U_0, V_0 are two finite-rank operators and if $\varepsilon > 0$ is given then, for all large enough $n \in \mathbf{N}$, one can find a finite-rank operator U_n such that $\|U_n - U_0\| < \varepsilon$ and $\|T^n U_n - V_0\| < \varepsilon$. Let us fix U_0, V_0 and ε.

We set $H := \mathrm{Ker}(U_0) \cap \mathrm{Ker}(V_0)$. Then H is a closed finite-codimensional subspace of X. Let $E \subset X$ be any finite-dimensional subspace such that $X = E \oplus H$, and let us denote by $\pi_E : X \to E$ the corresponding projection.

Since E is finite-dimensional, the map $W \mapsto W\pi_E$ is $(SOT, \|\cdot\|)$-continuous on $\mathfrak{L}(X)$. Indeed, writing $\pi_E = \sum_1^N e_i \otimes e_i^*$, where $e_i \in X$ and $e_i^* \in X^*$, we get $W\pi_E(x) = \sum_1^N \langle e_i^*, x \rangle W(e_i)$. It follows that

$$\|W\pi_E\| \le C \max(\|W(e_1)\|, \ldots, \|W(e_N)\|),$$

for some finite constant C.

Therefore, one can find SOT-open sets $\mathcal{U}, \mathcal{V} \subset \mathfrak{L}(X)$ with $U_0 \in \mathcal{U}$ and $V_0 \in \mathcal{V}$ such that the following implications hold:

$$U \in \mathcal{U} \implies \|U\pi_E - U_0\| < \varepsilon,$$
$$V \in \mathcal{V} \implies \|V\pi_E - V_0\| < \varepsilon.$$

Since $\mathbf{L}_T$ is $\mathbf{N}$-mixing for the strong operator topology, one can find an $n_0 \in \mathbf{N}$ such that $\mathbf{L}_T^n(\mathcal{U}) \cap \mathcal{V} \neq \varnothing$ for all $n \in \mathbf{N}$, $n \ge n_0$. In other words, for each $n \ge n_0$ in $\mathbf{N}$ there exists an operator $\widetilde{U}_n \in \mathcal{U}$ with $T^n \widetilde{U}_n \in \mathcal{V}$. Then $U_n := \widetilde{U}_n \pi_E$ is a finite-rank operator such that $\|U_n - U_0\| < \varepsilon$ and $\|T^n U_n - V_0\| < \varepsilon$. This concludes the proof of (ii) $\implies$ (iii).

(iii) $\implies$ (i): Assume that (iii) holds. Fix $x, y \in X$ with $x \neq 0$, and let $\varepsilon > 0$. Let us choose two finite-rank operators $U, V \in \mathfrak{L}(X)$ such that $U(x) = x$ and $V(x) = y$. By assumption, one can find $n_0 \in \mathbf{N}$ such that the following holds: for any $n \in \mathbf{N}$, $n \ge n_0$, there exists a finite-rank operator W_n such that $\|W_n - U\| < \varepsilon$ and $\|T^n W_n - V\| < \varepsilon$. Setting $z_n = W_n(x)$, we get immediately

$$\|z_n - x\| < \varepsilon\|x\| \quad \text{and} \quad \|T^n(z_n) - y\| < \varepsilon\|x\|$$

for any $n \in \mathbf{N}$, $n \ge n_0$. This shows that T is $\mathbf{N}$-mixing.   □

From now on, we will assume that the Banach space X is separable (and infinite-dimensional).

PROPOSITION 8.8   *Let $T \in \mathfrak{L}(X)$. Assume that T is $\mathbf{N}$-mixing, for some infinite set $\mathbf{N} \subset \mathbb{N}$. Then the operator $\mathbf{L}_T$ is $\mathbf{N}$-hypercyclic for the strong operator topology.*

PROOF   Since the Banach space X is separable, its dual space X^* is w^*-separable. By Lemma 8.5, it follows that $\mathfrak{L}(X)$ is SOT-separable. More precisely, there exists a countable SOT-dense set $\mathcal{D} \subset \mathfrak{L}(X)$ consisting of finite-rank operators. For each pair $(V, \varepsilon) \in \mathcal{D} \times \mathbb{Q}^+$, set

$$\mathcal{O}_{V,\varepsilon} := \left\{ U \in \mathcal{FIN}; \, \exists n \in \mathbf{N}: \|\mathbf{L}_T^n(U) - V\| < \varepsilon \right\}.$$

Each $\mathcal{O}_{V,\varepsilon}$ is open in $\mathcal{FIN}$ with respect to the operator norm topology. Moreover, since $(\mathbf{L}_T)_{|\mathcal{FIN}}$ is $\mathbf{N}$-mixing, $\mathcal{O}_{V,\varepsilon}$ is also norm-dense in $\mathcal{FIN}$ by Lemma 8.7. By the Baire category theorem applied to $(\mathcal{FIN}, \| \cdot \|)$, it follows that $\bigcap_{V,\varepsilon} \mathcal{O}_{V,\varepsilon}$ is non-empty. In other words, there exists some $U \in \mathcal{FIN}$ such that each $V \in \mathcal{D}$ is in the norm closure of $\{\mathbf{L}_T^n(U);\ n \in \mathbf{N}\}$. Since $\mathcal{D}$ is SOT-dense in $\mathfrak{L}(X)$, this shows that $\mathbf{L}_T$ is $\mathbf{N}$-hypercyclic for the strong operator topology, with hypercyclic vector U. □

As a simple corollary, we obtain another, equivalent, form of the Hypercyclicity Criterion. This can be compared with the Bès–Peris theorem, of course.

COROLLARY 8.9   *Let $T \in \mathfrak{L}(X)$. Then T satisfies the Hypercyclicity Criterion if and only if the associated operator $\mathbf{L}_T : \mathfrak{L}(X) \to \mathfrak{L}(X)$ is hypercyclic for the strong operator topology.*

PROOF   The direct implication is exactly the content of Proposition 8.8, since T satisfies the Hypercyclicity Criterion iff it is $\mathbf{N}$-mixing for some infinite $\mathbf{N} \subset \mathbb{N}$. Conversely, suppose that $\mathbf{L}_T$ is hypercyclic, and let us show that $T \oplus T$ is also hypercyclic. Let $U \in \mathfrak{L}(X)$ be a hypercyclic vector for $\mathbf{L}_T$, and let $(x, y) \in X^2$ be linearly independent. We claim that $Ux \oplus Uy$ is a hypercyclic vector for $T \oplus T$. Indeed, for any $(a, b) \in X^2$ there exists $V \in \mathfrak{L}(X)$ such that $Vx = a$ and $Vy = b$. Since U is $\mathbf{L}_T$-hypercyclic, we may find an $n \in \mathbb{N}$ such that

$$\|T^n Ux - Vx\| < \varepsilon \quad \text{and} \quad \|T^n Uy - Vy\| < \varepsilon,$$

which concludes the proof. □

SECOND PROOF OF THEOREM 8.1   This is now very easy. By assumption, the operator T is $\mathbf{N}$-mixing, where $\mathbf{N} = \{n_k;\ k \in \mathbb{N}\}$. By Proposition 8.8, the operator $\mathbf{L}_T$ is $\mathbf{N}$-hypercyclic. Let us choose $U \in \mathfrak{L}(X)$ such that the set $\{T^n U;\ n \in \mathbf{N}\}$ is SOT-dense in $\mathfrak{L}(X)$. Then, for any non-zero $x \in X$, the vector $U(x)$ is $\mathbf{N}$-hypercyclic for T. Now, let $V := U + \alpha I$, where α is any positive number with $\alpha > \|U\|$. Then V is invertible, so that $Z := V(E)$ is a closed infinite-dimensional subspace of X. Moreover, if $z = U(x) + \alpha x$ is a non-zero vector in Z then z is $\mathbf{N}$-hypercyclic for T, because $U(x)$ is $\mathbf{N}$-hypercyclic and $T^n(x) \to 0$ as $n \to \infty$, $n \in \mathbf{N}$. □

REMARK   It follows from Corollary 8.9 that using Chan's approach one cannot hope to prove a strengthened version of Theorem 8.1 where the assumption "T satisfies the Hypercyclicity Criterion" is replaced by "T is hypercyclic".

The techniques developed in this section may also be used to obtain *common* hypercyclic subspaces (see Chapter 7). Suppose that $(T_\lambda)_{\lambda \in \Lambda}$ is a family of hypercyclic operators on X and that each single operator T_λ admits a hypercyclic subspace. It is natural to ask whether this family admits a common hypercyclic subspace. This is false in general, even for 2-operators (see [10] or Exercise 8.2). However, it becomes true under quite natural assumptions.

THEOREM 8.10 *Let* $(T_\lambda)_{\lambda \in \Lambda} \subset \mathfrak{L}(X)$ *be a family of operators parametrized by some* K_σ *topological space* Λ, *in such a way that the map* $(\lambda, x) \mapsto T_\lambda(x)$ *is continuous on* $\Lambda \times X$. *Assume that the following properties hold.*

(1) *For any* $K \geq 1$, *the set*

$$\bigcap_{\lambda \in \Lambda} HC(\underbrace{T_\lambda \times \cdots \times T_\lambda}_{K \text{ times}})$$

is dense in $X \times \cdots \times X$;

(2) *There exists a closed infinite-dimensional subspace* $E \subset X$ *such that* $T_\lambda^n(x) \to 0$ *for each* $\lambda \in \Lambda$ *and all* $x \in E$.

Then the operators T_λ *have a common hypercyclic subspace.*

PROOF We write $\Lambda = \bigcup_{k \in \mathbb{N}} \Lambda_k$, where the sets Λ_k are compact. The proof is a slight modification of the proofs of Lemma 8.7 and Proposition 8.8. It relies on the following two facts. Recall that we denote by $\mathcal{FIN} \subset \mathfrak{L}(X)$ the norm closure of the set of all finite-rank operators and by $\mathbf{L}_T : \mathfrak{L}(X) \to \mathfrak{L}(X)$ the left-multiplication-by-T operator associated with a given $T \in \mathfrak{L}(X)$.

FACT 1 For each finite-rank operator $V \in \mathfrak{L}(X)$ and each $\varepsilon > 0$, the set

$$\mathcal{O}_{V,\varepsilon} := \{U \in \mathcal{FIN}; \ \forall \lambda \in \Lambda \ \exists n \in \mathbb{N} : \ \|\mathbf{L}_{T_\lambda}^n(U) - V\| < \varepsilon\}$$

is dense in $\mathcal{FIN}$ with respect to the operator norm topology.

PROOF OF FACT 1 The proof is very similar to that of Lemma 8.7. Let us fix a finite-rank operator V_0 and $\varepsilon_0 > 0$.

We first show that for any SOT-neighbourhood $\mathcal{V} = \mathcal{N}_{\overline{y}, \varepsilon}(V_0)$ of V_0, see (8.1), the set

$$\widetilde{\mathcal{V}} := \{U \in \mathfrak{L}(X); \ \forall \lambda \in \Lambda \ \exists n \in \mathbb{N} : \ \mathbf{L}_{T_\lambda}^n U \in \mathcal{V}\}$$

is SOT-dense in $\mathfrak{L}(X)$. Let $\mathcal{U}$ be any non-empty SOT-open set in $\mathfrak{L}(X)$ of the form $\mathcal{U} = \mathcal{N}_{\overline{x}, \varepsilon'}(U_0)$. Adding some vectors to $\overline{x}$ or $\overline{y}$, we may assume that $\overline{x}$ and $\overline{y}$ have the same length. We write $\overline{x} = (x_1, \dots, x_K)$ and $\overline{y} = (y_1, \dots, y_K)$.

Since $\bigcap_{\lambda \in \Lambda} HC(T_\lambda \times \cdots \times T_\lambda)$ is dense in $X \times \cdots \times X$, one can find $z_1, \dots, z_K \in X$ with $\|z_i - U_0(x_i)\| < \varepsilon'$ for $i = 1, \dots, K$ such that, for any $\lambda \in \Lambda$, there exists $n \in \mathbb{N}$ satisfying $\|T_\lambda^n(z_i) - V_0(y_i)\| < \varepsilon$ for $i = 1, \dots, K$. Since $x_1, \dots, x_K$ are linearly independent, one can find an operator $U \in \mathfrak{L}(X)$ such that $U(x_i) = z_i$ for $i = 1, \dots, K$. Then $U \in \mathcal{U} \cap \widetilde{\mathcal{V}}$.

To prove Fact 1, let us now fix an arbitrary finite-rank operator U_0 and $\varepsilon > 0$. As in the proof of Lemma 8.7, one can find a finite-dimensional projection π and two SOT-open sets $\mathcal{U}_0 \ni U_0$ and $\mathcal{V}_0 \ni V_0$ such that $U \in \mathcal{U}_0 \implies \|U\pi - U_0\| < \varepsilon$ and $V \in \mathcal{V}_0 \implies \|V\pi - V_0\| < \varepsilon_0$. By what we have already proved, one can choose $\widetilde{U} \in \mathcal{U}_0 \cap \widetilde{\mathcal{V}_0}$. Then $U := \widetilde{U}\pi$ is in $\mathcal{O}_{V_0, \varepsilon_0}$ and $\|U - U_0\| < \varepsilon$. $\qquad \square$

FACT 2 Let us fix $k \in \mathbb{N}$. Then each map $(\lambda, U) \mapsto \mathbf{L}_{T_\lambda}^n (U)$ is continuous from $\Lambda_k \times (\mathcal{FIN}, \| \cdot \|)$ into $(\mathcal{FIN}, \| \cdot \|)$.

PROOF OF FACT 2 First, we note that the family $(T_\lambda)_{\lambda \in \Lambda_k}$ is bounded in $\mathfrak{L}(X)$. Indeed, for each $x \in X$ we have $\sup_{\lambda \in \Lambda_k} \|T_\lambda(x)\| < \infty$ by the compactness of Λ_k, because the map $\lambda \to T_\lambda(x)$ is continuous; therefore, we may apply the Banach–Steinhaus theorem.

Let us fix $(\lambda_0, U_0) \in \Lambda_k \times \mathcal{FIN}$. For any $(\lambda, U) \in \Lambda_k \times \mathcal{FIN}$, we have

$$\left\| \mathbf{L}_{T_\lambda}^n (U) - \mathbf{L}_{T_{\lambda_0}}^n (U_0) \right\| \leq \|T_\lambda^n (U - U_0)\| + \|(T_\lambda^n - T_{\lambda_0}^n) U_0\|$$
$$\leq C_{n,k} \|U - U_0\| + \sup_{z \in K_0} \|T_\lambda^n (z) - T_{\lambda_0}^n (z)\|,$$

where $C_{n,k} = \sup_{\lambda \in \Lambda_k} \|T_\lambda^n\|$ and $K_0 = \overline{U_0(B_X)}$. Since $U_0 \in \mathcal{FIN}$, the set K_0 is compact. Since the map $(\lambda, x) \mapsto T_\lambda(x)$ is continuous, we know that $T_\lambda^n (z) \xrightarrow{\lambda \to \lambda_0} T_{\lambda_0}^n (z)$ pointwise on K_0. Since $\sup_\lambda \|T_\lambda^n\| < \infty$, this convergence is uniform on the compact set K_0, which concludes the proof. $\square$

Using these two facts, we proceed as in the proof of Proposition 8.8 to show that the family $(\mathbf{L}_{T_\lambda})_\lambda$ has a common SOT-hypercyclic vector. For each triple $(V, \varepsilon, k) \in \mathcal{D} \times \mathbf{Q}^+ \times \mathbb{N}$, put

$$\mathcal{O}_{V,\varepsilon,k} := \left\{ U \in \mathcal{FIN}; \ \forall \lambda \in \Lambda_k \ \exists n \in \mathbb{N} \colon \ \|\mathbf{L}_T^n (U) - V\| < \varepsilon \right\}.$$

By Fact 2 and since Λ_k is compact, each set $\mathcal{O}_{V,\varepsilon,k}$ is $\| \cdot \|$-open in $\mathcal{FIN}$. Moreover, these sets $\mathcal{O}_{V,\varepsilon,k}$ are also $\| \cdot \|$-dense in $\mathcal{FIN}$ by Fact 1. By the Baire category theorem, $\bigcap_{V,\varepsilon,k} \mathcal{O}_{V,\varepsilon,k}$ is non-empty. Any $U \in \bigcap_{V,\varepsilon,k} \mathcal{O}_{V,\varepsilon,k}$ is a common $\| \cdot \|$-hypercyclic vector for the family $((\mathbf{L}_{T_\lambda})_{|\mathcal{FIN}})_{\lambda \in \Lambda}$, and hence a common SOT-hypercyclic vector for the family $(\mathbf{L}_{T_\lambda})_{\lambda \in \Lambda}$.

Having found a common SOT-hypercyclic vector for $(\mathbf{L}_{T_\lambda})$, we may now proceed exactly as in the second proof of Theorem 8.1 to get a common hypercyclic subspace for the operators T_λ. $\square$

It is easy to check that when (T_λ) satisfies the assumptions of one of the criteria of Chapter 7 then conditions (1), (2) of Theorem 8.10 are fulfilled. In particular, Theorem 8.10 improves a result from [20]. It also implies the following result of R. M. Aron, J. P. Bès, F. León-Saavedra and A. Peris [10].

COROLLARY 8.11 *Let $(T_\lambda)_{\lambda \in \Lambda}$ be a countable family of operators on X satisfying the Hypercyclicity Criterion. Suppose moreover that there exists a closed infinite-dimensional subspace $E \subset X$ such that $T_\lambda^n(x) \xrightarrow{n \to \infty} 0$ for each $\lambda \in \Lambda$ and all $x \in E$. Then the operators T_λ have a common hypercyclic subspace.*

PROOF Since the family (T_λ) is countable and each T_λ is weakly mixing, it follows from the Baire category theorem that condition (1) in Theorem 8.10 is satisfied. And, of course, condition (2) is also satisfied, by assumption. $\square$

8.3 Hypercyclic subspaces and the essential spectrum

In this section, we prove Theorem 8.2. The following terminology will be useful.

DEFINITION 8.12 *Let $\mathcal{K}$ be a family of compact operators from X into some Banach space Y. We say that a sequence $(e_i) \subset X$ is $\mathcal{K}$-**null** if $\lim_{i \to \infty} \|K(e_i)\| = 0$ for every $K \in \mathcal{K}$.*

The proof of Theorem 8.2 requires several lemmas. The first one is quite well known.

LEMMA 8.13 *Let Y be a Banach space, and let F be an infinite-dimensional closed subspace of X. Then, given a sequence of compact operators $(K_n) \subset \mathfrak{L}(X, Y)$ and a sequence of positive numbers (ε_n), one can find a decreasing sequence (F_n) of finite-codimensional subspaces of F such that $\|(K_n)_{|F_i}\| \leq \varepsilon_i$ whenever $i \geq n$.*

PROOF Let us fix (K_n) and (ε_n). Since K_0 is compact the adjoint operator K_0^* is also compact, so one can find $x_1^*, \ldots, x_N^* \in X^*$ such that $K_0^*(B_{Y^*}) \subset \bigcup_j B(x_j^*, \varepsilon_0)$. Then, for any $x \in X$, we have $\|K_0(x)\| = \sup\{|\langle x^*, x\rangle|;\ x^* \in K_0^*(B_{Y^*})\} \leq \max_j |\langle x_j^*, x\rangle| + \varepsilon_0 \|x\|$. Therefore, $F_0 := F \cap \bigcap_j \operatorname{Ker}(x_j^*)$ is a finite-codimensional subspace of F such that $\|(K_0)_{|F_0}\| \leq \varepsilon_0$. If $F_0, \ldots, F_i$ have been constructed, one obtains F_{i+1} in the same way, by considering a finite ε_{i+1}-covering of $\bigcup_{n=0}^{i+1} K_n^*(B_{Y^*})$. □

COROLLARY 8.14 *If $\mathcal{K} \subset \mathfrak{L}(X, Y)$ is a countable family of compact operators then any closed infinite-dimensional subspace of X contains a normalized $\mathcal{K}$-null basic sequence.*

PROOF Let us write $\mathcal{K} = \{K_n;\ n \in \mathbb{N}\}$, and let us fix an infinite-dimensional closed subspace $Z \subset X$. By Lemma 8.13, one can find a decreasing sequence (Z_n) of finite-codimensional subspaces of Z such that $\|(K_n)_{|Z_i}\| \leq 2^{-i}$ whenever $n \leq i$. Then Mazur's construction provides a normalized basic sequence $(e_i)_{i \in \mathbb{N}} \subset Z$ such that $e_i \in Z_i$ for all $i \in \mathbb{N}$. This sequence (e_i) has the required properties. □

In what follows, we denote by $\sigma_e(T)$ the essential spectrum of an operator $T \in \mathfrak{L}(X)$, i.e. the spectrum of $[T]_{\mathfrak{L}/\mathcal{K}}$ in the Calkin algebra $\mathfrak{L}(X)/\mathcal{K}(X)$. We recall that an operator $R \in \mathfrak{L}(X)$ is invertible modulo the compact operators if and only if it is a **Fredholm operator**, i.e. $\operatorname{Ran}(R)$ is closed and $\dim \operatorname{Ker}(R), \operatorname{codim} \operatorname{Ran}(R) < \infty$. Thus, $\sigma_e(T)$ is also the set of all complex numbers λ such that $T - \lambda$ is not Fredholm.

LEMMA 8.15 *Let $T \in \mathfrak{L}(X)$.*

(a) *Let $\lambda \in \sigma_e(T)$, and assume that $T - \lambda$ has dense range. Then, for any $\varepsilon > 0$, there exists an infinite-dimensional closed subspace $F_\varepsilon \subset X$ and a compact operator K_ε such that $\|K_\varepsilon\| < \varepsilon$, F_ε is invariant under $T - K_\varepsilon$ and $(T - K_\varepsilon)_{|F_\varepsilon} = \lambda I_{F_\varepsilon}$.*

(b) *If $\sigma_e(T) \cap \overline{\mathbb{D}} = \varnothing$ then one can find $\lambda > 1$, $n_0 \in \mathbb{N}$ and a countable family of compact operators $\mathcal{K}_0 \subset \mathfrak{L}(X)$ such that the following holds: for any normalized $\mathcal{K}_0$-null sequence $(e_i) \subset X$, one has*

$$\liminf_{i \to \infty} \|T^n(e_i)\| \geq \lambda^n \quad \text{for each } n \geq n_0.$$

PROOF   To prove (a), we note that, since the operator $T - \lambda$ is not Fredholm and has dense range, it has either an infinite-dimensional kernel or a non-closed range. In other words, $T - \lambda$ is not *left-Fredholm*; so the result follows at once from Proposition D.3.4 in Appendix D.

To prove (b), assume that $\sigma_e(T) \cap \overline{\mathbb{D}} = \varnothing$. Then $[T]_{\mathfrak{L}/\mathcal{K}}$ is invertible in the Calkin algebra and $\sigma([T]_{\mathfrak{L}/\mathcal{K}}^{-1}) \subset \mathbb{D}$. By the spectral radius formula, it follows that one can find a $\lambda > 1$ and an $n_0 \in \mathbb{N}$ such that $\|[T]_{\mathfrak{L}/\mathcal{K}}^{-n}\| < \lambda^{-n}$ for all $n \geq n_0$. Thus, for each $n \geq n_0$ we may choose an operator $A_n \in \mathfrak{L}(X)$ such that $\|A_n\| < \lambda^{-n}$ and $K_n := A_n T^n - I$ is compact. We put $\mathcal{K}_0 := \{K_n; \ n \geq n_0\}$.

Let (e_i) be any normalized $\mathcal{K}_0$-null sequence in X. Since

$$\|(I + K_n)(e_i)\| = \|A_n T^n(e_i)\| \leq \lambda^{-n} \|T^n(e_i)\|,$$

we have $\liminf_{i \to \infty} \|T^n(e_i)\| \geq \lambda^n \liminf_{i \to \infty} \|e_i + K_n(e_i)\| = \lambda^n$ for each $n \geq n_0$. This concludes the proof.   □

The next two lemmas are the main steps in the proof of Theorem 8.2. The first sets out what remains to be done for proving the implications (iv) $\implies$ (ii) and (iv) $\implies$ (i).

LEMMA 8.16   *Let $T \in \mathfrak{L}(X)$. Assume that there exist some dense set $\mathcal{D} \subset X$ and some infinite set $\mathbf{M} \subset \mathbb{N}$ such that $T^n(x) \to 0$ for all $x \in \mathcal{D}$ as $n \to \infty$, $n \in \mathbf{M}$. Assume also that there exist some compact operator $K \in \mathfrak{L}(X)$ and an infinite-dimensional closed subspace $F \subset X$ such that $\sup_{n \in \mathbf{M}} \|(T - K)^n_{|F}\| < \infty$. Then one can find an infinite-dimensional closed subspace $E \subset X$ and an increasing sequence $(n_k) \subset \mathbf{M}$ such that $T^{n_k}(x) \to 0$ for all $x \in E$.*

PROOF   For each $n \in \mathbb{N}$, one can write

$$T^n = (T - K)^n + K_n,$$

where K_n is a compact operator. This is clear since $[T]_{\mathfrak{L}/\mathcal{K}} = [T - K]_{\mathfrak{L}/\mathcal{K}}$.

By Lemma 8.13, one can find a decreasing sequence (F_n) of finite-codimensional closed subspaces of F such that

$$\|(K_n)_{|F_n}\| \leq 1 \tag{8.2}$$

for all $n \in \mathbb{N}$. Then Mazur's construction produces a normalized basic sequence $(f_n) \subset F$ such that $f_n \in F_n$ for all $n \in \mathbb{N}$.

Now, let (ε_n) be a summable sequence of positive numbers to be specified later, with $\sum_0^\infty \varepsilon_n \leq 1$. Since $\mathcal{D}$ is dense in X, one can find a sequence $(e_n) \subset \mathcal{D}$ such that $\|e_n - f_n\| < \varepsilon_n$ for all n. We assume that the ε_n are small enough to ensure that $\sum_0^\infty \|f_n^*\| \|e_n - f_n\| < 1$, where (f_n^*) is the sequence of coordinate functionals associated with (f_n). Then, any subsequence (e_{n_k}) of (e_n) is a semi-normalized basic sequence equivalent to (f_{n_k}).

Since $T^n(e) \to 0$ as $n \to \infty$ along $\mathbf{M}$ for any $e \in \mathcal{D}$, one can construct inductively an increasing sequence $(n_k) \subset \mathbf{M}$ such that

$$\|T^{n_k}(e_{n_j})\| < 2^{-k}\varepsilon_{n_j} \quad \text{whenever } j < k.$$

We will show that the subspace $E := \overline{\text{span}}\{e_{n_k}; \ k \in \mathbb{N}\}$ has the required property.

We note that since the basic sequence (e_{n_k}) is semi-normalized there exists some finite constant C such that $|x_k| \leq C\|x\|$ for any $x = \sum_0^\infty x_k e_{n_k} \in E$ and all $k \in \mathbb{N}$. Now, if $x = \sum_j x_j\,e_{n_j} \in E$ and if $k \in \mathbb{N}$, one can estimate $\|T^{n_k}(x)\|$ as follows:

$$\|T^{n_k}(x)\| \leq \sum_{j<k}|x_j|\,\|T^{n_k}(e_{n_j})\|$$

$$+ \sum_{j \geq k}|x_j|\,\|T^{n_k}(e_{n_j} - f_{n_j})\| + \left\|T^{n_k}\left(\sum_{j \geq k}x_j f_{n_j}\right)\right\|$$

$$\leq C\|x\|\left(2^{-k} + \|T^{n_k}\|\sum_{j \geq k}\varepsilon_{n_j}\right) + \left\|T^{n_k}\left(\sum_{j \geq k}x_j f_{n_j}\right)\right\|.$$

Moreover, we have

$$\left\|T^{n_k}\left(\sum_{j \geq k}x_j f_{n_j}\right)\right\| \leq \left\|(T - K)^{n_k}\left(\sum_{j \geq k}x_j f_{n_j}\right)\right\| + \left\|K_{n_k}\left(\sum_{j \geq k}x_j f_{n_j}\right)\right\|$$

$$\leq (M+1)\left\|\sum_{j \geq k}x_j f_{n_j}\right\|,$$

where we have put $M := \sup_{n \in \mathbf{M}}\|(T-K)^n_{|F}\|$ and used the inequality (8.2). Since (f_{n_j}) and (e_{n_j}) are equivalent, it follows that if the sequence (ε_n) is suitably chosen then

$$\|T^{n_k}(x)\| \leq \alpha_k\|x\| + A\|\pi_k(x)\|$$

for all $x \in E$, where π_k is the canonical projection from E onto $\overline{\text{span}}\{e_{n_j}; \ j \geq k\}$, A is some finite constant and $\lim_{k \to \infty}\alpha_k = 0$. This concludes the proof. $\qquad \square$

Let us now take a look at the implication (ii) $\implies$ (iv) in Theorem 8.2. By Corollary 8.14 and Lemma 8.15 we know that if $\sigma_e(T) \cap \overline{\mathbb{D}} = \varnothing$, we can find in any closed infinite-dimensional subspace Z of X a normalized basic sequence (e_i) such that $\liminf_{i \to \infty}\|T^n(e_i)\| \geq \lambda^n$ for all $n \in \mathbb{N}$ and some $\lambda > 1$. To contradict (ii), we need to find some vector $x \in Z$ such that $\lim_{n \to \infty}\|T^n(x)\| = \infty$. That this can indeed be done is the content of the next lemma.

LEMMA 8.17 *Let $(T_n)_{n \in \mathbb{N}}$ be a sequence in $\mathcal{L}(X, Y)$, where Y is a Banach space. Then there exist a Banach space $\widetilde{Y}$ and a countable family of compact operators*

$\mathcal{K}_1 \subset \mathfrak{L}(X, \widetilde{Y})$ such that the following holds for any normalized $\mathcal{K}_1$-null sequence $(e_i) \subset X$: given any summable sequence of positive numbers (α_n), there exists some vector $x \in \overline{\mathrm{span}}\{e_i; \ i \in \mathbb{N}\}$ such that

$$\|T_n(x)\| \geq \alpha_n \limsup_{i \to \infty} \|T_n(e_i)\| \quad \text{for each } n \in \mathbb{N}.$$

PROOF   Replacing Y by the closed linear span of $\bigcup_n \mathrm{Ran}(T_n)$, we may assume that Y is separable. Then Y embeds isometrically into a Banach space $\widetilde{Y}$ with a Schauder basis; one may take e.g. $\widetilde{Y} := \mathcal{C}([0,1])$. Let us fix a (bounded) sequence of finite-rank operators $(\pi_k)_{k \in \mathbb{N}}$ such that $\pi_k(y) \to y$ for all $y \in \widetilde{Y}$. We put $C := \sup \|\pi_k\|$. Finally, we denote by $\mathcal{K}_1$ the family of all operators $K \in \mathfrak{L}(X, \widetilde{Y})$ of the form $K = \pi_k T_n$, where $k, n \in \mathbb{N}$.

Now, let us fix a normalized $\mathcal{K}_1$-null sequence $(e_i) \subset X$ and a summable sequence of positive numbers (α_n). We set $A_n := \limsup_i \|T_n(e_i)\|$ and assume that $A_n > 0$ for all $n \in \mathbb{N}$.

Let (ε_n) be a sequence of positive numbers to be specified later, and put $k_{-1} := 0$. We construct by induction two increasing sequences of integers $(i_n)_{n \geq 0}$ and $(k_n)_{n \geq 0}$ in such a way that the following properties hold for each $n \in \mathbb{N}$:

(i) $\|(\pi_{k_n} - \pi_{k_{n-1}}) T_n(e_{i_n})\| > A_n / 2$;

(ii) $\|(\pi_{k_n} - \pi_{k_{n-1}}) T_n(e_{i_l})\| < \varepsilon_n A_n$ for all $l < n$;

(iii) $\|(\pi_{k_n} - \pi_{k_{n-1}}) T_n(e_i)\| < \varepsilon_n A_n$ for all $i \geq i_{n+1}$.

Since $\lim_{i \to \infty} \|\pi_0 T_0(e_i)\| = 0$, one can find an i_0 such that $\|\pi_0 T_0(e_{i_0})\| < A_0/4$. Moreover, by the definition of A_0 we may also require $\|T_0(e_{i_0})\| > 3A_0/4$. Then, one can choose k_0 such that $\|\pi_{k_0} T_0(e_{i_0})\| > 3A_0/4$ and also $\|(\pi_k - \pi_{k_0}) T_1(e_{i_0})\| < \varepsilon_1 A_1$ for all $k > k_0$. Then (i) holds for $n = 0$, and (ii) will hold for $n = 1$ whatever the choice of $k_1 > k_0$ may be. Of course, (ii) is vacuously satisfied for $n = 0$.

Since $\lim_{i \to \infty} \|(\pi_{k_0} - \pi_0) T_0(e_i)\| = 0 = \lim_{i \to \infty} \|\pi_{k_0} T_1(e_{i_1})\|$, one can find $i_1 > i_0$ such that $\|(\pi_{k_0} - \pi_0) T_0(e_i)\| < \varepsilon_0 A_0$ for all $i \geq i_1$ and $\|\pi_{k_0} T_1(e_{i_1})\| < A_1/4$. In particular, (iii) is satisfied for $k = 0$. We can also require that $\|T_1(e_{i_1})\| > 3A_1/4$, by the definition of A_1. Next, one can choose $k_1 > k_0$ such that $\|\pi_{k_1} T_1(e_{i_1})\| > 3A_1/4$ and $\|(\pi_k - \pi_{k_1}) T_2(e_{i_l})\| < \varepsilon_2 A_2$ for $l = 0, 1$ and all $k \geq k_1$. Then (i) holds for $n = 1$ and (ii) will be satisfied for $n = 2$ and any choice of $k_2 > k_1$. The inductive process should now be clear.

We set $x := M \sum_0^\infty \alpha_l e_{i_l}$, for some positive constant M to be specified. For each $n \in \mathbb{N}$, we have

$$\|T_n(x)\| \geq \frac{1}{2C} \|(\pi_{k_n} - \pi_{k_{n-1}}) T_n(x)\|$$

$$\geq \frac{M}{2C} \left(\alpha_n \|(\pi_{k_n} - \pi_{k_{n-1}}) T_n(e_{i_n})\| - \left\| \sum_{l \neq n} \alpha_l (\pi_{k_n} - \pi_{k_{n-1}}) T_n(e_{i_l}) \right\| \right)$$

$$\geq \frac{M}{4C} \alpha_n A_n - \varepsilon_n A_n \frac{M}{2C} \sum_{l \neq n} \alpha_l.$$

Thus, if we take $M := 8C$ and if the positive numbers ε_n are small enough, we get $\|T_n(x)\| \geq \alpha_n A_n$ for all n. □

PROOF OF THEOREM 8.2 Let us fix an increasing sequence of integers (m_k) such that T satisfies the Hypercyclicity Criterion with respect to (m_k).

First, we note that part (ii) of the theorem implies part (iii), by the Banach–Steinhaus theorem.

Assume that $\sigma_e(T)$ does not intersect $\overline{\mathbb{D}}$. Let $\lambda > 1$, $n_0 \in \mathbb{N}$ and the countable family of compact operators $\mathcal{K}_0$ be given by Lemma 8.15. Also, let $\mathcal{K}_1$ be the countable family of compact operators given by Lemma 8.17 applied with $T_n := T^n$. Finally, let us put $\mathcal{K} := \mathcal{K}_0 \cup \mathcal{K}_1$. By Corollary 8.14, any closed infinite-dimensional subspace $Z \subset X$ contains a normalized $\mathcal{K}$-null basic sequence (e_i). Applying Lemma 8.17 with any summable sequence (α_n) such that $\lim_{n\to\infty} \alpha_n \lambda^n = \infty$, we see that one can find a vector $x \in Z$ such that $\lim_{n\to\infty} \|T^n(x)\| = \infty$. This shows that (i) implies (iv) and that (iii) implies (iv) as well.

Now, assume that $\sigma_e(T) \cap \overline{\mathbb{D}} \neq \varnothing$. Since T is hypercyclic, $T - \lambda$ has dense range for any $\lambda \in \mathbb{C}$ (see Chapter 1). Therefore, one can apply Lemma 8.15 to get an infinite-dimensional closed subspace $F \subset X$ and a compact operator K such that $\|(T-K)^n_{|F}\| \leq 1$ for all $n \in \mathbb{N}$. By Lemma 8.16 applied with $\mathbf{M} := \{m_k;\ k \in \mathbb{N}\}$, one can find a subsequence (n_k) of (m_k) and an infinite-dimensional closed subspace $E \subset X$ such that $T^{n_k}(x) \to 0$ for all $x \in E$. This shows that (iv) implies (ii) and that, since T satisfies the Hypercyclicity Criterion with respect to (n_k), (iv) implies (i) by Theorem 8.1.

To summarize, we have shown that, in Theorem 8.2, (i) $\iff$ (iv) and (iii) $\implies$ (iv) $\implies$ (ii) $\implies$ (iii). This concludes the proof. □

REMARK 8.18 It follows from the above proof that if T satisfies the Hypercyclicity Criterion with respect to some increasing sequence of integers (m_k) and has a hypercyclic subspace then part (ii) of Theorem 8.2 holds also for some subsequence (n_k) of (m_k). Therefore, condition (2) of Theorem 8.1 is essentially optimal. Incidentally, we have no direct proof that if part (ii) of Theorem 8.2 holds then it holds also for some subsequence of (m_k).

It follows also from the above proof that $\sigma_e(T) \cap \overline{\mathbb{D}} \neq \varnothing$ as long as there exists an infinite-dimensional closed subspace $Z \subset X$ such that $\liminf_{n\to\infty} \|T^n(x)\| < \infty$ for all $x \in Z$, without assuming that T satisfies the Hypercyclicity Criterion.

8.4 Examples

In this section, we give some natural examples of hypercyclic operators which admit (or do not admit) hypercyclic subspaces. All results are due to F. León-Saavedra and A. Montes-Rodríguez [167]. The following well-known fact will be needed. For a proof, see e.g. [81] or Corollary D.3.2 in Appendix D.

LEMMA 8.19 *Let X be a complex Banach space, and let $T \in \mathfrak{L}(X)$. Then any non-isolated point in $\partial\sigma(T)$ belongs to $\sigma_e(T)$.*

EXAMPLE 8.20 Any hypercyclic bilateral weighted shift on $\ell^2(\mathbb{Z})$ admits a hypercyclic subspace.

PROOF We first observe that any hypercyclic bilateral weighted shift $B_{\mathbf{w}}$ satisfies the Hypercyclicity Criterion (see the proof of Theorem 1.38). Moreover, it is well known that the spectrum of $B_{\mathbf{w}}$ is either a disk or an annulus (centred at 0), depending on whether $B_{\mathbf{w}}$ is invertible or not (see [193]).

If the spectrum is a disk then $B_{\mathbf{w}}$ cannot have a closed range, because it is not invertible even though it is one-to-one with dense range. Thus, 0 belongs to $\sigma_e(B_{\mathbf{w}})$, which shows that $B_{\mathbf{w}}$ admits a hypercyclic subspace.

If the spectrum is an annulus then by Theorem 1.18 we know that this annulus intersects the unit circle. In particular, its boundary intersects the closed unit disk and hence Lemma 8.19 gives the result, thanks to Theorem 8.2.   □

EXAMPLE 8.21 Let $B_{\mathbf{w}}$ be a unilateral backward weighted shift on $\ell^2(\mathbb{N})$, with weight sequence $\mathbf{w} = (w_n)_{n\geq 1}$. Then $B_{\mathbf{w}}$ admits a hypercyclic subspace if and only if

$$\limsup_{n\to\infty} \prod_{i=1}^n w_i = \infty \quad \text{and} \quad \sup_n \left(\inf_k \prod_{i=1}^n w_{k+i} \right)^{1/n} \leq 1. \qquad (8.3)$$

In particular, it follows that the operator $2B$ has no hypercyclic subspace, where B is the canonical backward shift.

Observe that, in (8.3), the first condition is Salas' necessary and sufficient condition for $B_{\mathbf{w}}$ to be hypercyclic (Theorem 1.40). We shall assume that it holds true. Then $B_{\mathbf{w}}$ satisfies the Hypercyclicity Criterion. Having said that, Example 8.21 is now an immediate consequence of the following lemma.

LEMMA 8.22 *Let $B_{\mathbf{w}}$ be a unilateral backward weighted shift on $\ell^2(\mathbb{N})$. The essential spectrum of $B_{\mathbf{w}}$ is the annulus $\{r \leq |z| \leq R\}$, where*

$$r := \sup_{n\geq 1} \left(\inf_{k\in\mathbb{N}} \prod_{i=1}^n w_{k+i} \right)^{1/n} \quad and \quad R := \inf_{n\geq 1} \left(\sup_{k\in\mathbb{N}} \prod_{i=1}^n w_{k+i} \right)^{1/n}.$$

PROOF We first note that if we put $r_n := \inf_k \prod_{i=1}^n w_{k+i}$ then the sequence (r_n) is super-multiplicative, i.e. $r_{p+q} \geq r_p r_q$ for all $p, q \in \mathbb{N}$. From that, a well-known argument shows that in fact $r = \lim_{n\to\infty} r_n^{1/n}$. Similarly, we have $R = \lim_{n\to\infty} R_n^{1/n}$, where $R_n := \sup_k \prod_{i=1}^n w_{k+i}$; since $R_n = \|B_{\mathbf{w}}^n\|$ for each $n \geq 1$, this means that R is the spectral radius of $B_{\mathbf{w}}$.

Next we observe that, since an operator T is Fredholm iff T^* is Fredholm, we have

$$\sigma_e(B_{\mathbf{w}}) = \{\overline{\lambda}; \ \lambda \in \sigma_e(B_{\mathbf{w}}^*)\}.$$

This allows us to concentrate on the adjoint operator $B_\mathbf{w}^*$, i.e. the forward shift defined by

$$B_\mathbf{w}^*(e_n) = w_{n+1}e_{n+1},$$

where $(e_n)_{n \in \mathbb{N}}$ is the canonical basis of $\ell^2(\mathbb{N})$.

It is easy to see that $B_\mathbf{w}^* - \lambda$ has dense range if $\lambda \neq 0$ and that $\overline{\text{Ran}(B_\mathbf{w}^* - \lambda)} = H_0 := \{x \in \ell^2(\mathbb{N}); \ x_0 = 0\}$ if $\lambda = 0$. In particular $\overline{\text{Ran}(B_\mathbf{w}^* - \lambda)}$ has finite codimension for any $\lambda \in \mathbb{C}$. Moreover, $B_\mathbf{w}^* - \lambda$ is always one-to-one, since $B_\mathbf{w}$ is assumed to be hypercyclic (see Chapter 1). Thus, we see that

$$\sigma_e(B_\mathbf{w}^*) = \{\lambda \in \mathbb{C}; \ B_\mathbf{w}^* - \lambda \text{ is not bounded below}\}.$$

We first show that if $|\lambda| < r$ then $\lambda \notin \sigma_e(B_\mathbf{w}^*)$. Let us fix $\delta > 0$, with $r - \delta > |\lambda|$, and choose $n \in \mathbb{N}^*$ such that $\prod_{i=1}^n w_{k+i} \geq (r - \delta)^n$ for all $k \in \mathbb{N}$. Then

$$\|B_\mathbf{w}^{*n}(x)\|^2 = \sum_{k=0}^\infty \prod_{i=1}^n w_{k+i}^2 |x_k|^2$$
$$\geq (r - \delta)^{2n} \|x\|^2,$$

for any $x = \sum_k x_k e_k \in \ell^2(\mathbb{N})$. From this we get

$$\|B_\mathbf{w}^{*n}(x) - \lambda^n x\| \geq \big((r - \delta)^n - |\lambda|^n\big)\|x\|,$$

which shows that $B_\mathbf{w}^{*n} - \lambda^n$ is bounded below. Since the operator $B_\mathbf{w}^{*n} - \lambda^n$ can be factorized as $B_\mathbf{w}^{*n} - \lambda^n = (B_\mathbf{w}^* - \lambda)(B_\mathbf{w}^{*(n-1)} + \lambda B_\mathbf{w}^{*(n-2)} + \cdots + \lambda^{n-1})$, it follows that $B_\mathbf{w}^* - \lambda$ is also bounded below, so that $\lambda \notin \sigma_e(B_\mathbf{w}^*)$.

So far, we have shown that $\sigma_e(B_\mathbf{w}^*) \subset \{|\lambda| \geq r\}$. Since $\sigma_e(B_\mathbf{w}^*)$ is also contained in the disk $\overline{D}(0, R)$ (because R is the spectral radius of $B_\mathbf{w}^*$), we will get the desired result if we are able to show that $\sigma_e(B_\mathbf{w}^*)$ contains the (possibly degenerate) annulus $\{r \leq |\lambda| \leq R\}$. And since $\sigma_e(B_\mathbf{w}^*)$ is closed, non-empty and *rotation invariant*, it is in fact enough to prove that $\sigma_e(B_\mathbf{w}^*)$ contains the (possibly empty) open interval (r, R). The rotational invariance of $\sigma_e(B_\mathbf{w}^*)$ follows from the fact that $B_\mathbf{w}$ is unitarily equivalent to $\omega B_\mathbf{w}$, for any $\omega \in \mathbb{T}$. Indeed, the operator $U : \ell^2(\mathbb{N}) \to \ell^2(\mathbb{N})$ defined by $U(e_n) := \omega^{-n} e_n$ is unitary, and it is easily checked that $U B_\mathbf{w} U^{-1} = \omega B_\mathbf{w}$.

We assume that $r < R$ and we fix $\lambda \in (r, R)$. Also, let $\varepsilon > 0$ and let a, b be such that $r < a < \lambda < b < R$. By the definition of R, we may find n and k such that $(\lambda/b)^n < \varepsilon$ and $w_{k+1} \cdots w_{k+n} > b^n$; by the definition of r, we may find n' and $k' > k$ such that $(a/\lambda)^{n'} < \varepsilon$ and $w_{k'+1} \cdots w_{k'+n'} < a^{n'}$. Thus, we have $w_{k+1} \cdots w_{k+n} > \lambda^n/\varepsilon$ and $w_{k'+1} \cdots w_{k'+n'} < \lambda^{n'} \varepsilon$.

Let $x \in \ell^2(\mathbb{N})$ be defined by

$$x := e_k + \sum_{i=k+1}^{k'+n'-1} \frac{w_{k+1} \cdots w_i}{\lambda^{i-k}} e_i.$$

Then, an easy computation reveals that

$$B_\mathbf{w}^*(x) - \lambda x = \frac{w_{k+1} \cdots w_{k'+n'}}{\lambda^{k'+n'-k-1}} e_{k'+n'} - \lambda e_k. \tag{8.4}$$

Now, one has

$$\frac{w_{k+1} \cdots w_{k'+n'}}{\lambda^{k'+n'-k-1}} = \frac{w_{k+1} \cdots w_{k'}}{\lambda^{k'-k}} \times \frac{w_{k'+1} \cdots w_{k'+n'}}{\lambda^{n'-1}}$$

$$= \lambda \, x_{k'} \frac{w_{k'+1} \cdots w_{k'+n'}}{\lambda^{n'}}$$

$$< \lambda \varepsilon \, \|x\|$$

and

$$\|x\| \geq x_{k+n} = \frac{w_{k+1} \cdots w_{k+n}}{\lambda^n} > \frac{1}{\varepsilon} \, .$$

Combining these estimates with (8.4), we get

$$\|B_\mathbf{w}^*(x) - \lambda x\| < 2\lambda \varepsilon \|x\| \, .$$

Since $\varepsilon > 0$ is arbitrary, this shows that $B_\mathbf{w}^* - \lambda$ is not bounded below, and therefore that $\lambda \in \sigma_e(B_\mathbf{w}^*)$. This concludes the proof. $\qquad\square$

EXAMPLE 8.23   Let Λ be the set of all automorphisms of $\mathbb{D}$ having $\alpha = 1$ as attractive fixed point. Then the composition operators C_ϕ, $\phi \in \Lambda$, have a common hypercyclic subspace.

PROOF   We apply Theorem 8.10. By the proof of Example 7.3, condition (1) in Theorem 8.10 is satisfied by the family $(C_\phi)_{\phi \in \Lambda}$. Hence, it is enough to find some closed infinite-dimensional subspace $E \subset H^2(\mathbb{D})$ such that $C_\phi^n(f) \to 0$ for each $\phi \in \Lambda$ and all $f \in E$. This will be achieved thanks to the following lemma.

LEMMA 8.24   *Let $f \in H^2(\mathbb{D})$, and let $\phi \in \Lambda$. Assume that f is continuous at $\alpha = 1$ with $f(1) = 0$. Then $\lim_{n \to \infty} \|C_\phi^n(f)\|_{H^2} = 0$.*

PROOF   The hypothesis means that the boundary value of f is a.e. equal to a function f^* which is continuous at 1 with $f^*(1) = 0$. We keep the notation of Example 7.3 by setting $\mathcal{H}^2(\mathbb{P}_+) := \{g \circ \omega^{-1}; \, g \in H^2(\mathbb{D})\}$, where $\omega : \mathbb{D} \to \mathbb{P}_+$ is the Cayley map, $\omega(z) = i(1+z)/(1-z)$. We have to prove that if $F \in \mathcal{H}^2$ and if $|F(t)| \to 0$ as $|t| \to \infty$, $t \in \mathbb{R}$, then $F \circ \psi_n \to 0$ in $\mathcal{H}^2$ when $\psi(s) = \lambda(s - s_0) + s_0$ ($\lambda > 1$, $s_0 \in \mathbb{R}$) or $\psi(s) = s + b$ ($b \in \mathbb{R} \backslash \{0\}$). Here, of course, $\psi_n = \psi \circ \cdots \circ \psi$.

Let $\varepsilon > 0$, and choose $A > 0$ such that $|t| \geq A \implies |F(t)| \leq \varepsilon$. For $\psi(s) = \lambda(s - s_0) + s_0$, we have

$$\|F \circ \psi_n\|_{\mathcal{H}^2}^2 \leq \int_{s_0 - A/\lambda^n}^{s_0 + A/\lambda^n} |F(\lambda^n(t - s_0) + s_0)|^2 \frac{dt}{1 + t^2} + \varepsilon^2 \pi$$

$$\leq \int_{-A}^{A} |F(\tau)|^2 \frac{d\tau}{\lambda^n \left(1 + (s_0 + (\tau - s_0)/\lambda^n)^2\right)} + \varepsilon^2 \pi \, .$$

By Lebesgue's theorem (or a straightforward estimate), the last integral vanishes as $n \to \infty$, which gives the result. When $\psi(s) = z + b$, where $b \neq 0$, a similar computation gives

$$\|F \circ \psi_n\|_{\mathcal{H}^2}^2 \leq \int_{-A-nb}^{A-nb} |F(t+nb)|^2 \frac{dt}{1+t^2} + \varepsilon^2 \pi$$

$$\leq \int_{-A}^{A} |F(\tau)|^2 \frac{d\tau}{1+(\tau-nb)^2} + \varepsilon^2 \pi,$$

and the result follows. $\qquad\square$

Let us now construct the subspace E. For each positive integer n, we define $f_n \in H^2(\mathbb{D})$ by

$$f_n(z) := \frac{(1-z)^n}{\|(1-z)^n\|_2} = \frac{(1-z)^n}{2^{n/2}}.$$

Clearly, one can find an $a > 0$ such that $|f_n(e^{it})| \leq 2^{-n}$ for $|t| \leq a$. By setting $P_n(z) := z^{k_n} f_n(z)$ for a sufficiently fast increasing sequence of integers (k_n), we obtain a sequence of polynomials (P_n) such that

$$P_n(1) = 0, \quad \langle P_n, P_m \rangle_{H^2} = \delta_{n,m} \quad \text{and} \quad |P_n(e^{it})| \leq 2^{-n} \text{ for } |t| \leq a.$$

In particular, (P_n) is an orthonormal sequence in $H^2(\mathbb{D})$.

Let $E \subset H^2(\mathbb{D})$ be the closed linear span of the P_n. The estimate $|P_n(e^{it})| \leq 2^{-n}$, $|t| \leq a$, ensures that for any bounded sequence of scalars $(\alpha_n)_{n \geq 1}$ the series $\sum \alpha_n P_n(z)$ is uniformly convergent on the arc $\{e^{it}; |t| \leq a\}$. It follows that any $f = \sum_{n \geq 1} \alpha_n P_n \in E$ is continuous at 1 with $f(1) = 0$. By Lemma 8.24, E is the desired closed infinite-dimensional subspace. $\qquad\square$

8.5 Algebras of hypercyclic functions

When the underlying vector space X is an algebra, it is natural to ask whether the set of hypercyclic vectors for a given hypercyclic operator $T \in \mathfrak{L}(X)$ also contains a non-trivial algebra (except the zero vector). Such an algebra is called a **hypercyclic algebra** for T. The basic example here is $X = H(\mathbb{C})$. Perhaps surprisingly, it turns out that the situation is completely different if one considers translation operators or the derivative operator.

Concerning translation operators, the following result was proved by R. M. Aron, J. A. Conejero, A. Peris and J. B. Seoane-Sepúlveda in [12].

PROPOSITION 8.25 *Let p be a positive integer, and let $f \in H(\mathbb{C}) \backslash \{0\}$. Also, let T be a non-trivial translation operator on $H(\mathbb{C})$. If a non-constant function $g \in H(\mathbb{C})$ belongs to the closure of $O(f^p, T)$ then the order of each zero of g is a multiple of p.*

PROOF The operator T is given by $Tu(z) = u(z+a)$, for some $a \neq 0$. Let z_0 be a zero of g and choose $r > 0$ such that g has no other zero in the closed disk $\overline{D}(z_0, r)$. Assuming that $g \in \overline{O(f^p, T)}$, let (n_j) be a sequence of natural numbers such that $f^p(z + n_j a) \to g(z)$ as $j \to \infty$, uniformly on $\overline{D}(z_0, r)$. By Hurwitz's theorem, one can find a $j \geq 0$ such that $f^p(z + n_j a)$ and $g(z)$ have the same number of zeros (counted with their multiplicity) in $\overline{D}(z_0, r)$. Therefore, the order of $z_0 \in Z(g)$ is a multiple of p. $\qquad\square$

This result shows that there is no hypercyclic algebra for any translation operator T on $H(\mathbb{C})$. However, we have

THEOREM 8.26 *The derivative operator* $D : H(\mathbb{C}) \rightarrow H(\mathbb{C})$ *has a hypercyclic algebra.*

PROOF We first need a lemma, which is also taken from [12].

LEMMA 8.27 *Let* (U, V) *be a pair of non-empty open subsets of* $H(\mathbb{C})$, *and let* m *be a positive integer. Then one can find* $P \in U$ *and* $q \in \mathbb{N}$ *such that* $D^q(P^j) = 0$ *for* $j < m$ *and* $D^q(P^m) \in V$.

Observe that when $m = 1$ the lemma just says that T is topologically transitive. The stronger property given by the lemma will ensure the existence of an algebra of hypercyclic functions.

We postpone the (technical) proof of Lemma 8.27 to give that for Theorem 8.26. For any $f \in H(\mathbb{C})$, $m \in \mathbb{N}^*$ and $\alpha \in \mathbb{C}^m$, let us put $f_\alpha := \alpha_1 f + \cdots + \alpha_m f^m$. Also, let $(V_k)_{k \geq 1}$ be a countable basis of open sets for $H(\mathbb{C})$. We introduce the sets

$$\mathcal{A}(k, s, m) := \{ f \in H(\mathbb{C}); \ \forall \alpha \in \mathbb{C}^m \text{ with } \alpha_m = 1 \text{ and } \sup_i |\alpha_i| \leq s,$$
$$\exists q \in \mathbb{N} : \ D^q(f_\alpha) \in V_k \},$$

where k, s and m range over the positive integers. We claim that if one can find some function f in $\mathcal{A} := \bigcap_{k,s,m} \mathcal{A}(k, s, m)$ then the algebra $A(f)$ generated by f satisfies the theorem. Indeed, let $g = \alpha_1 f + \cdots + \alpha_m f^m$ be any non-zero vector of $A(f)$ with $\alpha_m \neq 0$. Since a vector is hypercyclic iff any non-zero multiple of it is hypercyclic, we may assume that $\alpha_m = 1$. Then $g = f_\alpha$ is clearly hypercyclic for D since $f \in \mathcal{A}$.

To show that $\mathcal{A}$ is non-empty, it is enough to check that each $\mathcal{A}(k, s, m)$ is open and dense in $H(\mathbb{C})$: once this is done, one may apply the Baire category theorem. So let us fix (k, s, m).

The set $\mathcal{A}(k, s, m)$ is clearly open in $H(\mathbb{C})$. Indeed, its complement is the projection of the closed set $C := \{(\alpha, f); \ \forall q \in \mathbb{N} \ D^q(f_\alpha) \notin V_k\} \subset \mathbb{C}^m \times H(\mathbb{C})$ along the compact set $K := \{\alpha \in \mathbb{C}^m; \ \alpha_m = 1 \text{ and } \sup_i |\alpha_i| \leq s\}$.

To prove that $\mathcal{A}(k, s, m)$ is dense, we use Lemma 8.27. Let U be any non-empty open subset of $H(\mathbb{C})$. By the lemma, one can find an entire function P and an integer q such that $P \in U$, $D^q(P^m) \in V_k$ and $D^q(P^j) = 0$ for all $j < m$. Then P belongs to $\mathcal{A}(k, s, m)$ by linearity, which concludes the proof. $\qquad\square$

PROOF OF LEMMA 8.27 It is enough to show that, for any pair of polynomials (A, B), one can find a sequence of polynomials R_n and a sequence of natural numbers (q_n) such that $D^{q_n}((A + R_n)^j) = 0$ for all $j < m$, $D^{q_n}((A + R_n)^m) = B$ and $R_n \rightarrow 0$ uniformly on compact sets. Indeed, having chosen A, B with $A \in U$ and $B \in V$, the polynomial $P := A + R_n$ and the integer $q := q_n$ will have the required properties if n is large enough. So let us fix A and B, with $\deg(B) = p$. The key point is the following

FACT Let n be a positive integer, and set $q := mn + (m-1)p$. Then there exists a polynomial R of the form $R(z) = z^n \sum_{i=0}^{p} c_i z^i$ such that $D^q(R^m) = B$. Moreover, writing $c_i = c_i(n)$ one has $c_i(n) = O\left(n^{p-i}/[(m(n+p))!]^{1/m}\right)$ as $n \to \infty$, for each $i \in \{0, \ldots, p\}$.

Granting this fact, we conclude the proof as follows. For each $n \in \mathbb{N}^*$, set $q_n := mn + (m-1)p$ and let R_n be the polynomial given by the fact. For any $k < m$, the polynomial R_n^k has degree at most $(m-1)(n+p) = q_n - n$. By the binomial theorem, it follows that if n is large enough then $D^{q_n}((A+R_n)^j) = 0$ for all $j < m$ and $D^{q_n}((A+R_n)^m) = D^{q_n}(R_n^m) = B$. Finally, $R_n \to 0$ uniformly on compact sets because $M^n c_i(n) \to 0$ for any $M > 0$. Thus, the sequences (R_n) and (q_n) have the required properties. $\qquad\square$

PROOF OF THE FACT We are looking for a polynomial of the form $R(z) = z^n \sum_{i=0}^{p} c_i z^i$. Then the polynomial R^m has degree (at most) $m(n+p) = q + p$. For each $i \in \{0, \ldots, p\}$, let us denote by $\mathbf{d}_i(c_0, \ldots, c_p)$ the coefficient of z^{q+i} in $R^m(z)$; in other words, the coefficient of $z^{mp-(p-i)}$ in $\left(\sum_{i=0}^{p} c_i z^i\right)^m$. Explicitly,

$$\mathbf{d}_i(c_0, \ldots, c_p) = \sum_{\gamma \in \Gamma_i} u_\gamma c_0^{\gamma_0} \cdots c_p^{\gamma_p},$$

where

$$\Gamma_i := \left\{ \gamma \in \mathbb{N}^{p+1}; \ \sum_{j=0}^{p} j\gamma_j = mp - (p-i) \text{ and } \sum_{j=0}^{p} \gamma_j = m \right\}$$

and u_γ is the multinomial coefficient

$$\frac{m!}{\gamma_0! \cdots \gamma_p!}.$$

We note that u_γ *does not depend on* n.

By the definition of the coefficients $\mathbf{d}_i(c_0, \ldots, c_p)$, the polynomial $D^q(R^m)$ is given by

$$D^q(R^m) = \sum_{i=0}^{p} \frac{(q+i)!}{i!} \mathbf{d}_i(c_0, \ldots, c_p) z^i.$$

Thus, writing $B(z) = \sum_{i=0}^{p} b_i z^i$, we have to solve the system of equations

$$\mathbf{d}_i(c_0, \ldots, c_p) = \frac{i! b_i}{(q+i)!} \quad (i = 0, \ldots, p). \tag{8.5}$$

This system is readily seen to be upper-triangular in the unknowns $c_0, \ldots, c_p$, so it can indeed be solved. We now check by (finite) induction on j that any solution of the system satisfies

$$c_{p-j}(n) = O\left(\frac{n^j}{[(m(n+p))!]^{1/m}}\right) \qquad (8.6)$$

as $n \to \infty$, for each $j \in \{0, \ldots, p\}$.

For $i = p$, (8.5) reduces to

$$c_p^m = \frac{p!b_p}{(m(n+p))!}.$$

Therefore, (8.6) is true for $j = 0$. Assume that it has been proved for all $j \leq l-1$, where $1 \leq l \leq p$. Looking at (8.5) for $i = p - l$, we get

$$\frac{(m(n+p)-l)!}{(p-l)!}\left(mc_{p-l}c_p^{m-1} + \sum_{\gamma \in \Gamma'_{p-l}} u_\gamma c_{p-l+1}^{\gamma_{p-l+1}} \cdots c_p^{\gamma_p}\right) = b_{p-l}, \qquad (8.7)$$

where $\Gamma'_{p-l} = \Gamma_{p-l}\backslash\{(0, \ldots, 1, 0, \ldots, 0, m-1)\}$, the number 1 being located at the $(p-l)$th position.

By the induction hypothesis, $c_{p-j}(n) = O(n^j/[(m(n+p))!]^{1/m})$ as $n \to \infty$, for all $j \leq l - 1$. Moreover, if $\gamma \in \Gamma'_{p-l}$ then $\sum_{j=0}^{l-1} j\gamma_{p-j} = \sum_{j=0}^{p}(p-j)\gamma_j = mp - (mp - l) = l$ and $\sum_{j=0}^{l-1} \gamma_{p-j} = \sum_{j=0}^{p} \gamma_j = m$. Hence, we get

$$(m(n+p)-l)! \sum_{\gamma \in \Gamma'_{p-1}} u_\gamma c_{p-l+1}^{\gamma_{p-l+1}} \cdots c_p^{\gamma_p} = O\left(\frac{(m(n+p)-l)!}{(m(n+p))!}n^l\right) = O(1).$$

(The coefficients u_γ have been absorbed into the "O" symbol since they do not depend on n.)

Inserting this estimate into (8.7) and remembering the exact value

$$c_p^m = \frac{p!b_p}{(m(n+p))!},$$

we finally arrive at

$$c_{p-l} = O\left(\frac{[(m(n+p))!]^{(m-1)/m}}{(m(n+p)-l)!}\right) = O\left(\frac{n^l}{[(m(n+p))!]^{1/m}}\right).$$

This concludes the proof of the fact.   □

REMARK 8.28   The reader may have noticed that, in the proof of Theorem 8.26, only Lemma 8.27 requires certain specific properties of the derivative operator. The remaining part of the proof is purely topological. Thus, one can extract the following statement. *Let T be a continuous operator on some separable F-algebra X. Assume that, for any pair (U, V) of non-empty open sets in X, for any open neighbourhood O of zero in X and for any positive integer m, one can find $u \in U$ and an integer q such that $T^q(u^j) \in O$ for all $j < m$ and $T^q(u^m) \in V$. Then T has a hypercyclic algebra.* In fact, the proof of Theorem 8.26 shows that there is a dense G_δ set of vectors $f \in X$ generating a hypercyclic algebra for T.

8.6 Comments and exercises

To our knowledge, hypercyclic subspaces were considered for the first time by L. Bernal-González and A. Montes-Rodríguez in [44], in a Fréchet space context. It is proved there that any non-trivial translation operator on $H(\mathbb{C})$ has a hypercyclic subspace. One may also consider the problem of the existence of a *supercyclic* subspace for a given operator; see [184], [125] and the survey paper [185].

The idea of considering hypercyclic operators on $\mathfrak{L}(X)$ is used in [70] and [74]. Another approach to hypercyclicity in the operator algebra, using tensor product techniques, can be found in [178] and in [58]. In particular, Theorem 8.1 and Corollary 8.9 are proved there in the context of a separable Fréchet space admitting a continuous norm.†

The idea of using the Baire category theorem to produce algebras of queer functions comes from [33]. As already stated, Proposition 8.25 and Lemma 8.27 appear in [12]. Let us mention the following question, asked by R. M. Aron: is it possible to find a hypercyclic algebra for D which is not finitely generated?

EXERCISE 8.1 Show that every (separable, infinite-dimensional) Banach space supports a hypercyclic operator which has a hypercyclic subspace.

EXERCISE 8.2 ([10]) Let $T \in \mathfrak{L}(\ell^2(\mathbb{N}))$ be a hypercyclic operator which satisfies the Hypercyclicity Criterion and has a hypercyclic subspace. We consider $T_1 := T \oplus (2B)$ and $T_2 := (2B) \oplus T$ acting on $\ell^2(\mathbb{N}) \oplus \ell^2(\mathbb{N})$, where B is the usual backward shift.

1. Show that T_1 and T_2 both have a hypercyclic subspace.
2. Show that T_1 and T_2 do not have any common hypercyclic subspace. (*Hint*: Show that if Z were such a subspace then at least one of the projections $\pi_1(Z)$, $\pi_2(Z)$ would contain an infinite-dimensional closed subspace.)

EXERCISE 8.3 **Hypercyclic subspaces for adjoints of multipliers** ([167])
Let ϕ be a function in $H^\infty(\mathbb{D})$, and assume that ϕ is one-to-one. Let M_ϕ be the associated multiplication operator on $H^2(\mathbb{D})$.

1. Prove that $\sigma(M_\phi^*)$ is the closure of $\phi(\mathbb{D})^*$, where $A^* = \{\bar{z};\ z \in A\}$.
2. Prove that for any $\lambda \in \phi(\mathbb{D})^*$ the operator $M_\phi^* - \lambda$ is Fredholm. (*Hint*: Factorize $\phi - \bar{\lambda}$ as $(z - z_0)^p h(z)$, where h does not vanish on $\mathbb{D}$.)
3. Show that M_ϕ^* has a hypercyclic subspace if and only if $\phi(\mathbb{D}) \cap \mathbb{T} \neq \varnothing$ and $\partial\phi(\mathbb{D}) \cap \overline{\mathbb{D}} \neq \varnothing$.

EXERCISE 8.4 ([58]) Let $X := \{(x_n)_{n \in \mathbb{Z}};\ (x_n)_{n < 0} \in \mathbb{C}^{\mathbb{N}}$ and $(x_n)_{n \geq 0} \in \ell^2(\mathbb{N})\}$, endowed with its natural topology. Let $T := 2B$, where B is the backward shift on X.

1. Show that there exists a closed infinite-dimensional subspace $E \subset X$ such that $T^n(x) \to 0$ for all $x \in E$.
2. Show that T has no hypercyclic subspace.

EXERCISE 8.5 **An algebra of hypercyclic vectors for** $2B$
Let $X := \ell^1(\mathbb{N})$ be endowed with the convolution product $(a * b)_n = \sum_{k=0}^n a_k b_{n-k}$. Show that the operator $T := 2B$ acting on X has a hypercyclic algebra. (*Hint*: Look at Remark 8.28 and try to imitate the proof of Lemma 8.27.)

---

† See also a similar result on Fréchet spaces by H. Petersson, Hypercyclic subspaces for Fréchet space operators. *J. Math. Anal. Applic.*, 319: 764–782, 2006.

9

Supercyclicity and the Angle Criterion

Introduction

Let X be a separable Banach space. The heart of this chapter is a very simple necessary condition for supercyclicity, the so-called **Angle Criterion**. In geometrical terms, this criterion says that if $x \in X$ is a supercyclic vector for some operator $T \in \mathcal{L}(X)$ then the set $\{T^n(x)/\|T^n(x)\|; \ n \in \mathbb{N}\}$ is a *norming set* for the dual space X^*. We start the chapter by proving the Angle Criterion and some useful consequences. Then we show that when the unit ball of X is "rotund" at sufficiently many points of the sphere, the Angle Criterion turns out to be also a sufficient condition for supercyclicity. In the remaining two sections, we illustrate the criterion by showing that the classical *Volterra operator* is not supercyclic and that composition operators associated with parabolic non-automorphisms of the disk are not supercyclic either.

9.1 The Angle Criterion

Throughout the chapter, the letter X stands for a real or complex separable Banach space. We denote by B_X the *closed* unit ball of X and by S_X the unit sphere.

The starting point is the following simple geometrical observation: if a vector $x \in X$ is a supercyclic vector for $T \in \mathcal{L}(X)$ then its orbit must visit any *cone* based on a non-empty open subset of X. Translating this into the language of linear functionals, we get a necessary condition for supercyclicity, the so-called Angle Criterion.

THEOREM 9.1 (ANGLE CRITERION) *Let* $T \in \mathcal{L}(X)$. *If a vector* $x \in X$ *is supercyclic for* T *then*

$$\limsup_{n \to \infty} \frac{|\langle x^*, T^n(x) \rangle|}{\|T^n(x)\| \|x^*\|} = 1 \tag{9.1}$$

for every non-zero $x^* \in X^*$.

PROOF Assume that x is supercyclic for T, and let $x^* \in X^* \backslash \{0\}$. Given $\varepsilon > 0$, one can find a $z \in S_X$ such that $\langle x^*, z \rangle \geq (1-\varepsilon)\|x^*\|$ and then a sequence $(\lambda_k) \subset \mathbb{K}$ and an increasing sequence of integers (n_k) such that $\lambda_k T^{n_k}(x) \to z$. Writing

$$\frac{|\langle x^*, T^{n_k}(x) \rangle|}{\|T^{n_k}(x)\| \|x^*\|} = \frac{|\langle x^*, \lambda_k T^{n_k}(x) \rangle|}{\|\lambda_k T^{n_k}(x)\| \|x^*\|},$$

we see that

$$\lim_{k \to \infty} \frac{|\langle x^*, T^{n_k}(x) \rangle|}{\|T^{n_k}(x)\| \|x^*\|} = \frac{|\langle x^*, z \rangle|}{\|x^*\|} \geq 1 - \varepsilon,$$

which gives the result since $\varepsilon > 0$ is arbitrary. $\qquad \square$

The Angle Criterion is often useful to prove that an operator is *not* supercyclic. For example, we get the following corollary, which can be viewed as an extension of Theorem 1.24.

COROLLARY 9.2 *Let $T \in \mathfrak{L}(X)$. Assume that $X = X_1 \oplus X_2 \oplus X_3$, where each X_i is a closed T-invariant subspace and $X_j \neq \{0\}$ for $j = 1, 2$. Suppose moreover that $\|T^n(x_1)\|/\|T^n(x_2)\| \to 0$ for any $(x_1, x_2) \in X_1 \times X_2$ with $x_2 \neq 0$. Then T is not supercyclic.*

PROOF Let us denote by $p_i : X \to X_i$ the projection onto X_i along $\oplus_{j\neq i} X_j$. Towards a contradiction, assume that T is supercyclic and let x be a supercyclic vector for T. Since $p_2 T = T p_2$, the vector $p_2(x)$ is then supercyclic for $T_{|X_2}$, hence $p_2(x) \neq 0$. Choose $x_1^* \in X_1^* \backslash \{0\}$, and define $x^* \in X^*$ by $\langle x^*, z \rangle = \langle x_1^*, p_1(z) \rangle$, that is, $x^* = p_1^*(x_1^*)$. Since $T p_1 = p_1 T$, we get

$$\frac{|\langle x^*, T^n(x) \rangle|}{\|T^n(x)\| \|x^*\|} = \frac{|\langle x_1^*, T^n(p_1(x)) \rangle|}{\|T^n(x)\| \|x^*\|}$$
$$\leq C \frac{\|T^n(p_1(x))\|}{\|T^n(p_2(x))\|}$$

for all $n \in \mathbb{N}$, where C is some finite constant. Therefore, by Theorem 9.1, this is a contradiction. □

In the same spirit, the following result shows that if an operator T has many orbits which grow faster than some orbit of T^* then T cannot be supercyclic.

COROLLARY 9.3 *Let $T \in \mathfrak{L}(X)$. Assume that, for all x in some non-empty open set $U \subset X$, one can find a non-zero linear functional $x^* \in X^*$ such that*

$$\lim_{n \to \infty} \frac{\|T^{*n}(x^*)\|}{\|T^n(x)\|} = 0$$

for every $x \in U$. Then T is not supercyclic.

PROOF Clearly, no vector $x \in U$ satisfies the Angle Criterion (9.1). Hence T cannot be supercyclic, since the set of supercyclic vectors for T is either empty or dense in X. □

9.2 About the converse

The following two examples show that the converse of the Angle Criterion can fail without additional assumptions on the Banach space X.

EXAMPLE 9.4 Let X be the real two-dimensional space $\ell^1(2)$, and let $T \in \mathfrak{L}(X)$ be a rotation by an angle $\pi/2$. Then T is not supercyclic. However, for any non-zero $x^* \in \ell^\infty(2)$, we have

$$\frac{|\langle x^*, T^n(e_0)\rangle|}{\|T^n(e_0)\|\|x^*\|} = 1$$

either for all $n \in 2\mathbb{N}$ or for all $n \in 2\mathbb{N}+1$.

EXAMPLE 9.5   Let B be the backward shift on $X = c_0(\mathbb{N})$. There exists an $x \in c_0(\mathbb{N})$ which is not a supercyclic vector for B and yet satisfies

$$\limsup_{n \to \infty} \frac{|\langle x^*, B^n(x)\rangle|}{\|B^n(x)\|\|x^*\|} = 1 \tag{9.2}$$

for every non-zero linear functional $x^* \in \ell^1(\mathbb{N})$.

PROOF   Let $(x_k^*)_{k \in \mathbb{N}}$ be a dense sequence in $\ell^1(\mathbb{N}) \setminus \{0\}$ made up of finitely supported vectors. For each $k \in \mathbb{N}$, choose a positive integer l_k, such that x_k^* is supported on $[0, l_k)$, and a vector $v_k \in c_0(\mathbb{N})$ supported on $[0, l_k)$ satisfying $\langle x_k^*, v_k \rangle = \|x_k^*\|$ and $|v_k(n)| = 1$ for all $n < l_k$. Finally, let $(n_k) \subset \mathbb{N}$ be defined by $n_0 = 0$ and $n_k = n_{k-1} + l_{k-1}$ for $k \geq 1$. Now, define $x \in c_0(\mathbb{N})$ by

$$x := \sum_{p=0}^{\infty} \frac{1}{2^p} S^{n_p}(v_p),$$

where, as usual, S is the forward shift. Then $\langle x_k^*, B^{n_k}(x)\rangle = 2^{-k}\langle x_k^*, v_k \rangle$ and $\|B^{n_k}(x)\| = 2^{-k}$, so that

$$\frac{|\langle x_k^*, B^{n_k}(x)\rangle|}{\|B^{n_k}(x)\|\|x_k^*\|} = 1$$

for all $k \in \mathbb{N}$. Since the sequence (x_k^*) is dense in $\ell^1(\mathbb{N}) \setminus \{0\}$, it follows at once that (9.2) is satisfied for every $x^* \in \ell^1(\mathbb{N})\setminus\{0\}$. However, x cannot be supercyclic for B. Indeed, since $|x(n+1)|/|x(n)| \geq 1/2$ for all $n \in \mathbb{N}$, we see that $\|\lambda B^n(x) - e_0\| \geq \max(|\lambda x_n - 1|, |\lambda x_{n+1}|) \geq 1/3$ for any $\lambda \in \mathbb{K}$.   □

One common feature of $\ell^1(2)$ and $c_0(\mathbb{N})$ is that their unit balls are not "rotund". We now show that when the unit ball of X is rotund at sufficiently many points of the sphere, the converse of the Angle Criterion does hold true. The precise definition that we need is the following.

DEFINITION 9.6   *A point $x \in S_X$ is said to be **strongly exposed** if there exists a linear functional $x^* \in S_{X^*}$ such that $\langle x^*, x \rangle = 1$ and the following property holds: whenever (x_n) is a sequence in B_X such that $\langle x^*, x_n \rangle \to 1$, it follows that $\|x_n - x\| \to 0$. The linear functional x^* is then called an **exposing functional** for x.*

It is important to note that this definition depends on the given norm $\|\cdot\|$ on X and not merely on the topology of X. In particular, it is well known that any separable Banach space can be equipped with an equivalent norm with respect to which every point of the unit sphere is strongly exposed. A proof of this result is outlined in Exercise 9.4.

We also note that if $x \in S_X$ is strongly exposed by some linear functional $x^* \in S_{X^*}$ then x is the only point $z \in B_X$ such that $\langle x^*, z \rangle = 1$: to see this, just apply the definition to the constant sequence $x_n = z$. It follows that if x is a strongly exposed point of S_X then x must be an **extreme point** of the (closed!) unit ball B_X. Indeed, if $x = (u+v)/2$ where $u, v \in B_X$ and if $x^* \in S_{X^*}$ is an exposing functional for x, then we must have $\langle x^*, u \rangle = 1 = \langle x^*, v \rangle$, because $1 = \langle x^*, x \rangle = (\langle x^*, u \rangle + \langle x^*, v \rangle)/2$; hence $u = x = v$.

EXAMPLE 9.7 For any positive measure μ and $1 < p < \infty$, every point of $S_{L^p(\mu)}$ is strongly exposed.

This follows from the **uniform convexity** of the L^p norm (see e.g. [95], or Exercise 9.3).

EXAMPLE 9.8 The unit sphere of $c_0(\mathbb{N})$ does not have any strongly exposed point.

PROOF It is well known that B_{c_0} has no extreme point; we recall here the simple proof of this fact. If $x = (x(k))_{k \in \mathbb{N}} \in B_{c_0}$, one can find $k_0 \in \mathbb{N}$ such that $|x(k_0)| < 1$. Then one can write $x(k_0) = (\alpha + \beta)/2$, with $|\alpha|, |\beta| \leq 1$ and $\alpha, \beta \neq x(k_0)$. Denoting by $(e_k)_{k \in \mathbb{N}}$ the canonical basis of $c_0(\mathbb{N})$ it follows that $x = (u+v)/2$, where $u = \alpha e_{k_0} + \sum_{k \neq k_0} x(k) e_k$ and $v = \beta e_{k_0} + \sum_{k \neq k_0} x(k) e_k$. This shows that x is not an extreme point of B_{c_0}. $\square$

EXAMPLE 9.9 A point $x \in S_{\ell^1(\mathbb{N})}$ is strongly exposed iff it has the form $x = \lambda e_p$, where $|\lambda| = 1$ and $p \in \mathbb{N}$. Here, $(e_k)_{k \in \mathbb{N}}$ is the canonical basis of ℓ^1.

PROOF We first show that each point $x = \lambda e_p$ is strongly exposed by the linear functional $x^* := \bar{\lambda} e_p^* \in \ell^\infty(\mathbb{N})$, where $(e_k^*)_{k \in \mathbb{N}} \subset \ell^\infty(\mathbb{N})$ is the sequence of coordinate functionals associated with (e_k). Indeed, if (x_n) is a sequence in S_{ℓ^1} such that $\langle x^*, x_n \rangle = \lambda x_n(p) \to 1$ then, writing

$$1 = \sum_{k=0}^\infty |x_n(k)| = |\lambda x_n(p)| + \sum_{k \neq p} |x_n(k)|,$$

we see that $\sum_{k \neq p} |x_n(k)| \to 0$. This shows that $\|x_n - x\|_1 \to 0$.

Conversely, if $x \in S_{\ell^1}$ does not have the form $x = \lambda e_p$ then x has at least two non-zero coordinates, say $x(0)$ and $x(1)$. It follows that x is not an extreme point of B_{ℓ^1}. Indeed, assuming that we are in the real case and setting $r := |x(0)| + |x(1)|$, the point $z := (x(0), x(1))$ is an interior point of a face of the sphere $S_r = \{\|\xi\|_1 = r\}$ in the two-dimensional space $\ell^1(2)$. Thus, one can write $z = (\alpha + \beta)/2$ where $\alpha, \beta \in S_r$ and $\alpha, \beta \neq z$, so that $x = (u+v)/2$, where $u := \alpha(0) e_0 + \alpha(1) e_1 + \sum_{k \geq 2} x(k) e_k$ and $v := \beta(0) e_0 + \beta(1) e_1 + \sum_{k \geq 2} x(k) e_k$ are in S_{ℓ^1} and $u, v \neq x$. $\square$

The following very simple lemma explains the relevance of strongly exposed points in the study of supercyclicity. Recall that a set $D \subset S_X$ is said to be **norming** for X^* if for every $x^* \in X^*$, one has

$$\|x^*\| = \sup\{|\langle x^*, z \rangle|; \ z \in D\}.$$

LEMMA 9.10 *Let $D \subset S_X$ be a norming set for X^*. Then every strongly exposed point of S_X is in the $\| \cdot \|$-closure of the set $\{\lambda z; \ z \in D, \ |\lambda| = 1\}$.*

PROOF   Let x be a strongly exposed point of S_X, and let $x^* \in S_{X^*}$ be an exposing functional for x. Since D is norming for X^*, one can find a sequence $(z_n) \subset D$ and a sequence of scalars (λ_n) with $|\lambda_n| = 1$ for all n such that $\langle x^*, \lambda_n z_n \rangle \to 1$. Then $\|\lambda_n z_n - x\| \to 0$. □

We can now state the following converse to the Angle Criterion. The reader should compare it with the counter-examples given in the previous section.

PROPOSITION 9.11   *Assume that the strongly exposed points of S_X are dense in S_X. If $T \in \mathfrak{L}(X)$ and if $x \in X$ satisfies (9.1) in Theorem 9.1 for every non-zero $x^* \in X^*$ then x is a supercyclic vector for T.*

PROOF   If (9.1) holds then $D := \{T^n(x)/\|T^n(x)\|; \ n \in \mathbb{N}\}$ is a norming set for X^*. By Lemma 9.10 it follows that $\overline{\mathbb{K} \cdot O_T(x)}$ contains every strongly exposed point of S_X, so that x is supercyclic for T if the strongly exposed points of S_X are dense in S_X. □

COROLLARY 9.12   *Let $T \in \mathfrak{L}(X)$. Then a vector $x \in X$ is supercyclic for T iff (9.1) holds for any equivalent norm $\| \cdot \|$ on X.*

PROOF   As mentioned above, there exists an equivalent norm on X with respect to which all points of the unit sphere are strongly exposed. □

In the next two sections, we give two applications of the Angle Criterion.

9.3 The Volterra operator

The **Volterra operator** $V : L^2([0,1]) \to L^2([0,1])$ is defined by the formula

$$V f(x) = \int_0^x f(t)\, dt.$$

The operator V is clearly bounded on $L^2([0,1])$. It is also cyclic, the constant function **1** being a cyclic vector by the Weierstrass theorem. In this section, we show that this cannot be very much improved.

THEOREM 9.13   *The Volterra operator is not supercyclic.*

PROOF   We will argue by contradiction; so, let us assume that V is supercyclic. We first claim that one can find a continuous function $f \in \mathcal{C}([0,1])$ which is a supercyclic vector for V and satisfies $f(1/2) \neq 0$. Indeed, since the operator V maps $L^2([0,1])$ continuously into $\mathcal{C}([0,1])$, the set $H := \{u \in L^2([0,1]); \ Vu(1/2) = 0\}$ is well defined, and H is a closed hyperplane in $L^2([0,1])$. As such, H is nowhere dense in $L^2([0,1])$ and, since the supercyclic vectors for V are dense in $L^2([0,1])$, it follows that one can find a $u \in SC(V)$ such that $f := Vu$ (is continuous and) does not vanish at $1/2$. We are going to show that the Angle Criterion is not satisfied by this supercyclic vector f, which will yield the desired contradiction.

We first give a lower estimate for $\|V^n(f)\|$, $n \in \mathbb{N}$. To this end, let us define $f_n \in L^2([0,1])$ as $f_n(t) = t^n(1-t)^n$ and set $L_n := d^n f_n/dt^n$. The polynomials L_n are the classical **Legendre polynomials**.

Integrating by parts, we see that the adjoint operator V^* is given by

$$V^* g(x) = \int_x^1 g(t)\, dt.$$

Therefore, we have

$$V^{*n}(L_n) = (-1)^n f_n$$

for all $n \in \mathbb{N}$.

We also note that, after a suitable normalization, the sequence (f_n) is a summability kernel at $1/2$. Indeed, a simple computation reveals that

$$\int_0^1 f_n(t)\, dt = \frac{(n!)^2}{(2n+1)!};$$

and since the function $f_n(1/2+s)$ is even on $[-1/2, 1/2]$ and decreasing on $[0, 1/2]$, we also have $\int_{|t-1/2|\geq\delta} f_n(t)\, dt \leq f_n(1/2+\delta) = (1/4-\delta^2)^n$ for any $\delta \in (0, 1/2)$. Thus, setting $h_n := ((2n+1)!/(n!)^2)f_n$, we get $\|h_n\|_{L^1} = 1$ and

$$\int_{|t-1/2|\geq\delta} |h_n(t)|\, dt \leq \left(\frac{1}{4}-\delta^2\right)^n \frac{(2n+1)!}{(n!)^2}$$

for any $\delta \in (0, 1/2)$. By Stirling's formula, $(2n+1)!/(n!)^2$ behaves asymptotically as $\sqrt{n}\, 4^n$, which gives immediately

$$\int_{|t-1/2|\geq\delta} h_n(t)\, dt \xrightarrow{n\to\infty} 0.$$

Hence, (h_n) is a summability kernel at $1/2$. Since f is continuous, it follows that $\langle f, h_n\rangle_{L^2} = \int_0^1 f(t)\, h_n(t)\, dt$ tends to $f(1/2)$ as $n \to \infty$, so that

$$\langle f, f_n\rangle = \frac{(n!)^2}{(2n+1)!}\, f\left(\frac{1}{2}\right)(1+o(1)).$$

Moreover, repeated integration by parts yields

$$\begin{aligned}
\|L_n\|^2 &= \int_0^1 \frac{d^n}{dt^n}\left(t^n(1-t)^n\right) \frac{d^n}{dt^n}\left(t^n(1-t)^n\right) dt \\
&= (2n)! \int_0^1 t^n(1-t)^n dt \\
&= \frac{(n!)^2}{2n+1}.
\end{aligned}$$

This gives the following lower estimate for $\|V^n(f)\|$:

$$\|V^n(f)\| \geq \frac{|\langle V^n(f), L_n \rangle|}{\|L_n\|} = \frac{|\langle f, V^{*n}(L_n) \rangle|}{\|L_n\|} = \frac{|\langle f, f_n \rangle|}{\|L_n\|}$$

$$\geq \frac{n!\sqrt{2n+1}}{(2n+1)!} \left| f\left(\frac{1}{2}\right) \right| (1 + o(1)).$$

Using Stirling's formula again, we conclude that on the one hand

$$\|V^n(f)\| \geq \frac{c}{4^n n!} \qquad (9.3)$$

for some constant $c > 0$ and every $n \in \mathbb{N}$. On the other hand, let $g := \mathbf{1}_{(0,1/5)} \in L^2([0,1])$ and recall that f is continuous. Straightforward induction yields

$$|V^n f(x)| \leq \frac{x^n}{n!} \|f\|_\infty$$

for any $x \in [0,1]$, so that

$$|\langle V^n(f), g \rangle| \leq \|f\|_\infty \int_0^{1/5} \frac{x^n}{n!} \, dx = \frac{\|f\|_\infty}{5^{n+1}(n+1)!} . \qquad (9.4)$$

From (9.3) and (9.4), we can now conclude that the presumed supercyclic vector f does not satisfy the Angle Criterion. $\qquad \Box$

9.4 Parabolic composition operators

In this section, we turn back to the dynamics of parabolic composition operators, which we have already encountered in Chapter 1. Our aim is to give a simple proof of the following beautiful theorem, which is due to E. A. Gallardo Guttiérez and A. Montes-Rodríguez [114].

THEOREM 9.14   *Let $\phi \in LFM(\mathbb{D})$ be a parabolic non-automorphism. Then the composition operator C_ϕ is not supercyclic on $H^2(\mathbb{D})$.*

PROOF   As in Chapter 1, it will be more convenient to replace the unit disk $\mathbb{D}$ by the upper half-plane $\mathbb{P}_+ = \{\mathrm{Im}(w) > 0\}$. Let $\omega(z) = i(1+z)/(1-z)$ be the Cayley transform, which maps $\mathbb{D}$ conformally onto $\mathbb{P}_+$. We denote by $\psi = \omega \circ \phi \circ \omega^{-1}$ the sister of ϕ living on $\mathbb{P}_+$. We may assume that ϕ has the form (see the proof of Theorem 1.47)

$$\phi(z) = \frac{(2-a)z + a}{-az + 2 + a},$$

where $\mathrm{Re}(a) > 0$. Then its sister ψ is given by

$$\psi(w) = w + ia.$$

Let us denote by $\mathcal{H}^2$ the image of $H^2(\mathbb{D})$ under ω, i.e.

$$\mathcal{H}^2 = \left\{ f \circ \omega^{-1}; \ f \in H^2(\mathbb{D}) \right\},$$

with norm

$$\|h\|_{\mathcal{H}^2}^2 = \frac{1}{\pi} \int_{\mathbb{R}} |h(x)|^2 \frac{dx}{1+x^2}.$$

The space $\mathcal{H}^2$ is not the usual Hardy space on $\mathbb{P}_+$; the latter is defined as

$$H^2(\mathbb{P}_+) := \left\{ F \in H(\mathbb{P}_+); \|F\|_{H^2(\mathbb{P}_+)}^2 := \sup_{y>0} \int_{\mathbb{R}} |F(x+iy)|^2 \, dx < \infty \right\}.$$

We denote by $J : \mathcal{H}^2 \to H^2(\mathbb{P}_+)$ the natural isometry from $\mathcal{H}^2$ onto $H^2(\mathbb{P}_+)$, i.e.

$$Jh(w) = \frac{1}{\sqrt{\pi}(w+i)} h(w) \cdot$$

The reason for introducing $H^2(\mathbb{P}_+)$ is that we are going to use the the **Paley–Wiener theorem**, according to which the Fourier transform is an isometry from $L^2(0,\infty)$ onto $H^2(\mathbb{P}_+)$. Here, we define the Fourier transform $\mathcal{F} : L^2(0,\infty) \to H^2(\mathbb{P}_+)$ by

$$\mathcal{F}\varphi(w) = \frac{1}{\sqrt{2\pi}} \int_0^\infty e^{itw} \varphi(t) \, dt.$$

Our intention is to apply Corollary 9.3. To this end, we fix some real number $\tau > 0$ and adopt the following notation. Given $f \in H^2(\mathbb{D})$, we put $F := J(f \circ w^{-1}) \in H^2(\mathbb{P}_+)$ and $\widehat{f} := \mathcal{F}^{-1}(F) \in L^2(0,\infty)$. Finally, we denote by $\widehat{f}_\tau$ the orthogonal projection of $\widehat{f}$ onto $L^2(0,\tau)$, i.e. $\widehat{f}_\tau = \mathbf{1}_{(0,\tau)}\widehat{f}$.

Let $T := J \circ C_\psi \circ J^{-1}$. The operator T acts on $H^2(\mathbb{P}_+)$, and it is is easy to check that if $f \in H^2(\mathbb{D})$ then

$$T^n F(w) = \frac{w+i(1+na)}{w+i} F(w+ina)$$

for all $n \in \mathbb{N}$. Now, since $\mathrm{Re}(a) > 0$ there exists $\delta > 0$ such that

$$\left| \frac{x+i(1+na)}{x+i} \right|^2 = \frac{(x-n\,\mathrm{Im}(a))^2 + (1+n\,\mathrm{Re}(a))^2}{x^2+1} \geq \delta^2$$

for every $x \in \mathbb{R}$ and all $n \in \mathbb{N}$. In particular, this gives

$$\|T^n(F)\|_{H^2(\mathbb{P}_+)} \geq \delta \, \|F(\cdot + ina)\|_{H^2(\mathbb{P}_+)}$$

for each $f \in H^2(\mathbb{D})$ and all $n \in \mathbb{N}$. To estimate the left-hand side, we use the Fourier transform and observe that

$$F(\cdot + ina) = \mathcal{F}\left(e^{-nat}\widehat{f}\right).$$

Thus, we get

$$\|F(\cdot + ina)\| = \|e^{-nat}\widehat{f}\|_{L^2(0,\infty)}$$
$$\geq \|e^{-nat}\widehat{f}_\tau\|_{L^2(0,\tau)}$$
$$\geq e^{-n\mathrm{Re}(a)\tau} \|\widehat{f}_\tau\|_{L^2(0,\tau)}.$$

Let $U := \{f \in H^2(\mathbb{D}); \; \hat{f}_\tau \neq 0\}$. Then U is a non-empty open subset of $H^2(\mathbb{D})$, and the above inequalities show that for any $f \in U$ there exists a positive constant $c = c_{f,\tau}$ such that

$$\|C_\phi^n(f)\| \geq c\, e^{-n\mathrm{Re}(a)\tau}$$

for all $n \in \mathbb{N}$.

In order to apply Corollary 9.3, we now just need to find some non-zero vector $g \in H^2(\mathbb{D})$ such that $\|C_\phi^{*n}(g)\| = o(e^{-n\mathrm{Re}(a)\tau})$ as $n \to \infty$. We will make use of the following lemma, which gives a simple explicit formula for the adjoint of the composition operator induced by a linear fractional map. This result can be found in C. C. Cowen's paper [89].

LEMMA 9.15 *Let* $\varphi(z) = (sz + t)/(uz + v) \in LFM(\mathbb{D})$. *The adjoint of the composition operator C_φ is given by*

$$C_\varphi^* = M_\gamma C_\sigma M_\theta^*,$$

where $\sigma(z) = (\bar{s}z - \bar{u})/(-\bar{t}z + \bar{v})$, $\gamma(z) = 1/(-\bar{t}z + \bar{v})$ *and* $\theta(z) = uz + v$. *Here M_α denotes the operator corresponding to multiplication by the bounded function α.*

We will postpone the proof of Lemma 9.15 in order to finish that of Theorem 9.14. In our case, we have $C_\phi^n = C_{\phi_n}$, where

$$\phi_n(z) = \frac{(2 - na)z + na}{-naz + (2 + na)}.$$

Thus $C_\phi^{*n} = M_{\gamma_n} C_{\sigma_n} M_{\theta_n}^*$ where

$$\theta_n(z) = -naz + (2 + na),$$
$$\gamma_n(z) = \frac{1}{-n\bar{a}z + (2 + n\bar{a})},$$
$$\sigma_n(z) = \frac{(2 - n\bar{a})z + n\bar{a}}{-n\bar{a}z + (2 + n\bar{a})}.$$

The key point is that σ_n is itself a parabolic non-automorphism of $\mathbb{D}$, so that one has considerable amount of information on its spectrum (see e.g. [88]). For our purposes, we just need to know that for each $s > 0$ the function $e_s \in H^2(\mathbb{D})$ defined by $e_s(z) = \exp\left(s\,(z + 1)/(z - 1)\right)$ is an eigenfunction of C_{σ_n} with associated eigenvalue $e^{-n\bar{a}s}$. This is easy to check, for example by looking at the sister ψ_n of σ_n living on $\mathbb{P}_+$ (since $\psi_n(w) = w + in\bar{a}$, the associated eigenfunction for C_{ψ_n} is the map $w \mapsto \exp(isw)$).

Now we set $g(z) := ze_s(z)$, where $s > 0$ has yet to be specified. Then

$$M_{\theta_n}^*(g) = -n\bar{a}e_s + (2 + n\bar{a})ze_s$$

for all $n \in \mathbb{N}$, so that

$$(C_\phi^*)^n(g) = \left(-n\bar{a} + (2 + n\bar{a})\sigma_n\right)e^{-n\bar{a}s}\gamma_n e_s.$$

To conclude the proof, we observe that $\|\gamma_n\|_\infty = O(1)$, which can be checked easily by writing

$$\frac{1}{\gamma_n(z)} = -n\bar{a}\left(z - \left(1 + \frac{2}{n\bar{a}}\right)\right)$$

and by observing that the modulus of $1 + 2/(n\bar{a})$ is greater than $1 + C/n$ for some constant $C > 0$. It follows that

$$\left\|C_\phi^{*n}(g)\right\| = O\left(ne^{-n\mathrm{Re}(a)s}\right).$$

Choosing $s > \tau$, Corollary 9.3 now shows that C_ϕ is not supercyclic. $\square$

PROOF OF LEMMA 9.15   We first observe that σ maps $\mathbb{D}$ into itself, so that the composition operator C_σ is well defined on $H^2(\mathbb{D})$. Indeed, the linear fractional map φ may be viewed as an automorphism of the Riemann sphere $\widehat{\mathbb{C}}$, and a simple computation shows that

$$\sigma(z) = \frac{1}{\psi^{-1}(1/z)},$$

where $\psi(z) = \overline{\varphi(\bar{z})}$.

For any $w \in \mathbb{D}$, let $k_w(z) = (1 - \bar{w}z)^{-1}$ be the reproducing kernel at w. Then $C_\varphi^*(k_w) = k_{\varphi(w)}$ and $M_\theta^*(k_w) = \overline{\theta(w)}k_w$ (see Example 1.11 and subsection 1.4.3). Hence

$$
\begin{aligned}
M_\gamma C_\sigma M_\theta^* k_w(z) &= \overline{\theta(w)}\gamma(z)C_\sigma(k_w)(z) \\
&= (\bar{u}\bar{w} + \bar{v})\frac{1}{-\bar{t}z + \bar{v}}\frac{1}{1 - \bar{w}((\bar{s}z - \bar{u})/(-\bar{t}z + \bar{v}))} \\
&= \frac{\bar{u}\bar{w} + \bar{v}}{-\bar{t}z + \bar{v} - \bar{w}\bar{s}z + \bar{u}\bar{w}} \\
&= \frac{1}{1 - \overline{\varphi(w)}z} = k_{\varphi(w)}(z) = C_\varphi^*(k_w)(z).
\end{aligned}
$$

Since the reproducing kernels k_w span a dense linear subspace of $H^2(\mathbb{D})$, this concludes the proof of the lemma. $\square$

9.5 Comments and exercises

The Angle Criterion appears for the first time in [184], where the authors use it to construct non-supercyclic vectors for certain weighted shifts. The criterion was formally stated in a Hilbert space setting in [114], and its extension to Banach spaces was studied in [117]. In [117], the converse part of the Angle Criterion is proved for *locally uniformly convex* Banach spaces (see the definition in Exercise 9.2). A more general formulation involving strongly exposed points may perhaps be useful. For example, it is proved in [36] that every normalized polynomial is strongly exposed in the unit sphere of the Bergman space $A^1(\mathbb{D})$. Therefore, the converse part of the Angle Criterion holds true in $A^1(\mathbb{D})$.

Besides the Angle Criterion, there is another method for showing that an operator cannot be supercyclic. The method works for operators which preserve positivity, typically for a convolution operator associated with some positive function. It can be roughly described as follows.

Let T be an operator on a complex Banach function space X. Suppose that $\sigma_p(T^*) = \varnothing$ and that T preserves real functions, i.e. $T(f)$ is real whenever $f \in X$ is. Now let $f \in X$, and assume that there exists an operator $R \in \mathfrak{L}(X)$ with dense range such that $TR = RT$ and $R(f)$ is a real function. Using the positive supercyclicity theorem (Corollary 3.4), it is then not hard to show that f cannot be supercyclic. This method was suggested by F. León-Saavedra and his co-authors; see e.g. [170] or [171]. Exercise 9.5 shows how it works for the Volterra operator; see also Exercise 9.6.

Theorem 9.13 was obtained in [116], and we have followed very closely the proof given therein. A. Montes-Rodríguez and S. Shkarin showed in [187] that the Volterra operator is in fact not even *weakly* supercyclic on any space $L^p([0,1])$, $1 \le p < \infty$. Further results on the Volterra operator can be found e.g. in [186].

Theorem 9.14 was obtained first in [114] (see also [115]). Besides its simplicity, the interest of the proof given here is that it works in various contexts (H^p spaces with $p \neq 2$, Bergman spaces, several complex variables, ...; see e.g. [23]). An alternative approach, using so-called *decomposable operators*, can be found in [221].

EXERCISE 9.1   Use the Angle Criterion to prove Theorem 1.24.

EXERCISE 9.2   **Three notions of convexity**
Let $(X, \| \cdot \|)$ be a Banach space. The norm $\| \cdot \|$ is said to be **uniformly convex** if, given any $\varepsilon > 0$, there exists $\delta > 0$ such that $\|(x + y)/2\| < 1 - \delta$ whenever $x, y \in S_X$ satisfy $\|x - y\| \ge \varepsilon$. It is said to be **locally uniformly convex** if, for each $x \in X$ and any sequence $(x_n) \subset X$ with $\|x\| = 1 = \|x_n\|$, the condition $\|x + x_n\| \to 2$ implies $x_n \to x$. It is said to be **strictly convex** if S_X contains no non-trivial segment, i.e. if, whenever $x, y \in S_X$ satisfy $\|x + y\| = 2$, it follows that $x = y$.

1. Prove the following chain of implications: uniformly convex $\implies$ locally uniformly convex $\implies$ strictly convex.
2. Prove that if $\| \cdot \|$ is locally uniformly convex then every point of S_X is strongly exposed.

EXERCISE 9.3   **Uniform convexity of L^p**
Let (Ω, μ) be a measure space, and let $p \in (1, \infty)$. We intend to show that the natural norm of $L^p(\Omega, \mu)$ is uniformly convex (following [95]).

1. Show that $|x - y|^p \le 2^p(|x|^p + |y|^p)$ for all $x, y \in \mathbb{C}$.
2. Prove that, for any $\eta > 0$, there exists $\delta > 0$ such that the following holds: if $x, y \in \mathbb{C}$ satisfy $|x - y| \ge \eta \max(|x|, |y|)$ then
$$\left| \frac{x+y}{2} \right|^p \le (1 - \delta) \frac{|x|^p + |y|^p}{2}$$
   (*Hint*: Prove it for $x = 1$ and $|y| \le 1$.)
3. Let $x, y \in L^p(\Omega, \mu)$ with $\|x\| = \|y\| = 1$ and $\|x - y\| \ge \varepsilon > 0$. Set
$$M := \left\{ \omega \in \Omega; \ \varepsilon^p \left(|x(\omega)|^p + |y(\omega)|^p \right) \le 4|x(\omega) - y(\omega)|^p \right\}.$$
   (a) Show that there exists $\delta > 0$ such that
$$\left| \frac{x(\omega) + y(\omega)}{2} \right|^p \le (1 - \delta) \frac{|x(\omega)|^p + |y(\omega)|^p}{2}$$
   for every $\omega \in M$.
   (b) Prove that $\int_M |x(\omega) - y(\omega)|^p d\mu(\omega) \ge \varepsilon^p / 2$.
   (c) Show that
$$\int_M \left[\frac{|x(\omega)|^p + |y(\omega)|^p}{2} - \left| \frac{x(\omega) + y(\omega)}{2} \right|^p \right] d\mu(\omega) \ge \frac{\delta \varepsilon^p}{2^{p+2}}.$$
4. Conclude that $L^p(\Omega, \mu)$ is uniformly convex.

EXERCISE 9.4   **Renormings of a separable Banach space**
Let X be a separable (real) Banach space, with given norm $\| \cdot \|$. We intend to show that X admits a locally uniformly convex renorming (following [122]).

1. Let (f_n) be a sequence in B_{X^*} which separates the points of X. Show that the formula

$$|x|^2 := \|x\|^2 + \sum_{n=1}^{+\infty} 2^{-n} f_n^2(x)$$

defines an equivalent norm on X and that $| \cdot |$ is strictly convex.
2. Let (X_n) be an increasing sequence of finite-dimensional subspaces such that $X_0 \neq \{0\}$ and $\bigcup_n X_n$ is dense in X. Let d_n be the seminorm defined by

$$d_n(x) = \mathrm{dist}_{|\cdot|}(x, X_n).$$

We define a new (equivalent) norm N on X by the formula

$$N(x)^2 = \sum_{n=0}^{\infty} 2^{-n} d_n(x)^2.$$

Finally, let $(x_k) \subset X$, $x \in X$ be such that $N(x_k) = N(x) = 1$ and $N(x_k + x) \to 2$.
   (a) Prove that $d_n(x_k) \to d_n(x)$ for every $n \in \mathbb{N}$.
   (b) Let $\varepsilon > 0$, and choose $n \in \mathbb{N}$ such that $d_n(x) < \varepsilon$. Writing $x_k = u_k + v_k$ with $u_k \in X_n$ and $|v_k|$ small, prove that (x_k) admits a subsequence (y_k^ε) such that $|y_p^\varepsilon - y_q^\varepsilon| < 3\varepsilon$ for all $p, q \in \mathbb{N}$.
   (c) Prove that (x_k) has a subsequence converging to some point $z \in X$ such that $N(z) = 1$.
   (a) Prove that $z = x$.
   (d) Conclude that N is a locally uniformly convex norm.

EXERCISE 9.5   **Non-supercyclicity of the Volterra operator revisited** ([171])
Let $p \in [1, \infty)$. We denote by V the Volterra operator acting on the complex space $L^p([0, 1])$. For any $a \in L^1(\mathbb{R})$, we denote by $R_a : L^p([0, 1]) \to L^p([0, 1])$ the operator defined by

$$R_a f(x) = \int_0^x \overline{a(t)} f(x - t)\, dt.$$

1. Show that $R_a V = V R_a$ for any $a \in L^1$.
2. Let $a \in L^p$ and assume that a is a cyclic vector for V. Show that R_a has dense range. (*Hint*: Compute $R_a(V^n \mathbf{1})$, $n \in \mathbb{N}$.)
3. Show that $R_{\bar{a}}(a)$ is a real function for any $a \in L^p$.
4. Show that $\sigma_p(V^*) = \varnothing$.
5. Show that V is not supercyclic on $L^p([0, 1])$. (*Hint*: Use the positive supercyclicity theorem, Corollary 3.4.)

EXERCISE 9.6   **The Cesáro operator on $\mathcal{C}([0, 1])$**
Let $T : \mathcal{C}([0, 1]) \to \mathcal{C}([0, 1])$ be the Cesáro operator, $Tf(x) = x^{-1} \int_0^x f(t) dt$. Show that T is not supercyclic on $\mathcal{C}([0, 1])$. (*Hint:* Use the positive supercyclicity theorem.)

10

Linear dynamics and the weak topology

Introduction

This chapter is devoted to hypercyclicity and supercyclicity with respect to the weak topology of a given Banach space. Let X be a separable infinite-dimensional Banach space. An operator $T \in \mathfrak{L}(X)$ is said to be **weakly hypercyclic** if it is hypercyclic when considered as an operator on the topological vector space (X, w), in other words, if there exists some vector $x \in X$ whose T-orbit $O(x, T)$ is weakly dense in X. Such a vector x is of course called a weakly hypercyclic vector for T. One defines in the same way **weakly supercyclic** operators and weakly supercyclic vectors.

One unpleasant fact immediately comes to mind when considering these definitions. Up to now, we have almost exclusively concentrated on hypercyclicity or supercyclicity for linear operators acting on completely metrizable topological vector spaces; however, the weak topology of an infinite-dimensional Banach space is neither Baire nor metrizable. This means first that we have no Hypercyclicity Criterion at our disposal and second that we must be careful with *sequences*, since a vector $z \in X$ may belong to the weak closure of some T-orbit $O(x, T)$ without being the weak limit of a sequence $(T^{n_k}(x))$.

Having said that, the first natural question is whether weak hypercyclicity and supercyclicity really make sense, i.e. whether there exist weakly hypercyclic or supercyclic operators which are not already hypercyclic or supercyclic with respect to the norm topology. Fortunately the answer is positive, but this is a non-trivial result.

To consider a still more basic question, one may ask what can be said of weakly dense sequences which are not dense with respect to the norm topology.

In any infinite-dimensional Banach space Z, one can find a sequence (z_n) such that $\|z_n\| \to \infty$ and 0 belongs to the weak closure of the set $\{z_n; \ n \in \mathbb{N}\}$. More precisely, given any sequence of positive numbers (a_n) such that $\sum_0^\infty a_n^{-2} = \infty$ (e.g. $a_n = \sqrt{n+1}$), one can find a sequence $(z_n) \subset Z$ such that $\|z_n\| = a_n$ for all n and $0 \in \overline{\{z_n; \ n \in \mathbb{N}\}}^w$.

(To see this, one can proceed as follows. Let (λ_n) be a sequence of positive numbers such that $\sum_0^\infty \lambda_n^2 < \infty$ and $\sum_0^\infty \lambda_n a_n^{-1} = \infty$. By the **Dvoretzky–Rogers theorem** (see [97]) there exists a normalized sequence $(u_n) \subset Z$ such that the series $\sum \lambda_n u_n$ is unconditionally convergent. Then $\sum_0^\infty \|R(\lambda_n u_n)\| < \infty$ for any finite-rank operator $R : Z \to \mathbb{K}^N$. Putting $z_n := a_n u_n$, this implies that 0 is in the weak closure of the sequence (z_n). Indeed, otherwise one could find a finite-rank operator $R : Z \to \mathbb{K}^N$ such that $\|R(a_n u_n)\| \geq 1$ for all n, so that we would have $\sum_n \lambda_n a_n^{-1} \leq \sum_n \lambda_n \|R(u_n)\| < \infty$.)

From the existence of such a sequence (z_n), it is easy to construct a weakly dense sequence $(x_n) \subset X$ such that $\|x_n\| \to \infty$ (see Exercise 10.1). However, it turns out that $\|x_n\|$ *cannot grow too fast*. This statement can be made precise by means of a natural parameter depending on the geometry of the Banach space X; we will discuss this in some detail in Section 10.1. Regarding weak linear dynamics, the main consequence will be that we get some conditions ensuring that a vector cannot be weakly hypercyclic or supercyclic for a given operator T.

After this general discussion, Section 10.2 is devoted to weighted backward shifts. We first show that for *unilateral* shifts acting on $\ell^p(\mathbb{N})$, weak and norm hypercyclicity are equivalent. The same is true for bilateral shifts acting on $\ell^p(\mathbb{Z})$ when $p < 2$, and likewise for supercyclicity when $p \le 2$. Then, we show in particular: that there exist weakly hypercyclic shifts on $\ell^2(\mathbb{Z})$ which are not hypercyclic; that, unlike norm-hypercyclicity, the weak hypercyclicity of weighted shifts on $\ell^p(\mathbb{Z})$ depends on p when $p > 2$; and that the unweighted shift B is weakly supercyclic on $\ell^p(\mathbb{Z})$ if (and only if) $p > 2$.

In Section 10.3, we consider unitary operators. Perhaps unexpectedly, it turns out that unitary operators can be weakly supercyclic. In other words, there exist probability measures μ on $\mathbb{T}$ for which the multiplication operator M_z is weakly supercyclic on $L^2(\mu)$. Clearly, this has much to do with harmonic analysis.

We conclude the chapter by saying a few words about weakly *sequentially* hypercyclic operators. This will point out the highly non-metrizable nature of the weak topology.

10.1 Weakly closed and weakly dense sequences

In this section, we take X to be a separable infinite-dimensional Banach space. We are going to prove several precise results centred around the following idea: if (x_n) is a sequence in X such that $\|x_n\|$ tends "rapidly" to infinity then the set $\{x_n;\ n \in \mathbb{N}\}$ is weakly closed in X and hence cannot be weakly dense. We start with a simple but already non-trivial example.

PROPOSITION 10.1 *Let* $(x_n)_{n \in \mathbb{N}}$ *be a sequence in* X *such that* $\|x_n\| \ge aC^n$ *for all* $n \in \mathbb{N}$ *and some constants* $a > 0$ *and* $C > 1$. *Then the set* $\{x_n;\ n \in \mathbb{N}\}$ *is weakly closed in* X.

PROOF   If z is any vector in $X \setminus \{x_n;\ n \in \mathbb{N}\}$ and if we put $\tilde{x}_n := x_n - z$ then the sequence $(\tilde{x}_n)$ satisfies $\|\tilde{x}_n\| \ge \tilde{C}^n$ for large enough n, say $n \ge n_0$, and some constant $\tilde{C} > 1$. Moreover, z is in the weak closure of the sequence (x_n) iff 0 is in the weak closure of the set $\{\tilde{x}_n;\ n \ge n_0\}$. Therefore, it is in fact enough to show that if (x_n) is a sequence in X such that $\|x_n\| \ge C^n$ for all n and some constant $C > 1$ then 0 is not in the weak closure of $\{x_n;\ n \in \mathbb{N}\}$. Finally, replacing x_n by $C^n x_n / \|x_n\|$, we may assume that $\|x_n\| = C^n$ for all n.

We first suppose that $C \geq 4$. In this case we shall find a single linear functional $x^* \in X^*$ such that $|\langle x^*, x_n \rangle| \geq 1/3$ for all $n \in \mathbb{N}$.

We construct by induction a sequence $(x_n^*) \subset X^*$ such that

$$|\langle x_n^*, x_n \rangle| \geq 1 \quad \text{and} \quad \|x_n^* - x_{n-1}^*\| \leq \frac{2}{C^n}$$

for all $n \in \mathbb{N}$, with the convention that $x_{-1}^* = 0$. Assume that x_{n-1}^* has already been constructed. If $|\langle x_{n-1}^*, x_n \rangle| \geq 1$ we put $x_n^* = x_{n-1}^*$. If $|\langle x_{n-1}^*, x_n \rangle| < 1$ we choose $y_n^* \in X^*$ such that $\|y_n^*\| = 1$ and $|\langle y_n^*, x_n \rangle| = C^n$, and we put $x_n^* := x_{n-1}^* + (2/C^n)y_n^*$. Then x_n^* clearly satisfies the above requirements.

Now, let $x^* := \lim_{n \to \infty} x_n^*$ and observe that

$$|\langle x^*, x_n \rangle| \geq |\langle x_n^*, x_n \rangle| - |\langle x^* - x_n^*, x_n \rangle|$$

$$\geq 1 - \sum_{j > n} \frac{2}{C^j} \, C^n \geq \frac{1}{3} \, .$$

For the general case, we choose some positive integer N such that $C^N \geq 4$ and split (x_n) into N sequences $\mathbf{x}^i = (x_k^i)_{k \in \mathbb{N}}$, $0 \leq i \leq N - 1$ in the obvious way, i.e. setting $x_k^i = x_{Nk+i}$. By the previous case, one can find N linear functionals $x_0^*, \ldots, x_{N-1}^*$ such that $|\langle x_i^*, x_{Nk+i} \rangle| \geq 1/3$ for each $i \in \{0, \ldots, N-1\}$ and all $k \in \mathbb{N}$. Then $\max_i |\langle x_i^*, x_n \rangle| \geq 1/3$ for all $n \in \mathbb{N}$, which shows that $0 \notin \overline{\{x_n; \, n \in \mathbb{N}\}}^w$. $\qquad\square$

10.1.1 Ball's theorem and the weak closure exponent

Proposition 10.1 says that a sequence $(x_n) \subset X$ tending to infinity exponentially fast is as far as possible from being weakly dense in X. This already has interesting consequences regarding weak linear dynamics; see e.g. [98] or Exercise 10.2.

However, the exponential rate of growth is clearly not optimal. Most people would surmise that the weaker condition $\sum_0^\infty \|x_n\|^{-1}$ already forces the set $\{x_n; \, n \in \mathbb{N}\}$ to be weakly closed and that an even weaker assumption should suffice if the space X is e.g. a Hilbert space. This is indeed true. However, it seems to be a non-trivial fact, depending as it does on deep theorems due to K. Ball ([16], [17]). We state Ball's results as a single theorem, to which we will refer as *Ball's theorem*.

THEOREM 10.2 (BALL'S THEOREM) *Let* $(z_n)_{n \in \mathbb{N}}$ *be a sequence of vectors in* X *such that* $\inf_n \|z_n\| > 0$, *and let* (w_n) *be a sequence of positive real numbers. Assume that either* $\sum_0^\infty w_n < \infty$ *or* X *is a* complex *Hilbert space and* $\sum_0^\infty w_n^2 < \infty$. *Then one can find a linear functional* $z^* \in X^*$ *such that* $|\langle z^*, z_n \rangle| \geq w_n$ *for all* $n \in \mathbb{N}$.

From this theorem, it is not hard to deduce

COROLLARY 10.3 *Let* $(x_n)_{n \in \mathbb{N}}$ *be a sequence of non-zero vectors in* X. *In each of the following two cases, the set* $\{x_n; \, n \in \mathbb{N}\}$ *is weakly closed in* X:

(a) X is arbitrary and $\sum_0^\infty \|x_n\|^{-1} < \infty$;

(b) X is a real or complex Hilbert space and $\sum_0^\infty \|x_n\|^{-2} < \infty$.

PROOF OF COROLLARY 10.3   We first note that in each case it is enough to show that the weak closure of the x_n does not contain 0. Indeed, once this is done one can replace x_n by $\widetilde{x}_n := x_n - z$ where $z \in X \setminus \{x_n;\ n \in \mathbb{N}\}$ is arbitrary and observe that the sequence $(\widetilde{x}_n)$ satisfies the same assumption as (x_n).

To prove (a), put $z_n := \|x_n\|^{-1} x_n$ and $w_n = \|x_n\|^{-1}$. Applying Ball's theorem, find $z^* \in X^*$ such that $|\langle z^*, x_n \rangle| \geq 1$ for all $n \in \mathbb{N}$. Thus, 0 is not in the weak closure of $\{x_n;\ n \geq 0\}$.

The proof of (b) when X is a complex Hilbert space is exactly the same. If X is real, we consider its complexification $X_{\mathbb{C}} = X \oplus iX$. From the complex case, one can find $z^* \in X_{\mathbb{C}}^*$ such that $|\langle z^*, x_n \rangle| \geq 1$ for all $n \in \mathbb{N}$. Writing $z^* = z_1^* \oplus i z_2^*$ with $z_1^*, z_2^* \in X^*$, we get $\max(|\langle z_1^*, x_n \rangle|, |\langle z_2^*, x_n \rangle|) \geq 1/2$ for all $n \in \mathbb{N}$. This concludes the proof.   □

The above results suggest that we should associate with the Banach space X the following two parameters:

- the **weak closure exponent** of X, defined by

$$w(X) := \sup \Big\{ a > 0;\ \text{for any sequence } (x_n) \subset X \text{ with } \sum_n \|x_n\|^{-a} < \infty,$$
$$\text{the set } \{x_n;\ n \in \mathbb{N}\} \text{ is weakly closed in } X \Big\}$$

$$= \inf \Big\{ b > 0;\ \text{there exists } (x_n) \subset X \text{ such that } \sum_n \|x_n\|^{-b} < \infty$$
$$\text{and } \{x_n;\ n \in \mathbb{N}\} \text{ is not weakly closed} \Big\};$$

- The **weak density exponent** of X, defined by

$$w_{\text{dens}}(X) := \sup \Big\{ a > 0;\ \forall (x_n) \subset X \text{ with } \sum_n \|x_n\|^{-a} < \infty,$$
$$\text{the set } \{x_n;\ n \in \mathbb{N}\} \text{ is not weakly dense in } X \Big\}$$

$$= \inf \Big\{ b > 0;\ \exists (x_n) \subset X \text{ such that } \sum_n \|x_n\|^{-b} < \infty$$
$$\text{and } \{x_n;\ n \in \mathbb{N}\} \text{ is weakly dense} \Big\}.$$

These definitions call for some comment. First, one may replace "$\{x_n;\ n \in \mathbb{N}\}$ is weakly closed" by "0 is not in the weak closure of $\{x_n;\ n \in \mathbb{N}\}$" in the definition of $w(X)$, according to the remark at the beginning of the proof of Corollary 10.3. From this and Ball's theorem, we see that $w(X)$ is well defined (i.e. the set over which the supremum is taken is non-empty), with $w(X) \geq 1$ for any X and $w(X) \geq 2$ if X is a Hilbert space. Moreover, it follows from the Dvoretzky–Rogers theorem that $w(X) \leq 2$ (see the argument given in the introduction to the chapter). Thus, we have

$$1 \leq w(X) \leq 2$$

for any Banach space X, and $w(X) = 2$ when X is a Hilbert space.

Second, unlike in Ball's theorem, the exponents $w(X)$ and $w_{\text{dens}}(X)$ are not sensitive to the scalar field, i.e. any general statement concerning them holds for all real Banach spaces iff it holds for all complex Banach spaces. Indeed, one can pass from the complex case to the real case as in the proof of Corollary 10.3, and from the real case to the complex case by considering a complex Banach space as a real Banach space.

Finally, the two exponents are in fact the same (!), as the next lemma shows. This observation essentially comes from the paper [11] by R. Aron, D. García and M. Maestre.

LEMMA 10.4   *We have $w(X) = w_{\mathrm{dens}}(X)$.*

PROOF   The relation $w(X) \leq w_{\mathrm{dens}}(X)$ is clear, since a weakly dense sequence cannot be weakly closed. Conversely let $b > w(X)$, so that one can find $(x_n) \subset X$ with $\sum_n \|x_n\|^{-b} < \infty$ and $0 \in \overline{\{x_n;\ n \in \mathbb{N}\}}^w$. We have to find another sequence $(z_n) \subset X$ such that $\sum_0^\infty \|z_n\|^{-b} < \infty$ and $\{z_n;\ n \in \mathbb{N}\}$ is weakly dense in X. In other words, we are looking for some countable weakly dense set $Z \subset X$ such that $\sum_{z \in Z} \|z\|^{-b} < \infty$.

We fix a sequence of linear functionals $(x_n^*) \subset B_{X^*}$ such that $\langle x_n^*, x_n \rangle = \|x_n\|$ for all $n \in \mathbb{N}$. Also, let $(y_k)_{k \geq 1}$ be a norm-dense sequence in X. For $n \in \mathbb{N}$ and $k \geq 1$, we put

$$\gamma_{n,k} := \begin{cases} \frac{\langle x_n^*, y_k \rangle}{|\langle x_n^*, y_k \rangle|} & \text{if } \langle x_n^*, y_k \rangle \neq 0, \\ 1 & \text{otherwise.} \end{cases}$$

Then

$$\begin{aligned} \|y_k + \gamma_{n,k} x_n\| &\geq |\langle x_n^*, y_k + \gamma_{n,k} x_n \rangle| \\ &= |\langle x_n^*, y_k \rangle| + \|x_n\| \\ &\geq \|x_n\|. \end{aligned}$$

Let us choose a sequence of positive integers (M_n) tending to infinity such that $\sum_0^\infty M_n \|x_n\|^{-b} < \infty$. For each $n \in \mathbb{N}$, put $Z_n := \{y_k + \gamma_{n,k} x_n;\ 1 \leq k \leq M_n\}$, and let $Z := \bigcup_n Z_n$. Then

$$\begin{aligned} \sum_{z \in Z}^\infty \|z\|^{-b} &= \sum_{n=0}^\infty \sum_{k \leq M_n}^\infty \|y_k + \gamma_{n,k} x_n\|^{-b} \\ &\leq \sum_{n=0}^\infty M_n \|x_n\|^{-b} < \infty. \end{aligned}$$

However, the set Z is weakly dense in X. Indeed, let U be a non-empty weakly open set in X. Since the sequence (y_k) is dense in X, one can find a $k \geq 1$ such that $U \supset y_k + O$, where O is a circled weak neighbourhood of 0. Since all vectors x_n are non-zero and $0 \in \overline{\{x_n;\ n \in \mathbb{N}\}}^w$, there exists $n \in \mathbb{N}$ such that $M_n \geq k$ and $x_n \in O$. Then $\gamma_{n,k} x_n \in O$ because $|\gamma_{n,k}| = 1$, so that $z := y_k + \gamma_{n,k} x_n \in U \cap Z$. This concludes the proof. $\square$

Our objective in this section is to compute the value of $w(X)$ for a large class of Banach spaces X. The inequalities $1 \leq w(X) \leq 2$ and the exact value $w(X) = 2$

in the Hilbertian case strongly suggest that $w(X)$ depends on the geometry of the Banach space X. In fact, the important thing turns out to be the type of the dual space X^*. See Appendix C for the definition of type and cotype.

For any Banach space Z, let us put

$$p(Z) := \sup\{p \in [1,2];\ Z \text{ has type } p\},$$
$$q(Z) := \inf\{q \in [2,\infty];\ Z \text{ has cotype } q\}.$$

We shall say that Z has a **non-trivial type** if $p(Z) > 1$, i.e. Z has type p for some $p > 1$.

Before stating the main result of this section, let us make some comments concerning the parameters $p(Z)$ and $q(Z)$. For any number $r \in [1,\infty]$, we shall denote by r^* the conjugate exponent of r, i.e. the number defined by

$$\frac{1}{r} + \frac{1}{r^*} = 1.$$

This notation will be used constantly throughout the chapter.

It is well known (and easily checked from the definitions of type and cotype) that if a Banach space Z has type $p \in [1,2]$ then Z^* has cotype p^*. Applying this to $Z = X^*$ and remembering that X embeds isometrically into X^{**}, we see that if X^* has type $p = q^*$ then X has cotype q. It follows that $q(X) \le p(X^*)^*$, i.e.

$$p(X^*) \le q(X)^*.$$

This inequality cannot be reversed, as shown by taking $X = \ell^1(\mathbb{N})$. Indeed, in that case $p(X^*) = 1$ and $q(X) = 2 = q(X)^*$. However, it was shown by G. Pisier that if X is assumed to have a non-trivial type then X has cotype $q \in [2,\infty]$ iff X^* has type q^* ([197]; see [97, Proposition 13.17]). From this, it follows that $p(X^*) = q(X)^*$ when X has a non-trivial type.

The following theorem was obtained by the first author in [22].

THEOREM 10.5 *The weak closure exponent satisfies* $p(X^*) \le w(X) \le q(X)^*$. *Therefore,* $w(X) = p(X^*)$ *if X has a non-trivial type.*

COROLLARY 10.6 *If (x_n) is a sequence in X such that $\sum_0^\infty \|x_n\|^{-a} < \infty$ for some $a < p(X^*)$ then the set $\{x_n;\ n \in \mathbb{N}\}$ is weakly closed in X.*

Since $\ell^r(\mathbb{N})$ has type exactly $\min(r,2)$ for any $r \in [1,\infty)$, we can immediately deduce the following result due to S. Shkarin [224]. Of course, this was the main motivation behind the search for a statement such as Theorem 10.5.

COROLLARY 10.7 *Let (x_n) be a sequence in $\ell^p(\mathbb{N})$, $1 < p < \infty$. Assume that $\sum_0^\infty \|x_n\|^{-a} < \infty$ for some $a < \min(2, p^*)$. Then the set $\{x_n;\ n \in \mathbb{N}\}$ is weakly closed in ℓ^p.*

10.1.2 Upper estimate for $w(X)$; q-sequences

In this subsection, we show that $w(X) \leq q(X)^*$. The key point is that, for a given $q \in [2, \infty]$, it is in fact not very hard to show that $w(X) \leq q^*$ if the space X contains ℓ^q spaces of arbitrarily large (finite) dimension. The key idea turns out to be the following (the terminology is that of [224]).

DEFINITION 10.8 *Let $q \in [1, \infty]$, and let C be a positive real number. A finite or infinite sequence $(g_i)_{i \in I} \subset X$ is said to be a q-**sequence** with constant C if*

$$\left\| \sum_{i \in I} a_i g_i \right\| \leq C \left(\sum_{i \in I} |a_i|^q \right)^{1/q}$$

for every finite sequence of scalars (a_i). When $q = \infty$, the right-hand side is to be interpreted as $C \sup_i |a_i|$.

In other words, (g_i) is a q-sequence iff there is an operator $L : \ell^q(I) \to X$ such that $L(e_i) = g_i$ for all $i \in I$, with $\|L\| \leq C$. Here (e_i) is the canonical basis of $\ell^q(I)$ (putting $\ell^q(I) = c_0(I)$ when $q = \infty$!). The next lemma explains why q-sequences are useful in our discussion.

LEMMA 10.9 *Let $(x_n)_{n \in \mathbb{N}}$ be a sequence in X, and let C be a positive real number. Assume that for each $A > 0$ one can find $q \in [1, \infty]$, a finite set $I \subset \mathbb{N}$ and a sequence of scalars $(\alpha_i)_{i \in I}$ such that $\|(\alpha_i)\|_{l^{q^*}} > A$ and $(g_i) := (\alpha_i x_i)$ is a q-sequence with constant C. Then the weak closure of $\{x_n; \ n \in \mathbb{N}\}$ in X contains 0.*

The statement of this lemma is arguably less than pleasing. The reader may look at Corollary 10.10 directly for a more friendly formulation.

PROOF   The proof relies on the following fact.

FACT   Let $(g_i)_{i \in I}$ be a finite q-sequence in X with constant C. For every linear functional $z^* \in X^*$, one has

$$\left(\sum_{i \in I} |\langle z^*, g_i \rangle|^{q^*} \right)^{1/q^*} \leq C \|z^*\|.$$

The left-hand side is to be interpreted as $\sup_i |\langle z^*, g_i \rangle|$ if $q = 1$.

PROOF OF THE FACT   Let $L : \ell^q(I) \to X$ be the linear operator sending each basis vector e_i to g_i. Then $\|L\| \leq C$, so the adjoint operator $L^* : X^* \to \ell^{q^*}(I)$ has norm $\leq C$ as well. Writing down the definition of L^*, this means exactly that for any $z^* \in X^*$ the sequence $(\langle z^*, g_i \rangle)_{i \in I}$ has ℓ^{q^*} norm at most $C \|z^*\|$.   □

Now, assume that 0 is not in the weak closure of $\{x_n; \ n \in \mathbb{N}\}$. Then one can find $z_1^*, \ldots, z_N^* \in B_{X^*}$ and a $\delta > 0$ such that

$$\max_{1 \leq j \leq N} |\langle z_j^*, x_n \rangle| \geq \delta$$

for all $n \in \mathbb{N}$. Let A be a large positive number, and let $(\alpha_i)_{i \in I}$ be a finite sequence of scalars such that $\|(\alpha_i)\|_{\ell^{q^*}} > A$ and $(g_i) := (\alpha_i x_i)$ is a q-sequence with constant C for some $q \in [1, \infty]$. By the above fact, we can then write

$$N \geq \left(\sum_{j=1}^N \|z_j^*\|^{q^*} \right)^{1/q^*}$$

$$\geq C^{-1} \left(\sum_{j=1}^N \sum_{i \in I} |\langle z_j^*, \alpha_i x_i \rangle|^{q^*} \right)^{1/q^*} \geq C^{-1} \delta \, \|(\alpha_i)\|_{\ell^{q^*}} > C^{-1} \delta A,$$

so that $A \leq \delta^{-1} CN$. Since A is arbitrary, this is a contradiction. $\qquad\square$

We point out the following immediate consequence of Lemma 10.9, which will be needed in Sections 10.2 and 10.3.

COROLLARY 10.10 *Let $q \in (1, \infty]$, and let $(g_i)_{i \in I}$ be an infinite q-sequence in X. Also, let $(\alpha_i)_{i \in I}$ be a sequence of non-zero scalars and assume that $\sum_{i \in I} |\alpha_i|^{q^*} = \infty$. Then the weak closure of the set $\{g_i / \alpha_i; \ i \in I\}$ contains 0.*

We now use Lemma 10.9 to prove the following result due to V. Kadets [149]. We state it for $q \geq 2$ only, since we already know that $w(X) \leq 2$.

PROPOSITION 10.11 *Let $q \in [2, \infty]$. Assume that, for some constant $C < \infty$, the Banach space X contains arbitrarily long normalized q-sequences with constant C. Then $w(X) \leq q^*$.*

PROOF Let $b > q^*$ and let us fix a sequence of positive numbers $(a_n)_{n \in \mathbb{N}}$ such that $\sum_n a_n^{-b} < \infty$ and $\sum_n a_n^{-q^*} = \infty$. It is enough to exhibit a sequence $(x_n) \subset X$ such that $\|x_n\| = a_n$ for all n and $0 \in \overline{\{x_n; \ n \in \mathbb{N}\}}^w$.

Let (I_k) be a partition of $\mathbb{N}$ into finite sets such that $\sum_{i \in I_k} a_i^{-q^*} \to \infty$ as $k \to \infty$. For each k, one can find in X a normalized q-sequence $(g_i)_{i \in I_k}$ with constant C. Putting $x_n := a_n g_n$, it follows at once from Lemma 10.9 that the sequence $(x_n)_{n \in \mathbb{N}}$ has the required properties. $\qquad\square$

To put this result in perspective, let us recall the important notion of finite representability. A Banach space Y is said to be **finitely representable** in some Banach space Z if, for any finite-dimensional subspace $E \subset Y$ and every $\varepsilon > 0$, one can find an embedding $J : E \to Z$ such that $\|J\| \, \|J^{-1}\| < 1 + \varepsilon$.

In particular, the finite representability of ℓ^q, $1 \leq q \leq \infty$, in X means that for any $M \in \mathbb{N}$ and every $\varepsilon \in (0, 1)$ one can find $f_1, \dots, f_M \in X$ such that

$$(1 - \varepsilon) \|(a_i)\|_{\ell^q} \leq \left\| \sum_{i=1}^M a_i f_i \right\| \leq (1 + \varepsilon) \|(a_i)\|_{\ell^q} \tag{10.1}$$

for any choice of scalars $a_1, \ldots, a_M$. Taking $\varepsilon := 1/2$, the sequence $(f_1, \ldots, f_M)$ is a q-sequence with constant 2 and $\|f_i\| \geq 1/2$, so we get a normalized q-sequence of length M with constant 4 if we put $g_i := f_i/\|f_i\|$. Therefore, we can state

COROLLARY 10.12 *Let $q \in [2, \infty]$ and assume that ℓ^q is finitely representable in X. Then $w(X) \leq q^*$.*

To apply this result, we now call on a deep and famous result from the local theory of Banach spaces, the so-called **Maurey–Pisier theorem**: *if Z is an infinite-dimensional Banach space then $\ell^{p(Z)}$ and $\ell^{q(Z)}$ are finitely representable in Z* (see [181] and also the books [97] or [173]).

This result says essentially that there is a canonical witness to the fact that a Banach space Z has a "bad" type or cotype. Indeed, $\ell^{p(Z)}$ has type exactly $p(Z)$ and $\ell^{q(Z)}$ has cotype exactly $q(Z)$. Of course, we should mention here the celebrated **Dvoretzky's theorem**, according to which ℓ^2 is finitely representable in *any* infinite-dimensional Banach space [103].

From Corollary 10.12 and the Maurey–Pisier theorem, we immediately get the required upper estimate for $w(X)$.

COROLLARY 10.13 *It holds that $w(X) \leq q(X)^*$.*

10.1.3 How to get a lower estimate

In this subsection we prove the following theorem, which gives at once the required lower bound for $w(X)$.

THEOREM 10.14 *Assume that X^* has type $p \geq 1$. Let $a \in (0, p)$, and let N be any integer such that $N(1 - a/p) \geq a$. Given any sequence $(x_n) \subset X$ such that $\sum_0^\infty \|x_n\|^{-a} < \infty$, one can find N linear functionals $z_1^*, \ldots, z_N^* \in X^*$ such that*

$$\max_{j=1,\ldots,N} |\langle z_j^*, x_n \rangle| \geq 1 \quad \textit{for all} \;\; n \in \mathbb{N}.$$

In particular, we see that if (x_n) is a sequence in X such that $\sum_0^\infty \|x_n\|^{-a} < \infty$ for some $a < p(X^*)$ then 0 is not in the weak closure of the x_n. Thus, we can state

COROLLARY 10.15 *It holds that $w(X) \geq p(X^*)$.*

PROOF OF THEOREM 10.14 Without loss of generality, we may assume that X is a *real* Banach space and that $\sum_0^\infty \|x_n\|^{-a} = 1$.

The linear functionals $x_1^*, \ldots, x_N^*$ will be found by a probabilistic argument. Here is the idea. For each $n \in \mathbb{N}$, choose $x_n^* \in B_{X^*}$ such that $\langle x_n^*, x_n \rangle = \|x_n\|$. We consider a random linear functional $z^*(\omega) = \sum_0^\infty g_n(\omega) r_n x_n^*$, where (g_n) is a sequence of independent Gaussian variables and the positive real numbers r_n are chosen in such a way that the series is almost surely convergent. We hope that, with large probability, $\langle z^*(\omega), x_n \rangle$ is large for every $n \in \mathbb{N}$. Unfortunately, this does not hold if we consider just one random linear functional z^*, so we have to duplicate (in fact, to "N-plicate") the construction.

Precisely, we fix a countable family $\{g_{i,n}; \ n \in \mathbb{N}, \ 1 \leq i \leq N\}$ of independent standard Gaussian variables defined on the same probability space $(\Omega, \mathcal{F}, \mathbb{P})$. For each $i \in \{1, \dots, N\}$, we define a random variable z_i^* with values in X^* by the formula

$$z_i^*(\omega) = \sum_{n \geq 0} g_{i,n}(\omega) \frac{x_n^*}{\|x_n\|^{a/p}} \, .$$

We first have to check that this formula makes sense, i.e. that the series is indeed convergent for almost every $\omega \in \Omega$, or, equivalently, that the series is convergent in $L^1(\Omega, X^*)$ (see Chapter 5). This follows easily from the assumption on X^*. Indeed,

$$\mathbb{E}\left(\left\| \sum_{n=N}^{M} g_{i,n}(\omega) \|x_n\|^{-a/p} x_n^* \right\|_{X^*} \right) \leq C \left(\sum_{n=N}^{M} \|x_n\|^{-a} \right)^{1/p}$$

for every $N, M \in \mathbb{N}$, by the very definition of type.

Let us fix $n \in \mathbb{N}$. We consider the N-dimensional Gaussian random vector ξ_n defined by

$$\xi_n(\omega) = (\langle z_1^*(\omega), x_n \rangle, \dots, \langle z_N^*(\omega), x_n \rangle) \, .$$

Let Q be the covariance matrix of ξ_n. Since the random variables $z_1^*, \dots, z_N^*$ are independent, the matrix Q is diagonal with diagonal entries

$$
\begin{aligned}
Q_{i,i} &= \mathbb{E}\left(\langle z_i^*(\omega), x_n \rangle^2 \right) \\
&= \sum_{m=0}^{\infty} \|x_m\|^{-2a/p} \langle x_m^*, x_n \rangle^2 \\
&:= \sigma_n^2 \geq \|x_n\|^{2(1-a/p)}.
\end{aligned}
$$

For each $c > 0$, let us denote by D_c the N-dimensional cube $\{x \in \mathbb{R}^N; \|x\|_\infty \leq c\}$, and let us put $B_{n,c} := \{\omega; \ \xi_n(\omega) \in D_c\}$. Then

$$
\begin{aligned}
\mathbb{P}(B_{n,c}) &= \left(\int_{-c}^{c} \frac{1}{\sqrt{2\pi}\sigma_n} \exp\left(\frac{-t^2}{2\sigma_n^2} \right) dt \right)^N \\
&= \left(\int_{-c/\sigma_n}^{c/\sigma_n} \frac{1}{\sqrt{2\pi}} \exp\left(\frac{-t^2}{2} \right) dt \right)^N \\
&\leq \left(\frac{c}{\sigma_n} \right)^N \\
&\leq c^N \|x_n\|^{-N(1-a/p)}.
\end{aligned}
$$

This holds for each $n \in \mathbb{N}$, so we get

$$\mathbb{P}\left(\bigcup_{n \in \mathbb{N}} B_{n,c} \right) \leq c^N \sum_{n=0}^{\infty} \|x_n\|^{-a},$$

where we have used the inequality $N(1 - a/p) \geq a$. Thus, when $c < 1$ we may pick $\omega_0 \in \Omega \backslash \bigcup_n B_{n,c}$. Then $\max_{1 \leq j \leq N} |\langle z_j^*(\omega_0), x_n \rangle| > c$ for all $n \in \mathbb{N}$, which concludes the proof of Theorem 10.14. $\square$

REMARK   Since the value $p = 1$ is allowed, Theorem 10.14 says something about *arbitrary* Banach spaces. The result obtained is weaker than Ball's theorem (Theorem 10.2) but still sufficient to show that $w(X) \geq 1$.

The reader may have noticed that Theorem 10.5 leaves open the following natural and somewhat irritating question.

OPEN QUESTION   What is the exact value of $w(\ell^1(\mathbb{N}))$?

10.1.4 Back to linear dynamics

Ball's theorem 10.2 and Theorem 10.14 have interesting applications to weak linear dynamics. More precisely, they can be helpful to show that a vector x is *not* weakly hypercyclic (or not weakly supercyclic) for a given operator T. The following proposition is the most obvious example.

PROPOSITION 10.16   *Let X be a separable Banach space, and let $T \in \mathfrak{L}(X)$. Also, let $x \in X$. In each of the following three cases, x is not a weakly hypercyclic vector for T:*

(a) $\sum_0^\infty \|T^n(x)\|^{-a} < \infty$ *for some $a < p(X^*)$;*
(b) $\sum_0^\infty \|T^n(x)\|^{-1} < \infty$;
(c) X *is a Hilbert space and* $\sum_0^\infty \|T^n(x)\|^{-2} < \infty$.

PROOF   In the first case, the set $\{T^n(x); \ n \in \mathbb{N}\}$ is weakly closed in X by Corollary 10.6. In the other two cases, the result follows from Ball's theorem. $\square$

In the same spirit, the next proposition is a kind of Angle Criterion for weak supercyclicity (see Chapter 9). We call it the **Weak Angle Criterion**.

PROPOSITION 10.17   *Let X be a separable Banach space, and let $T \in \mathfrak{L}(X)$. Also, let $x \in X$ and let $a > 0$. Assume that one can find some non-zero linear functional $x^* \in X^*$ such that*

$$\sum_{n=0}^\infty \left(\frac{|\langle x^*, T^n(x) \rangle|}{\|T^n(x)\|} \right)^a < \infty.$$

Then x is not a weakly supercyclic vector for T provided that $a < p(X^*)$, or $a = 1$, or $a \leq 2$ and X is a Hilbert space.

The proof relies on the following lemma.

LEMMA 10.18   *Let (x_n) be a sequence in X, and let $x \in X \setminus \{0\}$. Assume that x is in the weak closure of $\mathbb{K} \cdot \{x_n; \ n \geq 0\}$. Also, let $x^* \in X^*$ be a linear functional such that $\langle x^*, x \rangle = 1$. Then x is in the weak closure of*

$$\left\{ \frac{x_n}{\langle x^*, x_n \rangle}; \ n \in \mathbb{N}, \ \langle x^*, x_n \rangle \neq 0 \right\}.$$

PROOF   The point x is the weak limit of some net of the form $(\lambda_i x_{n_i})$, where $\lambda_i \in \mathbb{K}$ and $n_i \in \mathbb{N}$. Then $\langle x^*, x_{n_i} \rangle \neq 0$ eventually, and

$$\frac{x_{n_i}}{\langle x^*, x_{n_i} \rangle} = \frac{\lambda_i x_{n_i}}{\langle x^*, \lambda_i x_{n_i} \rangle} \overset{w}{\longrightarrow} x. \qquad \square$$

PROOF OF PROPOSITION 10.17   Choose any vector $z \in X \setminus (\mathbb{K} \cdot O(x, T))$ such that $\langle x^*, z \rangle = 1$ (this can be done since X is infinite-dimensional). If x were a weakly supercyclic vector for T then z would belong to the weak closure of the set

$$\left\{ \frac{T^n(x)}{\langle x^*, T^n(x) \rangle}; \ n \geq 0 \right\},$$

by Lemma 10.18. This is impossible, since the latter set is weakly closed and does not contain z. $\qquad \square$

10.2 Weak dynamics of weighted shifts

In this section, we consider in some detail weakly hypercyclic and weakly super-cyclic weighted shifts. By a **weight sequence**, we shall always mean a bounded sequence of positive numbers. If $\mathbf{w} = (w_n)_{n \in I}$ is a weight sequence ($I = \mathbb{N}^*$ or $\mathbb{Z}$), then the weighted backward shift $B_{\mathbf{w}}$ is well defined and bounded on any space $\ell^p(I)$, $1 \leq p < \infty$. In what follows, we will always denote by $(e_i)_{i \in I}$ the canonical basis of $\ell^p(I)$ and by (e_i^*) the associated sequence of coordinate functionals.

10.2.1 When weak is not so weak

As already stated in the introduction to this chapter, the first question that comes to mind is whether there exist weakly hypercyclic weighted shifts which are not hyper-cyclic. Examples of such operators will be given in the next subsection. However, there are some limitations, which we will discuss now.

First, we show that there is no hope of finding an example in the class of *unilateral* weighted shifts.

PROPOSITION 10.19   *Let $\mathbf{w} = (w_n)_{n \geq 1}$ be a weight sequence, and let $B_{\mathbf{w}}$ be the associated backward shift acting on $\ell^p(\mathbb{N})$, $1 \leq p < \infty$. Then $B_{\mathbf{w}}$ is weakly hypercyclic if and only if it is hypercyclic.*

PROOF Assume that $B_{\mathbf{w}}$ is weakly hypercyclic with weakly hypercyclic vector $x \in \ell^p(\mathbb{N})$. Since the set $\{z; \ |\langle e_0^*, z \rangle| > 1\}$ is weakly open in $\ell^p(\mathbb{Z})$, one can find an increasing sequence of integers (n_k) such that $|\langle B_{\mathbf{w}}^{n_k}(x), e_0 \rangle| > 1$ for all $k \in \mathbb{N}$. Then $w_1 \cdots w_{n_k} |x_{n_k}| > 1$ for all k, hence $w_1 \cdots w_{n_k} \to \infty$ because $x_{n_k} \to 0$. By Salas' characterization of hypercyclicity for unilateral shifts (Theorem 1.40) it follows that $B_{\mathbf{w}}$ is hypercyclic. $\qquad\square$

There are also limitations for bilateral weighted shifts acting on $\ell^p(\mathbb{Z})$, but here the situation is more involved because it depends on the exponent p. More precisely, we shall prove the following results, which are due to S. Shkarin [224]. Part (1) of the proposition should be compared with Corollary 10.28 below, and part (2) with Corollary 10.32.

PROPOSITION 10.20 *Let $B_{\mathbf{w}}$ be a weighted backward shift acting on $\ell^p(\mathbb{Z})$, $1 \le p < \infty$.*

(a) *If $B_{\mathbf{w}}$ is weakly hypercyclic and $p < 2$ then $B_{\mathbf{w}}$ is hypercyclic.*

(b) *If $B_{\mathbf{w}}$ is weakly supercyclic and $p \le 2$ then $B_{\mathbf{w}}$ is supercyclic.*

For the proof, we first need a lemma which gives an equivalent formulation of Salas' criteria for the supercyclicity or hypercyclicity of bilateral shifts (see Corollary 1.39).

LEMMA 10.21 *Let $B_{\mathbf{w}}$ be a bilateral weighted backward shift acting on $\ell^p(\mathbb{Z})$, with weight sequence $\mathbf{w} = (w_n)_{n \in \mathbb{Z}}$. Then $B_{\mathbf{w}}$ is hypercyclic iff*

$$\forall q \in \mathbb{N} \ : \ \liminf_{n \to +\infty} \max\left\{ (w_{q+1} \cdots w_{q+n})^{-1}, (w_q \cdots w_{q-n+1}) \right\} = 0 \,; \quad (10.2)$$

and $B_{\mathbf{w}}$ is supercyclic iff

$$\forall q \in \mathbb{N} \ : \ \liminf_{n \to +\infty} \left((w_{q+1} \cdots w_{q+n})^{-1} (w_q \cdots w_{q-n+1}) \right) = 0 \,. \quad (10.3)$$

PROOF If $q \in \mathbb{N}$ is given and if $n > q$ then we may write

$$w_{q+1} \cdots w_{q+n} = \frac{w_1 \cdots w_{q+n}}{w_1 \cdots w_q}$$

and similarly

$$w_q \cdots w_{q-n+1} = (w_0 \cdots w_{-n+q+1}) (w_q \cdots w_1).$$

From Salas' conditions, the result follows immediately. $\qquad\square$

For the hypercyclic part of Proposition 10.20, we will also need an elementary lemma on infinite matrices with positive coefficients.

LEMMA 10.22 *Let Λ be a countably infinite set, and let $(c_{\alpha,\beta})_{(\alpha,\beta)\in\Lambda^2}$ be a matrix with positive entries. Assume that $\max(c_{\alpha,\beta}, c_{\beta,\alpha}) \geq \delta > 0$ for every $(\alpha, \beta) \in \Lambda^2$ and some positive constant δ. Then*

$$\sum_{\alpha\in\Lambda}\left(\sum_{\beta\in\Lambda} c_{\alpha,\beta}\right)^{-r} < \infty \quad \text{for any } r > 1.$$

PROOF   We may assume that $\delta = 1$, so that $c_{\alpha,\beta} + c_{\beta,\alpha} \geq 1$ for any $\alpha, \beta \in \Lambda$. For each $\alpha \in \Lambda$, let us put $S_\alpha := \sum_{\beta\in\Lambda} c_{\alpha,\beta} \in (0,\infty]$.

We note that, for any positive integer $M > 0$, the set $\Lambda_M := \{\alpha;\ S_\alpha \leq M\}$ has cardinality at most $2M$. Indeed, if I is any finite subset of Λ_M then

$$M|I| \geq \sum_{\alpha\in I} S_\alpha \geq \sum_{\alpha,\beta\in I} c_{\alpha,\beta} = \frac{1}{2}\sum_{\alpha,\beta\in I}(c_{\alpha,\beta} + c_{\beta,\alpha}) \geq \frac{|I|^2}{2},$$

so that $|I| \leq 2M$. Using that with $M = 2^j$, $j \geq 0$, we get

$$\sum_{\alpha\in\Lambda} S_\alpha^{-r} \leq \sum_{S_\alpha \leq 1} S_\alpha^{-r} + \sum_{j=1}^{\infty}\ \sum_{2^{j-1} < S_\alpha \leq 2^j} S_\alpha^{-r}$$

$$\leq 2 + \sum_{j=1}^{\infty} 2 \times 2^j \times \frac{1}{2^{r(j-1)}},$$

which concludes the proof since $r > 1$.   □

PROOF OF PROPOSITION 10.20   We first make some reductions. By Salas' criteria, the norm hypercyclicity or supercyclicity of $B_{\mathbf{w}}$ on $\ell^p(\mathbb{Z})$ does not depend on p. Moreover, it is easily checked that weak hypercyclicity or supercyclicity on $\ell^{p_1}(\mathbb{Z})$ implies the same property on $\ell^{p_2}(\mathbb{Z})$ for any $p_2 \geq p_1$, since $\ell^{p_1} \subset \ell^{p_2}$ and $\ell^{p_2^*} \subset \ell^{p_1^*}$. Therefore it is enough to consider the case $p = 2$ to prove (a), and we may assume that $p > 1$ in (b).

To prove (b), assume that $B_{\mathbf{w}}$ is weakly supercyclic yet not supercyclic on $\ell^2(\mathbb{Z})$. By Lemma 10.21, one can find $\delta > 0$ and $q \in \mathbb{N}$ such that

$$(w_{q+1}\cdots w_{q+n})^{-1}(w_q\cdots w_{q-n+1}) \geq \delta$$

for all $n \geq 1$. Pick some weakly supercyclic vector $x \in \ell^2(\mathbb{Z})$ with $\langle e_q^*, x\rangle \neq 0$. Such a vector exists because the set of weakly supercyclic vectors is weakly dense in $\ell^2(\mathbb{Z})$ (the projective orbit of any weakly supercyclic vector is made up entirely of weakly supercyclic vectors, excluding the zero vector). Since $\|B_{\mathbf{w}}^n(x)\| \geq w_q\cdots w_{q-n+1}|x_q|$, we get

$$\frac{|\langle e_q^*, B_{\mathbf{w}}^n(x)\rangle|}{\|B_{\mathbf{w}}^n(x)\|} \leq \frac{w_{q+1}\cdots w_{q+n}\,|x_{q+n}|}{w_q\cdots w_{q-n+1}\,|x_q|} \leq \frac{|x_{q+n}|}{\delta\,|x_q|}$$

for all $n \geq 1$. Since $x \in \ell^2(\mathbb{Z})$, we conclude that

$$\sum_{n=0}^{\infty}\left(\frac{|\langle e_q^*, B_{\mathbf{w}}^n(x)\rangle|}{\|B_{\mathbf{w}}^n(x)\|}\right)^2 < \infty,$$

which contradicts Proposition 10.17.

To prove (a), assume that $p > 1$ and that $B_{\mathbf{w}}$ is weakly hypercyclic on $\ell^p(\mathbb{Z})$ but not hypercyclic. By Lemma 10.21, one can find $q \in \mathbb{N}$ and $\delta > 0$ such that

$$\max\left((w_{q+1}\cdots w_{q+k})^{-1}, w_q\cdots w_{q-k+1}\right) \geq \delta$$

for all $k \geq 1$. Let $x \in \ell^p(\mathbb{Z})$ be a weakly hypercyclic vector for $B_{\mathbf{w}}$ and put

$$\Lambda := \left\{n \geq 1;\ |\langle e_q^*, B_{\mathbf{w}}^n(x)\rangle| > 1\right\}.$$

Then on the one hand Λ is an infinite subset of $\mathbb{N}$ because $U := \{z;\ |\langle e_q^*, z\rangle| > 1\}$ is weakly open in $\ell^p(\mathbb{Z})$. Moreover, the set $\{B_{\mathbf{w}}^n(x);\ n \in \Lambda\}$ cannot be weakly closed in $\ell^p(\mathbb{Z})$, since it is dense in the weakly open set U.

On the other hand, the definition of Λ gives $w_{q+1}\cdots w_{q+m}|x_{q+m}| \geq 1$ for all $m \in \Lambda$. It follows that, for any $n \in \Lambda$,

$$\|B_{\mathbf{w}}^n(x)\|^p \geq \sum_{m\in\Lambda} (w_{q+m-n+1}\cdots w_{q+m})^p|x_{q+m}|^p$$

$$\geq \sum_{m\in\Lambda, m\leq n} (w_{q-(n-m)+1}\cdots w_q)^p + \sum_{m\in\Lambda, m>n} \frac{1}{(w_{q+1}\cdots w_{q+m-n})^p}.$$

Applying Lemma 10.22 to the infinite matrix $(c_{n,m})_{(n,m)\in\Lambda^2}$ defined by

$$\begin{cases} c_{n,m} := (w_{q-(n-m)+1}\cdots w_q)^p & \text{if } n \geq m, \\ c_{n,m} := (w_{q+1}\cdots w_{q+m-n})^{-p} & \text{if } m < n, \end{cases}$$

we infer that the series $\sum_{n\in\Lambda} \|B_{\mathbf{w}}^n(x)\|^{-rp}$ is convergent for any $r > 1$.

Now, since $p < 2$ one may choose $r > 1$ such that $a := rp < 2$; since $(\ell^p(\mathbb{Z}))^* = \ell^{p^*}(\mathbb{Z})$ has type 2 we may call on Theorem 10.5 to conclude that the set $\{B_{\mathbf{w}}^n(x);\ n \in \Lambda\}$ is weakly closed in ℓ^p, a contradiction. $\square$

10.2.2 Weakly hypercyclic weighted shifts

According to the results of the previous subsection, we should now concentrate on bilateral shifts acting on $\ell^p(\mathbb{Z})$, $p \geq 2$. In that setting, several conditions ensuring weak hypercyclicity have been discovered, by K. C. Chan and R. Sanders [72] and by S. Shkarin [224]. We provide here another sufficient set of conditions. Before stating the result, we introduce some terminology.

DEFINITION 10.23 *Let $(n_k)_{k\geq 0} \subset \mathbb{N}$ be an increasing sequence of integers.*

(a) *The sequence (n_k) is said to be a **Sidon sequence** if all sums $n_k + n_l$ with $k < l$ are distinct.*

(b) *Given a sequence of natural numbers $(\Delta_l)_{l\in\mathbb{N}}$, the sequence (n_k) is said to be (Δ_l)-**Sidon** if the sets of natural numbers*

$$\mathbf{J}_l := \mathbb{N} \cap \bigcup_{k\leq l} [n_k + n_l - \Delta_l, n_k + n_l + \Delta_l]$$

are pairwise disjoint.

Sidonicity is a very classical notion of lacunarity for sequences of integers, and (Δ_l)-Sidonicity is a rather natural quantitative strengthening. It should be clear from the beginning that any fast increasing sequence (n_k), e.g. $n_k = 3^k$, is (Δ_l)-Sidon, for some sequence (Δ_l) tending to infinity. For example, if $n_0 = 0$, $n_1 \geq 4$ and $n_{k+1} \geq 3n_k$ for all $k \geq 1$ then (n_k) is (l)-Sidon. Indeed, each set $\mathbf{J}_l$ is contained in $[n_l - l, 2n_l + l]$ and the latter intervals are pairwise disjoint because $n_l > 2l + 1$ for all $l \in \mathbb{N}$.

We also point out one less pleasing detail: by definition, a sequence (n_k) is (0)-Sidon iff the sets $\{n_k + n_l; \ k \leq l\}$ are pairwise disjoint. Thus, we do not recover exactly the classical definition of a Sidon sequence but a slightly stronger property. However, the definition of (Δ_l)-Sidon sequences will be needed as given above.

Here is now our main tool for producing weakly hypercyclic weighted shifts.

THEOREM 10.24 *Let* $\mathbf{w} = (w_n)_{n \in \mathbb{Z}}$ *be a weight sequence with* $\inf_n w_n > 0$, *and let* $p \geq 2$. *Assume that there exists an increasing sequence of integers* (n_k) *with* $n_0 = 0$ *such that*

(1) (n_k) *is* (Δ_l)-*Sidon, for some sequence* (Δ_l) *tending to infinity;*

(2) *the series* $\displaystyle\sum_{k \geq 1} (w_1 \cdots w_{n_k})^{-p}$ *is convergent;*

(3) $\displaystyle\sum_{l=0}^{\infty} \left(\sum_{k<l} (w_{-1} \cdots w_{-(n_l - n_k)})^p + \sum_{k>l} (w_1 \cdots w_{n_k - n_l})^{-p} \right)^{-p*/p} = \infty.$

Then $B_{\mathbf{w}}$ *is weakly hypercyclic on* $\ell^p(\mathbb{Z})$.

REMARK The assumption $\inf_n w_n > 0$ renders the shift $B_{\mathbf{w}}$ invertible. Moreover, condition (2) forces the convergence of all series $\sum_{k>l} (w_1 \cdots w_{n_k - n_l})^{-p}$, $l \in \mathbb{N}$, because $(w_1 \cdots w_{n_k - n_l})^{-p} \leq (\sup_n w_n)^{pn_l} (w_1 \cdots w_{n_k})^{-p}$. Therefore, condition (3) makes sense provided that (2) holds.

The proof of Theorem 10.24 relies on the following lemma.

LEMMA 10.25 *Let* $(z(l))_{l \in \mathbb{N}}$ *be a sequence of disjointly supported vectors in* $\ell^p(\mathbb{Z})$, $1 < p < \infty$. *Assume that* $\sum_0^{\infty} \|z(l)\|^{-p^*} = \infty$. *Then the weak closure of* $\{z(l); \ l \in \mathbb{N}\}$ *in* $\ell^p(\mathbb{Z})$ *contains* 0.

PROOF Any normalized sequence of disjointly supported vectors in $\ell^p(\mathbb{Z})$ is obviously a p-*sequence* in the sense of Definition 10.8. Thus, the result follows from Corollary 10.10 applied to $g_l := z(l)/\|z(l)\|$ and $\alpha_l := 1/\|z(l)\|$. $\square$

PROOF OF THEOREM 10.24 As pointed out in the above remark, condition (2) implies the convergence of all series $\sum_{k>l} (w_1 \cdots w_{n_k - n_l})^{-p}$, $l \in \mathbb{N}$. From this and condition (3), it is not hard to deduce the following fact. The detailed proof is rather tedious and does not add anything much, so we omit it.

FACT One can find a non-decreasing sequence of positive integers (M_k) tending to infinity such that

$$\sum_{k>l}^{\infty} M_k \left(w_1 \cdots w_{n_k - n_l}\right)^{-p} < \infty \quad \text{for all } l \geq 0$$

and

$$\sum_{l=0}^{\infty} \left(\sum_{k<l} M_k \left(w_{-1} \cdots w_{-(n_l - n_k)}\right)^p + \sum_{k>l} M_k \left(w_1 \cdots w_{n_k - n_l}\right)^{-p}\right)^{-p^*/p} = \infty.$$

Let us fix a sequence (M_k) satisfying the above properties. We may decompose $\mathbb{N}$ as $\mathbb{N} = \bigcup_{m=0}^{\infty} I_m$, where the I_m are infinite and pairwise disjoint and $\min(I_m) \geq m$ for all m, in such a way that

$$\sum_{l \in I_m} \left(\sum_{k<l} M_k \left(w_{-1} \cdots w_{-(n_l - n_k)}\right)^p + \sum_{k>l} M_k \left(w_1 \cdots w_{n_k - n_l}\right)^{-p}\right)^{-p^*/p} = \infty$$

for every $m \in \mathbb{N}$.

We may assume that $\Delta_l \geq 1$ for all l and that the sequence (Δ_l) is non-decreasing. Moreover, taking Δ_k smaller if necessary (which does not affect the property "(n_k) is (Δ_l)-Sidon"), we may also assume that $n_k \geq \Delta_k$ for all $k \geq 1$. Then, since $n_{k+1} > n_k$ and $[n_k - \Delta_k, n_k + \Delta_k]$ is contained in $\mathbf{J}_k$ we must have

$$n_{k+1} - \Delta_{k+1} > n_k + \Delta_k \qquad (10.4)$$

(recall that $n_0 = 0$).

Next, we choose an increasing sequence of integers (K_l) such that

$$\sum_{k>K_l} M_k \left(w_1 \cdots w_{n_k - n_l}\right)^{-p} \xrightarrow{l \to \infty} 0.$$

Having fixed (K_l), we choose a non-decreasing sequence of positive integers (d_k) tending to infinity such that the following implication holds for all $k, l \in \mathbb{N}$:

$$d_k \geq \frac{n_{l+1} - n_l}{2} \quad \Longrightarrow \quad k > K_l. \qquad (10.5)$$

Moreover, we also assume that $d_k \leq \Delta_k$ and that

$$C(\mathbf{w})^{d_k p} \leq M_k^{1/2}$$

for all $k \in \mathbb{N}$, where we have put $C(\mathbf{w}) := (\sup_n w_n)/(\inf_n w_n)$. Clearly, this is not a restriction since if (10.5) is true for some d_k then it is true for any smaller d_k.

Now, let $\mathbf{Q}$ be a countable dense subset of $\mathbb{K}$ and let $\mathcal{D} \subset \ell^p(\mathbb{N})$ be the the set of all finitely supported vectors with coordinates in $\mathbf{Q}$. We choose an enumeration $(x(l))_{l \in \mathbb{N}}$ of the countable set $\mathcal{D}$ satisfying the following properties:

(i) $l \mapsto x(l)$ is constant on each set I_m;
(ii) $x(l)$ is supported on $(-d_l, d_l)$;
(iii) $\|x(l)\|^p \leq M_l^{1/2}$.

Such a sequence $(x(l))$ can be easily obtained by first finding an enumeration $(x'(l))$ of $\mathcal{D}$ satisfying conditions (ii) and (iii) and then putting $x(l) := x'(m)$ whenever $l \in I_m$. This does not affect (ii) and (iii) because $\min(I_m) \geq m$ and the sequences (M_l), (d_l) are non-decreasing.

We define a sequence $(y(k))_{k \geq 1} \subset \ell^p(\mathbb{Z})$ by setting

$$y_i(k) := \begin{cases} (w_{i-n_k+1} \cdots w_i)^{-1} x_{i-n_k}(k) & \text{if } |i - n_k| < d_k, \\ 0 & \text{otherwise.} \end{cases}$$

In other words $y(k) = S_{\mathbf{w}}^{n_k} x(k)$, where $S_{\mathbf{w}} = B_{\mathbf{w}}^{-1}$ (this should look familiar). Thus, $B_{\mathbf{w}}^{n_k} y(k) = x(k)$ for all $k \geq 1$. Since the intervals $(n_k - d_k, n_k + d_k)$ are pairwise disjoint, the vectors $y(k)$ have disjoint supports. Moreover,

$$\|y(k)\|^p \leq \|x(k)\|^p C(\mathbf{w})^{d_k p} \frac{1}{(w_1 \cdots w_{n_k})^p} \leq \frac{M_k}{(w_1 \cdots w_{n_k})^p}$$

for all k. Thus, we may define

$$y := \sum_{k=1}^{\infty} y(k).$$

Now we will show that y is the weakly hypercyclic vector for which we are looking.

If $x \in \mathcal{D}$ and if we choose $m \in \mathbb{N}$ such that $x(l) = x$ when $l \in I_m$ then

$$\begin{aligned} B_{\mathbf{w}}^{n_l}(y) &= \sum_{k<l} B_{\mathbf{w}}^{n_l} y(k) + x + \sum_{k>l} B_{\mathbf{w}}^{n_l} y(k) \\ &= \sum_{k<l} B_{\mathbf{w}}^{n_l} y(k) + \sum_{\substack{k>l \\ d_k \leq (n_{l+1}-n_l)/2}} B_{\mathbf{w}}^{n_l} y(k) + x + \sum_{\substack{k>l \\ d_k > (n_{l+1}-n_l)/2}} B_{\mathbf{w}}^{n_l} y(k) \\ &:= z^-(l) + z^+(l) + x + u(l) \end{aligned}$$

for all $l \in I_m$. Thus, we just need to prove that, for any fixed $m \in \mathbb{N}$, the weak closure of the set $\{z^-(l) + z^+(l) + u(l); \ l \in I_m\}$ in $\ell^p(\mathbb{Z})$ contains 0. The key point is the following

FACT The vectors $z(l) := z^-(l) + z^+(l)$ are disjointly supported.

PROOF OF THE FACT We have to show that the vectors $z^-(l)$ are disjointly supported, as well as the vectors $z^+(l)$, and furthermore that $z^-(l)$ and $z^+(l')$ are disjointly supported for any l, l'. We note that $z^-(l)$ is supported on

$$E^-(l) := \bigcup_{k<l} (-n_l + n_k - d_k, -n_l + n_k + d_k)$$

and that $z^+(l)$ is supported on

$$E^+(l) := \bigcup_{k \in \Lambda_l} (-n_l + n_k - d_k, -n_l + n_k + d_k)$$

where $\Lambda_l = \{k > l; \ 2d_k \leq n_{l+1} - n_l\}$.

If $E^-(l) \cap E^-(l') \neq \varnothing$ then one can find $k < l$, $k' < l'$, $j \in (-d_l, d_l)$ and $j' \in (-d_{l'}, d_{l'})$ such that $n_l - n_k + j = n_{l'} - n_{k'} + j'$, i.e. $n_{l'} + n_k + j' = n_l + n_{k'} + j$. Assuming $l < l'$ and remembering that $d_l \leq \Delta_l$, it follows that (see Definition 10.23) $\mathbf{J}_{l'} \cap \mathbf{J}_{\max(l,k')} \neq \varnothing$, a contradiction.

Assume that $E^+(l) \cap E^+(l') \neq \varnothing$, with $l < l'$. Then one can find k, k', j, j' such that $n_k - n_l + j = n_{k'} - n_{l'} + j'$, with $k \in \Lambda_l$, $k' \in \Lambda_{l'}$, $|j| < d_k$ and $|j'| \leq d_{k'}$. Since (n_k) is (Δ_l)-Sidon, this is possible only if $k = k'$. Then we have $j' - j = n_{l'} - n_l$, so that $n_{l'} - n_l < d_k + d_{k'}$. But $d_k + d_{k'} \leq 2d_k \leq n_{l+1} - n_l \leq n_{l'} - n_l$ by definition of $\Lambda_{l'}$, so we have a contradiction.

Finally, if $k < l$ then $-n_l + n_k + d_k \leq -n_l + n_{l-1} + d_{l-1} < 0$, because of (10.4), and if $k > l'$ then $-n_{l'} + n_k - d_k > 0$. Thus, $E^-(l)$ is contained in $(-\infty, 0)$ and $E^+(l')$ is contained in $(0, +\infty)$, so that $z^-(l)$ and $z^+(l')$ are disjointly supported for any l, l'. □

We now show that $\|u(l)\| \to 0$ as $l \to \infty$ and that $\sum_{l \in I_m} \|z(l)\|^{-p^*} = \infty$. In view of Lemma 10.25, this will ensure that $0 \in \overline{\{z^-(l) + z^+(l) + u(l); \ l \in I_m\}}^w$, and the proof will be complete.

First, we observe that, for each fixed $l \in I_m$, the vectors $B_{\mathbf{w}}^{n_l} y(k)$ are disjointly supported. Therefore we get

$$
\|z^-(l)\|^p = \sum_{k<l} \sum_{|i|<d_k} (w_i \cdots w_{i+n_k-n_l+1})^p |x_i(k)|^p
$$
$$
\leq \sum_{k<l} C(\mathbf{w})^{d_k p} M_k^{1/2} (w_{-1} \cdots w_{-n_l+n_k})^p
$$
$$
\leq \sum_{k<l} M_k (w_{-1} \cdots w_{-(n_l-n_k)})^p .
$$

Similarly,

$$
\|z^+(l)\|^p \leq \sum_{k>l} M_k (w_1 \cdots w_{n_k-n_l})^{-p} .
$$

Thus, we have $\sum_{l \in I_m} \|z(l)\|^{-p^*} = \infty$, by the choice of the sequence (M_k). Finally, it follows from (10.5) that

$$
\|u(l)\|^p \leq \sum_{k>K_l} M_k (w_1 \cdots w_{n_k-n_l})^{-p}
$$

for all $l \in I_m$. By the choice of the sequence (K_l), this concludes the proof. □

We point out two perhaps more user-friendly consequences of Theorem 10.24.

COROLLARY 10.26 *Let $B_{\mathbf{w}}$ be an invertible weighted shift on $\ell^p(\mathbb{Z})$, $p \geq 2$. Assume that there exists a sequence of integers (n_k) with $n_0 = 0$ which is (Δ_l)-Sidon for some sequence (Δ_l) tending to infinity and is such that the following properties hold.*

(1) $\sup\limits_{l\geq 0}\sum_{k>l}(w_1\cdots w_{n_k-n_l})^{-p}<\infty;$

(2) $\sum\limits_{l=1}^{\infty}\left(\sum_{k<l}(w_{-1}\cdots w_{-(n_l-n_k)})^p\right)^{-p^*/p}=\infty.$

Then $B_{\mathbf{w}}$ is weakly hypercyclic.

Looking at the proof of Theorem 10.24, condition (1) of Corollary 10.26 says that the sequence $(z^+(l))$ considered therein is bounded and hence weakly convergent to 0 since the vectors $z^+(l)$ are disjointly supported.

COROLLARY 10.27 *Let $\mathbf{w}=(w_n)_{n\in\mathbb{Z}}$ be a weight sequence with $\inf_n w_n>0$, and let $p\geq 2$. Assume that $\limsup_{n\to+\infty}w_1\cdots w_n=\infty$ and $\sup_{n\geq 1}w_{-1}\cdots w_{-n}<\infty$. Then $B_{\mathbf{w}}$ is weakly hypercyclic on $\ell^p(\mathbb{Z})$.*

PROOF   It is not hard to construct by induction a sequence of integers (n_k) with $n_0=0$, $n_1=4$, $n_{k+1}\geq 3n_k$ for all k and $w_1\cdots w_{n_k-n_l}\geq 2^k$ whenever $l<k$. Indeed, if n_{k-1} has already been defined, it is enough to select a large integer n_k such that

$$w_1\cdots w_{n_k}\geq 2^k\|w\|_\infty^{n_{k-1}}.$$

If $l<k$ then we have

$$w_1\cdots w_{n_k-n_l}=\frac{w_1\cdots w_{n_k}}{w_{n_k-n_l+1}\cdots w_{n_k}}\geq\frac{2^k\|w\|_\infty^{n_{k-1}}}{\|w\|_\infty^{n_l}}\geq 2^k.$$

As observed just after the definition of a Sidon sequence, the sequence (n_k) is (l)-Sidon. Moreover, it is easy to verify assumptions (1), (2) of Corollary 10.26. Indeed,

$$\sum_{k>l}\frac{1}{(w_1\cdots w_{n_k-n_l})^p}\leq\sum_{k>l}\frac{1}{2^{kp}}\leq 2$$

for all $l\in\mathbb{N}$, and

$$\sum_{l=1}^{\infty}\left(\sum_{k<l}(w_{-1}\cdots w_{-(n_l-n_k)})^p\right)^{-p^*/p}\geq\sup_{n\geq 0}(w_{-1}\cdots w_{-n})^{-p^*}\sum_{l=1}^{\infty}l^{-p^*/p}=\infty,$$

since $p^*/p\leq 1$ when $p\geq 2$.   □

Corollary 10.27 allows us (finally!) to exhibit a weakly hypercyclic operator which is not hypercyclic. The existence of such operators was first proved by K. C. Chan and R. Sanders [72].

COROLLARY 10.28   *There exists a weakly hypercylic weighted shift $B_{\mathbf{w}}$ on $\ell^2(\mathbb{Z})$ which is not hypercyclic. For example, one may take $w_n:=2$ for $n\geq 1$ and $w_n:=1$ for $n\leq 0$.*

As another application of Theorem 10.24, we now intend to show that the class of weakly hypercyclic weighted shifts on $\ell^p(\mathbb{Z})$ does depend on p when $p>2$. We first need a lemma saying that $w_{-1} \cdots w_{-n}$ is controlled in some way by $w_1 \cdots w_n$ when the weighted shift $B_{\mathbf{w}}$ is weakly hypercyclic.

LEMMA 10.29   *Assume that the weighted shift $B_{\mathbf{w}}$ is weakly hypercyclic on $\ell^p(\mathbb{Z})$, $p \geq 2$. Moreover, assume that $w_{-1} \cdots w_{-n} \geq (w_1 \cdots w_n)^\alpha$ for some $\alpha > 0$ and all $n \geq 0$. Then $\alpha \leq p/p^*$.*

PROOF   Let us fix a weakly hypercyclic vector $x \in \ell^p(\mathbb{Z})$ for $B_{\mathbf{w}}$. Since the weakly hypercyclic vectors are weakly dense in $\ell^p(\mathbb{Z})$, we may assume that $x_0 \neq 0$.

Since x is weakly hypercyclic for $B_{\mathbf{w}}$, the basis vector e_0 belongs to the weak closure of the set $\{B_{\mathbf{w}}^n(x); \ n \in \mathbf{N}\}$, where

$$\mathbf{N} := \left\{n \in \mathbb{N}; \ |\langle e_0^*, B_{\mathbf{w}}^n(x)\rangle| > 1/2\right\}.$$

By Lemma 10.18, it follows that e_0 is also in the weak closure of the set

$$\left\{ \frac{B_{\mathbf{w}}^n(x)}{\langle e_0^*, B_{\mathbf{w}}^n(x)\rangle}; \ n \in \mathbf{N} \right\} := Z.$$

In particular, Z is not weakly closed in $\ell^p(\mathbb{Z})$. Hence, by Corollary 10.7 we know that

$$\sum_{n \in \mathbf{N}} \left(\frac{|\langle e_0^*, B_{\mathbf{w}}^n(x)\rangle|}{\|B_{\mathbf{w}}^n(x)\|} \right)^a = \infty$$

for any $a < p^*$. Since $\|B_{\mathbf{w}}^n(x)\| \geq w_{-1} \cdots w_{-n}|x_0|$ and $\langle e_0^*, B_{\mathbf{w}}^n x\rangle = w_1 \cdots w_n x_n$ for all $n \in \mathbb{N}$, this leads to

$$\infty = \sum_{n \in \mathbf{N}} \left(\frac{w_1 \cdots w_n}{w_{-1} \cdots w_{-n}} \right)^a |x_n|^a \leq \sum_{n \in \mathbf{N}} (w_1 \cdots w_n)^{a(1-\alpha)} |x_n|^a.$$

Putting $p = ra$ (so that $r > 1$), applying Hölder's inequality to the right-hand side and keeping in mind that $x \in \ell^p(\mathbb{Z})$, we arrive at

$$\sum_{n \in \mathbf{N}} (w_1 \cdots w_n)^{r^* a(1-\alpha)} = \infty. \tag{10.6}$$

However, we also have $|w_1 \cdots w_n x_n| \geq 1/2$ for all $n \in \mathbf{N}$. Since $x \in \ell^p(\mathbb{Z})$, it follows that

$$\sum_{n \in \mathbf{N}} (w_1 \cdots w_n)^{-p} < \infty. \tag{10.7}$$

Properties (10.6) and (10.7) are compatible only if $r^* a(\alpha - 1) < p$, i.e. $r^*(\alpha - 1) < r$ or, equivalently, $\alpha < r = p/a$. Since a can be chosen arbitrarily close to p^*, we conclude that $\alpha \leq p/p^*$.   □

We can now prove the following result, announced by S. Shkarin in [224].

PROPOSITION 10.30   *Assume that $2 \leq p < q$. Then there exists a weight sequence $\mathbf{w}$ such that $B_{\mathbf{w}}$ is weakly hypercyclic on $\ell^q(\mathbb{Z})$ but not weakly hypercyclic on $\ell^p(\mathbb{Z})$.*

PROOF   Let $(c_{i,j})_{0\leq i<j}$ be an infinite upper-triangular matrix with positive entries, to be specified later. Also, let $\alpha > 0$. If (n_k) is a rapidly increasing sequence of integers, it is possible to construct a weight sequence $\mathbf{w} = (w_n)_{n\in\mathbb{Z}}$ with $\inf_n w_n > 0$ such that

$$w_1 \cdots w_{n_j - n_i} = c_{i,j} \quad \text{whenever } i < j,$$
$$w_{-n} = w_n^\alpha \quad \text{for all } n \geq 1.$$

Indeed, let $(i,j) \mapsto \langle i,j \rangle$ be a bijection from $\Delta := \{(i,j) \in \mathbb{N}^2; \ i < j\}$ onto $\mathbb{N}$, and put $c_{\langle i,j \rangle} := c_{i,j}$. If (n_k) is a Sidon sequence, then the set $\mathbf{D} := \{n_j - n_i; (i,j) \in \Delta\}$ is enumerated in a one-to-one way by Δ. So the problem is to find some sequence of positive numbers $(u_n)_{n\geq 1}$ such that $0 < \inf_n u_{n+1}/u_n \leq \sup_n u_{n+1}/u_n < \infty$ and $u_{d_s} = c_s$ for all $s \in \mathbb{N}$, where (d_s) is the increasing enumeration of $\mathbf{D}$. Clearly, this can be done if $\mathbf{D}$ is sufficiently lacunary.

We fix (n_k) and also a weight sequence $\mathbf{w}$ as described above. Since (n_k) is rapidly increasing, we may assume that it is (Δ_l)-Sidon for some sequence (Δ_l) tending to infinity. Moreover, we may also assume that $n_0 = 0$.

By Lemma 10.29, $B_\mathbf{w}$ is not weakly hypercyclic on $\ell^p(\mathbb{Z})$ provided that $\alpha > p/p^*$. However, we now show that it is possible to choose $(c_{i,j})$ and $\alpha > p/p^*$ in such a way that $B_\mathbf{w}$ is weakly hypercyclic on $\ell^q(\mathbb{Z})$. By Lemma 10.29, this will conclude the proof.

By Theorem 10.24 and the choice of $\mathbf{w}$, it is enough to have

$$\sum_{k=1}^{\infty} c_{0,k}^{-q} < \infty$$

and

$$\sum_{l=0}^{\infty} \left(\sum_{k<l} c_{k,l}^{\alpha q} + \sum_{k>l} c_{l,k}^{-q} \right)^{-q^*/q} = \infty.$$

We now choose $\alpha := r/r^*$, where $r \in (p,q)$. Then $p/p^* < \alpha < q/q^*$. Let us look for a matrix of the form $c_{i,j} = a_i b_j$, where the sequences $(a_n)_{n\geq 0}$ and $(b_n)_{n\geq 0}$ satisfy $\sum_1^{\infty} b_n^{-q} < \infty$ and $\sum_0^{\infty} a_n^{\alpha q} < \infty$. Then

$$\sum_{k=1}^{\infty} c_{0,k}^{-q} = a_0^{-q} \sum_{n=1}^{\infty} b_n^{-q} < \infty$$

and

$$\sum_{k<l} c_{k,l}^{\alpha q} + \sum_{k>l} c_{l,k}^{-q} \leq b_l^{\alpha q} \sum_{n=0}^{\infty} a_n^{\alpha q} + a_l^{-q} \sum_{n=1}^{\infty} b_n^{-q}$$
$$\leq C(b_l^{\alpha q} + a_l^{-q}),$$

for some finite constant C. So we obtain the desired result provided that $\sum_0^{\infty} \left(b_l^{\alpha q^*} + a_l^{-q^*} \right)^{-1} = \infty$. Since $\alpha q^* < q$ and $q^* < \alpha q$, the latter condition is compatible with the convergence of the series $\sum b_l^{-q}$ and $\sum a_l^{\alpha q}$. For example, we may take $a_l := (l+1)^{-\beta}$ and $b_l := (l+1)^\beta$, where $\beta \alpha q^* \leq 1 < q\beta$. $\square$

10.2.3 Weak supercyclicity

The analogue of Theorem 10.24 for weak supercyclicity reads as follows.

THEOREM 10.31   *Let* $\mathbf{w} = (w_n)_{n \in \mathbb{Z}}$ *be a weight sequence with* $\inf_n w_n > 0$, *and let* $p \geq 2$. *Assume that there exists an increasing sequence of integers* $(n_k)_{k \in \mathbb{N}}$ *with* $n_0 = 0$ *and a sequence of positive real numbers* (λ_k) *such that:*

(1) (n_k) *is* (Δ_l)-*Sidon, for some sequence* (Δ_l) *tending to infinity;*

(2) *the series* $\sum_{k \geq 1} \lambda_k (w_1 \cdots w_{n_k})^{-p}$ *is convergent;*

(3) $\displaystyle\sum_{l=0}^{\infty} \left(\sum_{k<l} \frac{\lambda_k}{\lambda_l} (w_{-1} \cdots w_{-(n_l - n_k)})^p + \sum_{k>l} \frac{\lambda_k}{\lambda_l} (w_1 \cdots w_{n_k - n_l})^{-p} \right)^{-p^*/p} = \infty.$

Then $B_{\mathbf{w}}$ *is weakly supercyclic on* $\ell^p(\mathbb{Z})$.

PROOF   The proof closely imitates that of Theorem 10.24, so we will just indicate the necessary changes. We choose a non-decreasing sequence of positive integers (M_k) tending to infinity such that

$$\sum_{k>l}^{\infty} M_k \lambda_k (w_1 \cdots w_{n_k - n_l})^{-p} < \infty \quad \text{for all } l$$

and

$$\sum_{l=0}^{\infty} \left(\sum_{k<l} M_k \frac{\lambda_k}{\lambda_l} (w_{-1} \cdots w_{-(n_l - n_k)})^p + \sum_{k>l} M_k \frac{\lambda_k}{\lambda_l} (w_1 \cdots w_{n_k - n_l})^{-p} \right)^{-p^*/p} = \infty.$$

The sets of natural numbers I_m and the sequence $x((l))$ are the same as in the proof of Theorem 10.24, as well as the vectors $y(k)$. We still assume that $C(\mathbf{w})^{d_l p} \leq M_l^{1/2}$, which yields

$$\|y(k)\|^p \leq \frac{M_k}{(w_1 \cdots w_{n_k})^p} \cdot$$

Therefore, we may define

$$y := \sum_{k=1}^{\infty} \lambda_k^{1/p} y(k).$$

If $x \in \mathcal{D}$ and if $m \in \mathbb{N}$ is chosen in such a way that $x(l) = x$ for all $l \in I_m$ then we may write, as in the proof of Theorem 10.24

$$\frac{1}{\lambda_l^{1/p}} B_{\mathbf{w}}^{n_l}(y) = \sum_{k<l} \left(\frac{\lambda_k}{\lambda_l} \right)^{1/p} B_{\mathbf{w}}^{n_l} y(k) + x + \sum_{k>l} \left(\frac{\lambda_k}{\lambda_l} \right)^{1/p} B_{\mathbf{w}}^{n_l} y(k)$$

$$= z^-(l) + z^+(l) + x + u(l).$$

As before, one checks that $\|u(l)\| \to 0$ and that $0 \in \overline{\{z^-(l) + z^+(l)\}}^w$. Thus, any vector $x \in \mathcal{D}$ can be weakly approximated by vectors of the form $\lambda_l^{-1/p} B^{n_l}(y)$. This concludes the proof. $\square$

COROLLARY 10.32 *Let B be the unweighted bilateral weighted shift on $\ell^p(\mathbb{Z})$. Then B is weakly supercyclic if and only if $p > 2$.*

PROOF The unweighted shift B is not supercyclic, by Salas' theorem. When $p \leq 2$, it follows from Proposition 10.20 that B is not weakly supercyclic either.

Now assume that $p > 2$, i.e. $p^*/p < 1$. We will apply Theorem 10.31 with $n_0 = 0$, $n_k = 4^k$, $k \geq 1$, and any summable sequence of positive numbers (λ_k) such that $\sum_0^\infty \lambda_l^{p^*/p} = \infty$. Conditions (1), (2) are clearly satisfied, and (3) holds as well because $\sum_{k \neq l} \lambda_k/\lambda_l = O(1/\lambda_l)$. $\square$

REMARK We see in particular that on the one hand Banach space isometries can be weakly supercyclic. On the other hand, they cannot be supercyclic; see Exercise 1.4. Of course, an isometry cannot be *weakly hypercyclic* either, since it has bounded orbits.

To conclude this section, we point out that all the results we have proved remain true as stated for shifts acting on $\ell^p(\mathbb{Z})$ with $p = \infty$, provided that we interpret $\ell^\infty(\mathbb{Z})$ as its separable version $c_0(\mathbb{Z})$. For example, there exist weakly hypercyclic weighted shifts on $c_0(\mathbb{Z})$ which are not hypercyclic; the unweighted shift B is weakly supercyclic on $c_0(\mathbb{Z})$ as well.

10.3 Unitary operators

We have just seen that surjective isometries can be weakly supercyclic on $\ell^p(\mathbb{Z})$, $p > 2$. In this section, we show that unitary Hilbert space operators also can be weakly supercyclic.

We note that every weakly supercyclic operator is *cyclic*, since the weak closure of any linear subspace of a Banach space is equal to its norm closure. Now, by the spectral theorem any cyclic unitary operator is unitarily equivalent to the operator M_z corresponding to multiplication by the variable z acting on $L^2(\mu)$, for some Borel probability measure μ on $\mathbb{T}$. Therefore we are in fact interested in the following question: for which probability measures μ on $\mathbb{T}$ (if any) is the operator M_z weakly supercyclic on $L^2(\mu)$?

We start with two examples taken from [28].

EXAMPLE 10.33 Assume that μ has the following property: for each measurable set $\Omega \subset \mathbb{T}$ with $\mu(\Omega) = 1$, one can find a positive integer q and two distinct points $a, b \in \Omega$ such that $a^q = b^q$. Then M_z is not weakly supercyclic on $L^2(\mu)$. This holds in particular if μ is not singular with respect to the Lebesgue measure on $\mathbb{T}$.

PROOF The key point is the following simple observation.

FACT If $\phi \in L^\infty(\mu)$ and if the multiplication operator M_ϕ is cyclic on $L^2(\mu)$ then one can find a measurable set $\Omega \subset \mathbb{T}$ with $\mu(\Omega) = 1$ such that the function ϕ is one-to-one on Ω.

Indeed, assume that M_ϕ is cyclic with cyclic vector f. Then $f(z) \neq 0$ almost everywhere with respect to μ, and any function $g \in L^2(\mu)$ can be approximated in the L^2 norm by functions of the form $P(\phi)f$, where P is a polynomial. Since any L^2-convergent sequence has an almost everywhere convergent subsequence, it follows that one can find a sequence of polynomials (P_k) and a measurable set $\Omega \subset \mathbb{T}$ with $\mu(\Omega) = 1$ such that $f(z) \neq 0$ and $P_k(\phi(z))f(z) \to zf(z)$ for all $z \in \Omega$. Then ϕ is clearly one-to-one on Ω.

Now, assume that M_z is weakly supercyclic on $L^2(\mu)$. Then so are all powers of M_z, by the supercyclic version of Ansari's theorem (see Chapter 3). In particular, all powers of M_z are cyclic on $L^2(\mu)$. By the above fact, one can find $\Omega \subset \mathbb{T}$ with $\mu(\Omega) = 1$ such that all functions z^q, $q \in \mathbb{N}$, are one-to-one on Ω. Thus, μ does not have the property stated above.

To prove the last assertion, assume that μ is non-singular. Then any measurable set $\Omega \subset \mathbb{T}$ with $\mu(\Omega) = 1$ has positive Lebesgue measure. As is well known (see [157] or Exercise 10.13), the set $\Omega^{-1}\Omega$ is then a neighbourhood of the unit element $1 \in \mathbb{T}$. Hence, $\Omega^{-1}\Omega$ contains a rational number $\xi \neq 1$; in other words, one can find $a, b \in \Omega$ with $a \neq b$ such that $a^{-1}b$ is rational. This concludes the proof of Example 10.33. $\qquad\square$

Singularity of the measure μ is far from being a sufficient condition for the operator M_z to be weakly supercyclic. Indeed, one can find singular measures with the property described in Example 10.33, for example the canonical "Lebesgue" measure on the usual Cantor ternary set (considered as a subset of $\mathbb{T}$).

However, it turns out that if the support of μ is very small then M_z is indeed weakly supercyclic on $L^2(\mu)$.

Recall that a compact set $K \subset \mathbb{T}$ is said to be a **Kronecker set** if, for every continuous function $f : K \to \mathbb{T}$, one can find a sequence of positive integers (k_i) tending to infinity such that $z^{k_i} \to f(z)$ uniformly on K. A basic example of Kronecker sets is provided by **Kronecker's theorem**, according to which any finite independent set is Kronecker. (A set $E \subset \mathbb{T}$ is said to be **independent** if the relation $\omega_1^{n_1} \cdots \omega_d^{n_d} = 1$ with $n_j \in \mathbb{Z}$ and pairwise distinct $\omega_j \in E$ can hold only if $n_1 = \cdots = n_d = 0$; see Exercise 11.3 for two proofs of Kronecker's theorem.) From this result, it is easy to prove by Baire category arguments that there exist perfect Kronecker sets (see e.g. [157] or Exercise 10.12). Therefore, there exist *continuous* measures whose support is a Kronecker set (recall that a measure is said to be continuous if every single point has measure 0).

EXAMPLE 10.34   Assume that the measure μ is continuous. If the support of μ is a Kronecker set then M_z is weakly supercyclic on $L^2(\mu)$.

This result is a very simple consequence of the following lemma.

LEMMA 10.35   *If the measure μ is continuous then the measurable functions with constant modulus are weakly dense in $L^2(\mu)$.*

Indeed, assume that μ is continuous and supported on some Kronecker set E. By definition of a Kronecker set, the norm closure of $O(\mathbf{1}, M_z) = \{z^n; \ n \in \mathbb{N}\}$ in $L^2(\mu)$ contains all continuous functions $f : E \to \mathbb{C}$ with constant modulus 1, hence $\overline{\mathbb{C} \cdot O(\mathbf{1}, M_z)}^{\|\cdot\|_{L^2(\mu)}}$ contains all continuous functions with constant modulus. Now, the set E is totally disconnected (since Kronecker sets obviously have empty interior in $\mathbb{T}$), hence any Borel subset of E can be approximated in μ-measure by a *clopen* set. Moreover, every Borel function $f : E \to \mathbb{C}$ with constant modulus can be approximated in the $L^2(\mu)$ norm by a finite sum $\sum_{i=1}^p c_i \mathbf{1}_{Ai}$, where the A_i are Borel sets and the c_i have the same modulus. Approximating each A_i by a clopen set and observing that the characteristic function of a clopen set is continuous, one obtains an approximation of f by a continuous function with constant modulus. Thus the norm closure of $\mathbb{C} \cdot O(\mathbf{1}, M_z)$ in $L^2(\mu)$ contains in fact all Borel functions on E with constant modulus. By the lemma, this shows that M_z is weakly supercyclic with supercyclic vector $\mathbf{1}$.

PROOF OF LEMMA 10.35   Assume that μ is continuous. Then the measure space $(\mathbb{T}, \mu)$ is **non-atomic**, which means that every measurable set $A \subset \mathbb{T}$ with positive measure has a measurable subset B such that $0 < \mu(B) < \mu(A)$. It follows that for any measurable set $A \subset \mathbb{T}$, the range of $\mu_{|A}$ is the whole interval $[0, \mu(A)]$. This is the so-called **Liapounov convexity theorem** in its simplest form (see e.g. [210]).

Let us denote by $\mathcal{F}$ the set of all Borel functions $f : \mathbb{T} \to \mathbb{C}$ with constant modulus. It is enough to prove the following

FACT   Let $\varphi \in L^\infty(\mu)$. If $(A_1, \ldots, A_n)$ is a measurable partition of $\mathbb{T}$ then one can find $f \in \mathcal{F}$ such that $\|f\|_2 \leq 2\|\varphi\|_\infty$ and $\int_{A_i} f \, d\mu = \int_{A_i} \varphi \, d\mu$ for all i.

Indeed, once this is done, it follows that given any function $\varphi \in L^\infty(\mu)$ and any finite family of step functions $(h_1, \ldots, h_k)$, one can find $f \in \mathcal{F}$ with $\|f\|_2 \leq 2\|\varphi\|_\infty$ such that $\int f h_j \, d\mu = \int \varphi h_j \, d\mu$ for all j: just choose a partition $(A_1, \ldots, A_n)$ which is compatible with all functions h_j. Since we have a uniform estimate on $\|f\|_2$, this implies easily that each function $\varphi \in L^\infty(\mu)$ is in the weak closure of $\mathcal{F}$, and the lemma is proved.

To prove the fact, we may obviously assume that $\varphi \neq 0$ and that all sets A_i have positive measure. Put $\alpha_i := \int_{A_i} \varphi \, d\mu$ and $C := 2\|\varphi\|_\infty$. Then

$$\max\left\{\frac{|\alpha_i|}{\mu(A_i)}; \ 1 \leq i \leq n\right\} \leq \|\varphi\|_\infty < C,$$

so one can choose positive numbers $\delta_1, \ldots, \delta_n$ and complex numbers $C_1, \ldots, C_n$ such that $|C_i| = C$, $\delta_i < \mu(A_i)$ and $C_i \delta_i = \alpha_i$ for all i.

Let $i \in \{1, \ldots, n\}$. Since μ is continuous and $\delta_i < \mu(A_i)$, we may use the Liapounov convexity theorem to find a measurable partition $(B_i^+, \widetilde{A}_i, B_i^-)$ of A_i such that $\mu(\widetilde{A}_i) = \delta_i$ and $\mu(B_i^+) = \mu(B_i^-)$. Indeed, we can first choose $\widetilde{A}_i \subset A_i$ such that $\mu(\widetilde{A}_i) = \delta_i$ and next choose $B_i^+ \subset A_i \setminus \widetilde{A}_i$ such that $\mu(B_i^+) = (\mu(A_i) - \delta_i)/2$.

Now, let $f : \mathbb{T} \to \mathbb{C}$ be defined on each set A_i by $f(x) := C_i$ if $x \in \widetilde{A}_i$ and $f(x) := \pm C$ if $x \in B_i^{\pm}$. Then $f \in \mathcal{F}$, $\|f\|_2 = C = 2\|\varphi\|_\infty$ and $\int_{A_i} f \, d\mu = C_i \delta_i = \int_{A_i} \varphi \, d\mu$ for all i. This concludes the proof.

$\square$

REMARK   The continuity assumption on μ is necessary in Example 10.34. Indeed, it is not hard to check that M_z cannot be weakly supercyclic on $L^2(\mu)$ if the discrete part of μ is non-zero, unless μ is a point mass; see Exercise 10.11.

From the point of view of harmonic analysis, Kronecker sets are "very small" sets. At the other extreme, a compact set $K \subset \mathbb{T}$ is usually considered as "large" if it carries a **Rajchman measure**, i.e. a non-zero measure whose Fourier coefficients vanish at infinity. Thus, it is natural to ask whether M_z can be weakly supercyclic on $L^2(\mu)$ if μ is a Rajchman measure. This problem was raised in [28] and solved by S. Shkarin in [224].

THEOREM 10.36   *There exists a probability Rajchman measure μ on $\mathbb{T}$ such that M_z is weakly supercyclic on $L^2(\mu)$.*

For the proof, we need several lemmas.

LEMMA 10.37   *Let $(g_i)_{i \in I}$ be an infinite sequence of vectors in a Hilbert space H, and let $(\alpha_i)_{i \in I}$ be a family of non-zero complex numbers. Assume that the following properties hold.*

(1) *(g_i) is bounded and $\sum_{i \in I} \sum_{j \neq i} |\langle g_j, g_i \rangle|^2 < \infty$;*

(2) *$\sum_{i \in I} |\alpha_i|^2 = \infty$.*

Then 0 belongs to the weak closure of the set $\{\alpha_i^{-1} g_i; \ i \in I\}$.

PROOF   By Corollary 10.10, it is enough to show that (g_i) is a 2-sequence in H, in the sense of Definition 10.8. Thus, we look for an estimate of the form

$$\left\| \sum a_i g_i \right\|^2 \leq C \sum |a_i|^2,$$

for some constant C and all finite sequences of scalars (a_i). This follows easily from (1). Indeed,

$$\left\| \sum a_i g_i \right\|^2 = \sum_i |a_i|^2 \|g_i\|^2 + \sum_{i \neq j} \bar{a}_i a_j \langle g_i, g_j \rangle$$

$$\leq C \sum_i |a_i|^2 + \left(\sum_{i,j} |a_i|^2 |a_j|^2 \right)^{1/2} \left(\sum_{i \neq j} |\langle g_i, g_j \rangle|^2 \right)^{1/2}$$

$$\leq C \sum |a_i|^2$$

where, as usual, the constant C has changed between lines. $\square$

As a very simple consequence of this lemma, we now formulate a useful criterion for weak supercyclicity of the operator M_z. Let us fix a bijection $(p, q) \mapsto \langle p, q \rangle$ from $\mathbb{N} \times \mathbb{N}$ onto $\mathbb{N}$ and, for $i \in \mathbb{N}$, write $i = \langle p_i, q_i \rangle$.

COROLLARY 10.38 *Let* $\mathcal{H} = \{h_p; \; p \in \mathbb{N}\} \subset L^2(\mu)$ *be a countable set such that* $\mathbb{C} \cdot \mathcal{H}$ *is weakly dense in* $L^2(\mu)$. *Also, let* $(\alpha_i)_{i \in \mathbb{N}}$ *be a sequence of non-zero complex numbers such that* $\sum_{q=0}^{\infty} |\alpha_{\langle p, q \rangle}|^2 = \infty$ *for each* $p \in \mathbb{N}$, *and put* $f_i := \alpha_i h_{p_i}$. *Assume that one can find a sequence of positive integers* $(k_i)_{i \in \mathbb{N}}$ *such that* $\sum_i \sum_{j \neq i} |\langle g_i, g_j \rangle|^2 < \infty$, *where* $g_i(z) = z^{k_i} - f_i(z)$. *Then* M_z *is weakly supercyclic on* $L^2(\mu)$, *with supercyclic vector* **1**.

PROOF Putting $I_p := \{i \in \mathbb{N}; \; p_i = p\}$, it follows at once from Lemma 10.37 that each function h_p is in the weak closure of the set $\{\alpha_i^{-1} z^{k_i}; \; i \in I_p\} \subset \mathbb{C} \cdot O(\mathbf{1}, M_z)$. $\square$

From this result, it is clear that Theorem 10.36 follows from the next lemma: just put $\delta_i = 2^{-i}$ and $f_i = (2^{-p_i}/\sqrt{q_i}) h_{p_i}$, where $\mathcal{H} = (h_p)$ is any dense sequence in the unit ball of $\mathcal{C}(\mathbb{T})$.

LEMMA 10.39 *Let* $(f_i)_{i \in \mathbb{N}} \subset \mathcal{C}(\mathbb{T})$ *with* $\|f_i\|_\infty \leq 1$ *and* $\|f_i\|_\infty \to 0$. *Also, let* (δ_i) *be a sequence of positive numbers. Then one can find a Rajchman probability measure* μ *and a sequence of positive integers* (k_i) *such that* $|\langle g_i, g_j \rangle_{L^2(\mu)}| \leq \delta_i$ *whenever* $j < i$, *where* $g_i(z) = z^{k_i} - f_i(z)$.

For the proof of Lemma 10.39, one more lemma is needed. Let us denote by $\mathcal{M}(\mathbb{T})$ the space of all complex measures on $\mathbb{T}$. Since $\mathcal{M}(\mathbb{T})$ is the dual space of $\mathcal{C}(\mathbb{T})$, one can speak of the w^*-topology of $\mathcal{M}(\mathbb{T})$. By the separability of $\mathcal{C}(\mathbb{T})$, the unit ball of $\mathcal{M}(\mathbb{T})$ is (compact and) metrizable in its w^*-topology; let us fix some compatible metric d.

LEMMA 10.40 *Let* μ *be a probability Rajchman measure on* $\mathbb{T}$. *Also, let* $\varphi \in \mathcal{C}(\mathbb{T})$ *with* $\|\varphi\|_\infty \leq 1$ *and let* $N \in \mathbb{N}$. *Then one can find a probability Rajchman measure* ν *and an integer* $k \geq N$ *such that*

(1) ν *is* w^*-*close to* μ;
(2) $\|\widehat{\nu} - \widehat{\mu}\|_\infty \leq 2\|\varphi\|_\infty$;
(3) $z^k \nu$ *is* w^*-*close to* $\varphi \nu$.

PROOF The proof relies on the following two fairly well-known facts (see [157]).

FACT 1 Any measure absolutely continuous with respect to μ is Rajchman.

Indeed, since μ is Rajchman this is clear for measures of the form $P\mu$, where P is a trigonometric polynomial. So the result follows by the density of the trigonometric polynomials in $L^1(\mu)$.

FACT 2   Put $E := \mathrm{supp}(\mu)$, and let us denote by $\mathbf{P}(E) \subset \mathcal{M}(\mathbb{T})$ the set of all probability measures with support contained in E. The following sets are w^*-dense in $\mathbf{P}(E)$:

 (i) the set of all probability measures supported on a Kronecker set $K \subset E$;
(ii) the set of all probability Rajchman measures supported on E.

Part (i) follows because the finitely supported measures are w^*-dense in $\mathbf{P}(E)$ and any finite subset of the perfect set E can be approximated by a finite *independent* set contained in E, which is a Kronecker set by Kronecker's theorem. Part (ii) is proved by approximating any finitely supported measure $\sigma = \sum c_i \delta_{a_i} \in \mathbf{P}(E)$ by a measure of the form $\nu = \sum c_i \mu_{|V_i} / \mu(V_i)$, where V_i is a small neighbourhood of a_i. Such a measure ν is Rajchman by Fact 1.

Continuing to prove Lemma 10.40, we first choose a probability measure μ_1 which is w^*-close to μ and whose support is a Kronecker set K contained in $E = \mathrm{supp}(\mu)$; this is possible by Fact 2(i). By the definition of a Kronecker set, one can find an integer $k \geq N$ such that $z^k |\varphi| \mu_1$ is close to $\varphi \mu_1$ in the $\mathcal{M}(\mathbb{T})$-norm (just approximate $\varphi/|\varphi|$ by z^k on the set $\{z \in K; \ |\varphi(z)| \geq \varepsilon\}$, where $\varepsilon > 0$ is small enough). Moreover, since $(1 - |\varphi|)\mu$ is a Rajchman measure, we may assume additionally that $z^k(1 - |\varphi|)\mu$ is w^*-close to 0. Next, since $\mu_1 \in \mathbf{P}(E)$ we may use Fact 2(ii) to pick a probability Rajchman measure μ_2 which is w^*-close to μ_1, specifically, close enough to ensure that $z^k |\varphi| \mu_2$ is close to $z^k |\varphi| \mu_1$.

It is then easily checked that the measure $\nu := (1 - |\varphi|)\mu + |\varphi| \mu_2$ has the required properties. Indeed, ν is close to $(1 - |\varphi|)\mu + |\varphi| \mu_1$, which is close to $(1 - |\varphi|)\mu + |\varphi| \mu = \mu$, and $z^k \nu$ is close to $z^k |\varphi| \mu_2$, which is close to $z^k |\varphi| \mu_1$, which is close to $\varphi \mu_1$, which is close to $\varphi \mu$, which is close to $\varphi \nu$. Finally $\nu - \mu = |\varphi|(\mu_2 - \mu)$, so that $\|\widehat{\nu} - \widehat{\mu}\|_\infty \leq \||\varphi|(\mu_2 - \mu)\|_{\mathcal{M}(\mathbb{T})} \leq 2\|\varphi\|_\infty$. $\square$

PROOF OF LEMMA 10.39   We start with some sequence (ε_i) of positive numbers to be chosen later. We construct by induction two increasing sequences of integers (n_i) and (k_i), and a sequence of probability Rajchman measures (μ_i) such that the following conditions are fulfilled, where we have put $g_i(z) := z^{k_i} - f_i(z)$:

(a) $d(\mu_i, \mu_{i+1}) < \varepsilon_i$;
(b) $\|\widehat{\mu}_{i+1} - \widehat{\mu}_i\|_\infty \leq 2\|f_{i+1}\|_\infty$;
(c) $|\widehat{\mu}_{i+1}(n) - \widehat{\mu}_i(n)| < \varepsilon_i$ if $|n| \leq n_i$;
(d) $|\widehat{\mu}_i(n)| < \varepsilon_i$ if $|n| > n_i$;
(e) $|\langle \overline{g_j} g_{j'}, \mu_{i+1} - \mu_i \rangle| < \varepsilon_i$ for all $j, j' \leq i$;
(f) $|\int \overline{g_j} g_i \, d\mu_i| < \varepsilon_i$ for all $j < i$.

We start with any probability Rajchman measure μ_0, choose some positive integer n_0 such that (d) holds, and put $k_0 := 1$. If μ_i has been already defined then the measure μ_{i+1} and the integer k_{i+1} are given by Lemma 10.40 applied for $\mu := \mu_i$ and $\varphi := f_{i+1}$. More explicitly, conditions (a), (c) and (e) follow from (1) in Lemma 10.40, condition (b) follows from (2), and (f) follows from (3). Then one can choose n_{i+1} according to (d) since μ_{i+1} is Rajchman.

By (a), the sequence (μ_i) is w^*-convergent to some probability measure μ, provided that $\sum_i \varepsilon_i < \infty$. If $i \in \mathbb{N}$ and $n_i < |n| \le n_{i+1}$ then

$$|\widehat{\mu}(n)| \le |\widehat{\mu}_i(n)| + |\widehat{\mu}_{i+1}(n) - \widehat{\mu}_i(n)| + \sum_{j>i} |\widehat{\mu}_{j+1} - \widehat{\mu}_j(n)|$$

$$\le \varepsilon_i + 2\|f_{i+1}\|_\infty + \sum_{j>i} \varepsilon_j\,,$$

by conditions (b), (c) and (d). Hence, the measure μ is Rajchman. Moreover, if i, $j \in \mathbb{N}$ and $i < j$ then

$$|\langle g_i, g_j \rangle_{L^2(\mu)}| = \left| \int \overline{g_j} g_i \, d\mu \right|$$

$$\le \left| \int \overline{g_j} g_i \, d\mu_i \right| + \sum_{k \ge i} |\langle \overline{g_j} g_i, \mu_{k+1} - \mu_k \rangle|$$

$$\le \varepsilon_i + \sum_{k \ge i} \varepsilon_k$$

by (e) and (f). Thus, we see that the measure μ has the required properties if the ε_i are small enough. This concludes the proof of Lemma 10.39 and hence the proof of Theorem 10.36. $\qquad\qquad\qquad\qquad\qquad\qquad\qquad\qquad\qquad\qquad\qquad\square$

10.4 Weak sequential hypercyclicity and supercyclicity

Let X be a separable Banach space. There are at least two reasonable ways of defining the weak *sequential* density of a set $A \subset X$. First, one may declare that A is weakly sequentially dense if any vector $z \in X$ is the weak limit of a sequence $(x_k) \subset A$. Second, one may define the **weak sequential closure** of A as the smallest weakly sequentially closed subset of X containing A, and say that A is weakly sequentially dense if its weak sequential closure is the whole space X.

The second definition is of course weaker. In fact, one can describe the weak sequential closure of A as follows. For any set $B \subset X$, let us denote by $B^{(1)}$ the set of all weak limits of sequences from B. Put $A^{(0)} = A$, and let us define by induction a non-decreasing transfinite family of sets $(A^{(\xi)})_{\xi < \omega_1}$ by

$$A^{(\xi+1)} = \left(A^{(\xi)} \right)^{(1)}$$

and

$$A^{(\lambda)} = \bigcup_{\xi < \lambda} A^{(\xi)} \quad \text{if } \lambda \text{ is a limit ordinal.}$$

Here ω_1 is the first uncountable ordinal. Obviously, the weak sequential closure of A is equal to $\bigcup_{\xi < \omega_1} A^{(\xi)}$. Moreover, since X is separable it is not hard to see that the non-decreasing family $(A^{(\xi)})$ must stabilize at some countable ordinal $\xi(A)$, which may be called for example the *sequential closure index* of A (see Exercise 10.14). Thus, a set A is weakly sequentially dense in the first sense iff $A^{(1)} = X$, and weakly sequentially dense in the second sense iff $A^{(\xi)} = X$ for some countable ordinal ξ.

Each of the two above definitions has its own advantages, but we have to choose one. We choose the weaker, i.e. the second definition. Accordingly, we shall say that an operator $T \in \mathfrak{L}(X)$ is **weakly sequentially hypercyclic** if there exists some vector $x \in X$ such that the weak sequential closure of $O(x, T)$ is equal to X. Weakly sequentially *supercyclic* operators are defined in the same way.

A key observation here is that a weakly convergent sequence is necessarily *bounded*, by the uniform boundedness principle. It follows that if (x_n) is a sequence in X such that $\|x_n\| \to \infty$ then the set $\{x_n;\ n \in \mathbb{N}\}$ is weakly sequentially closed in X, and hence certainly not weakly sequentially dense (of course, this is to be compared with the discussion in Section 10.1). In particular, if $T \in \mathfrak{L}(X)$ and if $x \in X$ satisfies $\lim_{n \to \infty} \|T^n(x)\| = \infty$ then x cannot be a weakly sequentially hypercyclic vector for T.

In the same spirit, if $T \in \mathfrak{L}(X)$ is an operator such that $\|T(x)\| \geq \|x\|$ and $T^n(x) \to 0$ weakly for all $x \in X$ then T cannot be weakly sequentially supercyclic. Indeed, if (z_k) is a weakly convergent sequence of the form $z_k = \lambda_k T^{n_k}(x)$ then the sequence (λ_k) must be bounded since $\|T^{n_k}(x)\| \geq \|x\|$, so that in fact $z_k \xrightarrow{w} 0$ if the sequence (n_k) is unbounded. Therefore $\mathbb{K} \cdot O(x, T) \cup \{0\}$ is weakly sequentially closed for any $x \in X$.

For example, the unitary operator M_z acting on $L^2(\mu)$ cannot be weakly sequentially supercyclic if μ is a Rachjman measure on $\mathbb{T}$. However, it is not hard to show that M_z is weakly sequentially supercyclic if μ is continuous and supported on a Kronecker set; see Exercise 10.15.

There is at least one case where weak sequential hypercyclicity and supercyclicity have no interest at all, namely when every weakly convergent sequence in X is norm-convergent. Indeed, in that case the weak sequential closure of any set $A \subset X$ is equal to its norm closure; hence weak sequential hypercyclicity or supercyclicity is equivalent to (norm) hypercyclicity or supercyclicity. A typical example is $X = \ell^1$: this is a classical result proved in 1920 by J. Schur (see any book on Banach space theory). Since then, Banach spaces in which every weakly convergent sequence is norm-convergent are said to have the **Schur property**.

Very few results are known concerning weakly sequentially hypercyclic or supercyclic operators. We will prove two of them in this section. The first is a kind of angle criterion which lies between the original Angle Criterion and the Weak Angle Criterion 10.17. Like many results in this chapter, it is due to S. Shkarin [224].

PROPOSITION 10.41   *Let X be a Banach space, and let $T \in \mathfrak{L}(X)$. Also, let $x \in X$. Assume that $T^n(x) \neq 0$ for all $n \in \mathbb{N}$ and that one can find a non-zero linear functional $x^* \in X^*$ such that $\langle x^*, T^n(x) \rangle / \|T^n(x)\| \to 0$. Then x is not a weakly sequentially supercyclic vector for T.*

PROOF   It is enough to find some weakly sequentially closed set $Z \subset X$ such that $Z \neq X$ and $\mathbb{K} \cdot O(x, T) \subset Z$. The most obvious candidate is

$$Z := \mathbb{K} \cdot O(x, T) \cup \{z \in X;\ \langle x^*, z \rangle = 0\}.$$

Let $(z_k) \subset Z$ be a sequence converging weakly to some $z \in X$. If $z_k \in \operatorname{Ker}(x^*)$ for infinitely many k then $z \in \operatorname{Ker}(x^*) \subset Z$; so we may assume that $z_k \in \mathbb{K} \cdot O(x, T)$ for all k. Let us write $z_k = \lambda_k T^{n_k}(x)/\|T^{n_k}(x)\|$, where $\lambda_k \in \mathbb{K}$ and $n_k \in \mathbb{N}$. Then the sequence λ_k is bounded because the weakly convergent sequence (z_k) is. Thus, we may assume that (λ_k) converges to some $\lambda \in \mathbb{K}$. If $n_k \to \infty$ then $\langle x^*, z_k \rangle \to 0$ by assumption on x^* and x, so that $z \in \operatorname{Ker}(x^*)$. Otherwise, some subsequence of (n_k) is a constant, equal say to $n \in \mathbb{N}$, and then $z = \lambda T^n(x)/\|T^n(x)\| \in \mathbb{K} \cdot O(x, T)$. In either case, we get $z \in Z$. $\qquad\square$

REMARK In Shkarin's paper [223], an operator $T \in \mathfrak{L}(X)$ is said to be **antisupercyclic** if $\langle x^*, T^n(x) \rangle/\|T^n(x)\| \to 0$ for all $x \in X$ with $T^n(x) \neq 0$ and *every* linear functional $x^* \in X^*$, or, in other words, if $T^n(x)/\|T^n(x)\| \to 0$ weakly for all admissible x. Such operators fail the Angle Criterion in the strongest possible way. It is proved in [223] that the Volterra operator is antisupercyclic.

The second result we prove concerns weighted shifts (see [224] and [48]).

PROPOSITION 10.42 *Every weakly sequentially hypercyclic weighted shift on $\ell^p(\mathbb{Z})$ is hypercyclic, for any $p \in [1, \infty)$. Similarly, any weakly sequentially supercyclic weighted shift is supercyclic.*

PROOF Let $B_{\mathbf{w}}$ be a non-hypercyclic weighted shift on $\ell^p(\mathbb{Z})$, and let us show that $B_{\mathbf{w}}$ is not weakly sequentially hypercyclic. We assume that $B_{\mathbf{w}}$ is weakly hypercyclic, otherwise there is nothing to prove.

Let x be any weakly hypercyclic vector for $B_{\mathbf{w}}$. By the proof of Proposition 10.20, one can find $q \in \mathbb{N}$ such that

$$\|B_{\mathbf{w}}^n(x)\| \xrightarrow[n \in \Lambda]{n \to \infty} \infty \,,$$

where $\Lambda = \{n \in \mathbb{N}; \; |\langle e_q^*, B_{\mathbf{w}}^n(x) \rangle| > 1\}$. Therefore, the set $Z := \{B_{\mathbf{w}}^n(x); \; n \in \Lambda\}$ is weakly sequentially closed. Since the vector $2e_q$ belongs neither to Z nor to the weak closure of $\{B_{\mathbf{w}}^n(x); \; n \notin \Lambda\}$, it follows that $2e_q$ is not in the weak sequential closure of $O(x, B_{\mathbf{w}})$. Since x is an arbitrary weakly hypercyclic vector for $B_{\mathbf{w}}$, this shows that $B_{\mathbf{w}}$ is not weakly sequentially hypercyclic.

We now sketch the supercyclic case. The reader can show in Exercise 10.5 that if $p > 1$ then every non-supercyclic weighted shift on $\ell^p(\mathbb{Z})$ is in fact antisupercyclic. Therefore, the result follows from Proposition 10.41 if $p > 1$. The case $p = 1$ is trivial since $\ell^1(\mathbb{Z})$ has the Schur property. $\qquad\square$

10.5 Comments and exercises

Weak linear dynamics is still at its very beginning. Many interesting questions can be raised. Here are three.

OPEN QUESTION 1 Does there exist a weakly hypercyclic operator on ℓ^1 which is not hypercyclic? Is this true for every Banach space?

OPEN QUESTION 2 Does there exist any weakly sequential hypercyclic operator which is not hypercyclic?

OPEN QUESTION 3 Is it possible to characterize in a simple way the following "thinness" property of a compact set $E \subset \mathbb{T}$: for any continuous probability measure μ supported on E, the operator M_z is weakly supercyclic on $L^2(\mu)$?

As already mentioned, the study of the weak closure exponent $w(X)$ was motivated by results from S. Shkarin's paper [224], where $w(\ell^p(\mathbb{N}))$ is in effect determined for any $p \in (1, \infty)$. The inequalities $1 \le w(X) \le 2$ and the exact value $w(X) = 2$ in the Hilbertian case can be found in [187]. Ball's theorem is not used there, but the proofs make extensive use of Gaussian measures. Weakly dense sequences which are not dense are also considered in [11]. For another application of Ball's theorem to orbits of linear operators, see [189].

The technical assumptions in Theorem 10.24 are (again) inspired by Shkarin's paper [224]. The idea of using Sidon sequences comes from [31].

Surprisingly enough, it is unknown how quickly a (classical) Sidon sequence must increase. It is not too hard to show that the rate of growth of any Sidon sequence (n_k) is at least k^2 (see [138]). The best known admissible upper bound is $n_k = O(k^{1/(\sqrt{2}-1)+\varepsilon})$, for any given $\varepsilon > 0$; this is due to I. Ruzsa [213]. There is a simple and efficient method for producing Sidon sequences with polynomial growth, the so-called *greedy algorithm*; see [138] or Exercise 10.9.

EXERCISE 10.1 Let X be a separable Banach space. Construct a weakly dense sequence $(x_n) \subset X$ such that $\|x_n\| \to \infty$. (*Hint:* Start with a sequence $(z_i) \subset X$ such that $\|z_i\| \to \infty$ and $0 \in \overline{\{z_i;\ i \in \mathbb{N}\}}^w$, with $\|z_i\| \ge 2$. Consider the set $\{p(z_i + d_k);\ p, i \in \mathbb{N},\ k \le i\}$, where (d_k) is a norm-dense sequence in B_X.)

EXERCISE 10.2 ([98]) Let T be a weakly hypercyclic operator. Show that every connected component of $\sigma(T)$ intersects the unit circle. (*Hint:* Proceed as in Chapter 1, using Proposition 10.1.)

EXERCISE 10.3 Let X be a Banach space, and let $T \in \mathfrak{L}(X)$ be a weakly hypercyclic operator which is not hypercyclic. What can be said of the operator $I \oplus T$ acting on $\mathbb{K} \oplus X$?

EXERCISE 10.4 Let ϕ be a parabolic non-automorphism of the disk. Show that the composition operator C_ϕ is not weakly supercyclic on $H^2(\mathbb{D})$.

EXERCISE 10.5 ([224]) Let $p \in (1, \infty)$ and let $B_{\mathbf{w}}$ be a weighted shift on $\ell^p(\mathbb{Z})$ which is not supercyclic. Show that $B_{\mathbf{w}}$ is antisupercyclic. (*Hint:* Look at the proof of Proposition 10.20 and show that

$$\frac{\langle e_k^*, B_{\mathbf{w}}^n(x)\rangle}{\|B_{\mathbf{w}}^n(x)\|} \xrightarrow{n \to \infty} 0$$

for any $k \in \mathbb{Z}$. Conclude by a density argument.)

EXERCISE 10.6 Let $T = 2B$, where B is the backward shift on $\ell^2(\mathbb{N})$. Show that there exists a vector $x \in \ell^2(\mathbb{N})$ whose T-orbit is weakly dense in ℓ^2 but not norm-dense.

EXERCISE 10.7 Prove the converse of Lemma 10.25 when the supports of the vectors $z(l)$ have bounded cardinality.

EXERCISE 10.8 Show that Corollary 10.27 remains true even if $B_{\mathbf{w}}$ is not invertible, if we suppose now that $\sup_{1 \le m \le n} w_{-m} \cdots w_{-n} < \infty$. (*Hint:* Follow the proof of Theorem 10.24.)

EXERCISE 10.9 **The greedy algorithm**

1. Let $\alpha > 0$, and let $0 = n_0 < n_1 < \cdots < n_{k-1}$ be a finite increasing sequence of integers. Put

$$A_k := \{n_l + n_j + u - (n_m + v);\ j \le l \le k-1,\ u \le [l^\alpha],\ m \le k-1,\ v \le [k^\alpha]\},$$
$$B_k := \{n_l + n_j + u - v;\ j \le l \le n-1,\ u \le [l^\alpha],\ v \le [k^\alpha]\}.$$

Show that the cardinal numbers of A_k and of B_k are $O(n^{3+2\alpha})$.

2. Let $\varepsilon > 0$ and suppose that $2\alpha < \varepsilon$. Show that there exists an increasing sequence of integers (n_k) which is $([l^\alpha])$-Sidon and such that $n_k = O(k^{3+\varepsilon})$. (*Hint*: Proceed by induction, putting $n_k := \min\left(\mathbb{N} \setminus (A_k \cup B_k/2)\right)$.)

3. *Application.* Show that if $\alpha > 0$ is small enough and $p > 2$ then the weighted shift $B_\mathbf{w}$ defined by $w_n = 2$ and $w_{-1} \cdots w_{-n} = n^\alpha$ for $n \geq 0$ is weakly hypercyclic on $\ell^p(\mathbb{Z})$.

EXERCISE 10.10   **Some weakly hypercyclic weighted shifts**

1. Find a weakly hypercyclic weighted shift $B_\mathbf{w}$ on $\ell^2(\mathbb{Z})$ such that $w_{-1} \cdots w_{-n} \to \infty$ as $n \to +\infty$.

2. Find an invertible weighted shift $B_\mathbf{w}$ such that both $B_\mathbf{w}$ and $B_\mathbf{w}^{-1}$ are weakly hypercyclic on $\ell^2(\mathbb{Z})$, yet $B_\mathbf{w}$ is not hypercyclic. (*Hint* : Pick a sequence of integers (m_k) such that $m_k - m_{k-1} > m_{k-1}$ for all $k \geq 1$, and define $\mathbf{w}$ as follows:

$$\begin{cases} w_n = 2, & n \in [m_{2k} - m_{2k-1} - k, m_{2k} - m_{2k-1}] \\ & \qquad \cup \left[-m_{2k+1} + m_{2k}, -m_{2k+1} + m_{2k} + k\right] \\ w_n = 1/2, & n \in [m_{2k}, m_{2k} + k] \cup [-m_{2k+1} - k, -m_{2k+1}]. \end{cases}$$

Apply Theorem 10.24 with either $n_k = m_{2k}$ or $n_k = m_{2k+1}$).

EXERCISE 10.11   Let μ be a Borel probability measure on $\mathbb{T}$. Assume that the multiplication operator M_z is weakly supercyclic on $L^2(\mu)$, with supercyclic vector f. Assume also that μ is not a continuous measure, and fix $a \in \mathbb{T}$ with $\mu(\{a\}) > 0$.

1. Show that $f(z) \neq 0$ almost everywhere with respect to μ. In particular, $f(a) \neq 0$.

2. Observe that the map $g \mapsto g(a)$ is a well-defined continuous linear functional on $L^2(\mu)$. Deduce that any function $g \in L^2(\mu)$ is in the weak closure of some *bounded* subset of $\mathbb{C} \cdot O(f, M_z)$. Then show that M_z is weakly sequentially supercyclic with index 1.

3. Show that, for any $g \in L^2(\mu)$ with $g(a) \neq 0$, one can find a sequence of integers (n_k) and some complex number λ such that $|\lambda| = |g(a)|/|f(a)|$ and $z^{n_k} \to \lambda^{-1} g(z)/f(z)$ almost everywhere with respect to μ.

4. Show that μ is the point mass δ_a.

EXERCISE 10.12   ([157])   Let $\mathcal{K}(\mathbb{T})$ be the space of all compact subsets of $\mathbb{T}$, equipped with the Hausdorff metric, and let $\mathbf{K}$ be the set of all Kronecker sets in $\mathbb{T}$.

1. Show that $\mathbf{K}$ is dense in $\mathcal{K}(\mathbb{T})$. (*Hint*: Finite independent sets are Kronecker.)

2. Show that $\mathbf{K}$ is a G_δ subset of $\mathcal{K}(\mathbb{T})$.

3. Show that there exist perfect Kronecker sets.

EXERCISE 10.13   Let G be a locally compact abelian group with Haar measure m. Show that if $Z \subset G$ is a measurable set such that $m(Z) > 0$ then $Z - Z$ is a neighbourhood of 0 in G. (*Hint*: Assuming $m(Z) < \infty$, the function $f := \mathbf{1}_Z * \mathbf{1}_Z$ is continuous.)

EXERCISE 10.14   Let X be a separable Banach space.

1. Let $(F_\xi)_{\xi < \omega_1}$ be a non-decreasing transfinite family of closed subsets of X. Show that (F_ξ) is stationary, i.e. it stabilizes at some countable ordinal ξ. (*Hint*: Pass to complements and use the Lindelöf property.)

2. Let A be an arbitrary subset of X, and define the sets $A^{(\xi)}$ as in Section 10.4. Show that the family $(A^{(\xi)})_{\xi < \omega_1}$ is stationary. (*Hint*: There is a closed set between $A^{(\xi)}$ and $A^{(\xi+1)}$.)

EXERCISE 10.15   Let μ be a continuous probability measure on $\mathbb{T}$ supported on a Kronecker set. Show that the multiplication operator M_z is weakly sequentially supercyclic on $L^2(\mu)$, with index 1. (*Hint*: Let $\mathcal{F} \subset L^2(\mu)$ be the set of all Borel functions with constant modulus. Show that any function $f \in L^2(\mu)$ is in the weak closure of a bounded subset of $\mathcal{F}$, and deduce that $\mathcal{F}^{(1)} = L^2(\mu)$.)

11

Universality of the Riemann zeta function

11.1 Voronin's theorem and how to prove it

We have encountered many hypercyclic operators in this book, most of which are quite classical and very simply defined: translation operators, weighted shifts, composition operators, adjoints of multipliers, ... However, the corresponding hypercyclic vectors are usually far from being well identified. Indeed, they are found by an appeal to the Baire category theorem, by a probabilistic argument or by writing down a series which depends on an enumeration of some countable dense subset of the underlying space. As a rule, we have little information on *individual* hypercyclic vectors and are unable to show that any particular vector is hypercyclic.

There is, however, at least one remarkable exception: as shown by S. Voronin in [234], the famous **Riemann zeta function** is universal à la Birkhoff. Recall that the zeta function is usually defined first on the half-plane $\{\mathrm{Re}(s) > 1\}$ by the formula

$$\zeta(s) := \sum_{n \geq 1} \frac{1}{n^s}$$

and then extended to a meromorphic function on the whole complex plane, with one (simple) pole at $s = 1$.

To state Voronin's result, we need to fix some notation and to recall some terminology. Throughout this chapter, we denote by Ω the critical strip of the zeta function,

$$\Omega := \{s \in \mathbb{C};\ 1/2 < \mathrm{Re}(s) < 1\}$$

and we put

$$H^*(\Omega) = \{f \in H(\Omega);\ f \text{ has no zeros in } \Omega\}.$$

Here, we should of course recall the *Riemann hypothesis*, which is the assertion that ζ has no zeros in the critical strip Ω, i.e. $\zeta \in H^*(\Omega)$.

Since the strip Ω is invariant under imaginary translations, there is a well-defined **translation semigroup** $(T_t)_{t>0}$ acting on the space of holomorphic functions $H(\Omega)$ according to the formula

$$T_t f(s) = f(s + it).$$

Voronin's theorem asserts that one can approximate any function $f \in H^*(\Omega)$ by translates of the zeta function, and in fact by many such translates. Recall that the **lower density** of a (measurable) set of real numbers $D \subset \mathbb{R}_+$ is defined by

$$\underline{\text{dens}}(D) = \liminf_{T \to +\infty} \frac{1}{T} \int_0^T \mathbf{1}_D(t)\, dt\,.$$

We say that D *has a density* if $T^{-1} \int_0^T \mathbf{1}_D(t)\, dt$ has a limit as $T \to \infty$, and we denote this limit by $\text{dens}(D)$ (the density of D).

VORONIN'S THEOREM   *Given $f \in H^*(\Omega)$, $\varepsilon > 0$ and a compact set $K \subset \Omega$, one can find positive real numbers t such that*

$$|\zeta(s + it) - f(s)| < \varepsilon \quad \text{for every } s \in K\,.$$

Moreover, the set of all such real numbers t has positive lower density.

REMARK   With the notion of lower density at hand, one can define **frequently hypercyclic** (one-parameter) continuous semigroups of operators, exactly as we did in Chapter 6 for frequently hypercyclic operators (the discrete case). Then Voronin's theorem leads to the following rather pedantic statement: *if the Riemann hypothesis is true then the zeta function is a frequently hypercyclic vector for the translation semigroup (T_t) acting on the invariant subset $H^*(\Omega) \subset H(\Omega)$.*

The proof of Voronin's theorem is a truly beautiful piece of mathematics, mixing several tools which appear elsewhere in the book: Banach space geometry, complex approximation and ergodic theory. However, it can be read almost independently of the rest of the book.

Throughout this chapter, we shall denote by $\mathcal{P} = \{p_n;\ n \geq 1\}$ the set of all prime numbers, enumerated as an increasing sequence. We recall the **Euler product expansion** of the zeta function,

$$\zeta(s) = \prod_{n=1}^{\infty} \left(\frac{1}{1 - p_n^{-s}} \right)\,.$$

By the classical convergence criterion for infinite products, this expansion is valid in the half-plane $\{\text{Re}(s) > 1\}$; but it is by no means obvious that the formula holds in any reasonable sense in the strip Ω. Nevertheless, one can still define in Ω the partial products

$$\zeta_n(s) := \prod_{j=1}^{n} \left(\frac{1}{1 - p_j^{-s}} \right)\,,$$

and think of them as presumed approximations of ζ.

Voronin's theorem is proved using the following two results. The first states that the finite Euler products ζ_n are in some sense good approximations of ζ in the strip Ω.

THEOREM 11.1   *Let K be a compact subset of Ω. Given $\varepsilon, \alpha > 0$, one can find $n_0 \in \mathbb{N}$ such that*

$$\underline{\text{dens}}\left(\left\{ t \in \mathbb{R}_+;\ \sup_{s \in K} |\zeta(s + it) - \zeta_n(s + it)| < \varepsilon \right\} \right) > 1 - \alpha$$

for every $n \geq n_0$.

The second result states that one can approximate any function $f \in H^*(\Omega)$ by translates of the Euler products ζ_n, with an additional (and important) Cauchy-like condition.

THEOREM 11.2   *Let $f \in H^*(\Omega)$, and let K be a compact subset of Ω. Given $\varepsilon, \delta > 0$, one can find a set $D \subset \mathbb{R}_+$ with positive density and $N \geq 1$ such that*

(1) $\sup_{s \in K} |\zeta_N(s + it) - f(s)| < \varepsilon$ *for all $t \in D$;*
(2) *for any $n \geq N$,*

$$\underline{\mathrm{dens}} \left(\left\{ t \in D; \; \sup_{s \in K} |\zeta_n(s + it) - \zeta_N(s + it)| < \varepsilon \right\} \right) > (1 - \delta) \, \mathrm{dens}(D).$$

It should be clear that Voronin's theorem follows easily from these two results. To prove this in detail, let us fix $f \in H^*(\Omega)$, $\varepsilon > 0$ and a compact set $K \subset \Omega$.

By Theorem 11.2 applied for $\delta := 1/3$, one can find a set $D \subset \mathbb{R}_+$ with positive density and $N \geq 1$ such that

$$|\zeta_N(s + it) - f(s)| < \varepsilon \text{ on } K \text{ for each } t \in D;$$

then, for any $n \geq N$, the set

$$A_n := \left\{ t \in D; \; \sup_{s \in K} |\zeta_n(s + it) - \zeta_N(s + it)| < \varepsilon \right\}$$

has lower density greater than $2 \, \mathrm{dens}(D)/3$.

Next, we may apply Theorem 11.1 to $\alpha := \mathrm{dens}(D)/3$ to get an integer n_0 such that for every $n \geq n_0$, the set

$$B_n := \left\{ t \in \mathbb{R}_+; \; \sup_{s \in K} |\zeta(s + it) - \zeta_n(s + it)| < \varepsilon \right\}$$

has lower density greater than $1 - \mathrm{dens}(D)/3$.

Now, we fix an integer $n \geq \max(N, n_0)$ and put $C := A_n \cap B_n$. Then $\sup_{s \in K} |\zeta(s + it) - f(s)| < 3\varepsilon$ on C, by the triangle inequality. Moreover, the set C has lower density greater than $\mathrm{dens}(D)/3$, since it is easily checked that

$$\underline{\mathrm{dens}}(A \cap B) \geq \underline{\mathrm{dens}}(A) + \underline{\mathrm{dens}}(B) - 1$$

for any sets $A, B \subset \mathbb{N}$. This concludes the proof of Voronin's theorem.

Thus, our goal in this chapter is to prove Theorems 11.1 and 11.2. These two results have different natures. Theorem 11.1 really pertains to the theory of *Dirichlet series*, which is never far from analytic number theory. Its proof is developed in Sections 11.2 and 11.3 and can be omitted at a first reading. As regards Theorem 11.2, its proof is much more in the spirit of the present book. It can be divided into three steps, which are interesting for their own sake. The first step is an elegant result from Hilbert space geometry (Section 11.4). This result is used in the second step to prove the density of a set of Dirichlet polynomials in some suitable function space (Section 11.5). The third step uses ideas from ergodic theory and the classical Kronecker theorem on diophantine approximation (Section 11.6).

After these preliminaries, the proof of Theorem 11.2 becomes easy; it is given in Section 11.7.

We end this introduction by pointing out a functional-analytic tool that will be needed in the proofs of both Theorems 11.1 and 11.2. Instead of dealing with sup-norms $\| \cdot \|_{C(K)}$, it will be convenient to use some stronger L^2 norms. This leads us to introduce the well-known Bergman spaces.

Let U be a bounded domain in $\mathbb{C}$. The **Bergman space** $A^2(U)$ is the set of all holomorphic functions $f : U \to \mathbb{C}$ which are square integrable on U, endowed with its natural norm:

$$\|f\|_{A^2(U)}^2 := \int_U |f(s)|^2 \, dA(s).$$

Here (and elsewhere) dA is the area measure, i.e. the planar Lebesgue measure. The notation dA and $A^2(U)$ is conventional in that part of function theory, even though it seems hard to find a connection between the name "Bergman" and the letter "A".

Bergman spaces are extremely interesting objects of study; see e.g. [102] for a recent account of their properties. For our concerns, we will need just the following two results:

- $A^2(U)$ is a Hilbert space and convergence in $A^2(U)$ entails uniform convergence on compact subsets of U;
- if U is a smooth Jordan domain, i.e. the interior of some smooth Jordan curve, then the polynomials are dense in $A^2(U)$.

Proofs of these two facts are outlined in Exercise 11.1. The first one is crucial to establish the following lemma.

LEMMA 11.3 *Let K be a compact subset of Ω, and let U be a bounded domain such that $K \subset U \subset \Omega$. Given $\varepsilon, c > 0$, one can find $\eta > 0$ such that, whenever $D \subset \mathbb{R}_+$ has a positive density and $f, g \in H^*(\Omega)$ satisfy*

$$\limsup_{T \to +\infty} \frac{1}{T} \int_0^T \mathbf{1}_D(t) \|T_t(f) - T_t(g)\|_{A^2(U)}^2 dt < \eta \operatorname{dens}(D), \qquad (11.1)$$

it follows that

$$\underline{\operatorname{dens}} \left(\left\{ t \in D; \ \sup_{s \in K} |f(s + it) - g(s + it)| < \varepsilon \right\} \right) > (1 - c) \operatorname{dens}(D). \quad (11.2)$$

PROOF Let η be any positive number, and assume that $f, g \in H^*(\Omega)$ satisfy (11.1). Let us fix $T_0 \geq 0$ such that

$$\frac{1}{T} \int_0^T \mathbf{1}_D(t) \|T_t(f) - T_t(g)\|_{A^2(U)}^2 dt < \eta \operatorname{dens}(D)$$

for all $T \geq T_0$. Since convergence in $A^2(U)$ entails uniform convergence on compact sets, we may also choose some constant $C < \infty$ such that $\|h\|_{C(K)} \leq C\|h\|_{A^2(U)}$ for any $h \in A^2(U)$. Then

$$\frac{1}{T} \int_0^T \mathbf{1}_D(t) \sup_{s \in K} |f(s + it) - g(s + it)|^2 dt < C^2 \eta \operatorname{dens}(D)$$

for all $T \geq T_0$. By Chebyshev's inequality, it follows that the normalized Lebesgue measure of the set $\{t \in [0, T] \cap D; \ \sup_{s \in K} |f(s + it) - g(s + it)| < \varepsilon\}$ is greater than $1 - C^2 \eta \operatorname{dens}(D)/\varepsilon^2$ for any $T \geq T_0$. To conclude the proof, it is now sufficient to choose $\eta < \varepsilon^2 c/C^2$. $\qquad\qquad\qquad\qquad\qquad\qquad\qquad\qquad\square$

11.2 Dirichlet series and the zeta function

11.2.1 Approximation of the zeta function by Dirichlet polynomials

A *Dirichlet series* is a series of functions of the form

$$f(s) = \sum_{n \geq 1} a_n n^{-s},$$

where the coefficients a_n are complex numbers and $s \in \mathbb{C}$.

Using a suitable summation by parts, it is not hard to show that if a Dirichlet series f is convergent at some point s_0 then it is uniformly convergent throughout any angular region of the form $\{s \in \mathbb{C}; \ |\arg(s - s_0)| < \delta\}$, where $\delta \in (0, \pi/2)$ (see e.g. [232, Chapter IX]). The **abscissa of convergence** of $f(s) = \sum a_n n^{-s}$ is defined by

$$\sigma_c(f) := \inf \left\{ \operatorname{Re}(s); \ \sum a_n n^{-s} \text{ is convergent} \right\}.$$

Thus the Dirichlet series f converges for $\operatorname{Re}(s) > \sigma_c(f)$ and diverges for $\operatorname{Re}(s) < \sigma_c(f)$. Of course, the *function* f (i.e. the sum of the series) is analytic in the half-plane $\mathbb{C}_{\sigma_c(f)}$, where $\mathbb{C}_\sigma = \{s \in \mathbb{C}; \ \operatorname{Re}(s) > \sigma\}$.

Unlike in the case of a power series, the convergence of a Dirichlet series at some point s_0 does not entail absolute convergence at all points s such that $\operatorname{Re}(s) > \operatorname{Re}(s_0)$. Therefore, it makes sense to define the **abscissa of absolute convergence** of $f(s) = \sum a_n n^{-s}$ by

$$\sigma_a(f) := \inf \left\{ \sigma \in \mathbb{R}; \ \sum |a_n| n^{-\sigma} \text{ is convergent} \right\}.$$

Clearly $\sigma_c(f) \leq \sigma_a(f)$. Moreover, it is not difficult to see that $\sigma_a(f) \leq \sigma_c(f) + 1$, since $|a_n n^{-s}|$ is bounded when the series $\sum a_n n^{-s}$ is convergent at some point s.

Here, we are concerned with the particular Dirichlet series $\zeta(s) = \sum_{n \geq 1} n^{-s}$, which is used to define the zeta function in the half-plane $\mathbb{C}_1 := \{\operatorname{Re}(s) > 1\}$. In this case, the abscissae of convergence and absolute convergence are both equal to 1. Thus, the Dirichlet series $\sum n^{-s}$ is never convergent when $\operatorname{Re}(s) < 1$, but the zeta function can still be extended to the whole complex plane as a meromorphic function with one (simple) pole at $s = 1$.

The possibility of extending ζ to the whole complex plane is not a trivial result. However, it is much easier to obtain a meromorphic extension to the half-plane $\mathbb{C}_0 := \{\operatorname{Re}(s) > 0\}$. Indeed, it suffices to write (for $\operatorname{Re}(s) > 1$)

$$\zeta(s) - \frac{1}{s-1} = \sum_{n=1}^{\infty} \int_n^{n+1} \left(\frac{1}{n^s} - \frac{1}{t^s} \right) dt := \sum_{n=1}^{\infty} \varphi_n(s) \,,$$

and to note that the series appearing on the right-hand side is in fact uniformly convergent on compact subsets of $\mathbb{C}_0$ (by the mean-value theorem), thereby defining a holomorphic function $\varphi \in H(\mathbb{C}_0)$. An elaboration on this elementary argument leads to the following theorem, which is the simplest result on the approximation of ζ by partial sums of its defining Dirichlet series.

THEOREM 11.4   *Given $\sigma_0 > 0$, one can find some constant $M < \infty$ such that*

$$\left| \zeta(s) - \sum_{n \le x} n^{-s} + \frac{x^{1-s}}{1-s} \right| \le M x^{-\sigma}$$

whenever $x > 1$ and $s = \sigma + it$ with $0 < \sigma \le \sigma_0$, $|t| \le \pi x$ and $s \ne 1$.

The proof of this theorem relies heavily on the following classical inequality for exponential sums, which goes back to J. G. van der Corput [82].

LEMMA 11.5   *Let $a < b$ be two natural numbers and let $f, g : [a,b] \to \mathbb{R}$ be two functions of class C^2. Assume that*

(1) *f' is monotonic and $|f'| \le 1/2$;*
(2) *g is positive, non-increasing and convex.*

Then

$$\sum_{n=a}^{b} g(n) e^{2\pi i f(n)} = \int_a^b g(u) e^{2\pi i f(u)} \, du + O(g(a) + |g'(a)|) \,,$$

where the implied constant is absolute.

PROOF   We start with the following well-known summation formula.

FACT   Let $\phi \in C^1([a,b])$. Then

$$\sum_{n=a}^{b} \phi(n) = \int_a^b \phi(u) \, du + \int_a^b \left(u - [u] - \frac{1}{2} \right) \phi'(u) \, du + \frac{1}{2}\phi(b) + \frac{1}{2}\phi(a) \,, \quad (11.3)$$

where $[u]$ is the integer part of u.

PROOF OF THE FACT   For each $k \in \{a, \ldots, b-1\}$, integration by parts leads to

$$\int_k^{k+1} (u - k - 1/2)\phi'(u) \, du = \frac{1}{2}\phi(k+1) + \frac{1}{2}\phi(k) - \int_k^{k+1} \phi(u) \, du \,.$$

Summing from $k = a$ to $b - 1$, the result follows.   □

Applying the fact with $\phi(u) := g(u) e^{2\pi i f(u)}$, we get

$$\sum_{n=a}^{b} g(n) e^{2\pi i f(n)} = \int_a^b g(u) e^{2\pi i f(u)} \, du + \int_a^b \left(u - [u] - \frac{1}{2} \right) \phi'(u) \, du + O(g(a)) \,.$$

It remains to estimate the integral $\int_a^b (u - [u] - 1/2)\phi'(u)\, du$. To do this, we decompose the 1-periodic function $u \mapsto u - [u] - 1/2$ in Fourier series. After a simple computation, we get

$$u - [u] - 1/2 = \sum_{k=1}^{\infty} -\frac{\sin(2\pi ku)}{\pi k}, \quad u \in \mathbb{R} \setminus \mathbb{Z}.$$

It is well known (but not obvious) that the partial sums of the series appearing on the right-hand side are uniformly bounded on $\mathbb{R}$ (see e.g. [19, Vol. 1, p. 90] or Exercise 11.2). By Lebesgue's theorem, it follows that we can write

$$\int_a^b \left(u - [u] - \frac{1}{2}\right)\phi'(u)\, du = \sum_{k \geq 1} \frac{-1}{2i\pi k} \int_a^b (e^{2\pi iku} - e^{-2\pi iku})\phi'(u)\, du$$

$$= \frac{-1}{2i\pi} \sum_{k \in \mathbb{Z} \setminus \{0\}} \frac{1}{k} \int_a^b \phi'(u) e^{2\pi iku}\, du$$

$$= \frac{-1}{2i\pi} \sum_{k \neq 0} \frac{1}{k} \int_a^b \psi(u) e^{i\theta_k(u)}\, du,$$

where we have set $\theta_k(u) := 2\pi(ku + f(u))$ and $\psi = 2i\pi g f' + g'$. We write the last integral as

$$\int_a^b \psi(u) e^{i\theta_k(u)}\, du = \int_a^b \frac{\psi(u)}{i\theta_k'(u)} \left(e^{i\theta_k(u)}\right)'\, du$$

and integrate by parts. Since $|\theta_k'(u)| = 2\pi|k| + |f'(u)| \geq \pi k$, this gives

$$\int_a^b \psi(u) e^{i\theta_k(u)}\, du = O\left(\frac{|\psi(a)| + |\psi(b)|}{k} + \int_a^b \left|\left(\frac{\psi}{\theta_k'}\right)'(u)\right|\, du\right).$$

Since $|\psi| = O(|g'| + |f'g|)$, and by the assumptions on f and g, we see that $|\psi(a)| + |\psi(b)| = O(g(a) + |g'(a)|)$. Now, observe that

$$\left|\left(\frac{\psi}{\theta_k'}\right)'\right| = \left|\frac{\psi'}{\theta_k'} - \frac{\psi\theta_k''}{(\theta_k')^2}\right| = \frac{1}{k} O\left(|\psi'| + |\psi f''|\right),$$

since $\theta_k'' = 2\pi f''$ and $|\theta_k'| \geq |k| - 1/2$. Moreover, $|\psi'| = O(|g'f'| + |gf''| + |g'|) = O(|g'| + g|f''|)$. Using again the assumptions on f and g, it follows that

$$\int_a^b \left|\left(\frac{\psi}{\theta_k'}\right)'(u)\right|\, du = \frac{1}{k} O(g(a) + |g'(a)|).$$

Thus, we arrive at

$$\int_a^b \left(u - [u] - \frac{1}{2}\right)\phi'(u)\, du = O\left(\sum_{k \neq 0} \frac{1}{k^2}(g(a) + |g'(a)|)\right)$$

$$= O(g(a) + |g'(a)|). \qquad \square$$

PROOF OF THEOREM 11.4   Let us fix $x > 1$. Without loss of generality, we may assume that x in an integer. For $N > x$ and $\mathrm{Re}(s) > 1$, we may write

$$\zeta(s) = \sum_{n=1}^{N-1} n^{-s} + \sum_{n=N}^{\infty} n^{-s}.$$

Applying (11.3) with $\phi(u) := u^{-s}$, $a := N$ and $b \to \infty$, it follows that

$$\zeta(s) = \sum_{n=1}^{N-1} n^{-s} - \frac{N^{1-s}}{1-s} - s \int_{N}^{\infty} \frac{u - [u] - 1/2}{u^{s+1}} du + \frac{N^{-s}}{2}.$$

Since the integral defines an analytic function in $\mathbb{C}_0$, this identity remains valid for every $s \in \mathbb{C}_0\backslash\{1\}$. Hence, we may write

$$\zeta(s) = \sum_{n=1}^{N} n^{-s} - \frac{N^{1-s}}{1-s} + O\left(\frac{|s|/\sigma + 1}{N^{\sigma}}\right)$$

throughout $\mathbb{C}_0 \setminus \{1\}$; we have put $s = \sigma + it$.

Now assume that $s = \sigma + it$ with $0 < \sigma \le \sigma_0$ and $|t| \le \pi x$. We apply Lemma 11.5 to estimate the sum $\sum_{n=x}^{N} n^{-s}$, which has the required form with $g(u) := u^{-\sigma}$ and $f(u) := -t\log(u)/(2\pi)$ (observe that $|f'(u)| = |t|/(2\pi u) \le 1/2$ if $u \ge x$). Since $g(x) = x^{-\sigma}$ and $|g'(x)| = \sigma x^{-\sigma-1} \le \sigma_0 x^{-\sigma}$, and since $|s|/\sigma \ge 1$ (so that $|s|/\sigma + 1 = O(|s|/\sigma)$) this gives

$$\zeta(s) = \sum_{n=1}^{x} n^{-s} + \int_{x}^{N} u^{-s} du - \frac{N^{1-s}}{1-s} + O(x^{-\sigma}) + O\left(\frac{|s|}{\sigma N^{\sigma}}\right)$$

$$= \sum_{n=1}^{x} n^{-s} - \frac{x^{1-s}}{1-s} + O(x^{-\sigma}) + O\left(\frac{|s|}{\sigma N^{\sigma}}\right).$$

Since the O-constants are absolute, the required result follows if we let $N \to \infty$. □

11.2.2 A Parseval formula for Dirichlet series

Remembering that our first goal is to prove Theorem 11.1 and, looking at Lemma 11.3, our strategy is clear: we will in fact try to show that ζ_n is close to ζ with respect to some suitable Hilbertian norm. To this effect, we establish in this subsection a kind of Parseval formula for Dirichlet series.

We start with an elementary result. Recall that, for $\sigma \in \mathbb{R}$, we denote by $\mathbb{C}_\sigma$ the half-plane $\{\mathrm{Re}(s) > \sigma\}$.

PROPOSITION 11.6   *Let $f(s) = \sum a_n n^{-s}$ be a Dirichlet series with finite abscissa of absolute convergence σ_0. Then*

$$\lim_{T \to \infty} \frac{1}{T} \int_{0}^{T} |f(s+it)|^2 dt = \sum_{n=1}^{+\infty} |a_n|^2 n^{-2\mathrm{Re}(s)}$$

for every $s \in \mathbb{C}_{\sigma_0}$, uniformly on compact subsets of $\mathbb{C}_{\sigma_0}$.

PROOF   Let us denote by $AP(\mathbb{R})$ the class of all functions $\varphi : \mathbb{R} \to \mathbb{C}$ which are uniform limits of trigonometric polynomials, i.e. finite sums of the form $P(t) = \sum_n c_n e^{i\lambda_n t}$, where $\lambda_n \in \mathbb{R}$. This is the class of **almost periodic functions** in the sense of H. Bohr. It is well known, and easy to check, that any function $\varphi \in AP(\mathbb{R})$ has a well-defined (square) **mean** $M(\varphi)$ given by the formula

$$M(\varphi)^2 := \lim_{T \to \infty} \frac{1}{T} \int_0^T |\varphi(t)|^2 \, dt$$

and that the mean is continuous on $(AP(\mathbb{R}), \| \cdot \|_\infty)$. Moreover, if $P(t) = \sum_n c_n e^{i\lambda_n t}$ is a trigonometric polynomial then

$$M(P)^2 = \sum_n |c_n|^2 \, . \tag{11.4}$$

Indeed, since $\limsup_{T \to \infty} T^{-1} \int_0^T |\varphi(t)|^2 dt \le \|\varphi\|_\infty^2$ for all $\varphi \in AP(\mathbb{R})$, we need only check formula (11.4) to get the existence of $M(\varphi)$ for every $\varphi \in AP(\mathbb{R})$ and its continuity with respect to the norm $\| \cdot \|_\infty$. Now, expanding $|P(t)|^2$ and using linearity, (11.4) reduces to the orthogonality relation

$$\lim_{T \to \infty} \frac{1}{T} \int_0^T e^{i(\lambda-\mu)t} \, dt = \begin{cases} 0 & \text{if } \lambda \ne \mu, \\ 1 & \text{if } \lambda = \mu. \end{cases}$$

The connection with Proposition 11.6 should be clear. Indeed, the Dirichlet series $f(s) = \sum a_n n^{-s}$ is uniformly convergent on any closed half-space $\overline{\mathbb{C}}_\sigma$, $\sigma > \sigma_0$. This means that, for each $s \in \mathbb{C}_{\sigma_0}$, the function $f_s(t) := f(s+it)$ is the limit in the sense of $AP(\mathbb{R})$ of the trigonometric polynomials

$$P_N(s,t) := \sum_{n=1}^N a_n n^{-s} e^{-i \log(n) t} \, ,$$

the limit being uniform with respect to s on compact subsets of $\mathbb{C}_{\sigma_0}$. Therefore

$$M(f_s)^2 = \lim_{N \to \infty} M(P_N(s, \cdot))^2 = \sum_{n=1}^\infty |a_n|^2 n^{-2\mathrm{Re}(s)},$$

uniformly on compact subsets of $\mathbb{C}_{\sigma_0}$. $\qquad\qquad\square$

REMARK   The mean of an almost periodic function φ is usually defined via the two-sided "Cesàro means" $(2T)^{-1} \int_{-T}^T |\varphi(t)|^2 dt$. This is more natural in a group-theoretic context.

The proof of Proposition 11.6 relies heavily on the *absolute* convergence of the Dirichlet series for $\mathrm{Re}(s) > \sigma_0$. However, the formula can be extended to non-absolutely convergent Dirichlet series which are not too wildly growing. Here is the relevant definition.

DEFINITION 11.7   *Let f be an analytic function in some half-plane $\mathbb{C}_\alpha$, and let $\sigma > \alpha$. Then f is said to be **of finite order** in the closed half-plane $\overline{\mathbb{C}}_\sigma = \{\mathrm{Re}(s) \ge \sigma\}$ if $|f(x+it)| \le A + |t|^B$ throughout $\overline{\mathbb{C}}_\sigma$, for some finite constants A, B.*

The result that we would like to prove reads as follows. It is due to F. Carlson [69].

THEOREM 11.8   *Let $f(s) = \sum a_n n^{-s}$ be a Dirichlet series with finite abscissa of convergence. Also, let $\alpha, \beta \in \mathbb{R}$ with $\sigma_c(f) \leq \alpha < \beta$, and denote by $\Omega_{\alpha,\beta}$ the strip $\{s;\ \alpha < \mathrm{Re}(s) < \beta\}$. Suppose that*

(1) f is of finite order in $\overline{\mathbb{C}}_\sigma$, for any $\sigma > \alpha$;
(2) $\sup_{T>0} T^{-1} \int_0^T |f(\sigma + it)|^2 dt < \infty$ for any $\sigma \in (\alpha, \beta)$, with uniform bounds on compact sets.

Then

$$\lim_{T \to \infty} \frac{1}{T} \int_0^T |f(s + it)|^2 dt = \sum_{n \geq 1} |a_n|^2 n^{-2\mathrm{Re}(s)}$$

for every $s \in \Omega_{\alpha,\beta}$, uniformly on compact sets.

PROOF   The idea is (of course) to approximate f by absolutely convergent Dirichlet series, and then to use Proposition 11.6.

In what follows, we fix a compact set $K \subset \Omega_{\alpha,\beta}$ and put

$$\sigma_0 := \inf\{\mathrm{Re}(s);\ s \in K\}.$$

We also fix some real number $c > 0$ such that $\sigma_0 - c > \alpha$ and some positive real number $b > \sigma_a(f) - \sigma_0$ (recall that $\sigma_a(f) \leq \sigma_c(f) + 1 < +\infty$). Finally, we choose $\lambda > 0$ such that

$$-1 < \frac{-c}{\lambda} < 0 < \frac{b}{\lambda} < 1.$$

For each $\delta > 0$, we consider the Dirichlet series

$$g_\delta(s) := \sum_{n \geq 1} a_n e^{-(n\delta)^\lambda} n^{-s}.$$

We note that g_δ is absolutely convergent in the whole complex plane. Indeed, since $\sigma_c(f) < +\infty$ the coefficients a_n have at most polynomial growth. Moreover, the function g_δ should come close to f as $\delta \to 0$, so we have some hope of getting the result by applying Proposition 11.6 to g_δ. The crucial point is the following identity.

FACT   If $s \in \mathbb{C}$ satisfies $\mathrm{Re}(s) \geq \sigma_0$ then

$$g_\delta(s) - f(s) = \frac{1}{2i\pi\lambda} \int_{-c-i\infty}^{-c+i\infty} \Gamma\left(\frac{w}{\lambda}\right) f(s + w) \delta^{-w}\, dw. \tag{11.5}$$

Here, Γ is Euler's **gamma function**, which is usually defined on the half-plane $\{\mathrm{Re}(w) > 0\}$ by

$$\Gamma(w) = \int_0^{+\infty} e^{-x} x^{w-1}\, dx.$$

We recall that Γ extends to a meromorphic function on $\mathbb{C}$, with poles at the non-positive integers and residues $\mathrm{res}(\Gamma, -n) = (-1)^n/n!$. We also recall **Stirling's formula**:

$$\Gamma(s) = \sqrt{2\pi} s^{s-1/2} e^{-s} (1 + O(|s|^{-1})),$$

uniformly on any domain of the form $\mathcal{R}_\varepsilon = \{s \in \mathbb{C};\ |s| > \varepsilon \text{ and } |\arg(s)| < \pi - \varepsilon\}$, where $\varepsilon \in (0, \pi)$. These are well-established facts, a detailed exposition of which can be found e.g. in [188, Appendix 3].

From Stirling's formula, it is easy to infer that $|\Gamma(\sigma + it)| = O(e^{-\pi|t|/4})$ on any vertical line $\{\mathrm{Re}(s) = \sigma\}$, $\sigma \notin -\mathbb{N}$ (with a O-constant depending on σ). Since $-1 < -c/\lambda < 0 < b/\lambda < 1$, it follows that one can find some constants $C < \infty$ and $\kappa > 0$ such that

$$\left| \Gamma\left(\frac{x+iy}{\lambda}\right) \right| \leq C e^{-\kappa|y|} \tag{11.6}$$

on the lines $\{x = b\}$ and $\{x = -c\}$. In particular, formula (11.5) makes sense, bearing in mind that $\alpha < \sigma_0 - c \leq \mathrm{Re}(s) - c$ and that f is of finite order in $\overline{\mathbb{C}}_{\sigma_0-c}$.

PROOF OF THE FACT   We start with the identity

$$e^{-a} = \frac{1}{2\pi i} \int_{b-i\infty}^{b+i\infty} \Gamma(w) a^{-w} dw, \tag{11.7}$$

which holds for any positive number a. Thanks to Stirling's formula, this identity can be derived by applying the residue theorem in the square C_R whose right-hand side is the segment $[b - iR, b + iR]$ and letting $R \to \infty$; the details can be found in [188, Appendix 3].

We apply (11.7) with $a = (n\delta)^\lambda$, where $\lambda > 0$ is fixed (as well as b) and $\delta > 0$ is our parameter. After a change of variable, we get

$$e^{-(n\delta)^\lambda} = \frac{1}{2\pi i\lambda} \int_{b-i\infty}^{b+i\infty} \Gamma\left(\frac{w}{\lambda}\right) \frac{\delta^{-w}}{n^w} dw. \tag{11.8}$$

Now, let us fix $s \in \Omega_{\alpha,\beta}$ with $\mathrm{Re}(s) \geq \sigma_0$. Since $b > \sigma_a(f) - \sigma_0$, we may insert (11.8) in the definition of $g_\delta(s)$ and interchange the summation and integration, thanks to the estimate (11.6). This leads to the formula

$$g_\delta(s) = \frac{1}{2\pi i\lambda} \int_{b-i\infty}^{b+i\infty} \Gamma\left(\frac{w}{\lambda}\right) f(s+w)\delta^{-w} dw.$$

At this point, we apply the residue theorem to $\phi(w) := \Gamma(w/\lambda) f(s+w)\delta^{-w}$ between the lines $\{\mathrm{Im}(w) = b\}$ and $\{\mathrm{Im}(w) = -c\}$. By the choice of λ, the function ϕ has a unique pole at $w = 0$ between these two lines, with residue $\mathrm{res}(\phi, 0) = \lambda f(s)$. Using Stirling's formula together with the fact that f is of finite order in $\overline{\mathbb{C}}_{\sigma_0-c}$, we may estimate in a convenient way the integrals on the horizontal segments $[-c - iR, b - iR]$, $[-c + iR, b + iR]$. Letting $R \to \infty$, we get (11.5).   □

Using Stirling's formula again, we deduce from (11.5) and (11.6) that

$$|g_\delta(s+it) - f(s+it)| \leq C\delta^c \int_{-\infty}^{+\infty} e^{-\kappa|y|} |f(s-c+i(y+t))| dy$$

for all $s \in K$ and some constant $\kappa > 0$. Here and below, C is a constant whose value may change from line to line but remains independent of δ.

Since f has finite order in $\overline{\mathbb{C}}_{\sigma_0 - c}$ and since K is compact, one can write

$$|f(s - c + i(y + t))| \leq (A + |y + t|)^B,$$

for some finite constants A, B. It follows that if $T > 0$ and $t \in (0, T)$ then

$$\int_{|y| \geq 2T} e^{-\kappa|y|} |f(s - c + i(y + t))| \, dy \leq C e^{-\kappa T},$$

so that

$$|g_\delta(s + it) - f(s + it)| \leq C\delta^c \left(e^{-\kappa T} + \int_{-2T}^{2T} e^{-\kappa|y|} |f(s - c + i(y + t))| \, dy \right)$$

for all $s \in K$. Moreover,

$$\left(\int_{-2T}^{2T} e^{-\kappa|y|} |f(s - c + i(y + t))| \, dy \right)^2 \leq C \int_{-2T}^{2T} e^{-\kappa|y|} |f(s - c + i(y + t))|^2 \, dy,$$

by Schwarz's inequality applied with respect to the measure $e^{-\kappa|y|} dy$.

Collecting the estimates and integrating now with respect to $t \in (0, T)$, we get

$$\int_0^T |g_\delta(s + it) - f(s + it)|^2 dt$$

$$\leq C\delta^{2c} + C\delta^{2c} \int_{-2T}^{2T} e^{-\kappa|y|} \int_0^T |f(s - c + i(y + t))|^2 \, dt dy,$$

for any $T > 0$ and all $s \in K$.

Since K is compact and $\sigma_0 - c > \alpha$, we know that

$$\int_0^T |f(s - c + i(y + t))|^2 dt = O(T),$$

uniformly with respect to $(s, y) \in K \times [-2T, 2T]$. Thus

$$\frac{1}{T} \int_0^T |g_\delta(s + it) - f(s + it)|^2 \, dt \leq C\delta^{2c}.$$

Now, applying Proposition 11.6 to the absolutely convergent Dirichlet series g_δ and using the triangle inequality in $L^2(0, T)$, we deduce that for any $\delta > 0$ there exists $T_\delta > 0$ such that

$$\left| \left(\sum_{n=1}^{+\infty} |a_n|^2 n^{-2\mathrm{Re}(s)} e^{-2(n\delta)^\lambda} \right)^{1/2} - \left(\frac{1}{T} \int_0^T |f(s + it)|^2 dt \right)^{1/2} \right| \leq C\delta^c \quad (11.9)$$

for every $T \geq T_\delta$ and all $s \in K$.

Since the square means $T^{-1} \int_0^T |f(s + it)|^2 dt$ are uniformly bounded, we deduce first that for each $s \in K$ the sums $\Sigma_\delta(s) := \sum_{n=1}^\infty |a_n|^2 n^{-2\mathrm{Re}(s)} e^{-2(n\delta)^\lambda}$ are uniformly bounded with respect to $\delta \in (0, 1)$. By monotone convergence (letting $\delta \to 0$), it follows that the series $\sum |a_n|^2 n^{-2\mathrm{Re}(s)}$ is convergent for each $s \in K$, with $\Sigma(s) := \sum_1^\infty |a_n|^2 n^{-2\mathrm{Re}(s)} = \lim_{\delta \to 0} \Sigma_\delta(s)$. Moreover, the limit is in fact uniform with respect to $s \in K$ because $|\Sigma(s) - \Sigma_\delta(s)|$ is non-increasing with respect to $\mathrm{Re}(s)$. Choosing δ so that $\Sigma_\delta(s)$ is uniformly close to $\Sigma(s)$ on K and returning to (11.9), we now see that $T^{-1} \int_0^T |f(s + it)|^2 dt$ is uniformly close to $\sum_1^\infty |a_n|^2 n^{-2\mathrm{Re}(s)}$ when $T \geq T_\delta$. This concludes the proof of Theorem 11.8. $\qquad \square$

REMARK   An examination of the above proof shows that it is unnecessary to assume $\alpha \geq \sigma_c(f)$, i.e. the convergence of the Dirichlet series in $\mathbb{C}_\alpha$. It is enough to know that $f(s)$ extends analytically to $\mathbb{C}_\alpha$ and that f grows slowly on any vertical line.

11.3 The first half of the proof

We are now almost ready for the proof of Theorem 11.1. One more lemma is needed.

LEMMA 11.9   *Let $\sigma_0 \in [1/2, 1)$. Then, uniformly for $1/2 \leq \sigma \leq \sigma_0$,*

$$\sum_{0 < m < n \leq T} \frac{1}{m^\sigma n^\sigma \log(n/m)} = O(T^{2-2\sigma} \log(T)).$$

PROOF   Let us denote by Σ_1 the sum over all pairs (m, n) such that $m < n/2$, and by Σ_2 the remaining sum. Then

$$\Sigma_1 \leq \frac{1}{\log(2)} \sum_{0 < m < n \leq T} m^{-\sigma} n^{-\sigma} = O(T^{2-2\sigma}).$$

To estimate Σ_2, we write $m = n - r$ with $1 \leq r \leq n/2$, so that $\log(n/m) = -\log(1 - r/n) \geq r/n$. This yields

$$\Sigma_2 \leq \sum_{n \leq T} \sum_{r \leq n/2} \frac{(n-r)^{-\sigma} n^{-\sigma}}{r/n}$$

$$= O\left(\sum_{n \leq T} n^{1-2\sigma} \sum_{r \leq n} r^{-1} \right) = O(T^{2-2\sigma} \log(T)). \qquad \square$$

PROOF OF THEOREM 11.1   The idea is to apply Theorem 11.8 to $\zeta - \zeta_n$ for any $n \geq 1$, with $\alpha = 1/2$ and $\beta = 1$. This cannot be done directly, since $\zeta - \zeta_n$ has a pole at $s = 1$ and hence is not analytic in the half-plane $\mathbb{C}_{1/2}$. Fortunately, one can "kill" the pole $s = 1$ by considering $(1 - 2^{1-s})(\zeta(s) - \zeta_n(s))$ instead of $\zeta(s) - \zeta_n(s)$. The key point is the following

FACT   For any $n \geq 1$, the function $f_n(s) := (1 - 2^{1-s})(\zeta(s) - \zeta_n(s))$ satisfies the assumptions of Theorem 11.8 with $\alpha := 1/2$ and $\beta := 1$.

PROOF OF THE FACT   Since the Dirichlet series of $(1 - 2^{1-s})\zeta_n(s)$ is absolutely convergent in the half-plane $\mathbb{C}_0$, it suffices to show that $f(s) := (1 - 2^{1-s})\zeta(s)$ satisfies the required assumptions.

First, we note that $(1 - 2^{1-s})\zeta(s) = \sum_{n\geq 1}(-1)^n n^{-s}$, so that $f(s)$ is convergent in $\mathbb{C}_0$ (and hence $\sigma_c(f) \leq 1/2$). This was the only reason for introducing the factor $(1 - 2^{1-s})$.

Next, we observe that $|\zeta(\sigma + it)| \leq A + B|t|$ uniformly for $\sigma \geq \sigma_0 > 1/2$ and $|(\sigma+it)-1| > 1/4$. This follows from Theorem 11.4 applied for $x = |t| > 1$, thanks to the trivial estimate $|\sum_{n\leq|t|} n^{-s}| \leq |t|$. Since $(1 - 2^{1-s})$ is uniformly bounded in $\mathbb{C}_{1/2}$, it follows that f is of finite order in $\overline{\mathbb{C}}_{\sigma_0}$ for any $\sigma_0 > 1/2$.

Since (again) $(1 - 2^{1-s})$ is uniformly bounded in $\mathbb{C}_{1/2}$, the proof of the fact will be complete if we are able to show that $T^{-1}\int_0^T |\zeta(\sigma + it)|^2 dt$ is uniformly bounded on compact subsets of $\{1/2 < \sigma < 1\}$. Moreover, since ζ is bounded on compact subsets of Ω we may assume that $T > 1$ and replace the integral $\int_0^T$ by $\int_1^T$.

For any $t > 1$, we apply Theorem 11.4 with $s := \sigma + it$ and $x := t$, so that $|x^{1-s}/(1-s)| = O(t^{-\sigma})$. Integrating with respect to $t \in (1, T)$ gives

$$\int_1^T |\zeta(\sigma + it)|^2\, dt = \int_1^T \left| \sum_{n\leq t} n^{-(\sigma+it)} \right|^2 dt + O\left(\int_1^T t^{-2\sigma} dt \right). \qquad (11.10)$$

The first integral on the right-hand side looks rather unpleasant, since the range of summation depends on the variable of integration. However, expanding the square we may write

$$\int_1^T \left| \sum_{n\leq t} n^{-(\sigma+it)} \right|^2 dt = \sum_{n,m\leq T} (nm)^{-\sigma} \int_{\max(n,m)}^T \left(\frac{n}{m} \right)^{it} dt$$

$$= \sum_{n\leq T} n^{-2\sigma}(T - n)$$

$$+ \sum_{n\neq m} (nm)^{-\sigma} \frac{(n/m)^{iT} - (n/m)^{i\max(n,m)}}{i\log(n/m)}$$

$$= T \sum_{n\leq T} n^{-2\sigma} + O(T^{2-2\sigma}) + O(T^{2-2\sigma}\log(T)),$$

where we have used Lemma 11.9. It follows that $\int_1^T \left| \sum_{n\leq t} n^{-(\sigma+it)} \right|^2 dt = O(T)$, uniformly on compact subsets of $\{1/2 < \sigma < 1\}$. Inserting this estimate into (11.10), we obtain

$$\int_1^T |\zeta(\sigma + it)|^2 dt = O(T) + O(T^{-2\sigma+1}) = O(T).$$

□

We now return to the proof of Theorem 11.1. For any $n \geq 1$, let us write $\zeta(s) - \zeta_n(s) = \sum_{k\geq 1} b_{k,n} k^{-s}$, where $b_{k,n} \in \{0,1\}$ and $b_{k,n} = 0$ for all $k \leq p_n$. Then we may also write $(1 - 2^{1-s})(\zeta - \zeta_n(s)) = \sum_{k\geq 1} c_{k,n} k^{-s}$, where $|c_{k,n}| \leq 3|b_{k,n}|$. Since $|1-2^{1-s}|$ is uniformly bounded from below on compact subsets of Ω, it follows from the above fact that

$$\limsup_{T \to \infty} \frac{1}{T} \int_0^T |\zeta(s+it) - \zeta_n(s+it)|^2 \, dt = O\left(\sum_{k \geq 1} |c_k|^2 k^{-2\mathrm{Re}(s)} \right)$$

$$= O\left(\sum_{k \geq p_n} k^{-2\mathrm{Re}(s)} \right) \xrightarrow{n \to \infty} 0,$$

uniformly on compact subsets of Ω.

It is now easy to conclude the proof of Theorem 11.1, by going to some Bergman space $A^2(U)$ and then coming back. Let K be a compact subset of Ω, and choose a bounded domain U such that $K \subset U \subset \overline{U} \subset \Omega$. Let also $\eta > 0$. Applying the first part of the proof with the compact set $\overline{U} \subset \Omega$ and using Fubini's theorem, we find an integer $n_0 \geq 0$ such that

$$\limsup_{T \to \infty} \frac{1}{T} \int_0^T \int_U |\zeta(s+it) - \zeta_n(s+it)|^2 \, dA(s) dt < \eta,$$

for any $n \geq n_0$. By Lemma 11.3 (with $D = \mathbb{R}_+$), Theorem 11.1 follows immediately. $\qquad\qquad\square$

11.4 Some Hilbert space geometry

In this section, we are interested in the following question, which is seemingly unrelated to the previous sections: given a sequence $(x_n)_{n \geq 1}$ in a Hilbert space H, when is the set of all *unimodular* linear combinations of the vectors x_n dense in H? This is one aspect of the theory of conditionally convergent series in a Banach space; the interested reader may consult [148] to learn more on this subject.

We are going to prove the following result, which comes from [15].

PROPOSITION 11.10 *Let H be a Hilbert space, and let $(x_n)_{n \geq 1}$ be a sequence in H. Assume that the following properties hold.*

(1) $\sum_{n=1}^{\infty} |\langle x^*, x_n \rangle| = \infty$ *for any* $x^* \in H^* \backslash \{0\}$;
(2) $\sum_{n=1}^{\infty} \|x_n\|^2 < \infty$.

Then the set $\left\{ \sum_{j=1}^n a_j x_j; \; n \geq 1, \, |a_j| = 1 \right\}$ is dense in H.

We note that conditions (1) and (2) appear somewhat contradictory, since (1) says that the norm of x_n should be large whereas (2) says that it should be small! However, we will see in the next section that the sequence (n^{-s}), in a suitable Hilbert function space, satisfies these two conditions. For the time being, we will concentrate on the proof of Proposition 11.10. Two lemmas are needed.

LEMMA 11.11 *Let X be a locally convex topological vector space, and let $(x_n)_{n \geq 1}$ be a sequence in X. Assume that $\sum_1^{\infty} |\langle x^*, x_n \rangle| = \infty$ for any non-zero linear functional $x^* \in X^*$. Then, for any $N \in \mathbb{N}$, the set $\left\{ \sum_{j=N}^n a_j x_j; \; n \geq N, \; |a_j| \leq 1 \right\}$ is dense in X.*

PROOF Put $C := \{\sum_{j=N}^{n} a_j x_j; \; n \geq N, |a_j| \leq 1\}$. Then C is a convex subset of X. If C is not dense in X then, by the Hahn–Banach theorem, one can find a non-zero linear functional $x^* \in X^*$ and a real number c such that $\operatorname{Re} \langle x^*, z \rangle \leq c$ for any $z \in C$. Now let $M > N$, choose $a_N, a_{N+1}, \ldots, a_M \in \mathbb{T}$ such that $|\langle x^*, x_j \rangle| = a_j \langle x^*, x_j \rangle$ for each j and set $z := \sum_{j=N}^{M} a_j x_j$. Then $\sum_{j=N}^{M} |\langle x^*, x_j \rangle| = \operatorname{Re} \langle x^*, z \rangle \leq c$. Since $M > N$ is arbitrary, this contradicts the assumption $\sum_{1}^{\infty} |\langle x^*, x_n \rangle| = \infty$. $\qquad\square$

We now have to replace the coefficients a_j by unimodular coefficients.

LEMMA 11.12 *Let $x_1, \ldots, x_n$ be n vectors in a Hilbert space H and let $a_1, \ldots, a_n$ $\in \mathbb{K}$ with $|a_j| \leq 1$. Then one can find $b_1, \ldots, b_n \in \mathbb{K}$ with $|b_j| = 1$ for all j, such that*

$$\left\| \sum_{j=1}^{n} a_j x_j - \sum_{j=1}^{n} b_j x_j \right\|^2 \leq \sum_{j=1}^{n} \|x_j\|^2.$$

PROOF † Let $Z_1, \ldots, Z_n$ be a sequence of independent scalar-valued random variables, defined on the same probability space $(\Omega, \mathcal{F}, \mathbb{P})$, with distributions given by

$$\mathbb{P}\left(Z_j = \frac{a_j}{|a_j|} \right) = \frac{1 + |a_j|}{2} \quad \text{and} \quad \mathbb{P}\left(Z_j = -\frac{a_j}{|a_j|} \right) = \frac{1 - |a_j|}{2}.$$

Each variable Z_j has mean a_j. Hence, $Z_1 - a_1, \ldots, Z_n - a_n$ are independent centred random variables. Moreover, a straightforward computation gives $\mathbb{E}|Z_j - a_j|^2 = 1 - |a_j|^2$. Thus we have

$$\mathbb{E}\big((Z_j - a_j)(\overline{Z_k - a_k}) \big) = \begin{cases} 0 & \text{if } k \neq j, \\ 1 - |a_j|^2 & \text{if } k = j, \end{cases}$$

from which we infer that

$$\mathbb{E}\left(\left\| \sum_j a_j x_j - \sum_j Z_j x_j \right\|^2 \right) = \sum_j (1 - |a_j|^2) \|x_j\|^2 \leq \sum_j \|x_j\|^2.$$

Hence, at least one $\omega \in \Omega$ satisfies $\left\| \sum_j a_j x_j - \sum_j Z_j(\omega) x_j \right\|^2 \leq \sum_j \|x_j\|^2$, and we may take $b_j := Z_j(\omega)$. $\qquad\square$

PROOF OF PROPOSITION 11.10 Let us fix $z \in H$ and $\varepsilon > 0$. By (2), there exists $N \geq 1$ such that $\sum_{n \geq N} \|x_n\|^2 < \varepsilon^2$. By Lemma 11.11, one can find $n \geq N$ and $a_N, \ldots, a_n \in \overline{\mathbb{D}}$ such that

$$\left\| \sum_{j=N}^{n} a_j x_j - \left(-\sum_{j<N} x_j + z \right) \right\| < \varepsilon.$$

---

† This elegant proof was suggested to us by H. Queffélec.

Putting $a_j = 1$ for all $j < N$, we then have

$$\left\| \sum_{j=1}^{n} a_j x_j - z \right\| < \varepsilon .$$

By Lemma 11.12, we can now replace the coefficients $a_1, \ldots, a_n$ by unimodular coefficients $b_1, \ldots, b_n$, up to an error term with norm at most ε. Then we have $\| \sum_{j=1}^{n} b_j x_j - z \| < 2\varepsilon$, which concludes the proof. $\qquad\square$

REMARK   As one can easily imagine, Proposition 11.10 has a Banach space analogue. Namely, the conclusion remains true if the Hilbert space H is replaced by a
Banach space with type $p \in [1, 2]$ and (2) is replaced by $\sum_{n=1}^{\infty} \|x_n\|^p < \infty$. This
result is due to S. A. Chobanyan [75].

11.5 Density of Dirichlet polynomials

In this section, we return to Dirichlet series. Our aim is to prove the density of a
certain set of Dirichlet polynomials in some Bergman space $A^2(U)$. Recall that we
denote by $\mathcal{P}$ the set of all prime numbers, enumerated as an increasing sequence
$(p_n)_{n \geq 1}$.

PROPOSITION 11.13   *The set $\left\{ \sum_{j=1}^{n} b_j p_j^{-s}; \ n \geq 1, b_j \in \mathbb{T} \right\}$ is dense in $A^2(U)$*
for any smooth Jordan domain U such that $\overline{U} \subset \Omega$.

From now on, we will fix the Jordan domain U. If p is a prime number, we use
the symbol p^{-s} to denote the *function $s \mapsto p^{-s}$*, viewed as an element of $A^2(U)$.
In Proposition 11.10, the road to Proposition 11.13 is already traced: it is enough to
show that $\sum_{n=1}^{\infty} \|p_n^{-s}\|^2_{A^2(U)} < \infty$ and that $\sum_{n=1}^{\infty} |\langle \phi, p_n^{-s} \rangle_{A^2(U)}| = \infty$ for every
non-zero function $\phi \in A^2(U)$. One half of this is easy.

LEMMA 11.14   *We have $\sum_{1}^{\infty} \|p_n^{-s}\|^2_{A^2(U)} < \infty$.*

PROOF   Choose $\sigma_0 > 1/2$ such that $U \subset \{\text{Re}(s) \geq \sigma_0\}$. Then

$$\sum_{n=1}^{\infty} \|p_n^{-s}\|^2_{A^2(U)} = \sum_{n=1}^{\infty} \int_U |p_n|^{-2s} dA(z)$$
$$= \sum_{n=1}^{\infty} \text{Area}(U) \, p_n^{-2\sigma_0} < \infty . \qquad\square$$

The second half of the proof is more difficult and relies on the two forthcoming
lemmas. First, we recall the definition of the **Laplace transform** of a complex measure: if μ is a complex Borel measure on $\mathbb{C}$ with compact support K then its Laplace
transform is the entire function $\mathcal{L}\mu$ defined by

$$\mathcal{L}\mu(s) = \int_K e^{-sz} \, d\mu(z) .$$

Recall also that an entire function $f : \mathbb{C} \to \mathbb{C}$ is said to be of **exponential type** if $|f(z)| \leq Ae^{B|z|}$ for some finite constants A, B and all $z \in \mathbb{C}$ (see Appendix A). Clearly, the Laplace transform of any compactly supported complex measure is an entire function of exponential type.

LEMMA 11.15 *Let μ be a complex Borel measure on $\mathbb{C}$, with compact support K, and assume that its Laplace transform $\mathcal{L}\mu$ is not identically 0. Then*

$$\limsup_{x \to +\infty} \frac{\log |\mathcal{L}\mu(x)|}{x} \geq -\max\{\operatorname{Re}(z); \, z \in K\}.$$

PROOF We put $f := \mathcal{L}\mu$ and $\beta := \max\{\operatorname{Re}(z); \, z \in K\}$. Towards a contradiction, assume that

$$\limsup_{x \to +\infty} \frac{\log |f(x)|}{x} < -\beta.$$

Then one can find two constants $C < \infty$ and $\delta > 0$ such that on the one hand $|f(x)| \leq Ce^{-(\beta+2\delta)x}$ for all $x \geq 0$. On the other hand, if $x < 0$ then

$$|f(x)| \leq \int_K e^{-x\operatorname{Re}(z)}|d\mu(z)| \leq Ce^{-\beta x}.$$

From these two inequalities, it follows that

$$|e^{(\beta+\delta)x} f(x)| \leq Ce^{-\delta|x|} \tag{11.11}$$

for all $x \in \mathbb{R}$. In particular, $F(s) := e^{(\beta+\delta)s} f(s)$ is an entire function of exponential type whose restriction to the real axis belongs to $L^2(\mathbb{R})$. By the Paley–Wiener theorem (see Appendix B), its Fourier transform $\widehat{F}$ (defined on $\mathbb{R}$) has compact support. Now, $\widehat{F}$ is given by

$$\widehat{F}(\xi) = \int_{\mathbb{R}} F(x)e^{-ix\xi} \, dx.$$

By (11.11), this defines an analytic function in the strip $\{|\operatorname{Im}(\xi)| < \delta\}$. Therefore, $\widehat{F}$ cannot have compact support unless it is identically 0, i.e. $F = 0$ by analytic continuation. Since f is not identically 0, this is a contradiction. $\square$

REMARK The last part of the proof just says that a non-zero entire function of exponential type cannot converge to 0 at exponential speed along the real axis.

The relevance of Laplace transforms in our concerns can be seen from the following trivial but very important observation: if ϕ is any function in $A^2(U)$ and if μ is the complex measure defined by

$$d\mu(z) = \overline{\phi(z)} \, dA(z)$$

then

$$\langle \phi, p^{-s} \rangle_{A^2(U)} = \mathcal{L}\mu(\log(p)) \, ,$$

for any prime number p. Thus, we have to show that if $\phi \neq 0$ and if μ is defined as above, then the function $f := \mathcal{L}\mu$ satisfies $\sum_{p \in \mathcal{P}} |f(\log p)| = \infty$. Lemma 11.15

is a first step since it shows that $|f(x)|$ is sometimes rather large. The next lemma implies that this is already enough to get the desired conclusion.

LEMMA 11.16   *Let f be an entire function of exponential type. Suppose that*

$$\limsup_{x\to+\infty} \frac{\log|f(x)|}{x} > -1\,.$$

Then $\sum_{p\in\mathcal{P}} |f(\log(p))| = \infty$.

PROOF   We shall need two well-known and non-trivial results. The first is a basic estimate for the asymptotic distribution of prime numbers (see e.g. [188]):

HADAMARD–DE LA VALLÉE POUSSIN ESTIMATE   *For any $x > 1$, let us denote by $\Pi(x)$ the number of prime numbers $p \leq x$. Then*

$$\Pi(x) = \int_2^x \frac{du}{\log u} + O\!\left(xe^{-c\sqrt{\log x}}\right),$$

where $c > 0$ is an absolute constant.

The second result we need is a well-known Bernstein-type inequality relating the modulus of the derivative of a polynomial to the modulus of the polynomial itself (see e.g. [175]).

MARKOV'S INEQUALITY   *If P is a polynomial of degree $N \geq 0$ then*

$$\sup_{x\in[-1,1]} |P'(x)| \leq N^2 \sup_{x\in[-1,1]} |P(x)|\,. \tag{11.12}$$

We can now give the proof of Lemma 11.16. By assumption, one can find a $\delta \in (0,1)$ and an increasing sequence of positive real numbers (x_j) tending to infinity such that $|f(x_j)| \geq e^{-(1-\delta)x_j}$ for all $j \geq 0$. The key point is that $|f(x)|$ remains "large" in a not too small interval near x_j, because f is of exponential type.

FACT   There exist intervals $I_j = (w_j, w_j + \alpha_j) \subset [x_j - 1, x_j + 1]$ of length $\alpha_j \sim (2x_j^4)^{-1}$ (as $j \to \infty$) such that

$$|f(x)| \geq \frac{e^{-(1-\delta)x_j}}{4} \qquad \text{for all } x \in I_j\,,$$

if j is large enough.

PROOF OF THE FACT   Since f is an entire function of exponential type, we may write

$$f(x) = \sum_{n\geq 0} a_n x^n\,,$$

where the coefficients a_n satisfy $|a_n| \leq CR^n/n!$ for some constants C, R depending only on f. This enables one to approximate f by polynomials with a good error estimate. Indeed, if $N \in \mathbb{N}$ and $x \in [x_j - 1, x_j + 1]$ then

$$\sum_{n=N+1}^{\infty} |a_n||x^n| \leq C \sum_{n=N+1}^{\infty} \frac{R^n(x_j+1)^n}{n!}$$

$$\leq C \frac{R^{N+1}(x_j+1)^{N+1}}{(N+1)!} \sum_{n=0}^{\infty} \frac{R^n(x_j+1)^n}{n!}$$

$$\leq C \left(\frac{eR(x_j+1)}{N+1} \right)^{N+1} e^{R(x_j+1)},$$

where we have used Stirling's formula. Taking $N = N_j := ([x_j] + 2)^2$ and putting $P_j(x) := \sum_{n=0}^{N_j} a_n x^n$, this yields

$$|f(x) - P_j(x)| \leq C \left(\frac{eR\sqrt{N_j}}{N_j+1} \right)^{N_j+1} e^{R\sqrt{N_j}} = o\big(e^{-2\sqrt{N_j}}\big) = o(e^{-2x_j}),$$

where the estimate is uniform with respect to j and $x \in [x_j - 1, x_j + 1]$.

Now, let us choose a point $y_j \in [x_j - 1, x_j + 1]$ where $|f|$ attains its maximum value on $[x_j - 1, x_j + 1]$. Also, let $\alpha_j \in (0, 1]$, to be specified later. Then at least one of the two intervals $(y_j - \alpha_j, y_j]$, $(y_j, y_j + \alpha_j)$ is contained in $[x_j - 1, x_j + 1]$. We denote this interval by $I_j := (w_j, w_j + \alpha_j]$.

If $x \in I_j$ then, by the mean-value theorem and Markov's inequality (11.12), we may write

$$|f(x)| \geq |P_j(x)| - |f(x) - P_j(x)|$$

$$\geq |P_j(y_j)| - \sup_{w \in [x_j-1, x_j+1]} |P_j'(w)||x - y_j| + o(e^{-2x_j})$$

$$\geq |f(y_j)| - \alpha_j N_j^2 \sup_{w \in [x_j-1, x_j+1]} |P_j(w)| + o(e^{-2x_j})$$

$$\geq |f(y_j)| - \alpha_j N_j^2 |f(y_j)| + \alpha_j N_j^2 \, o(e^{-2x_j}) + o(e^{-2x_j}).$$

Taking $\alpha_j := \big(2N_j^2\big)^{-1}$, it follows that if j is large then

$$|f(x)| \geq \frac{|f(y_j)|}{4} \geq \frac{e^{-(1-\delta)x_j}}{4}$$

for all $x \in I_j$. $\qquad\qquad\qquad\qquad\qquad\qquad\qquad\qquad\qquad\qquad\qquad\qquad\square$

It is now easy to conclude the proof of the lemma. Indeed, the number of primes p such that $\log(p) \in I_j = (w_j, w_j + \alpha_j]$ is

$$\Pi(e^{w_j+\alpha_j}) - \Pi(e^{w_j}) = \int_{e^{w_j}}^{e^{w_j+\alpha_j}} \frac{dx}{\log x} + O\left(e^{w_j+\alpha_j} e^{-c\sqrt{w_j+\alpha_j}} \right)$$

$$\geq \frac{e^{w_j}(e^{\alpha_j} - 1)}{w_j + \alpha_j} + O\left(e^{w_j+\alpha_j} e^{-c\sqrt{w_j+\alpha_j}} \right).$$

Since we know that $w_j = x_j + O(1)$ and $\alpha_j \sim \left(2x_j^4\right)^{-1}$, it follows that if j is large, then at least $\kappa e^{x_j}/x_j^5$ prime numbers p satisfy $\log(p) \in I_j$, where $\kappa > 0$ is some absolute constant. Hence, we get

$$\sum_{\log(p) \in I_j} |f(\log(p))| \geq \kappa \, \frac{e^{x_j}}{x_j^5} \, \frac{e^{-(1-\delta)x_j}}{4} = \kappa \, \frac{e^{\delta x_j}}{4x_j^5},$$

for large enough j. Since the right-hand side tends to infinity, this concludes the proof. $\qquad\square$

PROOF OF PROPOSITION 11.13   By Lemma 11.14, we just have to show that if ϕ is a non-zero function in $A^2(U)$ then $\sum_{p \in \mathcal{P}} |\langle \phi, p^{-s} \rangle| = \infty$. Setting $d\mu(z) := \overline{\phi(z)} \, dA(z)$ and $f(s) := \mathcal{L}\mu(s) = \int_U e^{-sz} \, d\mu(z)$, this follows from Lemmas 11.15 and 11.16 provided that we are able to show that f is not identically zero.

Differentiating under the integral sign and evaluating $f^{(k)}$ at 0, we get

$$f^{(k)}(0) = (-1)^k \int_{\overline{U}} z^k \overline{\phi(z)} \, dA(z) = (-1)^k \langle \phi, z^k \rangle_{A^2(U)}$$

for all $k \in \mathbb{N}$. Since the polynomials are dense in $A^2(U)$ and $\phi \neq 0$, it follows that f cannot be 0. This concludes the proof. $\qquad\square$

We shall use Proposition 11.13 in the following form:

COROLLARY 11.17   *Let $f \in H^*(\Omega)$, let $\varepsilon > 0$, and let K be a compact subset of Ω. For any $n_0 \in \mathbb{N}$, there exist $N \geq n_0$ and $w_1, \ldots, w_N \in \mathbb{T}$ such that*

$$\sup_{s \in K} \left| f(s) - \prod_{j=1}^{N} \left(\frac{1}{1 - w_j p_j^{-s}} \right) \right| < \varepsilon.$$

PROOF   Since Ω is simply connected, there is a well-defined analytic determination of $\log f$ in Ω, say $g(s) = \log(f(s))$. For $(s, w) \in \Omega \times \mathbb{T}$, let us put

$$h_j(s, w) := \log\left(\frac{1}{1 - wp_j^{-s}} \right),$$

where $\log$ is the principal determination in the disk $\{|z - 1| < 1\}$. Finally, let $\delta > 0$.
Expanding $-\log(1 - u)$ in power series, we see that

$$|h_j(s, w) - wp_j^{-s}| \leq \sum_{k=2}^{\infty} \frac{|p_j^{-ks}|}{k}$$

$$\leq C_K p_j^{-2\sigma}$$

for all $s \in K$, where $\sigma = \inf\{\mathrm{Re}(z); \ z \in K\}$ and $C_K < \infty$ depends only on K. Since $\sigma > 1/2$, it follows that one can find $N_0 \geq n_0$ such that

$$\sup_{s \in K} \sum_{j=N_0+1}^{n} |h_j(s, w_j) - w_j p_j^{-s}| \leq \delta$$

for all $n \geq N_0$ and any choice of $w_{N_0+1}, \ldots, w_n \in \mathbb{T}$.

Now, let $U \subset \Omega$ be a smooth Jordan domain such that $K \subset U$ and $\overline{U} \subset \Omega$. By Proposition 11.13 and since convergence in $A^2(U)$ implies uniform convergence on K, there exist $N > N_0$ and $w_{N_0+1}, \ldots, w_N \in \mathbb{T}$ such that

$$\sup_{s \in K} \left| g(s) - \sum_{j=1}^{N_0} h_j(s,1) - \sum_{j=N_0+1}^{N} w_j p_j^{-s} \right| \leq \delta.$$

Setting $w_1 = \cdots = w_{N_0} := 1$, it follows that

$$\sup_{s \in K} \left| g(s) - \sum_{j=1}^{N} h_j(s, w_j) \right| \leq 2\delta.$$

Since δ is arbitrary, we get Corollary 11.17 by composing with the exponential map. $\qquad \square$

11.6 Unique ergodicity and the Kronecker flow

Let us fix $f \in H^*(\Omega)$, $\varepsilon > 0$ and a compact set $K \subset \Omega$. By the above results, we know that one can find an arbitrarily large N and $w_1, \ldots, w_N \in \mathbb{T}$ such that

$$\sup_{s \in K} \left| f(s) - \prod_{j=1}^{N} \left(\frac{1}{1 - w_j p_j^{-s}} \right) \right| < \varepsilon. \tag{11.13}$$

Since we are looking for an approximation of $f(s)$ by $\zeta_N(s + it)$ for some t, our first task is to replace each w_j by p_j^{-it}, for the *same* real number t. That this can indeed be done is the content of a famous theorem due to L. Kronecker (see e.g. [139]). To state it, we need to recall a definition.

DEFINITION 11.18 *A set of real numbers Θ is said to be $\mathbb{Q}$-linearly independent if the relation $c_1\theta_1 + \cdots + c_d\theta_d = 0$ with integer coefficients $c_1, \ldots, c_d$ and pairwise distinct $\theta_j \in \Theta$ holds only when the coefficients c_j are all zero.*

For example, the set $\{1\} \cup \{\log(p); \ p \in \mathcal{P}\}$ is $\mathbb{Q}$-linearly independent. This follows from the fundamental theorem of arithmetic and the fact that e^m is irrational for every non-zero integer m.

THEOREM 11.19 (KRONECKER'S THEOREM) *Let $\theta_1, \ldots, \theta_N \in \mathbb{R}$. Suppose that $1, \theta_1, \ldots, \theta_N$ are $\mathbb{Q}$-linearly independent, and put $\omega_j = e^{2i\pi\theta_j}$. Then, given arbitrary numbers $w_1, \ldots, w_N \in \mathbb{T}$ and $\varepsilon > 0$, one can find $m \in \mathbb{N}$ such that*

$$|\omega_j^m - w_j| < \varepsilon \quad \text{for all } j \in \{1, \ldots, N\}.$$

From Theorem 11.19 and the $\mathbb{Q}$-linear independence of $1, \log(p_1), \ldots, \log(p_N)$, we immediately deduce the following

COROLLARY 11.20 *The set $\{(p_1^{-it}, \ldots, p_N^{-it}); \ t \in \mathbb{R}_+\}$ is dense in $\mathbb{T}^N$ for any $N \geq 1$.*

Combining this result with (11.13), we can find at least one real number $t \geq 0$ such that $\sup_{s \in K} |f(s) - \zeta_N(s + it)| < \varepsilon$. However, Kronecker's theorem alone does not give any information on the size of the set of all such real numbers t. We need an additional tool, which comes from ergodic theory.

DEFINITION 11.21   *Let $(X, \mathcal{B})$ be a measurable space, and let $\mathcal{S} = (S_t)_{t>0}$ be a semigroup of measurable transformations, $S_t : X \to X$. The semigroup $\mathcal{S}$ is said to be* **uniquely ergodic** *if there exists exactly one $\mathcal{S}$- invariant measure μ on $(X, \mathcal{B})$.*

Of course, a measure μ is said to be $\mathcal{S}$-invariant if it is S_t-invariant for all t. As the terminology suggests, the unique ergodicity of a semigroup $\mathcal{S}$ implies ergodicity with respect to its invariant measure μ (see e.g. Exercise 5.3). And, indeed, we have the following strong form of Birkhoff's ergodic theorem (see [235] for the discrete version, which is often called *Oxtoby's theorem*).

THEOREM 11.22   *Let $\mathcal{S} = (S_t)_{t>0}$ be a semigroup of continuous transformation acting on a compact metric space X. Assume that $\mathcal{S}$ is uniquely ergodic, with invariant measure μ. Then*

$$\frac{1}{T} \int_0^T f(S_t x)\, dt \xrightarrow{T \to \infty} \int_X f\, d\mu$$

for every continuous function $f : X \to \mathbb{C}$ and all $x \in X$.

PROOF   Let us denote by $\mathbf{P}(X)$ the set of all Borel probability measures on X. By the Riesz representation theorem, we may identify $\mathbf{P}(X)$ with a w^*-compact subset of $\mathcal{C}(X)^*$.

Let us fix $x_0 \in X$. For each $T > 0$, let $\mu_T \in \mathbf{P}(X)$ be the measure defined by

$$\int_X f\, d\mu_T = \frac{1}{T} \int_0^T f(S_t x_0)\, dt , \quad f \in \mathcal{C}(X).$$

We have to show that $\mu_T \xrightarrow{w^*} \mu$ as $T \to \infty$.

By the compactness of $\mathbf{P}(X)$, it is enough to check that μ is the only possible w^*-cluster point of (μ_T) as $T \to \infty$. Now, for any given $u > 0$ we have

$$\int_X (f \circ S_u)\, d\mu_T - \int_X f\, d\mu_T = \frac{1}{T} \int_0^T f(S_{t+u} x_0)\, dt - \frac{1}{T} \int_0^T f(S_t x_0)\, dt$$

$$= \frac{1}{T} \left(\int_T^{T+u} f(S_t x_0)\, dt - \int_0^u f(S_t x_0)\, dt \right)$$

for every $f \in \mathcal{C}(X)$, so that

$$\left| \int_X (f \circ S_u)\, d\mu_T - \int_X f\, d\mu_T \right| \leq \frac{2u \|f\|_\infty}{T} \xrightarrow{T \to \infty} 0.$$

It follows that if ν is a w^*-cluster point of (μ_T) as $T \to \infty$ then ν is $\mathcal{T}$-invariant and hence $\nu = \mu$ by unique ergodicity. $\square$

COROLLARY 11.23   *Let X, S and μ be as above, and let $x \in X$. Then*

$$\frac{1}{T} \int_0^T \mathbf{1}_V(S_t x)\, dt \xrightarrow{T \to \infty} \mu(V)\,,$$

for any open set $V \subset X$ such that $\mu(\partial V) = 0$.

PROOF   Let us fix an open set $V \subset X$ with $\mu(\partial V) = 0$. Then, given $\varepsilon > 0$, one can find two real-valued continuous functions $f_\varepsilon, g_\varepsilon$ on X such that $f_\varepsilon \leq \mathbf{1}_V \leq g_\varepsilon$ and $\|f_\varepsilon - g_\varepsilon\|_{L^1(\mu)} < \varepsilon$. Indeed, the lower-semicontinuous function $\mathbf{1}_V$ is the pointwise limit of an increasing sequence of continuous functions, whereas the upper-semicontinuous $\mathbf{1}_{\overline{V}}$ is the limit of a decreasing sequence of continuous functions. Since $\mu(V) = \mu(\overline{V})$, the result follows.

Writing

$$\frac{1}{T} \int_0^T \mathbf{1}_V(S_t x)\, dt \leq \frac{1}{T} \int_0^T g_\varepsilon(S_t x)\, dt$$

$$\leq \mu(V) + \left| \frac{1}{T} \int_0^T g_\varepsilon(S_t x)\, dt - \int_X \mathbf{1}_V \, d\mu \right|$$

$$\leq \mu(V) + \left| \frac{1}{T} \int_0^T g_\varepsilon(S_t x)\, dt - \int_X g_\varepsilon \, d\mu \right| + \|g_\varepsilon - \mathbf{1}_V\|_{L^1(\mu)}$$

and applying Theorem 11.22, we see that

$$\limsup_{T \to \infty} \frac{1}{T} \int_0^T \mathbf{1}_V(S_t x)\, dt \leq \mu(V) + \varepsilon\,.$$

Similarly, $\liminf_{T \to \infty} T^{-1} \int_0^T \mathbf{1}_V(S_t x)\, dt \geq \mu(V) - \varepsilon$. Since $\varepsilon > 0$ is arbitrary, this concludes the proof.   □

The semigroup we are concerned with is the **Kronecker flow** $\mathcal{K} = (\mathcal{K}_t)_{t>0}$. It is defined on the polycircle $\mathbb{T}^N$ (for any $N \geq 1$) by

$$\mathcal{K}_t(z) := (p_1^{-it} z_1, \ldots, p_N^{-it} z_N)\,.$$

LEMMA 11.24   *The Kronecker flow is uniquely ergodic on $\mathbb{T}^N$ for any $N \geq 1$ with invariant measure λ^N, the Haar measure on $\mathbb{T}^N$.*

PROOF   The Haar measure is rotation-invariant, so it is $\mathcal{K}$-invariant. Conversely, if a Borel probability measure μ is $\mathcal{K}$-invariant then μ is invariant under *any* rotation of $\mathbb{T}^N$ by Kronecker's theorem and hence $\mu = \lambda^N$.   □

REMARK   There are proofs of Kronecker's theorem which give directly the strong conclusion of Theorem 11.22 for the Kronecker flow; see e.g. Exercise 11.3. However, it is more instructive to introduce the notion of unique ergodicity, which puts Kronecker's theorem inside a broader context.

11.7 The second half of the proof

We are now ready to give the proof of Theorem 11.2.

PROOF OF THEOREM 11.2   Let us fix $f \in H^*(\Omega)$, $\varepsilon, \delta > 0$ and a compact set $K \subset \Omega$. Also, let U be a smooth Jordan domain with $K \subset U$ and $\overline{U} \subset \Omega$. We choose a positive real number η according to Lemma 11.3, with $c = \delta$. Finally, let n_0 be a positive integer to be specified later.

By Corollary 11.17, one can find $N \geq n_0$ and $w_1, \ldots, w_N \in \mathbb{T}$ such that

$$\sup_{s \in \overline{U}} \left| f(s) - \prod_{j=1}^{N} \left(\frac{1}{1 - w_j p_j^{-s}} \right) \right| < \varepsilon. \tag{11.14}$$

In what follows, we denote by V an open cube of $\mathbb{T}^N$ made up of $(w_1, \ldots, w_N) \in \mathbb{T}^N$ satisfying (11.14). In particular, $\lambda^N(\partial V) = 0$.

Now let $n \geq N$. For $j \in \{1, \ldots n\}$ and $(s, z) \in K \times \mathbb{T}^N$, we set

$$F_j(s, z) := \prod_{k=1}^{j} \left(\frac{1}{1 - z_k p_k^{-s}} \right).$$

Then $F_j(s, \mathcal{K}_t \overline{1}) = \zeta_j(s + it)$, where $\overline{1} = (1, \ldots, 1) \in \mathbb{T}^n$. Thus F_j can be seen as the "image" of the Dirichlet product ζ_j on the polycircle $\mathbb{T}^n$. Finally, we put $V^n = \{z \in \mathbb{T}^n; \pi_N(z) \in V\}$, where $\pi_N : \mathbb{T}^n \to \mathbb{T}^N$ is the canonical projection, and we define

$$D := \{t \in \mathbb{R}_+; \mathcal{K}_t \overline{1} \in V^n\}.$$

It is important to note here that D *does not depend on* $n \geq N$, since only the first N coordinates of $\mathcal{K}_t \overline{1}$ are involved.

By Corollary 11.23 and the unique ergodicity of the Kronecker flow, the set D has density $\lambda^N(V) > 0$. Indeed,

$$\frac{1}{T} \int_0^T \mathbf{1}_D(t) \, dt = \frac{1}{T} \int_0^T \mathbf{1}_{V^n}(\mathcal{K}_t \overline{1}) \, dt \xrightarrow{T \to \infty} \lambda^n(V^n) = \lambda^N(V).$$

Moreover, condition (1) in Theorem 11.2 is satisfied by the very definition of D. Let us show that (2) holds true as well, provided that n_0 has been chosen large enough. We first apply Corollary 11.23:

$$\frac{1}{T} \int_0^T \mathbf{1}_D(t) \int_U |\zeta_n(s + it) - \zeta_N(s + it)|^2 \, dA(s) dt$$

$$= \frac{1}{T} \int_0^T \mathbf{1}_{V^n}(\mathcal{K}_t \overline{1}) \int_U |F_n(s, \mathcal{K}_t \overline{1}) - F_N(s, \mathcal{K}_t \overline{1})|^2 \, dA(s) dt$$

$$\xrightarrow{T \to \infty} \int_{\mathbb{T}^n} \mathbf{1}_{V^n}(z) \int_U |F_n(s, z) - F_N(s, z)|^2 \, dA(s) d\lambda^n(z).$$

Next, we note that if $z \in V^n$ and $s \in \overline{U}$ then

$$|F_n(s,z) - F_N(s,z)| = |F_N(s,z)| \times \left|1 - \frac{F_n(s,z)}{F_N(s,z)}\right|$$

$$\leq (\|f\|_{C(\overline{U})} + \varepsilon)\left|1 - \frac{F_n(s,z)}{F_N(s,z)}\right|.$$

Let us put

$$M := \int_{\mathbb{T}^n} \mathbf{1}_{V^n}(z)\left|1 - \frac{F_n(s,z)}{F_N(s,z)}\right|^2 d\lambda^n(z).$$

Since $1 - F_n(s,z)/F_N(s,z)$ does not depend on the first N coordinates of z, we have

$$M = \lambda^N(V) \int_{\mathbb{T}^{n-N}} \left|1 - \prod_{k=N+1}^{n} \frac{1}{1 - p_k^{-s}z_k}\right|^2 d\lambda^{n-N}(z).$$

Since D has density $\lambda^N(V)$, this leads to

$$\int_{\mathbb{T}^n} \mathbf{1}_{V^n}(z) \int_U |F_n(s,z) - F_N(s,z)|^2 \, dA(s)d\lambda^n(z)$$

$$\leq C\operatorname{dens}(D) \int_{\mathbb{T}^{n-N}} \left|1 - \prod_{k=N+1}^{n} \frac{1}{1 - p_k^{-s}z_k}\right|^2 d\lambda^{n-N}(z),$$

for some constant C which does not depend on n and N.

Now, we compute the last integral by expanding the product:

$$1 - \prod_{k=N+1}^{n} \frac{1}{1 - p_k^{-s}z_k} = -\sum_{\alpha_{N+1},\ldots,\alpha_n \geq 0} (p_{N+1}^{\alpha_{N+1}} \cdots p_n^{\alpha_n})z_{N+1}^{\alpha_{N+1}} \cdots z_n^{\alpha_n}.$$

By the orthogonality of the characters in $\mathbb{T}^{n-N}$, this yields

$$M \leq C\operatorname{dens}(D) \sum_{\alpha_{N+1},\ldots,\alpha_n \geq 0} (p_{N+1}^{\alpha_{N+1}} \cdots p_n^{\alpha_n})^{-2\operatorname{Re}(s)}$$

$$\leq C\operatorname{dens}(D) \sum_{k \geq n_0} \frac{1}{k^{2\sigma_0}},$$

where $\sigma_0 = \inf\{\operatorname{Re}(s); \; s \in K\}$. In particular, the quantity M is very small, say less than η/C if n_0 is large enough, regardless of $n \geq N$.

Assuming n_0 is now chosen in this way, we can summarize our discussion as follows: given $n \geq N$, one can find $T_0 > 0$ such that

$$\frac{1}{T} \int_0^T \mathbf{1}_D(t) \int_U |\zeta_n(s+it) - \zeta_N(s+it)|^2 \, dA(s)dt < \eta \operatorname{dens}(D)$$

for every $T \geq T_0$. By Lemma 11.3, this concludes the proof. $\qquad\square$

REMARK If we could take $n = \infty$ in Theorem 11.2, we would be able to dispense with proving Theorem 11.1(!) Unfortunately, the "image" of the zeta function in the infinite polycircle $\mathbb{T}^{\mathbb{N}}$ is not continuous, so it would appear to be difficult simply to replace ζ_n by ζ in the above proof without using something like Theorem 11.1.

11.8 Comments and exercises

The universality properties of the zeta function were first observed by H. Bohr in [56]. Bohr proved that for any $\sigma \in (1/2, 1)$, the function $\mathbb{R} \ni t \mapsto \zeta(\sigma + it)$ has dense range in $\mathbb{C}$. Kronecker's theorem and the idea of passing to the polycircle $\mathbb{T}^N$ are already present in [56].

Voronin's theorem was stated in [234] without the condition on the lower density, and with an approximation valid only on a disk $D(3/4, r)$, with $r < 1/4$. The result was improved in [205] and [15]. The book [163] presents several variants for some other Dirichlet series. For applications of Voronin's theorem, see Exercises 11.4 and 11.5.

EXERCISE 11.1   Let U be a bounded domain in $\mathbb{C}$.

1. Show that if $f \in A^2(U)$ and $z \in U$, then $|f(z)| \leq (\sqrt{\pi}\, d(z, \partial U))^{-1}\, \|f\|_{A^2(U)}$. (*Hint:* Express $f(z)$ as an integral over some suitable disk.) Deduce that $A^2(U)$ is a closed subspace of $L^2(U)$ (and hence a Hilbert space) and that convergence in $A^2(U)$ entails uniform convergence on compact sets.
2. Assume that U is the unit disk $\mathbb{D}$.

   (a) Show that $\|f\|_{A^2(\mathbb{D})}^2 = \pi \sum_{n=0}^{\infty} |c_n|^2/(n+1)$ for any $f \in A^2(\mathbb{D})$, $f(z) = \sum_0^{\infty} c_n z^n$. (*Hint:* Use polar coordinates and Parseval's formula.)

   (b) Deduce that the sequence $\left(\sqrt{(n+1)/\pi}\, z^n \right)_{n \in \mathbb{N}}$ is an orthonormal basis of $A^2(\mathbb{D})$.

   (c) For $a \in \mathbb{D}$, find the reproducing kernel $K_a \in A^2(\mathbb{D})$. (*Hint:* Compute $\langle K_a, z^n \rangle$ for $n \in \mathbb{N}$.)
3. Assume that U is a smooth Jordan domain. Then there exists some conformal map $\phi : \mathbb{D} \to U$ such that ϕ and ϕ^{-1} are smooth up to the boundaries of $\mathbb{D}$ and U (see [37]).

   (a) Let $f \in A^2(U)$, and let $\varepsilon > 0$. Show that one can find a function $\tilde{f}$ that is holomorphic in a neighbourhood of $\overline{U}$ such that $\|\tilde{f} - f\|_{L^2(U)} < \varepsilon$. (*Hint:* For $r \in (0, 1)$, put $f_r(z) := (f \circ \phi)(r\phi^{-1}(z))$. Consider first the case $U = \mathbb{D}$ and then use the change of variable formula.)

   (b) Show that the polynomials are dense in $A^2(U)$. (*Hint:* With the notation of (a), use Runge's theorem to approximate $\tilde{f}$ by a polynomial on $\overline{U}$.)

   Remark   The polynomials are dense in $A^2(U)$ for *any* Jordan domain U (not necessarily smooth) and in fact for a larger class of simply connected domains; see e.g. [177, III, 3.16] and [62].

EXERCISE 11.2   **Sums of sines**
Let (b_n) be a non-increasing sequence of positive numbers such that the sequence (nb_n) is bounded, say $nb_n \leq M$. We put $S_N(x) := \sum_{n=1}^{N} b_n \sin(nx)$. The aim of the exercise is to show that the sequence (S_N) is uniformly bounded on $\mathbb{R}$. Accordingly, we fix $x \in (0, \pi)$ and put $\nu := [\pi/x]$.

1. Let $D_n(x) := \sum_{k=1}^{n} \sin(kx)$. Show that $|D_n(x)| \leq \pi/x$ for all $n \geq 1$.
2. Show that if $N \leq \nu$ then $|S_N(x)| \leq M\pi$.
3. Show that if $N > \nu$ then $\left| \sum_{n=\nu+1}^{N} b_n \sin(nx) \right| \leq 2M$. (*Hint:* Use summation by parts.)
4. Conclude that $|S_N(x)| \leq M(\pi + 2)$ for all $N \geq 1$.

EXERCISE 11.3   **Kronecker's theorem**
In this exercise, we outline two proofs of Kronecker's theorem. So, we fix $\theta_1, \ldots, \theta_N \in \mathbb{R}$ and assume that $1, \theta_1, \ldots \theta_N$ are $\mathbb{Q}$-linearly independent. Putting $\omega_j := e^{i\theta_j}$, we have to show that $G := \{(\omega_1^m, \ldots, \omega_N^m);\ m \in \mathbb{N}\}$ is dense in $\mathbb{T}^N$.

1. *First proof* (Weyl's criterion)

(a) Show that if $f_1, \ldots, f_N : \mathbb{T} \to \mathbb{C}$ are arbitrary continuous functions then

$$\frac{1}{M} \sum_{m=0}^{M-1} \prod_{j=1}^{N} f_j(\omega_j^m) \xrightarrow{M \to \infty} \prod_{j=1}^{N} \int_{\mathbb{T}} f_j \, d\lambda \,,$$

where λ is the Lebesgue measure on $\mathbb{T}$. (*Hint:* Suppose first that $f_j(z) = z^{k_j}$, where $k_j \in \mathbb{Z}$.)

(b) Deduce the required result.

(c) Adapt the above proof to show directly that the Kronecker flow satisfies the conclusion of Theorem 11.22. (*Hint:* Replace the sums by integrals, and use the Stone–Weierstrass theorem.)

2. *Second proof* (Bohr compactification) Let us denote by Γ the group of all homomorphisms $\gamma : \mathbb{T} \to \mathbb{T}$ endowed with the topology of pointwise convergence. In other words, Γ is the character group of $\mathbb{T}_d$, the group $\mathbb{T}$ equipped with the discrete topology. Then $\mathbb{Z}$ is canonically identified with a subgroup of Γ, since any $m \in \mathbb{Z}$ gives rise to the character $\gamma_m(\xi) := \xi^m$. Moreover, $\mathbb{Z}$ is *dense* in Γ (*Bohr's theorem*; see [208, 1.8]).

(a) Prove *Dirichlet's theorem*, that any neighbourhood of 0 in Γ contains arbitrarily large positive integers. (*Hint:* Given a finite set $F \subset \mathbb{T}$, use the compactness of $\mathbb{T}$ to find a rapidly increasing sequence of integers (p_k) such that ξ^{p_k} has a limit for every $\xi \in F$.)

(b) Deduce that $\mathbb{N}$ is dense in Γ.

(c) Put $\Omega := \{\omega_1, \ldots, \omega_N\}$. Show that any map $f : \Omega \to \mathbb{T}$ can be extended to a character $\gamma \in \Gamma$. (*Hint:* Extend f to the subgroup of $\mathbb{T}$ generated by Ω. Then use Zorn's lemma.)

(d) Prove Kronecker's theorem.

EXERCISE 11.4 **Density of the values of** ζ ([233])

1. Let $a_0, \ldots, a_N \in \mathbb{C}$, with $a_0 \neq 0$. Show that one can find an entire function f without zeros such that $f^{(k)}(0) = a_k$ for $k = 0, \ldots, N$. (*Hint:* Look for a function of the form $f = e^P$, where P is a polynomial.)

2. Let $\sigma \in (1/2, 1)$. Show that the map $\mathbb{R} \ni t \mapsto (\zeta(\sigma + it), \ldots, \zeta^{(N)}(\sigma + it)) \in \mathbb{C}^{N+1}$ has dense range.

EXERCISE 11.5 **Functional independence of** ζ ([233])   Show that if $F_0, \ldots, F_N$ are continuous functions on $\mathbb{C}^{N+1}$ such that $\sum_{k=0}^{N} s^k F_k(\zeta(s), \zeta'(s), \ldots, \zeta^{(N)}(s)) = 0$ for all $s \in \mathbb{C} \backslash \{1\}$ then $F_0 = \cdots = F_N = 0$. (*Hint:* Argue by contradiction. Consider the largest k such that $F_k \neq 0$, and use the previous exercise with real numbers t tending to infinity.)

EXERCISE 11.6 **Disjoint hypercyclicity of** ζ
Let a, a' be two positive real numbers such that $a \log(n) \pm a' \log(m) \notin \mathbb{Z}$ for any integers $n, m \geq 2$.

1. Let N be a positive integer, and let $\mathcal{S} = (S_t)_{t>0}$ be the semigroup of measurable transformations on $\mathbb{T}^N \times \mathbb{T}^N$ defined by

$$S_t(z, z') = (p_1^{-ita} z_1, \ldots, p_N^{-ita} z_N, p_1^{-ita'} z_1', \ldots, p_N^{-ita'} z_N') \,.$$

Show that $\mathcal{S}$ is uniquely ergodic, and find the corresponding invariant measure.

2. Let $f, g \in H^*(\Omega)$. Show that for any compact set $K \subset \Omega$ and every $\varepsilon > 0$, one can find positive real numbers t such that

$$|\zeta(s + ita) - f(s)| < \varepsilon \quad \text{and} \quad |\zeta(s + ita') - g(s)| < \varepsilon \quad \text{for all } s \in K \,.$$

12

An introduction to Read-type operators

Introduction

In this final chapter, our aim is to give a short and gentle introduction to the kind of operator constructed by C. J. Read in the 1980s.

In his 1987 paper [202], Read solved in the negative the invariant subset problem for the space $\ell^1(\mathbb{N})$, and in fact for any separable Banach space containing a complemented copy of ℓ^1. In other words, he was able to produce on such a space an operator for which every non-zero vector is hypercyclic.

The construction carried out in [202] is something of a *tour de force*. Moreover, its understanding requires some familiarity with earlier constructions by the same author relating to the invariant *subspace* problem, which are already quite involved (see e.g. [201]). This convinced us that we should not be overly ambitious regarding the material presented in this chapter. Thus, we have chosen to concentrate on the simplest example of a "Read-type" operator, i.e. an operator T acting on $\ell^1(\mathbb{N})$ for which every non-zero vector x is *cyclic*. This is a counter-example to the invariant subspace problem for the space ℓ^1.

As already said, we have tried to give a helpful presentation, meaning that we have made some effort to explain the underlying ideas as we understand them, inserting heuristic comments whenever this seemed necessary. Some parts of the discussion are deliberately informal, but the construction is nevertheless complete and self-contained. We hope that this chapter will be useful for people interested in that kind of question.

12.1 The strategy

Throughout the chapter, we denote by $(e_j)_{j \in \mathbb{N}}$ the canonical basis of $X = \ell^1(\mathbb{N})$, and we put $c_{00} := \mathrm{span}\{e_j; \ j \in \mathbb{N}\}$. As a rule, an element of c_{00} will be denoted by the letters y or z, and an arbitrary vector in X will be denoted by $x. \ x \in X$.

We shall be working with a variable linear map $T : c_{00} \to c_{00}$. When T is bounded with respect to the ℓ^1 norm, we also denote by T its extension acting on the space $X = \ell^1(\mathbb{N})$. At the end of the chapter, T is the required operator, at which we will have arrived by successive approximations. Each approximation is denoted by the same generic letter T.

We shall *always* assume that T is an upper-triangular perturbation of a weighted forward shift; that is, each vector $T(e_j)$ has the form $\lambda_j e_{j+1} + f_j$, where $\lambda_j > 0$ and $f_j \in \mathrm{span}(e_0, \ldots, e_j)$. In that way, we have $\mathrm{span}(e_0, \ldots, T^N e_0) = \mathrm{span}(e_0, \ldots, e_N)$ for every $N \in \mathbb{N}$. In particular, the vectors $T^j e_0 \ (j \in \mathbb{N})$ span

c_{00}, so that e_0 is a cyclic vector for T if T is already known to be bounded. Thus, to show that a given vector $x \in X$ is cyclic for T, assuming that T is bounded, it is enough to approximate the single vector e_0 by vectors of the form $P(T)x$, where P is a polynomial. Hence, we are looking for a bounded upper-triangular perturbation of a weighted shift T such that, for any non-zero vector $x \in X$ and every $\varepsilon > 0$,

$$\exists P \text{ polynomial} : \quad \|P(T)x - e_0\| \le \varepsilon. \tag{12.1}$$

We will proceed as follows. First we explain how to ensure (12.1) for some fixed $\varepsilon > 0$ and all finitely supported vectors $x = y \ne 0$. Then we show how to get (12.1) for some fixed ε and every non-zero vector $x \in X$. Finally, we indicate how the construction can be modified in order to let ε tend to 0.

An important part of the proof will consist in showing that T and several related operators are bounded. To do this, the specificity of the ℓ^1 norm is crucial. Indeed, in principle it is very easy to show that a linear map $L : c_{00} \to c_{00}$ is bounded with respect to the ℓ^1 norm, since one just has to check that $\sup_j \|L(e_j)\| < \infty$. This is helpful from a practical point of view, but it is more than just a matter of convenience: after all, the invariant subspace problem remains open on ℓ^p for any $p > 1$.

Before we start the discussion, we will note that, since we are considering only upper-triangular perturbations of weighted shifts, our linear map $T : c_{00} \to c_{00}$ is perfectly well defined provided that we specify consistently the vectors $T^j e_0$, $j \in \mathbb{N}$. In other words, we may define T by the vectors $T^j e_0 := h_j$, provided that h_j has the required form ($h_j = \alpha_j e_j + k_j$ with $\alpha_j > 0$ and $k_j \in \mathrm{span}(e_i; i < j)$) and that the definition makes sense, i.e. h_j can be computed if we already know the vectors $T^i e_0$ for all $i < j$.

Now we settle some more notation and terminology. We will say that a vector $y = \sum_j y_j e_j \in c_{00}$ is *supported* on some set $J \subset \mathbb{N}$ if $y_j = 0$ for all $j \notin J$, and in that case we often write $y \subset J$. Likewise we will say that a polynomial P is supported on J if it can be written as $P(t) = \sum_{j \in J} p_j t^j$, and we will occasionally write $P \subset J$. The degree of a polynomial P is denoted by $\deg(P)$ and its valuation by $\mathrm{val}(P)$. Thus P is supported on the interval $[\mathrm{val}(P), \deg(P)]$. We denote by $|P|_1$ the ℓ^1 norm of P, that is, $|P|_1 = \sum_j |p_j|$ if $P(t) = \sum p_j t^j$. If y is any non-zero vector in c_{00}, we denote by $\mathrm{val}_T(y)$ the index of the first non-zero coefficient y_j when y is written as $\sum_j y_j T^j e_0$. For any $N \in \mathbb{N}$, we set

$$E_N := \mathrm{span}(e_0, \dots, e_N) = \mathrm{span}(e_0, \dots, T^N e_0).$$

Finally, the symbol $C_T(N)$ will denote some constant which depends only on the vectors $T^j e_0$, $j \le N$. As usual, the exact value of this constant $C_T(N)$ may change from line to line. We are now ready to start.

12.2 First step

The first step of the construction is rather natural. In order to ensure (12.1) for every finitely supported vector $x = y \ne 0$, it is convenient to assume that $T^a e_0$ is close to

e_0 for large integers a, say $T^a(e_0) = \varepsilon e_a + e_0$. Then, a combination of linear algebra and compactness arguments will do the job. What we need is the following observation, where $\varepsilon > 0$ is fixed. The condition below on the support of the polynomial P may look unimportant at this stage, but in fact later on it will turn out to be quite important.

OBSERVATION 1 *Assume that the linear map* $T : c_{00} \to c_{00}$ *satisfies*

$$T^a(e_0) = \varepsilon e_a + e_0$$

for some $a \in \mathbb{N}$. *Let* $\Delta > 0$, *and let* K *be a compact set depending only on the vectors* $T^j e_0$, $j < a + \Delta$, *with* $K \subset \{y \subset [0, a+\Delta); \; y \neq 0 \text{ and } \mathrm{val}_T(y) \leq a\}$. *Finally, put* $\mathrm{val}_T(K) := \max\{\mathrm{val}_T(y); \; y \in K\}$. *Then, for any* $y \in K$, *one can find a polynomial* P *supported on* $[a - \mathrm{val}_T(K), a + \Delta)$ *such that* $|P|_1 \leq C_T(a + \Delta - 1)$ *and*

$$\|P(T)y - e_0\| < 2\varepsilon + C_T(a + \Delta - 1) \times \max_{a+\Delta \leq j \leq 2(a+\Delta-1)} \|T^j e_0\| .$$

PROOF Let us denote by $T_{a+\Delta-1} : E_{a+\Delta-1} \to E_{a+\Delta-1}$ the operator T "truncated at the level $a + \Delta - 1$", that is, the operator defined by $T_{a+\Delta-1}(T^j e_0) = T^{j+1} e_0$ if $j < a + \Delta - 1$ and $T_{a+\Delta-1}(T^{a+\Delta-1} e_0) = 0$. If z is any non-zero vector in $E_{a+\Delta-1}$ then the vectors $z, T_{a+\Delta-1}z, \ldots, T_{a+\Delta-1}^{a+\Delta-1-\mathrm{val}_T(z)} z$ form a basis of $\mathrm{span}\{T^j e_0; \; \mathrm{val}_T(z) \leq j < a + \Delta\}$. Thus, if $\mathrm{val}_T(z) \leq a$, and in particular if $z \in K$, then one can find a polynomial P_z supported on $[a - \mathrm{val}_T(z), a + \Delta)$ such that $P_z(T_{a+\Delta-1})z = T^a e_0$, that is, $P_z(T_{a+\Delta-1})z = \varepsilon e_a + e_0$. Furthermore, one can find an open ball B_z centred at z such that $\|P_z(T_{a+\Delta-1})y - e_0\| < 2\varepsilon$ for all $y \in B_z$; the radius of this ball can be controlled by z and $T_{a+\Delta-1}$. By compactness, one can cover K by finitely many such balls $B_{z_1}, \ldots, B_{z_k}$, where the number k depends only on K. Since $T_{a+\Delta-1}$ and K depend only on the vectors $T^j e_0$, $j < a + \Delta$, we have proved the following: for any $y \in K$, one can find a polynomial P supported on $[a - \mathrm{val}_T(K), a + \Delta)$ such that $|P|_1 \leq C_T(a + \Delta - 1)$ and $\|P(T_{a+\Delta-1})y - e_0\| < 2\varepsilon$.

Now we replace the truncated operator $T_{a+\Delta-1}$ by T itself. If $y = \sum_k y_k T^k e_0 \in K$ and $P(t) = \sum_l p_l t^l$ are as above then, by the definition of $T_{a+\Delta-1}$ and since $\deg(P) \leq a + \Delta - 1$, we can write

$$P(T)y - P(T_{a+\Delta-1})y = \sum_{k+l \geq a+\Delta} y_k p_l T^{l+k} e_0$$

$$= \sum_{j=a+\Delta}^{2(a+\Delta-1)} \lambda_j T^j e_0 .$$

Moreover, we have $\sum_j |\lambda_j| \leq |P|_1 \sum_k |y_k| \leq C_T(a + \Delta - 1)$, because the norm $|\cdot|_T$ defined on $E_{a+\Delta-1}$ by $|\sum_k y_k T^k e_0|_T := \sum_k |y_k|$ is equivalent to the original norm of X up to a constant depending only on the vectors $T^j e_0$, $j < a + \Delta$. Thus we get

$$\|P(T)y - P(T_{a+\Delta-1})y\| \leq C_T(a + \Delta - 1) \times \max_{a+\Delta \leq j \leq 2(a+\Delta-1)} \|T^j e_0\| ,$$

which concludes the proof.   □

In view of Observation 1 (and replacing ε by 3ε), we see that (12.1) will be satisfied for a fixed $\varepsilon > 0$ and every non-zero $x = y \in c_{00}$ provided that we have at hand some $\Delta > 0$, a sequence of integers $(a_n)_{n \in \mathbb{N}}$ with $a_n < a_n + \Delta \ll a_{n+1}$ for all n and a sequence of compact sets (K_n) with

$$K_n \subset \{y \subset [0, a_n + \Delta); \ y \neq 0 \text{ and } \mathrm{val}_T(y) \leq a_n\}$$

such that

(i) $T^{a_n} e_0 = \varepsilon e_{a_n} + e_0$ for every $n \in \mathbb{N}$;
(ii) $\|T^j e_0\|$ is sufficiently small whenever $j \in [a_n + \Delta, 2(a_n + \Delta - 1)]$;
(iii) every non-zero vector $y \in c_{00}$ belongs to at least one K_n.

One way to ensure the existence of K_n and these conditions is to proceed as follows: given (a_n) and Δ, we define our first approximation T by

$$T^j e_0 := \begin{cases} \varepsilon e_j + T^{j-a_n} e_0 & j \in [a_n, a_n + \Delta), \\ \alpha_j e_j & j \in \mathbb{N} \setminus \bigcup_n [a_n, a_n + \Delta), \end{cases} \quad (12.2)$$

where the α_j are positive numbers to be specified. The reader may wonder why we are requiring that $T^j e_0 = \varepsilon e_j + T^{j-a_n} e_0$ for all $j \in [a_n, a_n + \Delta)$ and not just $T^{a_n} e_0 = \varepsilon e_{a_n} + e_0$. The reason is that, when looking at the boundedness of the operator T, an unpleasant factor $1/\varepsilon$ will appear, which we can keep under control thanks to the coefficient α_Δ; see below.

Now we put

$$K_n := \left\{ y \subset [0, a_n + \Delta); \ \|\tau_{a_n}(y)\| \geq 2^{-n} \text{ and } \|y\| \leq 2^n \right\}, \quad (12.3)$$

where $\tau_{a_n} : E_{a_n + \Delta - 1} \to E_{a_n - 1}$ is the projection defined by

$$\tau_{a_n}(T^j e_0) := \begin{cases} T^j e_0 & j < a_n, \\ 0 & j \geq a_n. \end{cases}$$

Clearly each K_n depends only on $T_{|E_{a_n + \Delta - 1}}$, and if $y \in c_{00}$ is non-zero then $y \in K_n$ for all but finitely many n, since τ_{a_n} is the identity on $E_{a_n - 1}$.

To make the first part of our plan work, we need to take the coefficients α_j as small for all $j \in [a_n + \Delta, 2(a_n + \Delta - 1)]$. More precisely, in view of the estimate given in Observation 1, we require that

$$\max_{a_n + \Delta \leq j \leq 2(a_n + \Delta - 1)} \alpha_j \leq \frac{\varepsilon}{C_T(a_n + \Delta - 1)}. \quad (12.4)$$

Regarding the boundedness of our operator T, we first note that T acts as a weighted forward shift on all basis vectors e_j for which $j + 1$ is not one of the critical indices a_n or $a_n + \Delta$. Explicitly, $T(e_j) = (\alpha_{j+1}/\alpha_j) e_{j+1}$ if $j \in \mathbb{N} \setminus \bigcup_n [a_n - 1, a_n + \Delta)$ and $T(e_j) = e_{j+1}$ if $j \in [a_n, a_n + \Delta - 1)$ for some n. Hence, this part of T defines a bounded operator if α_{j+1}/α_j remains bounded.

To compute $T(e_{a_n - 1})$, we write $T^{a_n}(e_0) = T(T^{a_n - 1} e_0) = \alpha_{a_n - 1} T(e_{a_n - 1})$, which yields

$$T(e_{a_n-1}) = \frac{1}{\alpha_{a_n-1}} \left(\varepsilon e_{a_n} + e_0 \right).$$

Likewise, writing $\alpha_{a_n+\Delta} e_{a_n+\Delta} = T^{a_n+\Delta} e_0 = T(\varepsilon e_{a_n+\Delta-1} + T^{\Delta-1} e_0)$, we get

$$T(e_{a_n+\Delta-1}) = \frac{\alpha_{a_n+\Delta}}{\varepsilon} e_{a_n+\Delta} - \frac{1}{\varepsilon} T^{\Delta}(e_0).$$

Thus, since we are working on $X = \ell^1$, we see that T is bounded by $\|T\| \le 2$ provided that

$$\begin{cases} \alpha_{j+1}/\alpha_j \le 2 & \text{whenever this is defined,} \\ \alpha_{a_n-1} \ge 1/\varepsilon & \text{for all } n, \\ \alpha_{a_n+\Delta} \le \varepsilon & \text{for all } n, \\ \|T^{\Delta} e_0\| \le \varepsilon. \end{cases} \tag{12.5}$$

The first three properties are compatible, since we can always reach a large $\alpha_{a_{n+1}-1}$ by starting with a small $\alpha_{2(a_n+\Delta-1)}$ and using "jumps" α_{j+1}/α_j less than 2 if a_{n+1} is large enough with respect to a_n. We also note that the requirement that $\alpha_{a_n+\Delta}$ should be small has already been met; see (12.4) above.

The last property, $\|T^{\Delta} e_0\| \le \varepsilon$, can be ensured by taking $a_0 > \Delta$ (so that $T^{\Delta} e_0 = \alpha_\Delta e_\Delta$) and $\alpha_\Delta \le \varepsilon$. This means in particular that Δ is in some way "attached" to ε, and this is the reason for requiring that the identities $T^j e_0 = \varepsilon e_{a_n} + T^{j-a_n} e_0$ hold in the whole interval $[a_n, a_n + \Delta)$. If we just considered $T^{a_n} e_0 = \varepsilon e_{a_n} + e_0$, i.e. if we took $\Delta = 1$, then we would have to control $1/\varepsilon$ by $Te_0 = \alpha_1 e_0$ rather than $T^{\Delta} e_0$. When ε is fixed this would cause no trouble but if we are letting ε go to 0 then we need a variable Δ.

At this point, it is important to be conscious that we have already used the specificity of the ℓ^1 norm. Indeed, the operator T is not bounded on any ℓ^p, $p > 1$, since infinitely many e_j satisfy $\pi_{[0,\Delta]} T(e_j) = \varepsilon^{-1} T^{\Delta} e_0$. Nevertheless, we have carried through the first part of our plan: if conditions (12.4) and (12.5) are satisfied then the operator T defined by (12.2) is bounded (with $\|T\| \le 2$) and satisfies (12.1) for 3ε and all finitely supported vectors $x = y \ne 0$.

REMARK 12.1 To conclude this section, a few more words must be added about the coefficients α_j. What comes out of the above discussion is that three things are required concerning these coefficients: the jumps α_{j+1}/α_j should be bounded; α_j has to be large near $j = a_n - 1$, i.e. at the end of the interval $[a_{n-1} + \Delta, a_n)$; α_j has to be small near $j = a_n + \Delta$, i.e. at the beginning of the interval $[a_n + \Delta, a_{n+1})$. The same properties will be needed later on, when the operator T becomes more complicated. We will even require that the jumps α_{j+1}/α_j are very close to 1.

The coefficients α_j are defined for all but some "critical" indices j. These critical indices j determine a countable family of maximal intervals $[A, B)$ where the α_j are defined. In such an interval $[A, B)$ the coefficients should increase very slowly from some "small" value α_A to some "large" value α_{B-1}. Thus, α_j should be small near $j = A$ and large near $j = B - 1$, and the jumps α_{j+1}/α_j should be very close to 1 in the intervals $[A, B - 1)$. These requirements are compatible because B is always much greater than A. For example, we may pass from $\alpha_A = 2^{-\sqrt{B}}$ to $\alpha_{B-1} = 2^{\sqrt{B}}$ with jumps $\alpha_{j+1}/\alpha_j \leq 16^{1/\sqrt{B}}$ if $B > 2A$.

12.3 Second step

We now come to the second step of our plan. That is, we would like to ensure (12.1) not only for all finitely supported vectors $x = y \neq 0$ but for every non-zero vector $x \in X$. At first sight, it is natural to write

$$x = \pi_{[0, a_n + \Delta)}(x) + x_n,$$

where $\pi_{[0, a_n + \Delta)} : X \to E_{a_n + \Delta - 1}$ is the canonical projection from X onto $E_{a_n + \Delta - 1}$, and to hope that the "tail" x_n will not cause too much trouble.

Although it may not be immediately apparent, we note that $\pi_{[0, a_n + \Delta)}(x) \in K_n$ for all but finitely many n. Indeed, we have $\tau_{a_n}(e_j) = e_j$ if $j < a_n$ (because $e_j \in \mathrm{span}(e_0, \ldots, T^j e_0)$) and

$$\tau_{a_n}(e_j) = \frac{1}{\varepsilon}\tau_{a_n}(T^j e_0 - T^{j-a_n} e_0) = -\frac{1}{\varepsilon}T^{j-a_n} e_0$$

if $j \in [a_n, a_n + \Delta)$. Thus we see that the projections τ_{a_n} are uniformly bounded, with $\|\tau_{a_n}\| \leq 2^\Delta/\varepsilon$. Since $\tau_{a_n} \pi_{[0, a_n + \Delta)} z = z$ eventually for any $z \in c_{00}$, it follows at once from (12.3) that $\pi_{[0, a_n + \Delta)}(x) \in K_n$ if n is large enough.

Choosing n such that $y_n := \pi_{[0, a_n + \Delta)}(x) \in K_n$, we may pick a polynomial P satisfying the conclusions of Observation 1 for $y = y_n$. Then, by (12.4), we get

$$\|P(T)x - e_0\| < 3\varepsilon + \|P(T)(I - \pi_{[0, a_n + \Delta)})x\|. \tag{12.6}$$

Thus we are essentially left with an estimate of $\|P(T)(I - \pi_{[0, a_n + \Delta)})\|$. For this to be small, we have to ensure that $\|P(T)e_j\|$ is small for all $j \geq a_n + \Delta$. Clearly this is hopeless if we know only that $P \subset [a_n - \mathrm{val}_T(K_n), a_n + \Delta)$ and $|P|_1 \leq C_T(a_n + \Delta - 1)$. So we have to do something else; and this means that our operator T is going to become more complicated.

One rather natural idea is to try to replace the polynomial P by another polynomial Q such that $|Q|_1$ is *small* rather than just controlled by some possibly large quantity $C_T(a_n + \Delta - 1)$. And yes, we can do this, as shown by the next observation.

OBSERVATION 2 *Let everything be as in Observation 1, and put* $v := a - \mathrm{val}_T(K)$. *Also, let* $b > a + \Delta$. *Assume that*

$$T^{b+v+i} e_0 = e_{b+v+i} + b T^{v+i} e_0 \quad \text{for all } i \in [0, a + \Delta - v).$$

Then, for any $y \in K$, *one can find a polynomial* $P \subset [v, a + \Delta)$ *such that* $|P|_1 \leq C_T(a + \Delta - 1)$ *and*

$$\left\| \frac{T^b}{b} P(T)y - e_0 \right\| < 2\varepsilon + \frac{C_T(a + \Delta - 1)}{b} \left(1 + \max_{b+a+\Delta \leq j \leq b+2(a+\Delta-1)} \|T^j e_0\| \right).$$

PROOF   The proof is similar to that of Observation 1. Given $y \in K$, we first find a polynomial $P \subset [v, a + \Delta)$ such that $\|P\|_1 \leq C_T(a + \Delta - 1)$ and

$$\|P(T_{a+\Delta-1})y - e_0\| < 2\varepsilon,$$

where $T_{a+\Delta-1} \colon E_{a+\Delta-1} \to E_{a+\Delta-1}$ is the operator T truncated at the level $a + \Delta - 1$. Next, we can write

$$P(T_{a+\Delta-1})y = T^v \sum_{i=0}^{a+\Delta-v-1} \lambda_i T^i e_0,$$

where $\sum_i |\lambda_i| \leq C_T(a + \Delta - 1)$. Since $T^{b+v+i} e_0 = e_{b+v+i} + b T^{v+i} e_0$ for all i, it follows that

$$\frac{T^b}{b} P(T_{a+\Delta-1})y = \sum_{i=0}^{a+\Delta-v-1} \frac{\lambda_i}{b} e_{b+v+i} + P(T_{a+\Delta-1})y,$$

whence

$$\left\| \frac{T^b}{b} P(T_{a+\Delta-1})y - P(T_{a+\Delta-1})y \right\| \leq \frac{C_T(a + \Delta - 1)}{b}.$$

Finally, we replace the truncated operator $T_{a+\Delta-1}$ by T itself, thanks to the estimate

$$\|T^b P(T)y - T^b P(T_{a+\Delta-1})y\| \leq C_T(a + \Delta - 1) \times \max_{b+a+\Delta \leq j \leq b+2(a+\Delta-1)} \|T^j e_0\|$$

(see the proof of Observation 1).   □

In view of Observation 2, we pick another increasing sequence of integers $(b_n)_{n \in \mathbb{N}}$, with $a_n < a_n + \Delta \ll b_n \ll a_{n+1}$ for all n. Temporarily disregarding the exact definition (12.3) of the compact sets K_n and putting $v_n := a_n - \mathrm{val}_T(K_n)$, we define our second approximation T as follows:

$$T^j e_0 = \begin{cases} \varepsilon e_j + T^{j-a_n} e_0 & j \in [a_n, a_n + \Delta), \\ e_j + b_n T^{j-b_n} e_0 & j \in [b_n + v_n, b_n + a_n + \Delta), \\ \alpha_j e_j & \text{otherwise.} \end{cases} \qquad (12.7)$$

The role of the numbers v_n will be explained below.

Since we have introduced new intervals, some care is needed regarding the boundedness of T. More precisely, we have to look at $T(e_{b_n+v_n-1})$ and $T(e_{b_n+a_n+\Delta-1})$. Now, straightforward computations (as above) give

$$T(e_{b_n+v_n-1}) = \frac{1}{\alpha_{b_n+v_n-1}} \left(e_{b_n+v_n} + b_n T^{v_n} e_0 \right)$$

and

$$T(e_{b_n+a_n+\Delta-1}) = \alpha_{b_n+a_n+\Delta} e_{b_n+a_n+\Delta} - b_n \alpha_{a_n+\Delta} e_{a_n+\Delta}.$$

It follows that the operator T remains bounded with $\|T\| \leq 2$ provided that

$$\alpha_{b_n+v_n-1} \geq b_n, \quad \alpha_{a_n+\Delta} \leq \frac{1}{b_n} \quad \text{and} \quad \alpha_{b_n+a_n+\Delta} \leq 1 \qquad (12.8)$$

for all $n \in \mathbb{N}$. These requirements are in accordance with Remark 12.1.

We also require that the coefficients α_j are less than 1 for all $j \in [b_n + a_n + \Delta, b_n + 2(a_n + \Delta - 1)]$ (which is again in accordance with Remark 12.1). Then, if b_n is large with respect to $C_T(a_n + \Delta - 1)$, the polynomial P from Observation 2 satisfies $\|(T^b/b)P(T)y - e_0\| < 3\varepsilon$.

Now let $x \in X$, and assume that $y_n = \pi_{[0,a_n+\Delta)}(x) \in K_n$ for some n. Then, by Observation 2, we obtain a polynomial $P \subset [v_n, a_n + \Delta)$ such that $|P|_1 \leq C_T(a_n + \Delta - 1)$ and

$$\left\| \frac{T^{b_n}}{b_n} P(T)x - e_0 \right\| < 3\varepsilon + \left\| \frac{T^{b_n}}{b_n} P(T)(I - \pi_{[0,a_n+\Delta)})x \right\|,$$

provided that b_n is large with respect to $C_T(a_n + \Delta - 1)$.

Write $P(t) = t^{v_n}Q(t)$ and $\widetilde{P}(t) := (t^{b_n}/b_n)P(t) = (t^{b_n+v_n}/b_n)Q(t)$. Then, since $|Q|_1 = |P|_1$, $\deg(Q) \leq a_n + \Delta$ and $\|T\| \leq 2$, we arrive at

$$\|\widetilde{P}(T)x - e_0\| < 3\varepsilon + \frac{C_T(a_n + \Delta - 1)2^{a_n+\Delta}}{b_n}\|T^{b_n+v_n}(I - \pi_n)\|\|x\|, \qquad (12.9)$$

where we have set $\pi_n := \pi_{[0,a_n+\Delta)}$.

Since b_n can be taken as much larger than a_n and $C_T(a_n + \Delta - 1)$, it follows that $\|\widetilde{P}(T)x - e_0\| < 4\varepsilon$ provided that n is large and we are able to show that $\|T^{b_n+v_n}(I - \pi_n)\|$ remains bounded. By the definition of π_n, the problem is then to estimate $\|T^{b_n+v_n}e_j\|$ for $j \geq a_n + \Delta$.

Looking back at (12.6), we see that we are now in a better situation for two reasons: we have to show that $\|T^{b_n+v_n}(I - \pi_n)\|$ is bounded, but not necessarily small, and, perhaps more importantly, we are dealing with a monomial $T^{b_n+v_n}$, which looks much more tractable than the presumably very complicated $P(T)$ appearing in (12.6).

At this point, we introduce a new increasing sequence of integers (s_n), with $s_n \ll a_n$ and $b_n \ll s_{n+1}$. Thus, we have

$$s_n \ll a_n < a_n + \Delta \ll b_n < b_n + a_n + \Delta \ll s_{n+1}.$$

The role of the integers s_n is a purely "visual" one: they are intended to separate clearly the zones $[a_n, b_n + a_n + \Delta)$ from each other, in order to make the whole picture look neater. Now, let us fix n and estimate $\|T^{b_n+v_n}e_j\|$ for $j \geq a_n + \Delta$.

We first assume that $j < s_{n+1}$, i.e. $a_n + \Delta \leq j < s_{n+1}$. If $j \geq b_n + a_n + \Delta$ then $\|T^{b_n+v_n}e_j\|$ is under control if a_{n+1} is large enough, because T acts as a weighted shift on the interval $[b_n + a_n + \Delta, a_{n+1})$ and we can require that the weights α_{j+1}/α_j are less than $1 + \eta$, for some $\eta > 0$ small enough that $(1 + \eta)^{b_n+v_n} \leq 2$ (see Remark 12.1).

If $a_n + \Delta \leq j < b_n + v_n$, then $e_j = \alpha_j^{-1}T^j(e_0)$ and hence $T^{b_n+v_n}(e_j) = \alpha_j^{-1}T^{j+b_n+v_n}(e_0)$. Now, $j + b_n + v_n$ is at least $b_n + a_n + \Delta$, so we have

$T^{j+b_n+v_n}(e_0) = \alpha_{j+b_n+v_n} e_{j+b_n+v_n}$. Thus $\|T^{b_n+v_n}(e_j)\|$ is under control in the interval $a_n+\Delta \leq j < b_n+v_n$ if we assume that $\alpha_{j+b_n+v_n} \leq \alpha_j$ for all such j, which is allowed since $j+b_n+v_n$ lives at the beginning of the interval $[b_n+a_n+\Delta, s_{n+1})$.

Finally, assume that $j \in [b_n+v_n, b_n+a_n+\Delta)$. Then the situation is a little more complicated. We have $e_j = T^j e_0 - b_n T^{j-b_n} e_0$, so that

$$T^{b_n+v_n}(e_j) = T^{j+b_n+v_n} e_0 - b_n T^{j+v_n} e_0.$$

The first term is under control, since $j+b_n+v_n \geq 2b_n > b_n+a_n+\Delta$ and hence $T^{j+b_n+v_n} e_0 = \alpha_{j+b_n+v_n} e_{j+b_n+v_n}$, which can be taken as very small since we are at the beginning of the interval $[b_n+a_n+\Delta, s_{n+1})$. For the second term, we note that $j+v_n \geq b_n+2v_n$. Thus, we will obtain the desired estimate if $2v_n \geq a_n+\Delta$, that is, v_n is large enough for $j+v_n$ to avoid the dangerous interval $[b_n+v_n, b_n+a_n+\Delta)$ and fall in the safe interval $[b_n+a_n+\Delta, s_{n+1})$. Indeed, in that case $T^{j+v_n} e_0 = \alpha_{j+v_n} e_{j+v_n}$, and α_{j+v_n} can be taken to be small enough to control b_n, because $j+v_n$ is at the beginning of $[b_n+a_n+\Delta, s_{n+1})$.

Unfortunately, things go wrong using the original definition of K_n. Indeed, from (12.3) it is clear that $\mathrm{val}_T(K_n) = a_n$ so that $v_n = 0$! Hence we need to modify the compact set K_n. From now on, we assume that $a_n \in 4\mathbb{N}$ and put

$$K_n := \left\{ y \subset [0, a_n+\Delta); \ \|y\| \leq 2^n \ \text{and} \ \|\tau_{a_n/4}(y)\| \geq 2^{-n} \right\}, \qquad (12.10)$$

where $\tau_{a_n/4} : E_{a_n+\Delta-1} \to E_{a_n/4-1}$ is defined exactly as τ_{a_n} is, but replacing a_n by $a_n/4$; that is,

$$\tau_{a_n/4}(T^j e_0) := \begin{cases} T^j e_0 & j < a_n/4, \\ 0 & j \geq a_n/4. \end{cases} \qquad (12.11)$$

In this way, we get $\mathrm{val}_T(K_n) = a_n/4$, so that $2v_n = 2 \times 3a_n/4 > a_n+\Delta$. Hence $\|T^{b_n+v_n}(e_j)\|$ is under control for all $j \in [b_n+v_n, b_n+a_n+\Delta)$ and no trouble arises from the indices $j \in [a_n+\Delta, s_{n+1})$.

For future reference, we quote here the true value of v_n:

$$v_n = 3a_n/4. \qquad (12.12)$$

We should also say something about the compact sets K_n, namely that if x is any non-zero vector in X then $\pi_{[0,a_n+\Delta)}(x) \in K_n$ for all but finitely many n. This is proved exactly as above, observing that the projections $\tau_{a_n/4}$ are uniformly bounded (see the start of section 12.3).

Now let us concentrate on $T^{b_n+v_n}(e_j)$ for $j \geq s_{n+1}$, that is, $j \in [s_p, s_{p+1})$ for some $p > n$.

If j and $j+b_n+a_n+\Delta$ both lie in one of the intervals $[s_p, a_p), [a_p+\Delta, b_p+v_p)$, $[b_p+v_p, b_p+a_p+\Delta)$ or $[b_p+a_p+\Delta, s_{p+1})$ then there will be no difficulty; in each of these intervals T acts as a weighted shift with weights equal to 1 or α_{j+1}/α_j, which can be taken less than $1+\eta$ for some $\eta > 0$ small enough to ensure that $(1+\eta)^{b_n+v_n} \leq 2$ (see Remark 12.1 again). So we just have to see what happens when we pass one of the critical indices $a_p, a_p+\Delta, b_p+v_p$ and $b_p+a_p+\Delta$.

Assume that $j < b_p + a_p + \Delta$ and $j + b_n + v_n \geq b_p + a_p + \Delta$. Then $j \geq b_p + v_p$ because the interval $[b_p + v_p, b_p + a_p + \Delta)$ has length $(a_p - v_p) + \Delta > 3a_p/4 > b_n + v_n$. Hence $e_j = T^j e_0 - b_p T^{j-b_p} e_0$, so that

$$T^{b_n+v_n}(e_j) = T^{j+b_n+v_n} e_0 - b_p T^{j+b_n+v_n-b_p} e_0$$

$$= \alpha_{j+b_n+v_n} e_{j+b_n+v_n} - \alpha_{j+b_n+v_n-b_p} e_{j+b_n+v_n-b_p}.$$

Both terms can be controlled because $j + b_n + v_n$ and $j + b_n + v_n - b_p$ live at the beginnings of the intervals $[b_p + a_p + \Delta, a_{p+1})$ and $[a_p + \Delta, b_p + v_p)$ respectively, and the coefficients α_k are known to be small in these regions.

Assume that $j < b_p + v_p$ and $j + b_n + v_n \geq b_p + v_p$. Then $j \in [a_p + \Delta, b_p + v_p)$ and $j + b_n + v_n \in [b_p + v_p, b_p + a_p + \Delta)$, because the intervals $[a_p + \Delta, b_p + v_p)$ and $[b_p + v_p, b_p + a_p + \Delta)$ have lengths greater than $b_n + v_n$. Thus, we have $e_j = \alpha_j^{-1} T^j e_0$ and

$$T^{b_n+v_n}(e_j) = \frac{1}{\alpha_j} T^{b_n+v_n+j} e_0$$

$$= \frac{1}{\alpha_j}\left(e_{b_n+v_n+j} + b_p T^{b_n+v_n+j-b_p} e_0\right).$$

Now $b_n + v_n + j - b_p < a_p + \Delta$ and $\|T\| \leq 2$, so $\|T^{b_n+v_n+j-b_p} e_0\| \leq 2^{a_p+\Delta}$. Moreover, j lives at the end of the interval $[a_p, b_p + v_p)$, which has length essentially equal to b_p, so, consistently with Remark 12.1, α_j can be taken larger than $b_p 2^{a_p+\Delta}$. Hence, in this case too we can obtain the desired estimate.

Assume that $j < a_p$ and $b_n + v_n + j \geq a_p$. Then $e_j = \alpha_j^{-1} T^j e_0$ and hence

$$T^{b_n+v_n}(e_j) = \frac{1}{\alpha_j} T^{b_n+v_n+j} e_0.$$

If $b_n + v_n + j \geq a_p + \Delta$ then $T^{b_n+v_n+j} e_0 = \alpha_{b_n+v_n+j} e_0$ is small because j lives at the beginning of the interval $[a_p + \Delta, b_p + v_p)$, and $T^{b_n+v_n}(e_j)$ is even smaller since α_j is large (j lives at the end of the interval $[s_p, a_p)$). If $b_n + v_n + j < a_p + \Delta$ then $T^{b_n+v_n+j} e_0 = \varepsilon e_{b_n+v_n+j} + T^{b_n+v_n+j-a_p} e_0$ has norm at most $\varepsilon + 2^\Delta$ (because $\|T\| \leq 2$), so again we are safe since α_j is large.

The only remaining case is when $a_p \leq j < a_p + \Delta$ and $b_n + v_n + j \geq a_p + \Delta$. Then we are in trouble. Indeed, in this case we have $e_j = \varepsilon^{-1}\left(T^j e_0 - T^{j-a_p} e_0\right)$, so that

$$T^{b_n+v_n}(e_j) = \frac{\alpha_{b_n+v_n+j}}{\varepsilon} e_{b_n+v_n+j} - \frac{1}{\varepsilon} T^{b_n+v_n+j-a_p} e_0. \qquad (12.13)$$

The first term is small because $b_n + v_n + j$ lives at the beginning of the interval $[a_p + \Delta, b_p + v_p)$, but the second one is not, if for example $j = a_p$. Indeed, we have $T^{b_n+v_n} e_{a_0} = e_{b_n+v_n} + b_n T^{v_n} e_0$, and we can say nothing more about $b_n T^{v_n} e_0$.

Thus something has to be done concerning the critical indices a_p and $a_p + \Delta$. The solution here is to simply "kill" these indices by taking more convenient (but more complicated) projections $\pi_n : X \to E_{a_n+\Delta-1}$ instead of the naive projections

$\pi_{[0,a_n+\Delta)}$. The definition of π_n is the obvious one if we want to be rid of that problem but nothing more. Hence, we set

$$\pi_n(e_j) := \begin{cases} e_j & j < a_n + \Delta, \\ \varepsilon^{-1}T^{j-a_p}e_0 & j \in [a_p, a_p + \Delta), \ p > n, \\ 0 & \text{otherwise.} \end{cases} \qquad (12.14)$$

Since we are working on $X = \ell^1$, these projections are indeed well defined and bounded, with $\|\pi_n\| \le 2^\Delta/\varepsilon$ for all n. Moreover, if $j \in [a_p, a_p + \Delta)$ for some $p > n$ and $b_n + v_n + j \ge a_p + \Delta$ then (12.13) gives

$$T^{b_n+v_n}(I - \pi_n)(e_j) = \frac{\alpha_{b_n+v_n+j}}{\varepsilon} e_{b_n+v_n+j},$$

which is indeed small because $b_n + v_n + j$ lives at the beginning of the interval $[a_p + \Delta, b_p + v_p)$ and nothing has changed for the other indices j.

Thus, we will have achieved our goal in this second step provided that we are able to show that if x is an arbitrary non-zero vector in X then $\pi_n(x) \in K_n$ for at least one $n \in \mathbb{N}$, where

$$K_n = \left\{y \subset [0, a_n + \Delta); \ \|y\| \le 2^n \text{ and } \|\tau_{a_n/4}(y)\| \ge 2^{-n}\right\}.$$

Let us show that $\pi_n(x) \in K_n$ for all but finitely many n. Since $\tau_{a_n/4}\pi_n(z) = z$ for large enough n if $z \in c_{00}$, it is clearly enough to check that the projections $\tau_{a_n/4}\pi_n$ are uniformly bounded. Now, the definitions of $\tau_{a_n/4}$ and π_n give

$$\tau_{a_n/4}\pi_n(e_j) = \begin{cases} e_j & j < a_n/4, \\ \varepsilon^{-1}T^{j-a_p}e_0 & j \in [a_p, a_p + \Delta), \ p \ge n, \\ 0 & \text{otherwise.} \end{cases} \qquad (12.15)$$

Indeed, this is clear if $j < a_n/4$ or $j \in [a_p, a_p + \Delta)$ for some $p > n$ and also if $j \ge a_n + \Delta$ and $j \notin \bigcup_{p>n}[a_p, a_p + \Delta)$. If $a_n/4 \le j < a_n$ then $\tau_{a_n/4}\pi_n(e_j) = \tau_{a_n/4}(e_j) = 0$ because $e_j = \alpha_j^{-1}T^j e_0$. Finally, if $j \in [a_n, a_n + \Delta)$ then $\pi_n(e_j) = e_j = \varepsilon^{-1}\left(T^j e_0 - T^{j-a_n}e_0\right)$, and hence $\tau_{a_n/4}\pi_n(e_j) = \varepsilon^{-1}T^{j-a_n}e_0$ because $\tau_{a_n/4}(T^j e_0) = 0$ and $T^{j-a_n}e_0 \subset [0, \Delta) \subset [0, a_n/4)$ (we are assuming that $a_0 \ge 4\Delta$).

Hence we see that $\|\tau_{a_n/4}\pi_n\| \le 2^\Delta/\varepsilon$ for all $n \in \mathbb{N}$, and this concludes the second step in our plan.

12.4 Third step

Having explained how to get (12.1) for every non-zero $x \in X$ and some *fixed* $\varepsilon > 0$, we now have to let ε go to 0. So, we will fix a decreasing sequence of positive numbers (ε_m) tending to 0 and explain how one can build the operator T in order to make the above arguments work for all ε_m. In spirit, this is just a carefully written diagonal argument, but there are some additional complications.

Before going into any detail, we note that it is enough to consider *normalized* vectors x (i.e. $\|x\| = 1$), since ε can be taken as arbitrarily small. Although not really crucial, this remark will be of some importance below.

When $\varepsilon > 0$ was fixed, in Section 12.3, there was a single positive integer Δ, which was there essentially to make $\|T\|$ independent of ε. However, we also considered two *sequences* of natural numbers (a_n) and (b_n). Thus, in having to deal with countably many ε_m, it is not hard to guess that we should consider a sequence of positive numbers (Δ_m) and two *double* sequences of natural numbers $(a_{n,m})$ and $(b_{n,m})$. Moreover, each number Δ_m will be restricted in some way to the scope of the sequence $(a_{n,m})_{n \in \mathbb{N}}$, so it is in fact natural to consider only those pairs $(n, m) \in \mathbb{N} \times \mathbb{N}$ with $n \geq m$. Thus, we consider double sequences $(a_{n,m})$ and $(b_{n,m})$ indexed by $\Lambda := \{(n, m) \in \mathbb{N} \times \mathbb{N}; \; n \geq m\}$. To arrange the terms of these two double sequences in a single increasing sequence, we need some ordering $\prec$ of Λ. We choose the following:

$$(n, 0) \prec (n + 1, n + 1) \prec (n + 1, n) \prec \cdots \prec (n + 1, 0) \prec \ldots$$

Accordingly, we assume that the sequences $(a_{n,m})$ and $(b_{n,m})$ are arranged in such a way that

$$b_{n,0} \ll a_{n+1,n+1} \ll b_{n+1,n+1} \ll a_{n+1,n} \ll \cdots$$

We also assume that $a_{n,m} \in 4\mathbb{N}$ for all pairs (n, m).

Next, we define the operator $T : c_{00} \to c_{00}$ as expected by consideration of (12.7) and (12.12):

$$T^j e_0 = \begin{cases} \varepsilon_m e_j + T^{j - a_{n,m}} e_0 & j \in [a_{n,m}, a_{n,m} + \Delta_m), \\ e_j + b_{n,m} T^{j - b_{n,m}} e_0 & j \in [b_{n,m} + v_{n,m}, b_{n,m} + a_{n,m} + \Delta_m), \\ \alpha_j e_j & \text{otherwise,} \end{cases}$$

$$(12.16)$$

where $v_{n,m} = 3a_{n,m}/4$.

Finally, we need a precise definition for Δ_m. This is obtained by induction, as follows: take $\Delta_0 := 1$ and

$$\Delta_m := b_{m-1,0} + a_{m-1,0} + \Delta_{m-1}, \quad m \geq 1. \tag{12.17}$$

In other words, Δ_m is taken as the first point of the interval between step $(m - 1, 0)$ and step (m, m) in the construction of T. In particular, the coefficient α_{Δ_m} can be small (less than $2^{-\sqrt{b_{m-1,0}}}$ if we so desire; see Remark 12.1).

One checks, exactly as above, that with a suitable choice of the positive numbers α_j the operator T is bounded, with $\|T\| \leq 2$. For example, we have

$$T(e_{a_{n,m} + \Delta_m - 1}) = \frac{1}{\varepsilon_m} \left(\alpha_{a_{n,m} + \Delta_m} e_{a_{n,m} + \Delta_m} + T^{\Delta_m} e_0 \right)$$

$$= \frac{\alpha_{a_{n,m} + \Delta_m}}{\varepsilon_m} e_{a_{n,m} + \Delta_m} + \frac{\alpha_{\Delta_m}}{\varepsilon_m} e_{\Delta_m},$$

so that $\|T(e_{a_{n,m}+\Delta_m-1})\| \leq 2$ if we take $\alpha_{a_{n,m}+\Delta_m} \leq \varepsilon_m$ and $\alpha_{\Delta_m} \leq \varepsilon_m$. Observe that the choice of Δ_m is compatible with this requirement.

To mimic what we did in Section 12.3, we now set

$$K_{n,m} := \left\{ y \subset [0, a_{n,m} + \Delta_m); \ \|y\| \leq C_{n,m} \ \text{and} \ \|\tau_{n,m}(y)\| \geq C_{n,m}^{-1} \right\}, \tag{12.18}$$

where $\tau_{n,m} : E_{a_{n,m}+\Delta_m-1} \to E_{a_{n,m}/4-1}$ is the projection defined by

$$\tau_{n,m}(T^j e_0) := \begin{cases} T^j e_0 & j < a_{n,m}/4, \\ 0 & \text{otherwise.} \end{cases} \tag{12.19}$$

Here, the parameters $C_{n,m}$ have to be specified. For the time being, the only restriction is that $C_{n,m}$ should depend only on the vectors $T^j e_0$, $j < a_{n,m} + \Delta_m$.

Assume that there are projections $\pi_{n,m} : X \to E_{a_{n,m}+\Delta_m-1}$ such that, for every normalized vector $x \in X$, one can find pairs $(n, m) \in \Lambda$ with arbitrarily large m such that $y_{n,m} := \pi_{n,m}(x) \in K_{n,m}$. Then the following holds: for every normalized vector $x \in X$ and arbitrarily large $m \in \mathbb{N}$, one can find a polynomial P such that

$$\|P(T)x - e_0\| < 3\varepsilon_m + \frac{C_T(a_{n,m} + \Delta_m - 1)2^{a_{n,m}+\Delta_m}}{b_{n,m}} \|T^{b_{n,m}+v_{n,m}}(I - \pi_{n,m})\|$$

for some $n \geq m$. This is proved exactly as (12.9), and so we omit the details.

Since $b_{n,m}$ can be taken as much larger than $a_{n,m}$ and $C_T(a_{n,m} + \Delta_m - 1)$, it follows that if $\sup_{n \geq m} \|T^{b_{n,m}+v_{n,m}}(I - \pi_{n,m})\| < \infty$ for each $m \in \mathbb{N}$ then every non-zero vector $x \in X$ is cyclic for T.

Thus, it remains to define the projections $\pi_{n,m} : X \to E_{a_{n,m}+\Delta_m-1}$ in such a way that for each $m \in \mathbb{N}$, the following properties hold true:

(i) $\sup_{n \geq m} \|T^{b_{n,m}+v_{n,m}}(I - \pi_{n,m})\| < \infty$ for each $m \in \mathbb{N}$;

(ii) given any normalized vector $x \in X$, one can find pairs $(m, n) \in \Lambda$ with arbitrarily large m such that $\pi_{n,m}(x) \in K_{n,m}$.

As in Section 12.3, the naive projections $\pi_{[0,a_{n,m}+\Delta_m)}$ almost fulfil property (i). The only difficulty comes from the vectors $T^{b_{n,m}+v_{n,m}}(e_j)$ when $a_{p,q} \leq j < a_{p,q} + \Delta_q$ and $b_{n,m} + v_{n,m} + j \geq a_{p,q} + \Delta_q$, for some $(p, q) \succ (n, m)$. In this case, we have $e_j = \varepsilon_q^{-1}\left(T^j e_0 - T^{j-a_{p,q}}e_0\right)$, so that

$$T^{b_{n,m}+v_{n,m}}(e_j) = \frac{\alpha_{b_{n,m}+v_{n,m}+j}}{\varepsilon_q} e_{b_{n,m}+v_{n,m}+j} - \frac{1}{\varepsilon_q} T^{b_{n,m}+v_{n,m}+j-a_{p,q}}e_0. \tag{12.20}$$

Since $b_{n,m} + v_{n,m} + j$ is at the beginning of the interval $[a_{p,q} + \Delta_q, b_{p,q} + v_{p,q})$ (because $(p, q) \succ (n, m)$, so that this interval is very large with respect to $b_{n,m} + v_{n,m}$), the coefficient $\alpha_{b_{n,m}+v_{n,m}+j}$ can be taken as smaller than ε_q and hence the first term in (12.20) is under control.

To estimate the second term, we note that if $q \geq n + 1$ then $b_{n,m} + v_{n,m} \leq b_{q-1,0} + v_{q-1,0}$, so that $\Delta_q \leq b_{n,m} + v_{n,m} + j - a_{p,q} < b_{q-1,0} + v_{q-1,0} + \Delta_q$. By (12.17), this gives

$$b_{q-1,0} + a_{q-1,0} + \Delta_{q-1} \le b_{n,m} + v_{n,m} + j - a_{p,q}$$
$$< 2b_{q-1,0} + v_{q-1,0} + a_{q-1,0} + \Delta_{q-1}.$$

Hence $b_{n,m} + v_{n,m} + j - a_{p,q}$ lives at the beginning of the interval $[b_{q-1,0} + a_{q-1,0} + \Delta_{q-1}, a_{q,q})$, so that

$$T^{b_{n,m}+v_{n,m}+j-a_{p,q}} e_0 = \alpha_{b_{n,m}+v_{n,m}+j-a_{p,q}} e_{b_{n,m}+v_{n,m}+j-a_{p,q}}$$

has norm less than ε_q, if we so desire. Thus, the second term in (12.20) is under control when $q \ge n + 1$.

However, we have no control over $T^{b_{n,m}+v_{n,m}+j-a_{p,q}} e_0$ if $q \le n$, since in this case $b_{n,m} + v_{n,m} + j - a_{p,q}$ can take some dangerous value beyond $a_{q,q}$. Therefore, as before we have to suppress those indices $(p, q) \succ (n, m)$ for which $q \le n$. Accordingly, we define the projections $\pi_{n,m} : X \to E_{a_{n,m}+\Delta_m-1}$ as follows:

$$\pi_{n,m}(e_j) := \begin{cases} e_j & j < a_{n,m} + \Delta_m, \\ -\varepsilon_q^{-1} T^{j-a_{p,q}} e_0 & j \in [a_{p,q}, a_{p,q} + \Delta_q), \ (p,q) \in \Lambda_{n,m}^*, \\ 0 & \text{otherwise,} \end{cases}$$

(12.21)

where $\Lambda_{n,m}^* := \{(p,q) \in \Lambda;\ (n,m) \prec (p,q) \text{ and } q \le n\}$.

Since we are working on ℓ^1, the operators $\pi_{n,m}$ are well defined and bounded, because for each fixed (n, m) there are only finitely many q such that $(p, q) \in \Lambda_{n,m}^*$ for some p. Moreover, $\pi_{n,m}$ is in fact a projection onto $E_{a_{n,m}+\Delta_m-1}$, since $T^{j-a_{p,q}} e_0 \subset [0, \Delta_q) \subset [0, a_{n,m} + \Delta_m)$ when $q \le n$. And, as should be clear from the above discussion, the definition of $\pi_{n,m}$ is the right one to ensure property (i) above, i.e. $\sup_{n \ge m} \|T^{b_{n,m}+v_{n,m}} (I - \pi_{n,m})\| < \infty$ for each $m \in \mathbb{N}$.

Let us now turn to property (ii). Recall the definition of the compact sets $K_{n,m}$:

$$K_{n,m} = \{y \subset [0, a_{n,m} + \Delta_m);\ \|y\| \le C_{m,n} \text{ and } \|\tau_{n,m}(y)\| \ge C_{n,m}^{-1}\}. \quad (12.22)$$

The constants $C_{n,m}$ are still undefined and can be arbitrarily chosen as long as $C_{n,m}$ depends only on $T^j e_0$, $j < a_{n,m} + \Delta_m$. We have to show that if the $C_{n,m}$ are suitably chosen then, for every normalized vector $x \in X$ and any $M \in \mathbb{N}$, one can find a pair $(n, m) \in \Lambda$ with $m \ge M$ such that $\pi_{n,m}(x) \in K_{n,m}$.

Note that $\|\pi_{n,m}(x)\| \le \|\pi_{n,m}\|$ and that $\|\pi_{n,m}\|$ depends only on $T^j e_0$, $j < a_{n,m} + \Delta_m$. So we may assume that $C_{n,m} \ge \|\pi_{n,m}\|$ for all pairs (n, m) and concentrate on the inequality $\|\tau_{n,m}\pi_{n,m}(x)\| \ge C_{n,m}^{-1}$ in (12.22). Thus, what we have to do is to show that if x is a normalized vector in X then $\|\tau_{n,m}\pi_{n,m}(x)\|$ is "not too small" for arbitrarily large m. Here, the guiding idea should be that this result is clearly true, even though the detailed proof is somewhat tedious. Indeed, by the definition of the projections $\tau_{n,m}$ and $\pi_{n,m}$, we have $\tau_{n,m}\pi_{n,m}(e_j) = e_j$ for all $j < a_{n,m}/4$, so that if $z \in c_{00}$ and if $m \in \mathbb{N}$ is fixed then $\tau_{n,m}\pi_{n,m}(z) = z$ for large enough n. Thus, if we knew that $\sup_{n \ge m} \|\tau_{n,m}\pi_{n,m}\| < \infty$ for infinitely many m

then we would have $\lim_{n\to\infty} \tau_{n,m}\pi_{n,m}(x) = x$ for each such m and every $x \in X$. Hence, it would be enough to take $C_{n,m} \geq 2$ to get the result.

Unfortunately, things are not so simple. Indeed, proceeding as for (12.15) above, the definitions of $\pi_{n,m}$ and $\tau_{n,m}$ give

$$\tau_{n,m}\pi_{n,m}(e_j) = \begin{cases} e_j & j < a_{n,m}/4, \\ -\varepsilon_q^{-1}T^{j-a_{p,q}}e_0 & j \in [a_{p,q}, a_{p,q} + \Delta_q), \ (p,q) \in \Lambda_{n,m}, \\ 0 & \text{otherwise}, \end{cases}$$

(12.23)

where

$$\Lambda_{n,m} = \{(p,q); \ (n,m) \preceq (p,q) \text{ and } q \leq n\} = \Lambda_{n,m}^* \cup \{(n,m)\}.$$

In particular, we see that $\|\tau_{n,m}\pi_{n,m}(e_{a_{p,q}})\| = \varepsilon_q^{-1}$ whenever $(p,q) \in \Lambda_{n,m}$, so that $\sup_{n\geq m}\|\tau_{n,m}\pi_{n,m}\| = \infty$ for every $m \in \mathbb{N}$.

However, we know exactly where the trouble comes from: $\|\tau_{n,m}\pi_{n,m}\|$ becomes large as $n \to \infty$ because $\Lambda_{n,m}$ contains pairs (p,q) with arbitrarily large q. This suggests the introduction of new projections $\mathbf{g}_{n,m} : X \to E_{a_{n,m}/4-1}$, defined as follows:

$$\mathbf{g}_{n,m}(e_j) := \begin{cases} e_j & j < a_{n,m}/4, \\ -\varepsilon_q^{-1}T^{j-a_{p,q}}e_0 & j \in [a_{p,q}, a_{p,q} + \Delta_q), \ (p,q) \in \mathbf{G}_{n,m}, \\ 0 & \text{otherwise}, \end{cases}$$

(12.24)

where

$$\mathbf{G}_{n,m} := \{(p,q) \in \Lambda_{n,m}; \ q \leq m\}.$$

Looking at (12.23) and (12.24), we see that $\mathbf{g}_{n,m}$ is obtained from $\tau_{n,m}\pi_{n,m}$ by putting $\mathbf{g}_{n,m}(e_j) = \tau_{n,m}\pi_{n,m}(e_j)$ if $j \notin J_{n,m}$ and $\mathbf{g}_{n,m}(e_j) = 0$ if $J_{n,m}$, where

$$J_{n,m} := \bigcup_{(p,q)\in\Lambda_{n,m}\setminus\mathbf{G}_{n,m}} [a_{p,q}, a_{p,q} + \Delta_q).$$

Hence, we may write

$$\tau_{n,m}\pi_{n,m} = \mathbf{g}_{n,m} + \tau_{n,m}\pi_{n,m}\pi_{J_{n,m}}$$ (12.25)

$$:= \mathbf{g}_{n,m} + \mathbf{b}_{n,m}.$$ (12.26)

Here and below, we denote by $\pi_J : X \to \text{span}\{e_j; \ j \in J\}$ the canonical projection from X onto $\text{span}\{e_j; \ j \in J\}$, for any set $J \subset \mathbb{N}$; that is, $\pi_J(e_j) = e_j$ if $j \in J$ and $\pi_J(e_j) = 0$ otherwise.

Now, the main difference between the projections $\mathbf{g}_{n,m}$ and $\tau_{n,m}\pi_{n,m}$ (and the reason for introducing $\mathbf{g}_{n,m}$) is that $\sup_{n\geq m}\|\mathbf{g}_{n,m}\| < \infty$ for each fixed $m \in \mathbb{N}$. Indeed, this is clear from (12.24) since $q \leq m$ whenever $(p,q) \in \mathbf{G}_{n,m}$. Since for any $z \in c_{00}$ and $m \in \mathbb{N}$, we clearly have $\mathbf{g}_{n,m}(z) = z$ if n is large enough, it follows that

$$\mathbf{g}_{n,m}(x) \xrightarrow{n\to\infty} x$$ (12.27)

for every $x \in X$ and each fixed $m \in \mathbb{N}$. In particular, if $\|x\| = 1$ then $\|\mathbf{g}_{n,m}(x)\| \geq 1/2$ for large enough n, so that $\|\mathbf{g}_{n,m}(x)\|$ is definitely "not too small". Thus, $\mathbf{g}_{n,m}$ and $\mathbf{b}_{n,m}$ should be viewed respectively as the good part and the bad part of the projection $\tau_{n,m}\pi_{n,m}$.

Given a normalized vector x and $M \in \mathbb{N}$, we now see that two cases can occur.

- If $\mathbf{g}_{N,M}(x)$ dominates $\mathbf{b}_{N,M}(x)$ for some large $N \geq M$ then we easily obtain that $\|\tau_{n,m}\pi_{n,m}(x)\|$ is not too small for $n := N$ and $m := M$.
- Otherwise, we have to find some other pair (n, m) with $m \geq M$ such that $\|\tau_{n,m}\pi_{n,m}(x)\|$ is not too small. This is done by observing that $\|\pi_{[a_{p,q},a_{p,q}+\Delta_q)}(x)\|$ has to be not too small for some $(p, q) \in \Lambda_{n,m} \setminus \mathbf{G}_{n,m}$ and by making use of the following strange identity, which holds true whenever $1 \leq q < p$:

$$\pi_{[a_{p,q},a_{p,q}+\Delta_q)}\tau_{p,q-1}\pi_{p,q-1} = \pi_{[a_{p,q},a_{p,q}+\Delta_q)} \, . \tag{12.28}$$

This identity can be checked as follows. First, we note that $(p, q - 1) \succ (p, q)$, and hence $a_{p,q} \ll a_{p,q-1}$. Therefore the interval $[a_{p,q}, a_{p,q} + \Delta_q)$ is contained in $[0, a_{p,q-1}/4)$, so that $\tau_{p,q-1}\pi_{p,q-1}(e_j) = e_j$ for all $j \in [a_{p,q}, a_{p,q} + \Delta_q)$, by (12.23). However, if $j \notin [a_{p,q}, a_{p,q} + \Delta_q)$ then either $j < a_{p,q} < a_{p,q-1}/4$ and $\tau_{p,q-1}\pi_{p,q-1}(e_j) = e_j$ or $j \in [a_{p',q'}, a_{p',q'} + \Delta_{q'})$ for some $(p', q') \in \Lambda_{p,q-1}$ and $\tau_{p,q-1}\pi_{p,q-1}(e_j) = -\varepsilon_{q'}^{-1}T^{j-a_{p',q'}}e_0$, or else $\tau_{p,q-1}\pi_{p,q-1}(e_j) = 0$. In any case this yields $\pi_{[a_{p,q},a_{p,q}+\Delta_q)}\tau_{p,q-1}\pi_{p,q-1}(e_j) = 0$, since in the first case $j < a_{p,q}$ and in the second case $T^{j-a_{p',q'}}e_0$ is supported on $[0, \Delta_{q'})$, with $\Delta_{q'} < a_{p,q}$ (observe that $(q' - 1, q' - 1) \prec (p, q)$). Thus, we see that $\pi_{[a_{p,q},a_{p,q}+\Delta_q)}\tau_{p,q-1}\pi_{p,q-1}(e_j) = 0$ for all $j \notin [a_{p,q}, a_{p,q}+\Delta_q)$, which concludes the proof of (12.28). It follows (and this is the important thing) that

$$\|\tau_{p,q-1}\pi_{p,q-1}(x)\| \geq \|\pi_{[a_{p,q},a_{p,q}+\Delta_q)}(x)\| \tag{12.29}$$

for all $x \in X$. Indeed, this is clear from (12.28) since $\|\pi_{[a_{p,q},a_{p,q}+\Delta_q)}\| = 1$.

Having said this, we start the actual proof of property (ii). So, let us fix $x \in X$ with $\|x\| = 1$ and $M \in \mathbb{N}$ and find $(n, m) \in \Lambda$ with $m \geq M$ such that $\|\tau_{n,m}\pi_{n,m}(x)\| \geq C_{n,m}^{-1}$, where the $C_{n,m}$ still have to be defined.

By (12.27), we can choose some integer $N \in \mathbb{N}$ such that $\|\mathbf{g}_{N,M}(x)\| \geq 1/2$. If $\|\mathbf{g}_{N,M}(x)\| \geq 2\|\mathbf{b}_{N,M}(x)\|$ then $\|\tau_{N,M}\pi_{N,M}(x)\| \geq \|\mathbf{g}_{N,M}(x)\|/2 \geq 1/4$. So we are done in that case, setting $m := M$ and $n := N$, provided only that $C_{n,m} \geq 4$ for all pairs (n, m).

Now assume that $\|\mathbf{g}_{N,M}(x)\| < 2\|\mathbf{b}_{N,M}(x)\|$, so that $\|\mathbf{b}_{N,M}(x)\| > 1/4$. By the definition of $\mathbf{b}_{N,M}$, i.e. (12.25), this means that

$$\|\tau_{N,M}\pi_{N,M}\pi_{J_{N,M}}(x)\| > 1/4 \, .$$

It follows that

$$\|\pi_{J_{N,M}}(x)\| \geq \frac{1}{4\|\tau_{N,M}\pi_{N,M}\|} := \eta_{N,M} \, ,$$

a quantity which depends only on $T_{|E_{N,M}}$. Now, by the definition of $J_{N,M}$ we may write

$$\pi_{J_{N,M}} = \sum_{(p,q)\in\Lambda_{N,M}\setminus \mathbf{G}_{N,M}} \pi_{[a_{p,q},a_{p,q}+\Delta_q)},$$

so that $\eta_{N,M} \leq \|\pi_{J_{N,M}}(x)\| \leq \sum_{p,q}\|\pi_{[a_{p,q},a_{p,q}+\Delta_q)}(x)\|$. Hence, one can find some pair $(p,q) \in \Lambda_{N,M}\setminus \mathbf{G}_{N,M}$ such that

$$\|\pi_{[a_{p,q},a_{p,q}+\Delta_q)}(x)\| \geq 2^{-p-q-2}\,\eta_{N,M}\,. \tag{12.30}$$

We note that $M < q \leq N$ and $(N,M) \prec (p,q)$, by the definitions of $\Lambda_{N,M}$ and $\mathbf{G}_{N,M}$. In particular, $N \leq p$. Moreover, $N = p$ is impossible since the inequality $(p,M) \prec (p,q)$ is equivalent to $q < M$. Hence, we have $M < q \leq N < p$.

Using (12.30) and (12.29), we arrive at

$$\|\tau_{p,q-1}\pi_{p,q-1}(x)\| \geq 2^{-p-q-2}\,\eta_{N,M}\,.$$

Thus, if we knew that $C_{p,q-1} \geq 2^{p+q+2}/\eta_{N,M}$ then we could conclude that $\pi_{n,m}(x) \in K_{n,m}$ with $n := p$ and $m := q - 1 \geq M$. We note here that $(n,m) \succ (N,M)$.

It should now be clear that we can choose the parameters $C_{n,m}$ in such a way that the long awaited property (ii) holds true. For example, we may take

$$C_{n,m} := \max\left(4, \|\pi_{n,m}\|, \frac{2^{n+m+3}}{\min\{\eta_{N,M};\ (N,M) \prec (n,m)\}}\right).$$

And that is the end of the story!

12.5 Comments and an exercise

The first example of a bounded operator without any non-trivial invariant subspace was found by P. Enflo around 1975. Enflo's construction is extremely difficult, and it was published only in 1987 [104]. In the meantime, it had been simplified by B. Beauzamy in [35], and other examples of operators without invariant subspaces had been given by C. J. Read. Our exposition is based on Read's paper [201], although the final step follows the more recent work [228] by G. Sirotkin. The construction is flexible enough to allow additional interesting properties for the operator T. For example, T can be made quasinilpotent (see [204] or [228]).

Besides ℓ^1, Read also exhibited a bounded operator without invariant subspaces on the space c_0 [203]. The basic idea of the construction is the same, but several modifications are needed. Coming back to the operator defined in the first step and looking at the boundedness of T, the real difficulty, when one is not working with the ℓ^1 norm, comes from the vectors $T(e_{a_n+\Delta-1})$. On c_0, this difficulty can be overcome by defining $T^j(e_0)$ as follows for $j \in [a_n, a_n + \Delta)$:

$$T^j(e_0) = \varepsilon e_j + T^{j-(a_n-a_{n-1})}e_0\,,$$

with the convention that $a_{-1} := 0$. By induction, it is easily checked that

$$T^{a_n}(e_0) = \varepsilon e_{a_n} + \varepsilon e_{a_{n-1}} + \cdots + \varepsilon e_{a_0} + e_0\,,$$

so that (and this is specific to the c_0 norm) we still get

$$\|T^{a_n}e_0 - e_0\|_\infty \leq \varepsilon\,.$$

However, we now have

$$T(e_{a_n+\Delta-1}) = \frac{\alpha_{a_n+\Delta}}{\varepsilon} e_{a_n+\Delta} - \frac{1}{\varepsilon}T^{a_{n-1}+\Delta}e_0$$

$$= \frac{\alpha_{a_n+\Delta}}{\varepsilon} e_{a_n+\Delta} - \frac{\alpha_{a_{n-1}+\Delta}}{\varepsilon} e_{a_{n-1}+\Delta}.$$

Thus, if we arrange (a_n) and (α_n) in such a way that $\sum_n \alpha_{a_{n-1}+\Delta}/\varepsilon < 1$ then T is bounded on c_0. Of course, many other details in the proof have to be modified (e.g. the definition of the projections $\pi_{n,m}$), but the ideas are similar. However, there are serious obstructions in trying to make this approach work on ℓ^p, $p \in (1, \infty)$, so serious that the invariant subspace problem remains unsolved for these spaces, in particular for the Hilbert space ℓ^2. In fact, the problem is still open for any reflexive Banach space.

A penetrating study of Read-type operators was carried out in a Hilbert space setting by S. Grivaux and M. Roginskaya in [131]. Among many other things, it is shown in [131] that there exists a Hilbert space operator T for which the set of non-hypercyclic vectors is extremely small; namely, $HC(T)^c$ is contained in a countable union of closed hyperplanes of the underlying Hilbert space H.

We should also mention that there exist Banach spaces on which every bounded operator does have a non-trivial invariant subspace. This follows from a recent work of S. A. Argyros and R. Haydon, who constructed an infinite-dimensional separable Banach space on which any operator has the form $T = \lambda I + K$, where K is a compact operator [9]. Such operators do have invariant subspaces, by a classical result of N. Aronszajn and K. T. Smith; see [2] for instance.

Finally, we note that life appears to be much easier if one is allowed to cheat: if *discontinuous* operators are permitted then it is not very difficult to find counter-examples to the invariant-subspace problem, even on a Hilbert space. This was first shown by A. L. Shields in [227]. A stronger result was proved by H. N. Salas in [217]. This is the topic of the only exercise in this chapter.

EXERCISE ([217]) Let X be a separable infinite-dimensional Banach space. In this exercise, our aim is to prove the following result: *there exists a linear map $T : X \to X$ for which every non-zero vector is hypercyclic*. In what follows, we denote by $\mathfrak{c}$ the cardinality of the continuum and fix a set Σ with cardinality $\mathfrak{c}$.

1. Show that one can find two linear subspaces $E, F \subset X$ with algebraic dimension $\mathfrak{c}$ such that $X = E \oplus F$ and F is dense in X.
2. Show that one can find a family $(B_\sigma)_{\sigma \in \Sigma}$ of pairwise disjoint subsets of F such that $B := \bigcup_\sigma B_\sigma$ is a Hamel basis for F and each B_σ is countable and dense in X.
3. Show that there exists a Hamel basis A for E with the following property: A can be written as $A = \bigcup_{\sigma \in \Sigma} A_\sigma$, where the A_σ are countably infinite and pairwise disjoint and each "sequence" A_σ tends to 0.
4. Let $\mathbf{D}$ be a countably infinite set. Show that one can find a family $(\mathbf{D}_\sigma)_{\sigma \in \Sigma}$ of infinite subsets of $\mathbf{D}$ such that $\mathbf{D}_\sigma \cap \mathbf{D}_\tau$ is finite whenever $\sigma \neq \tau$. In what follows, we set $\mathbf{D} := \{2^m; \ m \in \mathbb{N}\}$.
5. Enumerate each set B_σ as $(e_{j,\sigma})_{j \in \mathbf{D}_\sigma}$ and each set A_σ as $(e_{j,\sigma})_{j \in \mathbb{N} \setminus \mathbf{D}_\sigma}$. Thus the family $(e_{j,\sigma})_{(j,\sigma) \in \mathbb{N} \times \Sigma}$ is a Hamel basis for X, each set $\{e_{j,\sigma}; \ j \in \mathbf{D}_\sigma\}$ is dense in X and $e_{j,\sigma} \to 0$ as $j \to \infty$ with $j \notin \mathbf{D}_\sigma$. Now let $T : X \to X$ be the linear map defined on the basis vectors $e_{j,\sigma}$ by

$$T(e_{j,\sigma}) := e_{j+1,\sigma}.$$

(a) Let $(j, \sigma) \in \mathbb{N} \times \Sigma$, and let $(k, \tau) \neq (j, \sigma)$. Show that $T^n(e_{k,\tau}) \to 0$ as $n \to \infty$ with $j + n \in \mathbf{D}_\sigma$. (*Hint*: When $k \neq j$, use the lacunarity of $\mathbf{D}$.)
(b) Deduce that if $(j, \sigma) \in \mathbb{N} \times \Sigma$ and $z \in \text{span}\{e_{k,\tau}; \ (k, \tau) \neq (j, \sigma)\}$ then the set $\{T^n(e_{j,\sigma} + z); \ n \in \mathbb{N}\}$ is dense in X.
(c) Show that T has the required property.

Appendices

A Complex analysis

A.1 Runge's theorem

At several places in the book, we need the following classical and extremely useful approximation theorem (see e.g. [209]).

THEOREM A.1.1 (Runge's theorem) *Let K be a compact subset of $\mathbb{C}$, and let $S \subset \mathbb{C}$. Assume that S intersects every bounded connected component of $\mathbb{C} \setminus K$. Then any function f that is holomorphic in a neighbourhood of K can be uniformly approximated on K by rational functions with poles in S.*

When $\mathbb{C} \setminus K$ is connected, there are no bounded components, so $S = \varnothing$ is allowed. Thus, we get

COROLLARY A.1.2 *Let K be a compact subset of $\mathbb{C}$, and assume that $\mathbb{C} \setminus K$ is connected. Then, any function f that is holomorphic in a neighbourhood of K can be uniformly approximated on K by polynomial functions.*

For completeness (or just for interest), we outline one possible direct proof of Corollary A.1.2. So, assume that K is a compact subset of $\mathbb{C}$ with connected complement. We denote by $\mathcal{C}(K)$ the space of all continuous functions on K (endowed with the uniform norm) and by $\mathrm{P}(K) \subset \mathcal{C}(K)$ the closure of the set of all polynomial functions.

First, we note that any function f that is holomorphic in a neighbourhood of K can be uniformly approximated on K by rational functions without poles on K. This follows from Cauchy's formula. Indeed, for $z \in K$ one can write

$$f(z) = \frac{1}{2i\pi} \int_\Gamma \frac{f(w)}{w - z}\, dw \,,$$

where Γ is some suitable contour in $\mathbb{C} \setminus K$ (see e.g. [209, Theorem 13.5]) and then approximate the integral by Riemann sums. Therefore, it is enough to show that, for any complex number $a \in \mathbb{C} \setminus K$, the function $f_a(z) := (z - a)^{-1}$ belongs to $\mathrm{P}(K)$.

Let

$$A := \{a \in \mathbb{C} \setminus K; \ f_a \in \mathrm{P}(K)\}\,.$$

Clearly, A is a closed subset of $\mathbb{C} \setminus K$. Moreover, A is non-empty because $a \in A$ if $|a|$ is large enough, as can be seen by expanding

$$\frac{1}{z - a} = -\frac{1}{a} \times \frac{1}{1 - z/a}$$

in power series. Finally, A is also *open* in $\mathbb{C} \setminus K$. Indeed, if $a \in A$ and b is close to a then, writing

$$\frac{1}{z - b} = \frac{1}{z - a} \times \frac{1}{1 - (b - a)/(z - a)}\,,$$

we see that $f_b = \sum_0^\infty (b - a)^n f_a^{n+1}$, where the series is convergent in $\mathcal{C}(K)$. Since $f_a \in \mathrm{P}(K)$ and $\mathrm{P}(K)$ is a closed subalgebra of $\mathcal{C}(K)$, it follows that $f_b \in \mathrm{P}(K)$. By the connectedness of $\mathbb{C} \setminus K$ we conclude that $A = \mathbb{C} \setminus K$.

A.2 Entire functions of exponential type

An entire function $F : \mathbb{C} \to \mathbb{C}$ is said to be of **exponential type** if there exists some finite constant $B \geq 0$ such that $|F(z)| = O(e^{B|z|})$. The infimum of all such constants B is the **type** of the function F. In particular, F is of exponential type 0 iff, for each $\varepsilon > 0$, one can find some finite constant A_ε such that $|F(z)| \leq A_\varepsilon e^{\varepsilon|z|}$ for all $z \in \mathbb{C}$.

The following two simple and very well-known lemmas are all we need regarding entire functions of exponential type. The first says that exponential type can be detected by looking at the Taylor coefficients, and the second shows that the zero set of an entire function of exponential type cannot be too "concentrated". The reader may consult e.g. [177] or [172] for more on these functions.

LEMMA A.2.1 *Let $F(z) = \sum_0^\infty c_n z^n$ be an entire function. Then F is of exponential type if and only if $|c_n| \leq CR^n/n!$ for some finite constants C, R and all $n \in \mathbb{N}$.*

PROOF If $|F(z)| \leq Ae^{B|z|}$ then Cauchy's inequalities yield $|c_n| r^n \leq Ae^{Br}$ for any $r > 0$ and all $n \in \mathbb{N}$. For each fixed n, the quantity $r^{-n}e^{Br}$ is minimal for $r := n/B$. Thus, we get $|c_n| \leq A(n/B)^{-n}e^n \leq A(Be)^n/n!$. Conversely, if $|c_n| \leq AR^n/n!$ then $|F(z)| \leq Ae^{R|z|}$ by a straightforward computation. $\square$

LEMMA A.2.2 *Let F be a non-constant entire function of exponential type 0 and, for each $r > 0$, denote by $n_F(r)$ the number of zeros of F in the closed disk $\overline{D}(0,r)$. Then $n_F(r) = o(r)$ as $r \to \infty$.*

PROOF We may clearly assume that $F(0) \neq 0$ and that F has infinitely many zeros. We arrange these zeros into a sequence $(a_n)_{n \geq 1}$ tending to ∞ and assume that $|a_n|$ is non-decreasing. By Jensen's formula (see [209]), we then have

$$\sum_{n \leq n_F(s)} \log \left| \frac{s}{a_n} \right| = -\log |F(0)| + \int_{-\pi}^{\pi} \log |F(se^{i\theta})| \frac{d\theta}{2\pi}$$

for all $s > 0$. Since $\sum_{n \leq n_F(r)} \log |s/a_n| \leq \sum_{n \leq n_F(s)} \log |s/a_n|$ whenever $r \leq s$, and $|s/a_n| \geq |s/r|$ if $n \leq n_F(r)$, it follows (taking $s = 2r$ in Jensen's formula) that

$$n_F(r) \log 2 \leq -\log |F(0)| + \int_{-\pi}^{\pi} \log |F(2re^{i\theta})| \frac{d\theta}{2\pi}$$

for all $r > 0$. Since F is of exponential type 0, this yields the estimate

$$n_F(r) \leq C_\varepsilon + \varepsilon r$$

for any $\varepsilon > 0$, which concludes the proof. $\square$

B Function spaces

B.1 The space $H(\Omega)$

For any open set $\Omega \subset \mathbb{C}$, we denote by $H(\Omega)$ the space of all holomorphic functions on Ω. Then $H(\Omega)$ is a Fréchet space when endowed with the topology of uniform convergence on compact sets. Local convexity is clear since the topology of $H(\Omega)$ is generated by the continuous seminorms $\| \cdot \|_K$ defined by

$$\|f\|_K := \sup\{|f(z)|;\ z \in K\},$$

where K ranges over the compact subsets of Ω. Completeness is also clear, since the uniform Cauchy criterion entails uniform convergence. Finally, the space $H(\Omega)$ is metrizable because its topology is generated by the countable family of seminorms $(p_{K_n})_{n \in \mathbb{N}}$, where (K_n) is any increasing sequence of compact sets such that $\bigcup_n K_n = \Omega$ and, moreover, any compact set $K \subset \Omega$ is contained in some K_n. Explicitly, one defines a compatible translation-invariant (complete) metric on $H(\Omega)$ by setting

$$d(f,g) := \sum_{n=0}^{\infty} 2^{-n} \min(1, \|f - g\|_{K_n}).$$

The space $H(\Omega)$ is also *separable*. This is obvious for $\Omega = \mathbb{C}$, since the polynomial functions with rational coefficients are dense in $H(\mathbb{C})$. For an arbitrary open set Ω, this follows from Runge's theorem: the rational functions without poles in Ω and with rational coefficients are dense in $H(\Omega)$.

B.2 The Hardy space

The classical **Hardy space** of the unit disk $\mathbb{D}$ is the space $H^2(\mathbb{D})$ made up of all holomorphic functions $f : \mathbb{D} \to \mathbb{C}$ such that

$$\|f\|_{H^2}^2 := \sup_{r<1} \int_0^{2\pi} |f(re^{i\theta})|^2 \, \frac{d\theta}{2\pi} < \infty.$$

The space $H^2(\mathbb{D})$ is arguably the most important function space appearing in complex analysis and operator theory. Many excellent books provide detailed accounts of the beautiful theory of Hardy spaces. Just one of these is P. Duren's book [101]. In the present book, we use very few properties of $H^2(\mathbb{D})$, which we now summarize.

A holomorphic function $f : \mathbb{D} \to \mathbb{C}$ can be uniquely written as $f(z) = \sum_0^{\infty} c_n(f) z^n$. It is well known, and easy to check using Parseval's identity, that f is in $H^2(\mathbb{D})$ if and only if $\sum_0^{\infty} |c_n(f)|^2 < \infty$. Then $\|f\|_2^2 = \sum_0^{\infty} |c_n(f)|^2$. Conversely, if $(a_n)_{n \in \mathbb{N}}$ is any square-summable sequence of complex numbers then the formula $f(z) := \sum_0^{\infty} a_n z^n$ defines a function $f \in H^2(\mathbb{D})$. Thus $H^2(\mathbb{D})$ is an isometric copy of the ubiquitous space $\ell^2(\mathbb{N})$. In particular, $H^2(\mathbb{D})$ is a Hilbert space with orthonormal basis $(z^n)_{n \in \mathbb{N}}$.

This can be rephrased by saying that $H^2(\mathbb{D})$ is canonically isometric to a subspace of $L^2(\mathbb{T})$, namely $H^2(\mathbb{T}) := \{\varphi \in L^2(\mathbb{T}); \; \widehat{\varphi}(n) = 0 \text{ for all } n < 0\}$. Here $\widehat{\varphi}(n)$ is the nth Fourier coefficient of φ. If $f \in H^2(\mathbb{D})$, the associated function $f^* \in H^2(\mathbb{T})$ is called the **boundary value** of f.

It follows that the scalar product of $H^2(\mathbb{D})$ is given by

$$\langle f, g \rangle_{H^2} = \langle f^*, g^* \rangle_{L^2(\mathbb{T})} = \sum_{n=0}^{\infty} c_n(f) \overline{c_n(g)}.$$

It is easy to check that if $f \in H^2(\mathbb{D})$ then the functions $f_r(e^{i\theta}) := f(re^{i\theta})$ converge to f^* in the L^2 sense as $r \to 1^-$. It is also true, but much less elementary, that convergence holds *almost everywhere* with respect to the Lebesgue measure on $\mathbb{T}$. This is part of **Fatou's theorem**, which we are about to state.

Let us first introduce some terminology. For $\xi \in \mathbb{T}$ and $\alpha \in [0, \pi/2)$, we denote by $\Gamma(\xi, \alpha)$ the **Stolz angle** with vertex ξ and angle α, i.e.

$$\Gamma(\xi, \alpha) := \{z \in \mathbb{D}; \; |\arg(1 - \bar{\xi}z)| \leq \alpha\}.$$

A function $f : \mathbb{D} \to \mathbb{C}$ is said to admit a **non-tangential limit** at some point $\xi \in \mathbb{T}$ if there exists $w \in \mathbb{C}$ such that $f(z) \to w$ as $z \to w$ while remaining in any given Stolz angle $\Gamma(\xi, \alpha)$. If this holds then, in particular, w is the **radial limit** of f at ξ, i.e. $f(r\xi) \to w$ as $r \to 1^-$.

THEOREM B.2.1 (FATOU'S THEOREM) *If $f \in H^2(\mathbb{D})$ then $f(z)$ has a non-tangential limit $f^*(\xi)$ at almost every $\xi \in \mathbb{T}$.*

For any $a \in \mathbb{D}$, the evaluation map $f \mapsto f(a)$ is a continuous linear functional on $H^2(\mathbb{D})$. This follows from the estimate

$$|f(a)| \le \sum_{n=0}^{\infty} |c_n(f)| \, |a|^n$$

$$\le \left(\sum_{n=0}^{\infty} |c_n(f)|^2 \right)^{1/2} \left(\sum_{n=0}^{\infty} |a|^{2n} \right)^{1/2}$$

$$= \frac{1}{\sqrt{1 - |a|^2}} \, \|f\|_{H^2} \,.$$

Since $H^2(\mathbb{D})$ is a Hilbert space, this linear functional is given by the scalar product with some function $k_a \in H^2(\mathbb{D})$. In other words,

$$f(a) = \langle f, k_a \rangle \tag{12.31}$$

for all $f \in H^2(\mathbb{D})$. The function k_a is called the **reproducing kernel** at a.

Reproducing kernels exist in any functional Hilbert space where the evaluation maps are continuous. In $H^2(\mathbb{D})$, it is easy to compute them explicitly. Indeed, we have $\langle k_a, z^n \rangle = \overline{\langle z^n, k_a \rangle} = \bar{a}^n$ for all $n \in \mathbb{N}$. Since $(z^n)_{n \in \mathbb{N}}$ is an orthonormal basis of $H^2(\mathbb{D})$, it follows that k_a is given by the formula

$$k_a(z) = \sum_{n \ge 0} \bar{a}^n z^n = \frac{1}{1 - \bar{a}z} \,.$$

Thus, (12.31) is nothing more (and nothing less) than a rephrasing of Cauchy's formula.

We also note that convergence in $H^2(\mathbb{D})$ entails uniform convergence on compact sets: this is apparent from the above estimate of $|f(a)|$.

Finally, it is sometimes convenient to move from the unit disk $\mathbb{D}$ to the upper half-plane

$$\mathbb{P}_+ := \{ w \in \mathbb{C}; \; \mathrm{Im}(w) > 0 \}$$

in particular when dealing with composition operators. This can be done via the **Cayley map**

$$\omega(z) = i \, \frac{1 + z}{1 - z} \,,$$

which maps $\mathbb{D}$ conformally onto $\mathbb{P}_+$. We denote by $\mathcal{H}^2$ the image of $H^2(\mathbb{D})$ under the Cayley map, i.e.

$$\mathcal{H}^2 = \{ f \circ \omega^{-1}; \; f \in H^2(\mathbb{D}) \} \,,$$

with norm $\|h\|_{\mathcal{H}^2} := \|h \circ \omega\|_{H^2}$.

The inverse map ω^{-1} is given by $\omega^{-1}(w) = (w - i)/(w + i)$, whose derivative is $2i/(w + i)^2$. Writing $\omega^{-1}(x) = e^{i\theta}$ for $x \in \mathbb{R}$, we get $ie^{i\theta} d\theta = 2i \, dx/(x + i)^2$, so that $|d\theta| = 2 \, dx/(1 + x^2)$. Therefore the normalized Lebesgue measure on $\mathbb{T} = \partial \mathbb{D}$ corresponds to the measure $\pi^{-1} dx/(1 + x^2)$ on $\mathbb{R} = \partial \mathbb{P}_+$. It follows that if $h = f \circ \omega^{-1} \in \mathcal{H}^2$ and if we put $h^* := f^* \circ \omega^{-1}$ then

$$\|h\|_{\mathcal{H}^2}^2 = \frac{1}{\pi} \int_{\mathbb{R}} |h^*(x)|^2 \frac{dx}{1 + x^2} \,.$$

The space $\mathcal{H}^2$ is not the usual Hardy space of the upper half-plane, to be defined in the next subsection. However, one can pass from $\mathcal{H}^2$ to $H^2(\mathbb{P}_+)$ using the unitary operator J defined by

$$Jh(w) = \frac{1}{\sqrt{\pi}(w + i)} \, h(w) \,.$$

B.3 Paley–Wiener theorems

We state here two versions of the Paley–Wiener theorem. For any interval $I \subset \mathbb{R}$, denote by $L^2(I)$ the space of all functions $\varphi \in L^2(\mathbb{R})$ supported on I. We define the complex Fourier transform of a function $\varphi \in L^2(I)$ by

$$\mathcal{F}\varphi(z) := \frac{1}{\sqrt{2\pi}} \int_I e^{itz} \varphi(t)\, dt\,,$$

whenever this formula makes sense.

It is easily checked that if $\varphi \in L^2(\mathbb{R})$ is compactly supported then $\mathcal{F}\varphi(z)$ is well defined for all $z \in \mathbb{C}$ and $F = \mathcal{F}\varphi$ is an entire function of exponential type. Moreover, $F_{|\mathbb{R}} \in L^2(\mathbb{R})$ by Plancherel's theorem. The Paley–Wiener theorem gives the converse.

THEOREM B.3.1 *Let $F : \mathbb{C} \to \mathbb{C}$ be an entire function, and let $a > 0$. The following are equivalent.*

(i) $F(z) = O(e^{a|z|})$ *and* $F_{|\mathbb{R}} \in L^2(\mathbb{R})$.
(ii) F *is the Fourier transform of some function* $\varphi \in L^2([-a, a])$.

The second standard version of the Paley–Wiener theorem provides a canonical isomorphism between $L^2(0, \infty)$ and the usual Hardy space of the upper half-plane $\mathbb{P}_+ = \{\mathrm{Im}(w) > 0\}$, i.e.

$$H^2(\mathbb{P}_+) := \left\{ F \in H(\mathbb{P}_+); \|F\|^2_{H^2(\mathbb{P}_+)} := \sup_{y>0} \int_{\mathbb{R}} |F(x+iy)|^2 dx < \infty \right\}.$$

If $\varphi \in L^2(0, \infty)$ then $\mathcal{F}\varphi(w)$ is easily seen to be well defined for any $w \in \mathbb{P}_+$, and $\mathcal{F}\varphi$ is holomorphic in the half-plane $\mathbb{P}_+$. Moreover, it follows from Plancherel's formula that $\mathcal{F}\varphi \in H^2(\mathbb{P}_+)$ with $\|\mathcal{F}\varphi\|_{H^2} = \|\varphi\|_{L^2}$.

THEOREM B.3.2 *A function $F : \mathbb{P}_+ \to \mathbb{C}$ is in $H^2(\mathbb{P}_+)$ if and only if it is the Fourier transform of some $\varphi \in L^2(0, \infty)$. In other words, the Fourier transform is an isometric isomorphism from $L^2(0, \infty)$ onto $H^2(\mathbb{P}_+)$.*

Proofs of the Paley–Wiener theorems can be found in [155] or [209].

C Banach space theory

C.1 Basic sequences

We recall here some well-known facts concerning basic sequences in Banach spaces. This topic is well covered in many books, including [174], [96], [173] and [3].

Let X be a Banach space. A sequence $(e_n)_{n\in\mathbb{N}} \subset X$ is a **basic sequence** if each vector $x \in [(e_n)] := \overline{\mathrm{span}}\{e_n; \ n \geq 0\}$ can be uniquely written as the sum of a convergent series $\sum_0^\infty \alpha_n e_n$. The sequence (e_n) is a **Schauder basis** of X if (e_n) is a basic sequence and $[(e_n)] = X$.

If (e_n) is a basic sequence in X then the projections $\pi_m : [(e_n)] \to X$ defined by $\pi_m(\sum_0^\infty \alpha_n e_n) = \sum_0^m \alpha_n e_n$ are uniformly bounded, by a simple (but not trivial) application of Banach's isomorphism theorem. The finite constant $C := \sup_m \|\pi_m\|$ is called the **basis constant** of (e_n). Conversely, if $(e_n) \subset X$ is a linearly independent sequence and if the projections π_m are uniformly bounded on $\mathrm{span}\{e_n; \ n \in \mathbb{N}\}$ then (e_n) is a basic sequence.

If (e_n) is a basic sequence in X, we denote by (e_n^*) the associated sequence of coordinate functionals, defined on $[(e_n)]$ by

$$\left\langle e_n^*, \sum_k \alpha_k e_k \right\rangle := \alpha_n\,.$$

We will also denote by e_n^* any Hahn–Banach extension of e_n^* to the whole space X. Thus, we may consider (e_n^*) as a sequence in X^*.

It is not hard to check that $\|e_n\| \|e_n^*\| \le 2C$ for all $n \in \mathbb{N}$, where C is the basis constant of (e_n). In particular, if the sequence (e_n) is bounded below, i.e. $\inf_n \|e_n\| > 0$, then the sequence (e_n^*) is bounded.

By a classical result of S. Mazur, any infinite-dimensional Banach space contains a basic sequence. We use this result several times in the book, in the following form.

LEMMA C.1.1 (MAZUR'S CONSTRUCTION) *Let X be an infinite-dimensional Banach space, and let $(A_n)_{n \in \mathbb{N}}$ be a sequence of subsets of $X \setminus \{0\}$. Assume that $A_n \cap E \ne \varnothing$ for each n and every closed subspace $E \subset X$ with finite codimension. Then one can construct a basic sequence $(e_n)_{n \in \mathbb{N}} \subset X$ with $e_n \in A_n$ for all $n \in \mathbb{N}$.*

PROOF   The key point is the following

FACT   Let F be a finite-dimensional subspace of X, and let $n \in \mathbb{N}$. Given $\alpha \in (0,1)$, one can find a vector $e \in A_n$ such that $\|z + \lambda e\| \ge \alpha \|z\|$ for all $z \in F$ and every $\lambda \in \mathbb{K}$.

PROOF OF THE FACT   Put $\delta := 1 - \alpha$, and let $(z_1, \ldots, z_N)$ be a δ-net in the unit sphere S_F of F. Choose $z_1^*, \ldots, z_N^* \in X^*$ such that $\langle z_i^*, z_i \rangle = 1 = \|z_i^*\|$ for all $i \in \{1, \ldots, N\}$. Then $E := \bigcap_{i=1}^N \operatorname{Ker}(z_i^*)$ is a closed finite-codimensional subspace of X, so one can pick $e \in A_n \cap E$. If z is an arbitrary point of S_F, one can find i such that $\|z - z_i\| < \delta$. Then $\|z + \lambda e\| \ge \operatorname{Re} \langle z_i^*, z + \lambda e \rangle \ge 1 - \delta = \alpha$ for every $\lambda \in \mathbb{K}$. By homogeneity, this concludes the proof.   □

Now let $(\varepsilon_n)_{n \in \mathbb{N}}$ be any summable sequence of positive numbers. Starting with $e_0 \in A_0$ and using the above fact, one can construct inductively a sequence $(e_n)_{n \in \mathbb{N}} \subset X$ such that $e_n \in A_n$ for each n and $\|\sum_0^n \lambda_i e_i\| \le (1 + \varepsilon_{n+1}) \|\sum_0^n \lambda_i e_i + \lambda e_{n+1}\|$ for any $\lambda_0, \ldots, \lambda_n, \lambda \in \mathbb{K}$. Then (e_n) is a basic sequence with constant at most $\prod_0^\infty (1 + \varepsilon_i)$.   □

Two basic sequences (e_n) and (f_n) are said to be **equivalent** if the convergence of a series $\sum_n \alpha_n e_n$ is equivalent to that of $\sum_n \alpha_n f_n$, in other words, if there exists an isomorphism from $[(e_n)]$ onto $[(f_n)]$ sending e_n to f_n for each $n \in \mathbb{N}$. The following lemma is a simple but very useful perturbation result.

LEMMA C.1.2   *Let (e_n) be a basic sequence in X, and let (f_n) be a sequence in X such that $\sum_0^\infty \|e_n^*\| \|e_n - f_n\| < 1$. Then (f_n) is a basic sequence equivalent to (e_n).*

PROOF   By assumption, the formula $T(x) := \sum_0^\infty \langle e_n^*, x \rangle (e_n - f_n)$ defines a bounded operator on X with $\|T\| < 1$. Then $J := I - T$ is an invertible operator sending e_n to f_n for each $n \in \mathbb{N}$. This shows that (f_n) is basic and equivalent to (e_n).   □

C.2 Type and cotype

By a **Rademacher sequence** we mean any sequence $(\varepsilon_n)_{n \in \mathbb{N}}$ of independent random variables defined on some probability space $(\Omega, \mathcal{F}, \mathbb{P})$ and with the same Rademacher distribution, i.e.

$$\mathbb{P}(\varepsilon_n = 1) = \frac{1}{2} = \mathbb{P}(\varepsilon_n = -1).$$

Khinchine's inequalities assert that on the linear space generated by a Rademacher sequence all L^p norms are equivalent ($1 \le p < \infty$). Since $\|\sum_n \lambda_n \varepsilon_n\|_{L^2}^2 = \sum_n |\lambda_n|^2$ for any finite sequence of scalars (λ_n), this means that

$$\left\| \sum_n \lambda_n \varepsilon_n \right\|_{L^p} \sim \left(\sum_n |\lambda_n|^2 \right)^{1/2},$$

up to constants depending only on $p \in [1, \infty)$.

Khinchine's inequalities have been extended by J. P. Kahane to the case of vector-valued Rademacher sums: for each $p \in [1, \infty)$, there exist finite positive constants A_p, B_p such that

$$A_p \left\| \sum_n \varepsilon_n x_n \right\|_{L^2(\Omega, X)} \leq \left\| \sum_n \varepsilon_n x_n \right\|_{L^p(\Omega, X)} \leq B_p \left\| \sum_n \varepsilon_n x_n \right\|_{L^2(\Omega, X)}$$

for any Banach space X and every finite sequence $(x_n) \subset X$. These are the so-called **Kahane inequalities** (see [97], [173] or [3]).

It was shown by S. Kwapień that the best constant B_p in the upper Kahane inequality satisfies $B_p \leq C \sqrt{p}$, for some absolute constant C ([161]; see [173, p. 129]).

A Banach space X is said to have (Rademacher) **type** $p \in [1, 2]$ if

$$\left\| \sum_n \varepsilon_n x_n \right\|_{L^2(\Omega, X)} \leq C \left(\sum_n \|x_n\|^p \right)^{1/p},$$

for some constant $C < \infty$ and every finite sequence $(x_n) \subset X$. Similarly, X is said to have **cotype** $q \in [2, \infty]$ if

$$\left(\sum_n \|x_n\|^q \right)^{1/q} \leq C \left\| \sum_n \varepsilon_n x_n \right\|_{L^2(\Omega, X)},$$

for some constant C and every finite sequence $(x_n) \subset X$. When $q = \infty$, the left-hand side is to be interpreted as $\sup_n \|x_n\|$.

By Khinchine's inequalities, the restrictions $p \in [1, 2]$ and $q \in [2, \infty]$ are necessary. Moreover, it follows from Kahane's inequalities that one can replace the L^2 norm by any other L^p norm ($p < \infty$) in the definitions of type and cotype.

Any Banach space has type 1 and cotype ∞. Moreover, if a Banach space X has type p then it has type p' for all $p' \in [1, p]$, and if X has cotype q then it has cotype q' for all $q' \in [q, \infty]$. Any L^p space ($1 \leq p < \infty$) has type $\min(p, 2)$ and cotype $\max(p, 2)$, whereas $c_0(\mathbb{N})$ or $L^\infty([0, 1])$ have nothing better than type 1 and cotype ∞. By the orthogonality of the Rademacher variables, any Hilbert space has both type 2 and cotype 2; and by a famous result of S. Kwapień, this property characterizes Hilbert spaces up to isomorphism. Finally, if a Banach space X has type $p \in [1, 2]$ then X^* has cotype p^*, where p^* is the conjugate exponent $(p^{-1} + (p^*)^{-1} = 1)$.

One can also consider *Gaussian* random variables instead of Rademacher variables; i.e., one can replace the sequence (ε_n) by a sequence of independent standard Gaussian variables (g_n). Then the Khinchine–Kahane inequalities remain true. Moreover, it can be shown that the Gaussian versions of type and cotype are equivalent to their Rademacher counterparts: a Banach space has Gaussian type p iff it has Rademacher type p, and likewise for cotype. This is a non-trivial result due to B. Maurey and G. Pisier. More precisely, one-quarter of this result is hard, namely the implication that (Gaussian cotype q) $\implies$ (cotype q).

Proofs of the aforementioned results (and considerably more) can be found in [97], [3] or [173]. The reader may also consult B. Maurey's survey paper [180] and the references given therein.

D Spectral theory

D.1 Spectra

Let $\mathcal{A}$ be a complex Banach algebra with unit e. If $a \in \mathcal{A}$, the **spectrum** of a in $\mathcal{A}$ is the set of all complex numbers λ such that $a - \lambda e$ is not invertible in $\mathcal{A}$. It is denoted by $\sigma(a)$.

The spectrum of any element $a \in \mathcal{A}$ is a compact subset of $\mathbb{C}$ and is always *non-empty*. The *spectral radius* of a is the number $r(a) := \sup\{|\lambda|; \ \lambda \in \sigma(a)\}$. Then $r(a) \leq \|a\|$, and the **spectral radius formula** reads

$$r(a) = \lim_{n \to \infty} \|a^n\|^{1/n} = \inf_{n \in \mathbb{N}} \|a^n\|^{1/n}.$$

These basic facts and many others can be found in e.g. W. Rudin's functional analysis book [210].

When T is a bounded operator on some complex Banach space X, we denote by $\sigma(T)$ the spectrum of T in the algebra $\mathcal{L}(X)$. The **point spectrum** of T is the set of all eigenvalues of T; it is denoted by $\sigma_p(T)$.

The **essential spectrum** of an operator $T \in \mathcal{L}(X)$ is the spectrum of the equivalence class $[T]_{\mathcal{L}/\mathcal{K}}$ of T in the Calkin algebra $\mathcal{L}(X)/\mathcal{K}(X)$, where $\mathcal{K}(X) \subset \mathcal{L}(X)$ is the two-sided ideal of all compact operators on X. This makes sense only if X is infinite-dimensional, for otherwise $\mathcal{L}(X)/\mathcal{K}(X) = \{0\}$. The essential spectrum of T is denoted by $\sigma_e(T)$. We note that $\sigma_e(T)$ is a non-empty compact subset of $\mathbb{C}$ and that $\sigma_e(T) \subset \sigma(T)$. Finally, the essential spectrum can also be defined in terms of *Fredholm* operators; see below.

D.2 Functional calculus

At some places in the book, the reader is assumed to have some familiarity with the holomorphic functional calculus for linear operators. We do not recall here the definition and the basic properties of the functional calculus; see e.g. [210]. In fact, the only result we really need is the so-called **Riesz decomposition theorem**.

THEOREM D.2.1 (RIESZ DECOMPOSITION THEOREM) *Let $T \in \mathcal{L}(X)$, and assume that the spectrum of T can be decomposed as $\sigma(T) = \sigma_1 \cup \cdots \cup \sigma_N$, where the sets σ_i are closed and pairwise disjoint. Then one can write $X = X_1 \oplus \cdots \oplus X_N$, where each X_i is a closed T-invariant subspace and $\sigma(T_{|X_i}) = \sigma_i$ for each $i \in \{1, \ldots, N\}$.*

PROOF Let us choose pairwise disjoint open sets Ω_i such that $\sigma_i \subset \Omega_i$ for each i, and let $\Omega := \bigcup_1^N \Omega_i$. One defines holomorphic functions $\chi_i \in H(\Omega)$ by setting $\chi_i(z) = 1$ on Ω_i and $\chi_i(z) = 0$ elsewhere. Then the operators $p_i := \chi_i(T)$ are well-defined projections (since $\chi_i^2 = \chi_i$), which satisfy $p_i p_j = 0$ if $i \neq j$ (since $\chi_i \chi_j = 0$ if $i \neq j$) and $\sum_i p_i = I$ (since $\sum_i \chi_i = 1$). Thus we have $X = \oplus_i X_i$, where $X_i := \text{Ran}(p_i)$. Moreover, each subspace X_i is T-invariant because $p_i T = T p_i$ (which comes from the identity $\chi_i z = z \chi_i$). We set $T_i := T_{|X_i}$, and we have to show that $\sigma(T_i) = \sigma_i$ for all $i \in \{1, \ldots, N\}$. The key point is the following

FACT Let $i \in \{1, \ldots, N\}$. A complex number λ does not belong to $\sigma(T_i)$ iff one can find an operator $R_i(\lambda) \in \mathcal{L}(X)$ such that $R_i(\lambda)(T - \lambda) = p_i = (T - \lambda)R_i(\lambda)$.

PROOF OF THE FACT If $\lambda \notin \sigma(T_i)$ then one may set $R_i(\lambda) := (T_i - \lambda)^{-1} p_i$, considered as an operator from X into X: indeed, $(T - \lambda)R_i(\lambda) = (T_i - \lambda)(T_i - \lambda)^{-1} p_i = p_i$ and $R_i(\lambda)(T - \lambda) = (T_i - \lambda)^{-1} p_i (T - \lambda) = (T_i - \lambda)^{-1}(T_i - \lambda)p_i = p_i$. Conversely, if one can find such an operator $R_i(\lambda)$ then X_i is invariant under $R_i(\lambda)$ because $R_i(\lambda)$ commutes with $p_i = \chi_i(T)$, and $T_i - \lambda$ is invertible with inverse $R_i(\lambda)_{|X_i}$. □

So, let us prove that $\sigma(T) = \bigcup_i \sigma_i$. First assume that $\lambda \notin \bigcup_i \sigma(T_i)$. Then, since $\sum_i p_i = I$, it follows from the above fact that $T - \lambda$ is invertible with inverse $\sum_i R_i(\lambda)$. This shows that $\sigma(T) \subset \bigcup_i \sigma(T_i)$.

Conversely, if $\lambda \notin \sigma_i$ then one can find some holomorphic function f such that $(z - \lambda)f(z) = \chi_i(z)$ in a neighbourhood of $\sigma(T)$. Setting $R := f(T)$, we get $(T - \lambda)R = p_i = R(T - \lambda)$, so that $\lambda \notin \sigma(T_i)$ by the above fact. Thus, we have shown that $\sigma(T_i) \subset \sigma_i$ for each $i \in \{1, \ldots, N\}$. Since $\bigcup_i \sigma_i = \sigma(T) \subset \bigcup_i \sigma(T_i)$, this concludes the proof.

D.3 Fredholm operators

Let X be a Banach space. An operator $R \in \mathfrak{L}(X)$ is said to be a **Fredholm operator** if $\mathrm{Ker}(R)$ is finite-dimensional and $\mathrm{Ran}(R)$ has finite codimension in X; then R necessarily has closed range, by a simple application of Banach's isomorphism theorem (see e.g. [162, VII.1, Corollary 4]). Equivalently, R is Fredholm iff it has closed range and $\mathrm{Ker}(R)$, $\mathrm{Ker}(R^*)$ are finite-dimensional. The **Fredholm index** of R is defined by

$$\mathrm{ind}(R) := \dim \mathrm{Ker}(R) - \mathrm{codim}\,\mathrm{Ran}(R)$$
$$= \dim \mathrm{Ker}(R) - \dim \mathrm{Ker}(R^*)\,.$$

The basic theory of Fredholm operators is well covered in many books. For efficient and very readable accounts, we recommend [174, Vol. 1, 2.c] and [162]. We outline below the best known parts of the basic theory and then prove some equally important but perhaps less universally known facts.

It follows from the Riesz theory of compact operators that if $K \in \mathfrak{L}(X)$ is compact then $I + K$ is Fredholm with index 0. This example probably motivated the whole theory.

The set of all Fredholm operators is open in $\mathfrak{L}(X)$, and the Fredholm index is continuous, i.e. locally constant. Moreover, the product of two Fredholm operators is also Fredholm, with $\mathrm{ind}(R_1 R_2) = \mathrm{ind}(R_1) + \mathrm{ind}(R_2)$.

Fredholm operators can be defined equivalently as follows: an operator $R \in \mathfrak{L}(X)$ is Fredholm iff it is invertible modulo the compact operators, i.e. $[R]_{\mathfrak{L}/\mathcal{K}}$ is invertible in the Calkin algebra $\mathfrak{L}(X)/\mathcal{K}(X)$. When X is a complex Banach space, it follows that the essential spectrum of an operator $T \in \mathfrak{L}(X)$ is exactly the set of all complex numbers λ such that $T - \lambda$ is not Fredholm.

PROPOSITION D.3.1   *Let X be a Banach space, and let $R \in \mathfrak{L}(X)$ be a Fredholm operator. Then $\dim \mathrm{Ker}(R - \mu)$ and $\mathrm{codim}\,\mathrm{Ran}(R - \mu)$ do not depend on μ if $\mu \in \mathbb{K} \setminus \{0\}$ is close enough to 0.*

PROOF   Let us put $\widetilde{X} := \bigcap_{n \geq 1} \mathrm{Ran}(R^n)$. Since all operators R^n are Fredholm, $\widetilde{X}$ is a closed subspace of X and clearly $R(\widetilde{X}) \subset \widetilde{X}$. We claim that in fact $R(\widetilde{X}) = \widetilde{X}$. Indeed, let $x \in \widetilde{X}$. Since R is Fredholm, $F_x := R^{-1}(\{x\})$ is a finite-dimensional affine subspace of X. Moreover, we have $F_x \cap \mathrm{Ran}(R^n) \neq \varnothing$ for each $n \geq 1$, since $x \in \mathrm{Ran}(R^{n+1})$. Thus, the sequence $(F_x \cap \mathrm{Ran}(R^n))_{n \geq 1}$ is a non-increasing sequence of non-empty finite-dimensional affine subspaces of X. It follows that $F_x \cap \bigcap_{n \geq 1} \mathrm{Ran}(R^n) \neq \varnothing$, i.e. $x \in R(\widetilde{X})$.

Let $\widetilde{R} := R_{|\widetilde{X}}$. Then $\widetilde{R} \in \mathfrak{L}(\widetilde{X})$ is onto, hence Fredholm, because $\dim \mathrm{Ker}(\widetilde{R}) \leq \dim \mathrm{Ker}(R) < \infty$. It follows that any operator $\widetilde{S} \in \mathfrak{L}(\widetilde{X})$ close enough to $\widetilde{R}$ is also onto and Fredholm with the same index as $\widetilde{R}$, so that $\dim \mathrm{Ker}(\widetilde{S}) = \dim \mathrm{Ker}(\widetilde{R})$. Here, we use the fact that the set of onto operators is open in $\mathfrak{L}(\widetilde{X})$, which is best seen by passing to adjoints: an operator L is onto iff L^* is an embedding, and the latter condition is easily checked to define an open set.

Thus, we get in particular that $\dim \mathrm{Ker}(\widetilde{R} - \mu) = \dim \mathrm{Ker}(\widetilde{R})$ if $\mu \in \mathbb{K}$ is close enough to 0. But if $\mu \neq 0$ then $\mathrm{Ker}(\widetilde{R} - \mu) = \mathrm{Ker}(R - \mu)$. Indeed, if $x \in X$ satisfies $R(x) = \mu x$ then $x = R^n(\mu^{-n}x)$ for all $n \geq 1$, so that $x \in \widetilde{X}$. Hence, we have shown that if $\mu \in \mathbb{K}$ is close enough to 0 and $\mu \neq 0$ then $\dim \mathrm{Ker}(R - \mu)$ does not depend on μ.

Now, R is still Fredholm, so $R - \mu$ is Fredholm with the same index as R if $\mu \in \mathbb{K}$ is close enough to 0. Thus, $\mathrm{codim}\,\mathrm{Ran}(R - \mu)$ is also independent of μ if $\mu \neq 0$ is close to 0. $\square$

COROLLARY D.3.2   *Let X be a complex Banach space, and let $T \in \mathfrak{L}(X)$. If $\lambda \in \partial\sigma(T)$ and λ is not isolated in $\sigma(T)$ then $\lambda \in \sigma_e(T)$.*

PROOF   We have to show that if λ is a non-isolated point in $\sigma(T)$ and $\lambda \notin \sigma_e(T)$ then λ is in fact an interior point of $\sigma(T)$. Let us fix λ.

By the above proposition applied to the Fredholm operator $R := T - \lambda$, we know that if $\mu \neq \lambda$ is close to λ then $\dim \mathrm{Ker}(T - \mu)$ and $\mathrm{codim}\,\mathrm{Ran}(T - \mu)$ are independent of μ. Now, since λ is not an isolated point of $\sigma(T)$ one can find $\mu_0 \neq \lambda$ close to λ such that $T - \mu$ is not invertible, i.e. at least one of $\dim \mathrm{Ker}(T - \mu_0)$ and $\mathrm{codim}\,\mathrm{Ran}(T - \mu_0)$ is non-zero. It follows that either $T - \mu$ is not one-to-one for all $\mu \neq 0$ close to λ or $T - \mu$ is not onto for all such μ. In either case, λ is an interior point of $\sigma(T)$. $\qquad\square$

A Banach space operator R is said to be **left-Fredholm** if R has closed range and $\mathrm{Ker}(R)$ is finite-dimensional. This terminology is justified by the following lemma.

LEMMA D.3.3   *Let X be a Banach space, and let $R \in \mathfrak{L}(X)$. If $[R]_{\mathfrak{L}/\mathcal{K}}$ is left-invertible in the Calkin algebra $\mathfrak{L}(X)/\mathcal{K}(X)$ then R is left-Fredholm. The converse is true if $\mathrm{Ran}(R)$ is complemented in X, in particular when X is a Hilbert space.*

PROOF   Assume that $[R]_{\mathfrak{L}/\mathcal{K}}$ is left-invertible in $\mathfrak{L}(X)/\mathcal{K}(X)$, i.e. one can find an operator $A \in \mathfrak{L}(X)$ and a compact operator K such that $AR = I + K$. Then $\mathrm{Ker}(R) \subset \mathrm{Ker}(AR) = \mathrm{Ker}(I + K)$, hence $\mathrm{Ker}(R)$ is finite-dimensional because $I + K$ is Fredholm. Likewise, $R^*A^* = I + K^*$ is also Fredholm, hence $\mathrm{Ran}(R^*)$ has finite codimension in X^* because $\mathrm{Ran}(R^*) \supset \mathrm{Ran}(R^*A^*)$. This forces R^* to have closed range, so that R has closed range as well (see [210]). Thus, R is left-Fredholm.

Conversely, assume that R is left-Fredholm and that $\mathrm{Ran}(R)$ is complemented in X. Then $\mathrm{Ker}(R)$ and $\mathrm{Ran}(R)$ are both complemented in X; let E_0 and E_1 be closed subspaces of X such that $\mathrm{Ker}(R) \oplus E_0 = X = \mathrm{Ran}(R) \oplus E_1$. The operator $R_{|E_0}$ is an isomorphism from E_0 onto $\mathrm{Ran}(R)$, so one defines a bounded operator $A \in \mathfrak{L}(X)$ by setting $A(Ru) = u$ if $u \in E_0$ and $A(v) = 0$ if $v \in E_1$. Then $AR - I$ is a finite-rank operator, hence R is left-invertible modulo the compact operators. $\qquad\square$

We also note that if $R \in L(X)$ is left-Fredholm then the Fredholm index $\mathrm{ind}(R) = \dim \mathrm{Ker}(R) - \mathrm{codim}\,\mathrm{Ran}(R)$ is still well defined, except that it may be equal to $-\infty$. Moreover, the set of all left-Fredholm operators is open in $\mathfrak{L}(X)$ and the index is continuous, i.e. locally constant; see e.g. [154].

PROPOSITION D.3.4   *Let X be a Banach space, and let $R \in \mathfrak{L}(X)$. Assume that R is not left-Fredholm. Then, for any $\varepsilon > 0$, there exists an infinite-dimensional closed subspace $E_\varepsilon \subset X$ and a compact operator $K_\varepsilon \in \mathfrak{L}(X)$ such that $\|K_\varepsilon\| < \varepsilon$ and $R \equiv K_\varepsilon$ on E_ε.*

PROOF   Let us fix $\varepsilon > 0$. If $\mathrm{Ker}(R)$ is infinite-dimensional, one can take $E_\varepsilon := \mathrm{Ker}(R)$ and $K_\varepsilon = 0$. Therefore, we assume that $\mathrm{Ker}(R)$ is finite-dimensional. Then R does not have closed range since R is not left-Fredholm.

Let Z be a closed subspace of X such that $Z \oplus \mathrm{Ker}(R) = X$. If E is any closed finite-codimensional subspace of X then $E \cap Z$ has finite codimension as well, hence $R(E \cap Z)$ is not closed otherwise $\mathrm{Ran}(R)$ itself would be closed; therefore, one can find $e \in E$ such that $\|e\| = 1$ and $\|R(e)\|$ is arbitrarily small. By Mazur's construction, it follows that one can build a normalized basic sequence $(e_n) \subset X$ such that $\|R(e_n)\| \to 0$ (see Lemma C.1.1).

Let $(e_n^*) \subset X^*$ be a sequence of coordinate functionals associated with (e_n). Extracting a subsequence if necessary, we may assume that $\|R(e_n)\| < 2^{-n-1}C^{-1}\varepsilon$ for all $n \in \mathbb{N}$, where $C = \sup_n \|e_n^*\|$. Now, let us define $K_\varepsilon : X \to X$ by $K_\varepsilon(x) := \sum_0^\infty \langle e_n^*, x \rangle R(e_n)$. Then K_ε is a compact operator, $\|K_\varepsilon\| < \varepsilon$, and $K_\varepsilon(e_n) = R(e_n)$ for all $n \in \mathbb{N}$, so that $K_\varepsilon \equiv R$ on $E_\varepsilon := [(e_n)]$. This concludes the proof. $\qquad\square$

COROLLARY D.3.5   *Let H be a Hilbert space, let $T \in \mathfrak{L}(H)$ and let $\lambda \in \mathbb{K}$. Assume that $T - \lambda$ is not left-Fredholm. Then, for any $\varepsilon > 0$, one can find an operator $T_0 \in \mathfrak{L}(H)$ such that $\|T - T_0\| < \varepsilon$, E is invariant under T_0 and $(T_0)_{|E} = \lambda I_E$.*

PROOF   Let $E = E_\varepsilon$ be given by the previous proposition applied to $R := T - \lambda$, and set $F := E^\perp$. Let

$$T - \lambda = \begin{pmatrix} A & C \\ B & D \end{pmatrix}$$

be the block representation of $T - \lambda$ with respect to the decomposition $H = E \oplus F$. Then the operator T_0 defined by

$$T_0 - \lambda = \begin{pmatrix} 0 & C \\ 0 & D \end{pmatrix}$$

has the required properties. $\square$

PROPOSITION D.3.6 *Let X be an infinite-dimensional complex Banach space. Given any operator $T \in \mathcal{L}(X)$, one can find a complex number $\lambda \in \partial\sigma(T)$ such that $T - \lambda$ is not left-Fredholm.*

PROOF Since X is infinite-dimensional, the essential spectrum of T is a non-empty compact set. Choose $\lambda \in \sigma_e(T)$ with maximum modulus. We claim that λ has the required properties. Let us denote by C_λ the "corona" $\{z \in \mathbb{C}; \ |z| > |\lambda|\}$.

Since $\lambda \in \sigma(T)$ and $\sigma(T)$ is compact, any half-line starting at λ has to intersect $\partial\sigma(T)$. From this, it is easy to see that if λ were an interior point of $\sigma(T)$ then $\partial\sigma(T) \cap C_\lambda$ would contain some uncountable compact set (e.g. $\partial\sigma(T) \cap \{|z| \geq r\}$, for some $r > |\lambda|$); so one could find some non-isolated point $\mu \in \partial\sigma(T)$ with $|\mu| > |\lambda|$. But then we would have $\mu \in \sigma_e(T)$ by Corollary D.3.2, so we get a contradiction by our choice of λ. Thus, we have shown that $\lambda \in \partial\sigma(T)$.

Again by our choice of λ, we know that $T - \mu$ is Fredholm for every $\mu \in C_\lambda$. Moreover, $T - \mu$ is invertible if $|\mu|$ is large enough, so that $\mathrm{ind}(T - \mu) = 0$. Since the Fredholm index is continuous and the corona C_λ is connected, it follows that $\mathrm{ind}(T - \mu) = 0$ for every $\mu \in C_\lambda$. Now, the set of Fredholm operators with index 0 is relatively closed in the set of all left-Fredholm operators, by the continuity of the index. Since λ is a limit point of C_λ, it follows that if $T - \lambda$ were left-Fredholm then we would have $\mathrm{ind}(T - \lambda) = 0 > -\infty$ and hence $T - \lambda$ would be in fact Fredholm, which was excluded at the beginning since $\lambda \in \sigma_e(T)$. Therefore, $T - \lambda$ is not left-Fredholm. $\square$

References

[1] E. Abakumov and J. Gordon. Common hypercyclic vectors for multiples of back-ward shift. *J. Funct. Anal.*, 200: 494–504, 2003. (Cited on pp. 164, 165 and 192.)

[2] Y. A. Abramovich and C. D. Aliprantis. *An Invitation to Operator Theory*, volume 50 of *Graduate Studies in Mathematics*. American Mathematical Society, 2002. (Cited on p. 309.)

[3] F. Albiac and N. J. Kalton. *Topics in Banach Space Theory*, volume 233 of *Graduate Texts in Mathematics*. Springer-Verlag, 2006. (Cited on pp. 84, 196, 314 and 316.)

[4] E. Alfsen. *Compact Convex Sets and Boundary Integrals*, volume 57 of *Ergebnisse Der Mathematik Und Ihrer Grenzgebiete*. Springer-Verlag, 1971. (Cited on p. 100.)

[5] R. D. Anderson. Hilbert space is homeomorphic to the countable infinite product of real lines. *Bull. Amer. Math. Soc.*, 72: 515–519, 1966. (Cited on p. 16.)

[6] S. I. Ansari. Hypercyclic and cyclic vectors. *J. Funct. Anal.*, 128: 374–383, 1995. (Cited on pp. 28, 60, 61 and 72.)

[7] S. I. Ansari. Existence of hypercyclic operators on topological vector spaces. *J. Funct. Anal.*, 148: 384–390, 1997. (Cited on pp. 37 and 57.)

[8] S. I. Ansari and P. S. Bourdon. Some properties of cyclic operators. *Acta Sci. Math. (Szeged)*, 63: 195–207, 1997. (Cited on p. 28.)

[9] S. A. Argyros and R. G. Haydon. An $\mathcal{L}^\infty$ HI space solving the $\lambda I + K$ problem. Preprint, 2008. (Cited on pp. 31 and 151.)

[10] R. M. Aron, J. P. Bès, F. León-Saavedra and A. Peris. Operators with common hy-percyclic subspaces. *J. Operator Theory*, 54: 251–260, 2005. (Cited on pp. 202, 204 and 217.)

[11] R. M. Aron, D. García and M. Maestre. Construction of weakly dense, norm divergent sequences. *J. Convex Analysis*, 16: 2009. (Cited on pp. 234 and 262.)

[12] R. M. Aron, J. A. Conejero, A. Peris and J. B. Seoane-Sepúlveda. Powers of hypercyclic functions for some classical hypercyclic operators. *Integral Equations Operator Theory*, 58: 591–596, 2007. (Cited on pp. 213, 214 and 217.)

[13] C. Badea and S. Grivaux. Size of the peripheral point spectrum under power or resolvent growth conditions. *J. Funct. Anal.*, 246: 302–329, 2007. (Cited on p. 163.)

[14] C. Badea and S. Grivaux. Unimodular eigenvalues, uniformly distributed sequences and linear dynamics. *Adv. Math.*, 211: 766–793, 2007. (Cited on pp. 160, 162 and 163.)

[15] B. Bagchi. A joint universality theorem for Dirichlet L-functions. *Math. Z.*, 181: 319–334, 1982. (Cited on pp. 278 and 290.)

[16] K. Ball. The plank problem for symmetric bodies. *Invent. Math.*, 104: 535–543, 1991. (Cited on p. 232.)

[17] K. Ball. The complex plank problem. *Bull. London Math. Soc.*, 33: 433–442, 2001. (Cited on p. 232.)

[18] J. Banks, J. Brooks, G. Cairns, G. Davis and P. Stacey. On Devaney's definition of chaos. *Amer. Math. Monthly*, 99: 332–334, 1992. (Cited on pp. 137 and 162.)

[19] N. K. Bary. *A Treatise on Trigonometric Series*. Pergamon Press, 1964. (Cited on p. 270.)

[20] F. Bayart. Common hypercyclic subspaces. *Integral Equations Operator Theory*, 53: 467–476, 2005. (Cited on p. 204.)

[21] F. Bayart. Porosity and hypercyclic operators. *Proc. Amer. Math. Soc.*, 133 (11): 3309–3316, 2005. (Cited on p. 26.)

[22] F. Bayart. Weak-closure and polarization constant by Gaussian measures. Preprint, 2007. (Cited on p. 235.)

[23] F. Bayart. Parabolic composition operators on the ball. Preprint, 2008. (Cited on p. 228.)

[24] F. Bayart and T. Bermúdez. Semigroups of chaotic operators. *Math. Zeitschrifte:* 2009. (Cited on p. 162.)

[25] F. Bayart and S. Grivaux. Hypercyclicity and unimodular point spectrum. *J. Funct. Anal.*, 226: 281–300, 2005. (Cited on p. 130.)

[26] F. Bayart and S. Grivaux. Frequently hypercyclic operators. *Trans. Amer. Math. Soc.*, 358 (11): 5083–5117, 2006. (Cited on pp. 129, 130, 141, 142, 161 and 193.)

[27] F. Bayart and S. Grivaux. Invariant Gaussian measures for operators on Banach spaces and linear dynamics. *Proc. Lond. Math. Soc.*, 94: 181–210, 2007. (Cited on pp. 119, 130, 146, 156 and 161.)

[28] F. Bayart and É. Matheron. Hyponormal operators, weighted shifts, and weak forms of supercyclicity. *Proc. Edinb. Math. Soc.*, 49: 1–15, 2006. (Cited on pp. 253 and 256.)

[29] F. Bayart and É. Matheron. How to get common universal vectors. *Indiana Univ. Math. J.*, 56: 553–580, 2007. (Cited on pp. 93, 162, 178, 182, 184, 192 and 193.)

[30] F. Bayart and É. Matheron. Hypercyclic operators failing the Hypercyclicity Criterion on classical Banach spaces. *J. Funct. Anal.*, 250: 426–441, 2007. (Cited on pp. 83, 92 and 94.)

[31] F. Bayart and É. Matheron. (Non)-weakly mixing operators and hypercyclicity sets. *Ann. Inst. Four.:* 2009. (Cited on pp. 92, 161 and 262.)

[32] F. Bayart, É. Matheron and P. Moreau. Small sets and hypercyclic vectors. *Comment. Math. Univ. Carolinae*, 49: 53–65, 2008. (Cited on p. 26.)

[33] F. Bayart and L. Quarta. Algebrability of sets of queer functions. *Israel J. Math.*, 158: 285–296, 2007. (Cited on p. 217.)

[34] B. Beauzamy. Un opérateur, sur un espace de Banach, avec un ensemble non-dénombrable, dense, de vecteurs hypercycliques. In *Séminaire de géométrie des espaces de Banach*, volume 18 of *Publ. Math. Univ. Paris VII*, pp. 149–175. Université Paris VII, 1984. (Cited on p. 26.)

[35] B. Beauzamy. Un opérateur sans sous-espace invariant: simplification de l'exemple de P. Enflo. *Integral Equations Operator Theory*, 8: 314–384, 1985. (Cited on p. 308.)

[36] P. Beneker and F. Wiegerinck. Strongly exposed points in the ball of the Bergman space. *Integral Equations Operator Theory*, 52: 42–65, 2005. (Cited on p. 227.)

[37] C. A. Berenstein and R. Gay. *Complex Variables. An Introduction*, volume 125 of *Graduate Texts in Mathematics*. Springer-Verlag, 1991. (Cited on p. 290.)

[38] T. Bermúdez, A. Bonilla and A. Martinón. On the existence of chaotic and hypercyclic semigroups on Banach spaces. *Proc. Amer. Math. Soc.*, 131 (8): 2435–2441, 2003. (Cited on p. 57.)

[39] T. Bermúdez, A. Bonilla, and A. Peris. On hypercyclicity and supercyclicity criteria. *Bull. Austral. Math. Soc.*, 70: 45–54, 2004. (Cited on pp. 27 and 28.)

[40] L. Bernal-González. On hypercyclic operators on Banach spaces. *Proc. Amer. Math. Soc.*, 127 (4): 1003–1010, 1999. (Cited on pp. 37 and 57.)

[41] L. Bernal-González. Disjoint hypercyclic operators. *Studia Math.*, 182(2): 113–131, 2007. (Cited on p. 92.)

[42] L. Bernal-González and K.-G. Grosse-Erdmann. The Hypercyclicity Criterion for sequences of operators. *Studia Math.*, 157: 17–32, 2003. (Cited on p. 92.)

[43] L. Bernal-González and K.-G. Grosse-Erdmann. Existence and nonexistence of hypercyclic semigroups. (English summary) *Proc. Amer. Math. Soc.*, 135: 755–766, 2007. (Cited on p. 57.)

[44] L. Bernal-González and A. Montes-Rodríguez. Universal functions for composition operators. *Complex Variables Theory Appl.*, 27: 47–56, 1995. (Cited on p. 217.)

[45] J. P. Bès. Three problem's on hypercyclic operators. Ph.D. thesis, Bowling Green State University, Bowling Green, Ohio, 1998. (Cited on pp. 4 and 27.)

[46] J. P. Bès. Invariant manifolds of hypercyclic vectors for the real scalar case. *Proc. Amer. Math. Soc.*, 127 (6): 1801–1804, 1999. (Cited on p. 27.)

[47] J. P. Bès and K. C. Chan. Approximation by chaotic operators and by conjugate classes. *J. Math. Anal. Appl.*, 284: 206–212, 2003. (Cited on pp. 44 and 57.)

[48] J. P. Bès, K. C. Chan and R. Sanders. Every weakly sequentially hypercyclic shift is norm hypercyclic. *Math. Proc. R. Ir. Acad.*, 105A: 79–85, 2005. (Cited on p. 261.)

[49] J. P. Bès, K. C. Chan and S. M. Seubert. Chaotic unbounded differentiation operators. *Integral Equations Operator Theory*, 40: 257–267, 2001. (Cited on p. 27.)

[50] J. P. Bès and A. Peris. Hereditarily hypercyclic operators. *J. Funct. Anal.*, 167: 94–112, 1999. (Cited on p. 75.)

[51] J. P. Bès and A. Peris. Disjointness in hypercyclicity. *J. Math. Anal. Appl.*, 336: 297–315, 2007. (Cited on p. 92.)

[52] C. Bessaga and A. Pełczyński. *Selected Topics in Infinite-Dimensional Topology.* PWN, Warsaw, 1975. (Cited on p. 17.)

[53] G. D. Birkhoff. Surface transformations and their dynamical applications. *Acta Math.*, 43: 1–119, 1922. (Cited on pp. 2 and 26.)

[54] G. D. Birkhoff. Démonstration d'un théorème élémentaire sur les fonctions entières. *C. R. Acad. Sci. Paris*, 189: 473–475, 1929. (Cited on p. 3.)

[55] V. I. Bogachev. *Gaussian Measures*, volume 62 of *Mathematical Surveys and Monographs*. American Mathematical Society, 1998. (Cited on pp. 98, 99, 123, 127 and 131.)

[56] H. Bohr. Zur Theorie der Riemannschen Zetafunktion im kritischen Streifen. *Acta Math.*, 40: 67–100, 1915. (Cited on p. 290.)

[57] J. Bonet, F. Martínez-Giménez and A. Peris. A Banach space which admits no chaotic operator. *Bull. London Math. Soc.*, 33: 196–198, 2001. (Cited on pp. 151 and 162.)

[58] J. Bonet, F. Martínez-Giménez and A. Peris. Universal and chaotic multipliers on spaces of operators. *J. Math. Anal. Appl.*, 297: 599–611, 2004. (Cited on p. 217.)

[59] J. Bonet and A. Peris. Hypercyclic operators on non-normable Fréchet spaces. *J. Funct. Anal.*, 159: 587–595, 1998. (Cited on pp. 29, 37, 57 and 58.)

[60] A. Bonilla and K.-G. Grosse-Erdmann. On a theorem of Godefroy and Shapiro. *Integral Equations Operator Theory*, 56: 151–162, 2006. (Cited on pp. 146 and 161.)

[61] A. Bonilla and K.-G. Grosse-Erdmann. Frequently hypercyclic operators and vectors. *Ergodic Theory Dynamical Systems*, 27: 383–404, 2007. (Cited on pp. 138 and 161.)

[62] P. S. Bourdon. Density of the polynomials in Bergman spaces. *Pacific J. Math.*, 130: 215–221, 1987. (Cited on p. 290.)

[63] P. S. Bourdon. Invariant manifolds of hypercyclic vectors. *Proc. Amer. Math. Soc.*, 118 (3): 845–847, 1993. (Cited on p. 27.)

[64] P. S. Bourdon. The second iterate of a map with dense orbit. *Proc. Amer. Math. Soc.*, 124 (5): 1577–1581, 1996. (Cited on p. 72.)

[65] P. S. Bourdon. Orbits of hyponormal operators. *Michigan Math. J.*, 44: 345–353, 1997. (Cited on p. 29.)

[66] P. S. Bourdon and N. S. Feldman. Somewhere dense orbits are everywhere dense. *Indiana Univ. Math. J.*, 52: 811–819, 2003. (Cited on pp. 60, 69, 71 and 73.)

324   *References*

[67] P. Brunovský and J. Komorník. Ergodicity and exactness of the shift on $C([0, \infty])$ and the semiflow of a first-order partial differential equation. *J. Math. Anal. Appl.*, 104: 235–245, 1984. (Cited on p. 129.)

[68] G. Cantor. Beiträge zur Begründung der transfiniten Mengenlehre. *Math. Ann.*, 49: 207–246, 1897. (Cited on p. 41.)

[69] F. Carlson. Contributions à la théorie des séries de Dirichlet. *Ark. Mat.*, 16: 1–19, 1922. (Cited on p. 272.)

[70] K. C. Chan. Hypercyclicity of the operator algebra for a separable Hilbert space. *J. Operator Theory*, 42: 231–244, 1999. (Cited on pp. 196 and 217.)

[71] K. C. Chan. The density of hypercyclic operators on a Hilbert space. *J. Operator Theory*, 47: 131–143, 2001. (Cited on p. 57.)

[72] K. C. Chan and R. Sanders. A weakly hypercyclic operator that is not norm hypercyclic. *J. Operator Theory*, 52: 39–59, 2004. (Cited on pp. 244 and 249.)

[73] K. C. Chan and R. Sanders. Two criteria for a path of operators to have common hypercyclic vectors. *J. Operator Theory*, to appear, 2008. (Cited on pp. 170, 185, 189, 192 and 193.)

[74] K. C. Chan and R. D. Taylor Jr. Hypercyclic subspaces of a Banach space. *Integral Equations Operator Theory*, 41: 381–388, 2001. (Cited on pp. 196 and 217.)

[75] S. A. Chobanyan. The structure of the set of sums of a conditionally convergent series in a normed space. *Math. Sb.*, 56: 49–62, 1985. (Cited on p. 280.)

[76] S. A. Chobanyan and V. I. Tarieladze. Gaussian characterizations of certain Banach spaces. *J. Multivariate Anal.*, 7: 183–203, 1977. (Cited on p. 107.)

[77] S. A. Chobanyan, V. I. Tarieladze and N. N. Vakhania. *Probability distributions on Banach spaces*, volume 14 of *Mathematics and its Applications (Soviet Series.* Reidel, 1987. (Cited on pp. 98, 99, 102, 106, 109, 127, 128 and 131.)

[78] K. F. Clancey and D. D. Rogers. Cyclic vectors and seminormal operators. *Indiana Univ. Math. J.*, 27: 689696, 1978. (Cited on pp. 27 and 28.)

[79] J. A. Conejero, V. Müller and A. Peris. Hypercyclic behaviour of operators in a hypercyclic C_0-semigroup. *J. Funct. Anal.*, 244: 342–348, 2007. (Cited on pp. 60, 62, 161 and 193.)

[80] J. A. Conejero and A. Peris. Linear transitivity criteria. *Topology Appl.*, 153: 767–773, 2005. (Cited on p. 27.)

[81] J. B. Conway. *A Course in Functional Analysis*, volume 96 of *Graduate Texts in Mathematics*. Springer-Verlag, 1990. (Cited on p. 209.)

[82] J. G. van der Corput. Zahlentheoritische Abschätzungen. *Math. Annalen*, 84: 53–79, 1921. (Cited on p. 269.)

[83] G. Costakis. On a conjecture of D. Herrero concerning hypercyclic operators. *C. R. Acad. Sci. Paris*, 330: 179–182, 2000. (Cited on p. 72.)

[84] G. Costakis and A. Manoussos. J-class operators and hypercyclicity. Preprint, 2007. (Cited on pp. 29 and 33.)

[85] G. Costakis and A. Peris. Hypercyclic semigroups and somewhere dense orbits. *C. R. Acad. Sci. Paris*, 335: 895–898, 2002. (Cited on p. 72.)

[86] G. Costakis and M. Sambarino. Genericity of wild holomorphic functions and common hypercyclic vectors. *Adv. Math.*, 182: 278–306, 2004. (Cited on pp. 173 and 192.)

[87] G. Costakis and M. Sambarino. Topologically mixing hypercyclic operators. *Proc. Amer. Math. Soc.*, 132 (2): 385–389, 2004. (Cited on p. 93.)

[88] C. C. Cowen. Composition operators on H^2. *J. Operator Theory*, 9: 77–106, 1983. (Cited on p. 226.)

[89] C. C. Cowen. Linear fractional composition operators on $\mathcal{H}^2$. *Integral Equations Operator Theory*, 11: 151–160, 1988. (Cited on p. 226.)

[90] C. C. Cowen and B. MacCluer. *Composition Operators on Spaces of Analytic Functions*. Studies in Advanced Mathematics. CRC Press, 1995. (Cited on pp. 22 and 23.)

[91] M. De La Rosa and C. J. Read. A hypercyclic operator whose direct sum is not hypercyclic. *J. Operator Theory*, to appear, 2008. (Cited on pp. 75, 83 and 92.)

[92] M. Del Pilar Romero de La Rosa. Regular orbits. Preprint, 2008. (Cited on p. 74.)

[93] R. deLaubenfels, H. Emamirad and K.-G. Grosse-Erdmann. Chaos for semigroups of unbounded operators. *Math. Nachr.*, 261/262: 47–59, 2003. (Cited on pp. 27 and 161.)

[94] R. L. Devaney. *An Introduction to Chaotic Dynamical systems*. Addison-Wesley, 1989. (Cited on pp. 4 and 137.)

[95] J. Diestel. *Geometry of Banach Spaces – Selected Topics*, volume 485 of *Lecture Notes in Mathematics*. Springer-Verlag, 1975. (Cited on pp. 221 and 228.)

[96] J. Diestel. *Sequences and Series in Banach Spaces*, volume 92 of *Graduate Texts in Mathematics*. Springer-Verlag, 1984. (Cited on pp. 84, 138, 196, 197 and 314.)

[97] J. Diestel, H. Jarchow and A. Tonge. *Absolutely summing operators*, volume 43 of *Cambridge Studies in Advanced Mathematics*. Cambridge University Press, 1995. (Cited on pp. 105, 106, 107, 118, 138, 230, 235, 238, and 316.)

[98] S. Dilworth and V. Troitsky. Spectrum of a weakly hypercyclic operator meets the unit circle. In *Trends in Banach Spaces and Operator Theory*, volume 321, pp. 67–69. American Mathematical Society, 2006. (Cited on pp. 232 and 262.)

[99] R. G. Douglas and C. Pearcy. A note on quasitriangular operators. *Duke Math. J.*, 37: 177–188, 1970. (Cited on p. 54.)

[100] J. Dugundji. *Topology*. Allin and Bacon, 1966. (Cited on pp. 43, 66, 68 and 167.)

[101] P. L. Duren. *Theory of H^p spaces*, volume 38 of *Pure and Applied Mathematics*. Academic Press, 1970. (Cited on pp. 7 and 312.)

[102] P. L. Duren and A. Schuster. *Bergman Spaces*, volume 100 of *Mathematical Surveys and Monographs*. American Mathematical Society, 2004. (Cited on p. 267.)

[103] A Dvoretzky. A theorem on convex bodies and applications to Banach spaces. *Proc. Nat. Acad. Sci. USA*, 45: 223–226, 1959. (Cited on p. 238.)

[104] P. Enflo. On the invariant subspace problem for Banach spaces. *Acta Math.*, 158: 213–313, 1987. (Cited on pp. x and 308.)

[105] A. Fathi. Existence de systèmes dynamiques minimaux sur l'espace de Hilbert séparable. *Topology*, 22: 163–167, 1983. (Cited on pp. 16 and 27.)

[106] N. S. Feldman. Linear chaos? http://home.wlu.edu/~feldmann/research.html, 2001. (Cited on p. 49.)

[107] N. S. Feldman. Perturbations of hypercyclic vectors. *J. Math. Anal. Appl.*, 273: 67–74, 2002. (Cited on pp. 134, 136 and 161.)

[108] N. S. Feldman. Hypercyclic pairs of coanalytic Toeplitz operators. *Integral Equations Operator Theory*, 58: 153–173, 2007. (Cited on p. 62.)

[109] N. S. Feldman, V. G. Miller and T. L. Miller. Hypercyclic and supercyclic cohyponormal operators. *Acta Sci. Math. (Szeged)*, 68: 303–328, 2002. Corrected reprint: *Acta Sci. Math. (Szeged)*, 68: 965–990, 2002. (Cited on pp. 12 and 27.)

[110] P. A. Fillmore, J. G. Stampfli and J. P. Williams. On the essential numerical range, the essential spectrum, and a problem of Halmos. *Acta Sci. Math. (Szeged)*, 33: 179–192, 1972. (Cited on p. 58.)

[111] E. Flytzanis. Unimodular eigenvalues and linear chaos in Hilbert spaces. *Geom. Funct. Anal.*, 5: 1–13, 1995. (Cited on p. 129.)

[112] H. Furstenberg. Disjointness in ergodic theory, minimal sets, and a problem in diophantine approximation. *Math. Systems Theory*, 1: 1–49, 1967. (Cited on p. 92.)

[113] H. Furstenberg. *Recurrence in Ergodic Theory and Combinatorial Number Theory*. Princeton University Press, 1981. (Cited on p. 92.)

[114] E. Gallardo Gutiérrez and A. Montes-Rodríguez. The role of the angle in supercyclic behavior. *J. Funct. Anal.*, 203: 27–43, 2003. (Cited on pp. 141, 224, 227 and 228.)

[115] E. A. Gallardo Gutiérrez and A. Montes-Rodríguez. The role of the spectrum in the cyclic behavior of composition operators. *Mem. Amer. Math. Soc.*, 791: 2004. (Cited on p. 228.)

[116] E. A. Gallardo Gutiérrez and A. Montes-Rodríguez. The Volterra operator is not supercyclic. *Integral Equations Operator Theory*, 50: 211–216, 2004. (Cited on p. 228.)

[117] E. A. Gallardo Gutiérrez and J. R. Partington. Supercyclic vectors and the angle criterion. *Studia Math.*, 166: 93–99, 2005. (Cited on p. 227.)

[118] E. A. Gallardo Gutiérrez and J. R. Partington. Common hypercyclic vectors for families of operators. *Proc. Amer. Math. Soc.*, 136 (1): 119–126, 2008. (Cited on p. 193.)

[119] R. M. Gethner and J. H. Shapiro. Universal vectors for operators on spaces of holomorphic functions. *Proc. Amer. Math. Soc.*, 100 (2): 281–288, 1987. (Cited on pp. 4, 26, 27 and 145.)

[120] E. Glasner. *Ergodic Theory via Joinings*, volume 101 of *Mathematical Surveys and Monographs*. American Mathematical Society, 2003. (Cited on p. 92.)

[121] E. Glasner and B. Weiss. On the interplay between measurable and topological dynamics. *Handbook of Dynamical Systems*, 1B: 597–648, 2006. (Cited on p. 150.)

[122] G. Godefroy. Renorming of Banach spaces. In *Handbook of the Geometry of Banach Spaces, volume 1*, pp. 781–835. North-Holland, 2003. (Cited on p. 229.)

[123] G. Godefroy and J. H. Shapiro. Operators with dense, invariant, cyclic vector manifolds. *J. Funct. Anal.*, 98: 229–269, 1991. (Cited on pp. ix, 6, 20, 32, 137, 161 and 192.)

[124] R. Godement. Théorèmes taubériens et théorie spectrale. *Ann. Sci. École Norm. Sup.*, 64: 119–138, 1947. (Cited on p. 28.)

[125] M. González, F. León-Saavedra and A. Montes-Rodríguez. Semi-Fredholm theory: hypercyclic and supercyclic subspaces. *Proc. London Math. Soc.*, 81: 169–189, 2000. (Cited on pp. 195 and 217.)

[126] W. T. Gowers and B. Maurey. The unconditional basic sequence problem. *J. Amer. Math. Soc.*, 6: 851–874, 1993. (Cited on p. 151.)

[127] S. Grivaux. Construction of operators with prescribed behaviour. *Arch. Math. (Basel)*, 81: 291–299, 2003. (Cited on pp. 41 and 42.)

[128] S. Grivaux. Sums of hypercyclic operators. *J. Funct. Anal.*, 202: 486–503, 2003. (Cited on pp. 45, 50 and 54.)

[129] S. Grivaux. Hypercyclic operators, mixing operators, and the Bounded Steps Problem. *J. Operator Theory*, 54: 147–168, 2005. (Cited on pp. 28, 58, 75, 80, 92 and 93.)

[130] S. Grivaux. A probabilistic version of the frequent hypercyclicity criterion. *Studia Math.*, 176: 279–290, 2006. (Cited on p. 161.)

[131] S. Grivaux and M. Roginskaya. On Read's type operators on Hilbert spaces. *Int. Math. Res. Notices*, to appear, 2008. (Cited on pp. 26 and 309.)

[132] S. Grivaux and S. Shkarin. Non-mixing hypercyclic operators. Unpublished work, 2007. (Cited on pp. 31, 32, 36, 57 and 92.)

[133] K.-G. Grosse-Erdmann. Universal families and hypercyclic vectors. *Bull. Amer. Math. Soc.*, 36 (3): 345–381, 1999. (Cited on pp. ix, x and 27.)

[134] K.-G. Grosse-Erdmann. Hypercyclic and chaotic weighted shifts. *Studia Math.*, 139: 47–68, 2000. (Cited on pp. 20 and 161.)

[135] K.-G. Grosse-Erdmann. Dynamics of linear operators. In *Topics in Complex Analysis and Operator Theory*, pp. 41–84, University of Málaga, 2007 (Cited on pp. 111, 127 and 130.)

[136] K.-G. Grosse-Erdmann and A. Peris. Frequently dense orbits. *C. R. Acad. Sci. Paris*, 341: 123–128, 2005. (Cited on pp. 150 and 161.)

[137] D. Hadwin, E. Nordgren, H. Radjavi and P. Rosenthal. Most similarity orbits are strongly dense. *Proc. Amer. Math. Soc.*, 76 (2): 250–252, 1979. (Cited on p. 45.)

[138] H. Halberstam and K. F. Roth. *Sequences*. Second edition. Springer-Verlag, 1983. (Cited on p. 262.)

[139] G. H. Hardy and E. M. Wright. *An Introduction to the Theory of Numbers*. Clarendon Press, 1938. (Cited on p. 285.)

[140] F. Hausdorff. *Set Theory*. Chelsea, 1957. (Cited on p. 41.)

[141] D. A. Herrero. *Approximation of Hilbert Space Operators I*, volume 72 of *Pitman Research Notes in Mathematics*. Pitman, 1982. (Cited on pp. 47, 58 and 59.)

[142] D. A. Herrero. A metatheorem on similarity and approximation of operators. *J. London Math. Soc.*, 42: 535–554, 1990. (Cited on p. 58.)

[143] D. A. Herrero. Limits of hypercyclic and supercyclic operators. *J. Funct. Anal.*, 99: 179–190, 1991. (Cited on pp. 27, 58 and 75.)

[144] H. M. Hilden and L. J. Wallen. Some cyclic and non-cyclic vectors of certain operators. *Indiana Univ. Math. J.*, 23: 557–565, 1973/74. (Cited on pp. 26 and 28.)

[145] J. G. Hocking and G. S. Young. *Topology*. Second edition. Dover, 1988. (Cited on p. 12.)

[146] T. Hosokawa. Chaotic behavior of composition operators on the Hardy space. *Acta Sci. Math. (Szeged)*, 69: 801–811, 2003. (Cited on p. 161.)

[147] B. Jamison. Eigenvalues of modulus 1. *Proc. Amer. Math. Soc.*, 16: 375–377, 1965. (Cited on pp. 162 and 163.)

[148] M. Kadets and V. Kadets. *Series in Banach Spaces. Conditional and Unconditional Convergence.*, volume 94 of *Operator Theory: Advances and Applications*. Birkhäuser, 1997. (Cited on p. 278.)

[149] V. Kadets. Weak cluster points of a sequence and coverings by cylinders. *Mat. Fiz. Anal. Geom.*, 11: 161–168, 2004. (Cited on p. 237.)

[150] J.-P. Kahane. *Some Random Series of Functions*, volume 5 of *Cambridge Studies in Advanced Mathematics*. Cambridge University Press, 1993. (Cited on pp. 123.)

[151] G. Kalisch. On operators on separable Banach spaces with arbitrary prescribed point spectrum. *Proc. Amer. Math. Soc.*, 34: 207–208, 1972. (Cited on p. 130.)

[152] N. Kalton, J. van Neerven, M. Veraar and L. Weis. Embedding vector-valued Besov spaces into spaces of γ-radonifying operators. *Math. Nachr.*, 281: 238–252, 2008. (Cited on p. 130.)

[153] N. J. Kalton, N. T. Peck and J. W. Roberts. *An F-space sampler*, volume 89 of *London Mathematical Society Lecture Note Series*. Cambridge University Press, 1984. (Cited on p. 57.)

[154] T Kato. *Perturbation Theory for Linear Operators*. Classics in Mathematics. Springer-Verlag, 1995. (Cited on p. 319.)

[155] Y. Katznelson. *An Introduction to Harmonic Analysis*. John Wiley & Sons, 1968. (Cited on pp. 114, 133 and 314.)

[156] A. S. Kechris. *Classical Descriptive Set Theory*, volume 156 of *Graduate Texts in Mathematics*. Springer-Verlag, 1995. (Cited on pp. 113, 114, 154, 156 and 168.)

[157] A. S. Kechris and A. Louveau. *Descriptive Set Theory and the Structure of Sets of Uniqueness*, volume 128 of *London Mathematical Society Lecture Notes Series*. Cambridge University Press, 1987. (Cited on pp. 117, 254, 257 and 263.)

[158] C. Kitai. Invariant closed sets for linear operators. Ph.D. thesis, University of Toronto, Toronto, 1982. (Cited on pp. ix, 4, 11, 26, 27 and 29.)

[159] H. König. On the Fourier coefficients of vector-valued functions. *Math. Nachr.*, 152: 215–227, 1991. (Cited on p. 119.)

[160] S. Kwapień. On Banach spaces containing c_0. *Studia Math.*, 52: 187–188, 1974. (Cited on p. 106.)

[161] S. Kwapień. A theorem on Rademacher series with vector-valued coefficients, in volume 526 of *Lecture Notes in Mathematics*, pp. 157–158. Springer-Verlag, 1976. (Cited on pp. 182 and 316.)

[162] S. Lang. *Real Analysis*. Second edition. *Advanced Book Program*. Addison-Wesley, 1983. (Cited on p. 318.)

[163] A. Laurinčikas. *Limit Theorems for the Riemann Zeta Function*, volume 352 of *Mathematics and its Applications*. Kluwer Academic Press, 1996. (Cited on p. 290.)

[164] M. Ledoux and M. Talagrand. *Probability in Banach spaces. Isoperimetry and Processes*, volume 23 of *Ergebnisse Der Mathematik Und Ihrer Grenzgebiete*. Springer-Verlag, 1991. (Cited on pp. 98 and 180.)

[165] F. León-Saavedra. Notes about the hypercyclicity criterion. *Math. Slovaca*, 53: 313–319, 2003. (Cited on p. 92.)

[166] F. León-Saavedra and A. Montes-Rodríguez. Linear structure of hypercyclic vectors. *J. Funct. Anal.*, 148: 524–545, 1997. (Cited on p. 57.)

[167] F. León-Saavedra and A. Montes-Rodríguez. Spectral theory and hypercyclic subspaces. *Trans. Amer. Math. Soc.*, 353 (1): 247–267, 2001. (Cited on pp. 195, 196, 209 and 217.)

[168] F. León-Saavedra and V. Müller. Rotations of hypercyclic and supercyclic operators. *Integral Equations Operator Theory*, 50: 385–391, 2004. (Cited on pp. 60 and 61.)

[169] F. León-Saavedra and V. Müller. Hypercyclic sequences of operators. *Studia Math.*, 175: 1–18, 2006. (Cited on p. 197.)

[170] F. León-Saavedra and A. Piqueras-Lerena. Positivity in the theory of supercyclic operators. In *Perspectives in Operator Theory*, pp. 221–232, Banach Center Publication volume 75. Polish Academy of Science, Warsaw, 2007. (Cited on p. 230.)

[171] F. León-Saavedra and A. Piqueras-Lerena. Cyclic properties of Volterra operators II. Preprint, 2008. (Cited on pp. 228 and 229.)

[172] B. Ya. Levin. *Lectures on Entire Functions*, volume 150 of *Translations of Mathematical Monographs*. American Mathematical Society, 1996. (Cited on pp. 152 and 311.)

[173] D. Li and H. Queffélec. *Introduction à l'étude des espaces de Banach. Analyse et probabilités*, volume 12 of *Cours spécialisés*. Société Mathématique de France, 2004. (Cited on pp. 98, 105, 107, 123, 132, 180, 196, 238, 314 and 316.)

[174] J. Lindenstrauss and L. Tzafriri. *Classical Banach Spaces*, volume 12 of *Classics in Mathematics*. Springer-Verlag, 1997. (Cited on pp. 40, 107, 151, 175, 196, 314 and 318.)

[175] G. Lorentz. *Approximation of Functions*. Holt Rinehart and Winston, 1966. (Cited on p. 282.)

[176] G. R. MacLane. Sequences of derivatives and normal families. *J. Analyse Math.*, 2: 72–87, 1952/53. (Cited on p. 6.)

[177] A. I. Markushevich. *Theory of Functions of a Complex Variable*. Chelsea, 1977. (Cited on pp. 290 and 311.)

[178] F. Martínez-Giménez and A. Peris. Universality and chaos for tensor products of operators. *J. Approx. Theory*, 124: 7–24, 2003. (Cited on p. 217.)

[179] B. Maurey. Banach spaces with few operators. In *Handbook of the Geometry of Banach Spaces*, volume 2, pp. 1247–1297. North Holland, 2003. (Cited on p. 150.)

[180] B. Maurey. Type, cotype and K-convexity. In *Handbook of the Geometry of Banach Spaces*, volume 2, pp. 1299–1332. North Holland, 2003. (Cited on pp. 105 and 316.)

[181] B. Maurey and G. Pisier. Séries de variables aléatoires vectorielles indépendantes et propriétés géométriques des espaces de Banach. *Studia Math.*, 58: 45–90, 1976. (Cited on p. 238.)

[182] G. Metafune and V. B. Moscatelli. Dense subspaces with continuous norm in Fréchet spaces. *Bull. Polish. Acad. Sc.*, 37: 477–479, 1989. (Cited on p. 37.)

[183] A. Montes-Rodríguez. Banach spaces of hypercyclic vectors. *Michigan Math. J.*, 43: 419–436, 1996. (Cited on p. 195.)

[184] A. Montes-Rodríguez and H. N. Salas. Supercyclic subspaces: spectral theory and weighted shifts. *Adv. Math.*, 163: 74–134, 2001. (Cited on pp. 27, 92, 217 and 227.)

[185] A. Montes-Rodríguez and H. N. Salas. Supercyclic subspaces. *Bull. London Math. Soc.*, 35: 721–737, 2003. (Cited on p. 217.)

[186] A. Montes-Rodríguez and S. Shkarin. New results on a classical operator. In *Recent Advances in Operator-Related Function Theory*, volume 393, pp. 139–157. Amer. Math. Soc., 2006. (Cited on p. 228.)

[187] A. Montes-Rodríguez and S. Shkarin. Non-weakly supercyclic operators. *J. Operator Theory*, 58: 39–62, 2007. (Cited on pp. 93, 228 and 262.)

[188] H. L. Montgomery and R. C. Vaughan. *Multiplicative Number Theory. I. Classical Theory*, volume 97 of *Cambridge Studies in Advanced Mathematics*. Cambridge University Press, 2007. (Cited on pp. 274 and 282.)

[189] V. Müller and J. Vršovský. Orbits of linear operators tending to infinity. *Rocky Mountain J. Math*, to appear, 2008. (Cited on p. 262.)

[190] R. I. Ovsepian and A. Pełczyński. On the existence of a fundamental total and bounded biorthogonal sequence in every separable Banach space, and related constructions of uniformly bounded orthonormal systems in L^2. *Studia Math.*, 54: 159–159, 1975. (Cited on p. 40.)

[191] J. Pal. Zwei kleine Bemerkungen. *Tohoku Math J.*, 6: 42–43, 1914. (Cited on p. x.)

[192] A. Pazy. *Semigroups of Linear Operators and Applications to Partial Differential Equations*, volume 44 of *Applied Mathematical Sciences*. Springer-Verlag, 1983. (Cited on p. 126.)

[193] C. Pearcy ed. *Topics in Operator Theory*, volume 13 of *Mathematical Surveys*. American Mathematical Society, 1979. (Cited on p. 210.)

[194] A. Peris. Multi-hypercyclic operators are hypercyclic. *Math. Z.*, 236: 779–786, 2001. (Cited on p. 72.)

[195] A. Peris and L. Saldivia. Syndetically hypercyclic operators. *Integral Equations Operator Theory*, 51: 275–281, 2005. (Cited on pp. 80 and 92.)

[196] G. Pisier. Conditions d'entropie assurant la continuité de certains processus et applications à l'analyse harmonique. In *Séminaire d'Analyse Fonctionnelle, Ecole Polytechnique*, exposés 13–14, 1980. (Cited on p. 181.)

[197] G. Pisier. Holomorphic semigroups and the geometry of Banach spaces. *Ann. Math.*, 115: 375–392, 1980. (Cited on p. 235.)

[198] G. T. Prajitura. The density of hypercyclic operators in the strong operator topology. *Integral Equations Operator Theory*, 499: 559–560, 2004. (Cited on p. 57.)

[199] H. Radjavi and P. Rosenthal. *Invariant Subspaces*, volume 77 of *Ergebnisse der Mathematik und ihrer Grenzgebiete*. Springer-Verlag, 1973. (Cited on p. 47.)

[200] T. Ransford and M. Roginskaya. Point spectra of partially power-bounded operators. *J. Funct. Anal.*, 230: 432–445, 2006. (Cited on p. 162.)

[201] C. J. Read. A short proof concerning the invariant subspace problem. *J. London Math. Soc.*, 34: 335–348, 1986. (Cited on pp. 292 and 308.)

[202] C. J. Read. The invariant subspace problem for a class of Banach spaces. II. Hypercyclic operators. *Israel J. Math.*, 63: 1–40, 1988. (Cited on pp. x and 292.)

[203] C. J. Read. The invariant subspace problem on some Banach spaces with separable dual. *Proc. London Math. Soc.*, 58: 583–607, 1989. (Cited on p. 308.)

[204] C. J. Read. Quasinilpotent operators and the invariant subspace problem. *J. London Math. Soc.*, 56: 595–606, 1997. (Cited on p. 308.)

[205] A. Reich. Universelle Werteverteilung von Eulerproducten. *Nachr. Acad. Wiss. Göttingen*, 1: 1–17, 1977. (Cited on p. 290.)

[206] S. Rolewicz. On orbits of elements. *Studia Math.*, 32: 17–22, 1969. (Cited on pp. 1 and 6.)

[207] M. Rosenblum. On the operator equation $BX - XA = Q$. *Duke Math. J.*, 23: 263–269, 1956. (Cited on p. 47.)

[208] W. Rudin. *Fourier Analysis on Groups*. Interscience Publishers (John Wiley and sons), 1962. (Cited on p. 291)

[209] W. Rudin. *Real and Complex Analysis*. McGraw-Hill Series in Higher Mathematics. McGraw Hill, 1966. (Cited on pp. 4, 310, 311 and 314.)

[210] W. Rudin. *Functional analysis*. McGraw-Hill Series in Higher Mathematics. McGraw Hill, 1973. (Cited on pp. 2, 255, 317 and 319.)

[211] R. Rudnicki. Gaussian measure-preserving linear transformations. *Univ. Iagel. Acta Math.*, 30: 105–112, 1993. (Cited on p. 130.)

[212] I. Ruzsa. On difference-sequences. *Acta Arith.*, 25: 151–157, 1973/74. (Cited on p. 150.)

[213] I. Ruzsa. An infinite Sidon sequence. *J. Number Theory*, 68: 63–71, 1998. (Cited on p. 262.)

[214] H. N. Salas. A hypercyclic operator whose adjoint is also hypercyclic. *Proc. Amer. Math. Soc.*, 112 (3): 765–770, 1991. (Cited on p. 75.)

[215] H. N. Salas. Hypercyclic weighted shifts. *Trans. Amer. Math. Soc.*, 347 (3): 993–1004, 1995. (Cited on pp. 18, 20, 30, 31 and 32.)

[216] H. N. Salas. Supercyclicity and weighted shifts. *Studia Math.*, 135: 55–74, 1999. (Cited on pp. 9, 18 and 192.)

[217] H. N. Salas. Pathological hypercyclic operators. *Arch. Math.*, 86: 241–250, 2006. (Cited on p. 309.)

[218] H. N. Salas. Banach spaces with separable duals support dual hypercyclic operators. *Glasg. Math. J.*, 49: 281–290, 2007. (Cited on p. 57.)

[219] W. Seidel and J. L. Walsh. On approximation by euclidean and non-euclidean translations of an analytic function. *Bull. Amer. Math. Soc.*, 47: 916–920, 1941. (Cited on p. 30.)

[220] J. H. Shapiro. *Composition Operators and Classical Function Theory*. Universitext. Springer-Verlag, 1993. (Cited on pp. 10, 22 and 23.)

[221] J. H. Shapiro. Decomposability and the cyclic behavior of parabolic composition operators. In *Recent Progress in Functional Analysis (Proc. Conf. Valencia, 2000)*, volume 189 of *North-Holland Mathematics Studies*, pp. 143–157. North-Holland, Amsterdam, 2001. (Cited on p. 228.)

[222] J. H. Shapiro. Notes on the dynamics of linear operators. Available at the author's web page http://home.wlu.edu/~shapiro/Pubvit/LecNotes.html, 2001. (Cited on p. ix.)

[223] S. Shkarin. Antisupercyclic operators and orbits of the Volterra operator. *J. London Math. Soc.*, 73: 506–528, 2006. (Cited on p. 261.)

[224] S. Shkarin. Non-sequential weak supercyclicity and hypercyclicity. *J. Funct. Anal.*, 242: 37–77, 2007. (Cited on pp. 235, 236, 242, 244, 250, 256, 260, 261 and 262.)

[225] S. Shkarin. The Kitai Criterion and backward shifts. *Proc. Amer. Math. Soc.*, 136 (5): 1659–1670, 2008. (Cited on p. 57.)

[226] S. Shkarin. On the spectrum of frequently hypercyclic operators. Proc. Amer. Math. Soc. 137: 123–134, 2009.

[227] A. L. Shields. A note on invariant subspaces. *Michigan Math. J.*, 17: 231–233, 1970. (Cited on p. 309.)

[228] G. Sirotkin. A modification of Read's transitive operator. *J. Operator Theory*, 55: 153–167, 2006. (Cited on p. 308.)

[229] A. Sobczyk. Projection of the space m onto its subspace c_0. *Bull. Amer. Math. Soc.*, 47: 938–947, 1941. (Cited on p. 84.)

[230] M. Talagrand. *The Generic Chaining. Upper and Lower Bounds of Stochastic Processes*. Springer-Verlag, 2005. (Cited on p. 180.)

[231] M. Taniguchi. Chaotic composition operators on the classical holomorphic spaces. *Complex Variables Theory Appl.*, 49: 429–538, 2004. (Cited on pp. 138 and 161.)

[232] E. C. Titchmarsh. *The Theory of Functions*. Clarendon Press, 1932. (Cited on p. 268.)

[233] S. M. Voronin. The distribution of the non-zero values of the Riemann zeta function. *Trudy Mat. Inst. Steklov*, 128: 131–150, 1972. (Cited on p. 291.)

[234] S. M. Voronin. A theorem on the "universality" of the Riemann zeta function. *Math. USSR-Izv.*, 9: 443–453, 1975. (Cited on pp. 264 and 290.)

[235] P. Walters. *An Introduction to Ergodic Theory*, volume 79 of *Graduate Texts in Mathematics*. Springer-Verlag, 1982. (Cited on pp. 95, 108, 110, 132, 133 and 286.)

[236] J. Wengenroth. Hypercyclic operators on non-locally convex spaces. *Proc. Amer. Math. Soc.*, 131 (6): 1759–1761, 2003. (Cited on p. 27.)

[237] P. Y. Wu. Sums and products of cyclic operators. *Proc. Amer. Math. Soc.*, 122 (4): 1053–1063, 1994. (Cited on p. 58.)

Notation

Frequently used notation

$\mathbb{D}$: the unit disk, $\mathbb{D} = \{z \in \mathbb{C}; \; |z| < 1\}$.
$\mathbb{K}$: the scalar field, $\mathbb{K} = \mathbb{R}$ or $\mathbb{C}$.
$\mathbb{K}^*$: the *non-zero* scalars.
$\mathbb{N}$: the set of all natural numbers.
$\mathbb{N}^*$: the *positive* natural numbers.
$\mathbb{P}_+$: the upper half-plane, $\mathbb{P}_+ = \{w \in \mathbb{C}; \; \mathrm{Im}(w) > 0\}$.
$\mathbb{T}$: the unit circle, $\mathbb{T} = \{z \in \mathbb{C}; \; |z| = 1\}$.
$B_{\mathbf{w}}$: the weighted shift with weight sequence $\mathbf{w}$ (defined on pp. 9 and 18).
$c_{00}(\mathbb{N})$, $c_{00}(\mathbb{Z})$: the finitely supported sequences of scalars.
$\mathcal{C}(E)$: the set of all scalar-valued continuous functions on E.
$\mathcal{C}(E, F)$: the set of all continuous functions from E to F.
C_ϕ: the composition operator induced by ϕ (defined on p. 22).
$HC(T)$: the set of all hypercyclic vectors for T.
$\mathfrak{L}(X)$: the continuous linear operators on X.
$O(x, T)$: the T-orbit of x (defined on p. 1).

Function spaces

$AP(\mathbb{R})$: the class of almost periodic functions (defined on p. 272).
$A^2(U)$: the Bergman space of the domain U (defined on p. 267).
$H(\Omega)$: the space of all holomorphic functions on the open set $\Omega \subset \mathbb{C}$.
$\mathcal{H}^2$: the image of the Hardy space $H^2(\mathbb{D})$ under the Cayley map $\omega(z) = i(1 + z)/(1 - z)$ (defined on pp. 224, 165 and 313).
$H^2(\mathbb{D})$: the Hardy space of the disk $\mathbb{D}$ (defined on pp. 7 and 312).
$H^2(\mathbb{P}_+)$: the Hardy space of the upper half-plane (defined on pp. 225 and 314).
$H^\infty(\mathbb{D})$: the space of all bounded holomorphic functions on $\mathbb{D}$.

Names of sets

$SC(T)$: the set of all supercyclic vectors for T.
$Aut(\mathbb{D})$: the automorphisms of $\mathbb{D}$ (defined on p. 23).
$\mathbb{C}_\sigma$: the right half-plane $\{s \in \mathbb{C}; \; \mathrm{Re}(s) > \sigma\}$.
$D(z, r)$: the open disk in $\mathbb{C}$ with centre z and radius r.
$FHC(T)$: the frequently hypercyclic vectors for T (defined on p. 142).
$\mathcal{FIN} \subset \mathfrak{L}(X)$: the norm closure of the set of all finite-rank operators.
$GL(X)$: the invertible operators on X.
$\mathrm{Ker}^*(B)$: the "generalized kernel" of B (defined on p. 32).
$\mathbb{K}[T]$: *the set* $\{P(T); \; P \text{ polynomial}\}$.
$\mathbb{K}[T]x$: the set span $O(x, T) = \{P(T)x; \; P \text{ polynomial}\}$.
$J^{\mathrm{mix}}(R)$: defined on p. 33.
$LFM(\mathbb{D})$: the linear fractional maps of the disk $\mathbb{D}$ (defined on p. 23).
$\mathfrak{L}_{HC}(X)$: the hypercyclic operators on X.
$\Lambda^+(T), \Lambda^-(T)$: defined on p. 32.
$\mathbf{C}(U, V), \mathbf{N}(U, V)$: defined on p.78.
$\mathbf{N}(x, U)$: defined on p. 77.
$\mathcal{P}$: the set of all prime numbers.
$\mathrm{Per}(T)$: the periodic points of T (defined on p. 137).
$\sigma(T)$: the spectrum of the operator T.

$\sigma_e(T)$: the essential spectrum of T (defined on p. 317).
$\sigma_p(T)$: the point spectrum of T (defined on p. 11).

Names of operators

I: the identity operator.
$\mathbf{L}_T$: the left-multiplication-by-T operator (defined on p. 199).
M_ϕ: the multiplication-by-ϕ operator (defined on p. 8).
R_μ: the covariance operator of the Gaussian measure μ (defined on p. 99).
$u \otimes u^* \in \mathfrak{L}(X)$: the rank-1 operator defined by $u \otimes u^*(x) = \langle u^*, x \rangle\, u$.

Miscellaneous

$d\gamma_\sigma$: the centred Gaussian measure on $\mathbb{R}$ with variance σ^2 (defined on p. 97).
k_α: the reproducing kernel at $\alpha \in \mathbb{D}$ (defined on pp. 8 and 313).
$p(X)$, $q(X)$: the "maximal type" and "minimal cotype" of the Banach space X (defined on p. 235).
r^*: the conjugate exponent of $r \in [1, \infty]$, i.e. the number defined by $r^{-1} + (r^*)^{-1} = 1$
$r(T)$: the spectral radius of the operator T (defined on p. 317).
$\sigma_c(f)$, $\sigma_a(f)$: the abscissae of convergence and absolute convergence of the Dirichlet series f (defined on p. 268).
$\|u\|_E$: $\sup\{|u(z)|;\ z \in E\}$ (defined on p. 4).

Author index

Subject index